P9-EDJ-840

A SIERRA NEVADA FLORA

Norman F. Weeden, Ph.D.
Professor of Plant Genetics,
Cornell University

With new drawings by Amy David

WILDERNESS PRESS
BERKELEY

First edition 1975
Second edition 1981
Third edition 1986
FOURTH EDITION October 1996

Cover design by Larry B. Van Dyke

Library of Congress Card Catalog Number 96-32604
International Standard Book Number 0-89997-204-7

Manufactured in the United States of America

Published by Wilderness Press
2440 Bancroft Way
Berkeley, CA 94704
(800) 443-7227
FAX (510) 548-1355

Write, call or fax for a free catalog

Cover photo: *Dudleya Cymosa*

Library of Congress Cataloging-in-Publication Data

Weeden, Norman.
A Sierra Nevada flora / Norman F. Weeden. — 4th ed.
p. cm.
Includes bibliographical references and index.
ISBN 0-89997-204-7
1. Botany—Sierra Nevada (Calif. and Nev.) 2. Plants--Identification. I. Title.
QK149.W43 1990
582.13'09794'4—dc20 96-32604
CIP

FOREWORD

This book has been written for the outdoor enthusiast who becomes interested in the variety of plants that may be seen on a short walk or an extended trek along Sierran trails. I have attempted to present as practical and convenient a field guide as possible. The user of this guide should be able to identify most of the Sierran plants without the use of a hand lens or the need to dissect flower or fruit. Plant characteristics easily observed in the field and not necessarily those most acceptable to plant taxonomists have been selected for the keys. The use of specialized terminology in the keys and plant descriptions has also been minimized.

Many changes in the taxonomy of species currently found in the Sierra Nevada have occurred in the twenty years since this book first appeared under the title of *A Survival Handbook to Sierra Flora*. Recently, most of these taxonomic revisions have been summarized by highly regarded botanists in the excellent and monumental effort entitled *The Jepson Manual: Higher Plants of California*, edited by the late J.C. Hickman. It was the publication of this exhaustive treatment of the flora of California that stimulated me to undertake an extensive revision of my own small handbook. Clearly, science has marched forward and the new arrangements or names are based on a more detailed understanding of the phylogenic relationships within and among taxa. Although I will miss the recognition of taxa such as the Fumariaceae, Pyrolaceae, Amaryllidaceae, *Zauschneria*, and *Heterogaura* and occasionally still slip back to using old names such as the Compositae, Umbelliferae, *Osmaronia*, and *Peltiphyllum*, I have refrained from relying on any of my own predilections or prejudices and have strictly followed the nomenclature presented in *The Jepson Manual* as the basis for the scientific names. This decision has the advantage of my not having to present many of the details on each taxon normally expected in an extensive flora. Information on the authority for each name, where the taxon was first described, and what other synonyms have been used for the taxon can be found in the *The Jepson Manual*.

The revision has also given me the opportunity to improve many of the drawings used in the previous editions. I am particularly fortunate to be able to include over 100 new illustrations by Amy David. I am sure the reader will greatly appreciate these fine additions, identified by an "AD" notation.

Exploration of Sierran vegetation can be a continual source of stimulation and enjoyment, and, as in many other endeavors, the more one knows about the flora, the more one appreciates its existence and availability. It is critical that we recognize the vegetation of the Sierra as a resource of immense value, not just in direct economic criteria but also in terms of quality of life and value to future generations. This flora is now threatened by the seemingly inexorable pressure of civilization and burgeoning population. Should we fail to appreciate the long term advantages of so magnificent a flora at our doorstep, even for a generation, it will become irretrievably lost, and the grand forests of Yellow Pine, White Fir, Sugar Pine, Red Fir and Giant Sequoia will become like the cedars of Lebanon, a note in the history books and another example of man's difficulty in maintaining a long-term sustainable relationship with his environment.

There are several people who deserve special thanks for help and encouragement in getting this revision finished. First and foremost, I would like to thank my wife Cathy, without whose significant assistance and support this revision would never have seen the light of day. Special thanks are extended to Dr. Robert Patterson at San Francisco State University and his summer class of 1995 for field testing this revision. Thanks are also due the staff at the Jepson Herbarium, University of California, Berkeley, for furnishing me working space during my short sabbatical leave there and for allowing me access to early drafts of some sections of *The Jepson Manual*. Finally, I would like to thank the many people who have encouraged and aided in the production of this book or improved the quality of its contents. I hope that I am able to meet these and other users of this book on trails and at campsites in the Sierra Nevada.

TABLE OF CONTENTS

LIST OF ABBREVIATIONS AND SYMBOLS

alt	alternate	m	meter
c.	central	mdw	meadow
cm	centimeter	mly	mostly
cmpd	compound	mm	millimeter
Co. (cos.)	County (plural)	n.	north
dm	decimeter	opp	opposite
e.	east	recpt	receptacle
fl(s)	flower(s)	s.	south
fr(s)	fruit(s)	SN	Sierra Nevada
H	high	sp (spp)	species (plural)
infl	inflorescence	ssp (sspp)	subspecies (plural)
invol.	involucre	usu	usually
L	long	var.(s)	variety (plural)
lf (lvs)	leaf (plural)	w.	west
lft(s)	leaflet(plural)	W	wide

±	more or less
(-8)	numbers in parentheses indicate that the respective plant organ rarely contains that number of parts or grows to that size

INTRODUCTION

For years adventurers and vacationers have been drawn to the Sierra Nevada to enjoy its spectacular scenery and to briefly experience a more rugged life style away from the comforts of city life. Curiously, such a change has a refreshing and revitalizing effect on ones physical and mental state so that many have become addicted to this wilderness experience, returning as often as possible to bear further discomforts. Yet the Sierra is kind to these addicts, for they discover many of the secret charms and unique features of the range that usually pass unnoticed by the casual observer.

The vegetation of the Sierra Nevada is one of the range's most remarkable features, rarely overlooked on even a brief visit. The magnificent stands of pine and fir and the colorful meadows are appreciated by all. The vegetation's value as a source of food and timber and as a game refuge has always been recognized by the residents of California. Its value as an integral part of the Sierran landscape is becoming more generally acknowledged and better understood. The interrelationship between the vegetation and the other aspects of the landscape is so strong that virtually every variation in climate, soil composition or even historical events is expressed through the local flora. The diversity of habitats thus produced is unsurpassed elsewhere in the continental United States.

The vegetation described in this handbook is that generally found above 3500' on the western slope of the Sierra Nevada and above 8000' on its eastern flank. The northern limit is the Thousand Lakes Valley area just north of Lassen Peak. The southern limit is Walker Pass and approximately the Tulare County-Kern County border. The limits were selected to include the plants usually associated with the Black Oak, pine, fir and alpine belts. Excluded were species primarily of the grasslands, oak savanna or chaparral characteristic of the western foothills. Plants predominantly of Great Basin origin found on the lower eastern slopes were also excluded from this treatment. The book's major emphasis is on those areas most widely used for recreation by hikers, campers, sportsmen, and skiers. It should be remembered that the limits are only approximate and that the elevation limits are higher in the southern Sierra and lower in the north.

The species described within are those defined in *The Jepson Manual: Higher Plants of California*, edited by James C. Hickman. In keeping with the philosophy that this book is primarily a field guide, several genera, such as *Cryptantha* and *Carex*, which are very difficult or impossible to identify to species in the field, have not been treated at the species level. In a number of other genera containing many similar species (e.g. *Eriogonum* and *Poa*) only a key to the species has been included in order to save on space. However, no taxon has been intentionally omitted without mention of doing so.

Common names have been included in almost all instances in order to allow those unfamiliar with Latin names to feel more comfortable with the subject; however, one will soon find that familiarity with the "scientific" names makes the keys much easier. A key to the dicotyledon genera has been included as an additional aid in plant identification. This enlarged genera key parallels the key to the families and the genera key within each family but uses more easily observed plant characteristics. Especially if an unknown dicot is not in flower it will be most easily identified by using the Key to the Dicotyledon Genera.

One or two drawings are given for each genus. The drawings are intended to depict particularly important features for the identification of the genus or individual species within the genus. The user may find that for the larger genera, having only two drawings may be frustrating, because the drawings will provide

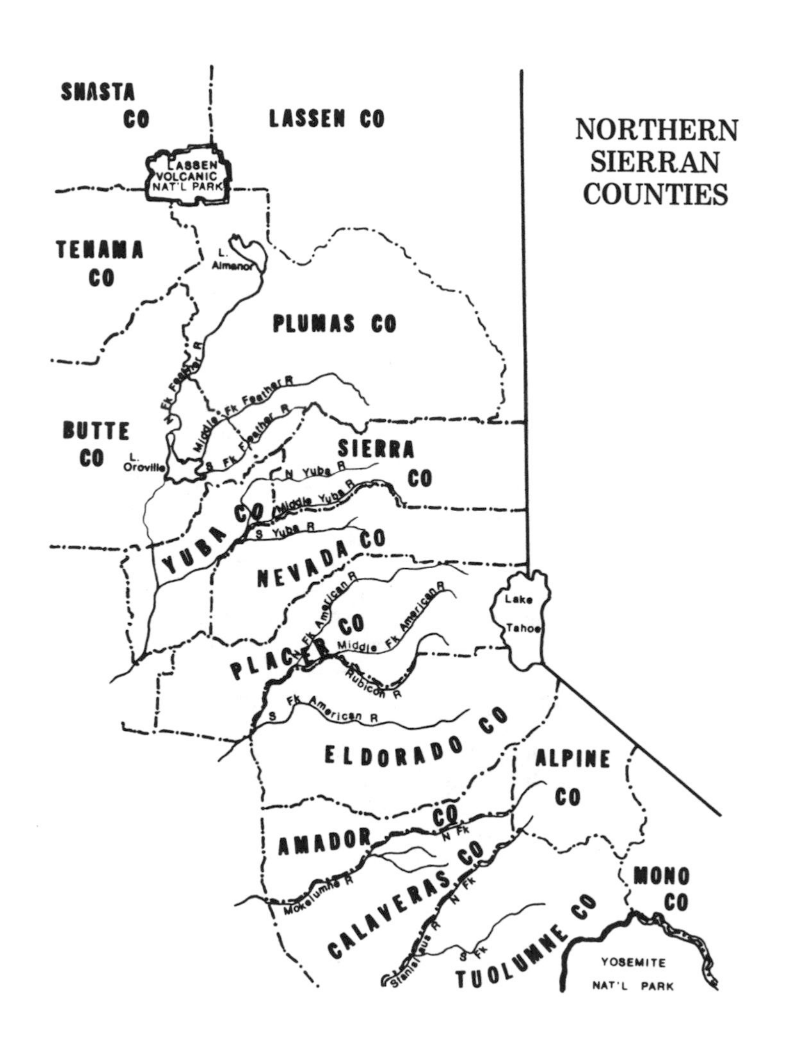
NORTHERN
SIERRAN
COUNTIES
SHASTA CO
LASSEN CO
LASSEN VOLCANIC NAT'L PARK
TEHAMA CO
L. Almanor
PLUMAS CO
N Fk Feather R
Middle Fk Feather R
S Fk Feather R
BUTTE CO
L. Oroville
SIERRA CO
N Yuba R
Middle Yuba R
S Yuba R
YUBA CO
NEVADA CO
N Fk American R
Middle Fk American R
Rubicon R
PLACER CO
Lake Tahoe
S Fk American R
ELDORADO CO
ALPINE CO
AMADOR CO
N Fk
Mokelumne R
CALAVERAS CO
N Fk
Stanislaus R
S Fk
TUOLUMNE CO
MONO CO
YOSEMITE NAT'L PARK

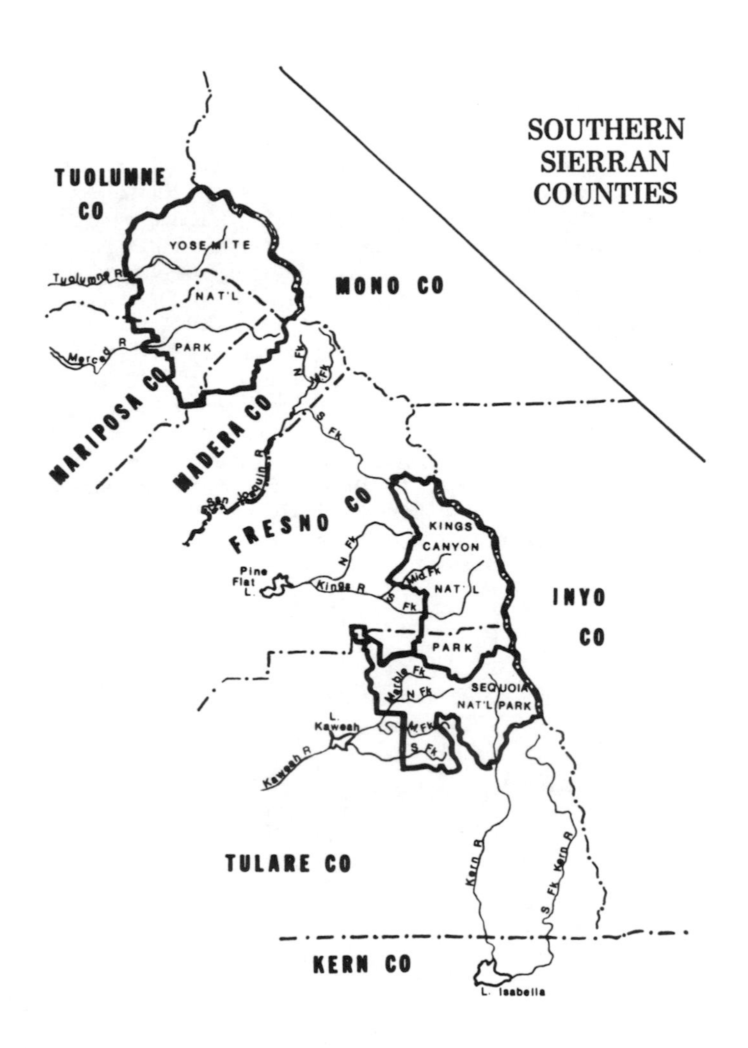

SOUTHERN
SIERRAN
COUNTIES
TUOLUMNE
CO
YOSEMITE
NAT'L
PARK
Tuolumne R
Merced R
MONO CO
MARIPOSA CO
MADERA CO
N Fk
S Fk
San Joaquin R
FRESNO CO
KINGS
CANYON
NAT'L
PARK
Mid Fk
N Fk
Pine
Flat
L.
Kings R
S Fk
INYO
CO
Marble Fk
N Fk
SEQUOIA
NAT'L PARK
L.
Kaweah
Kaweah R
S Fk
TULARE CO
Kern R
S Fk Kern R
KERN CO
L. Isabella

little help in distinguishing among the species. Unfortunately, the inclusion of sufficient drawings to present all the characters used in the keys would expand the book beyond a size convenient for field use. A brief caption is given underneath each drawing including (in parentheses) the length that the scale bar in the figure represents.

Notes on edibility have been included for most genera. Although most users of the book have no intention of eating Sierran vegetation, these notes on edibility hopefully will provide a unique and interesting reflection of the genetic diversity present in the flora. Except for those parts (such as berries) designed for dispersal of seeds by ingestion by animals, most plants possess some defense against being eaten. If a plant does not have obvious physical defenses (thorns, spines, thick bark, etc.) it probably contains toxic compounds or chemicals conferring an unpleasant taste. Despite such defenses, a surprising number of Sierran species can be safely eaten, particularly after some treatment such as leaching or cooking. However, a person should be very cautious about sampling plants. Not only are there very poisonous plants in the Sierra Nevada, but too much of any plant can cause digestive upset or worse. In addition, variation exists within a species. The reader may be aware that even in common crops such as squash (*Cucurbita pepo*) or potato (*Solanum tuberosum*) genetic or environmental variation can make the crop extremely toxic. In much the same way, selenium accumulators such as locoweed (*Astragalus* spp) can be poisonous on some soils and relatively innocuous on other soils. Although the author is not aware of specific instances where the toxicity of a Sierran species has been demonstrated to vary as a result of genetic differences, such variation certainly exists. Hence what might be eaten without problems in one location might be very unpleasant if sampled from another site. The author's own choice is to pack in his own food and sample an occasional berry or leaf of particularly flavorful or interesting species.

HOW TO BEGIN

In order to use this book most efficiently you must be familiar with the technique of using a dichotomous key. In such a key you are confronted with a series of choices. There are always two choices at each step, and you must decide which choice best fits the unknown plant. Start at the most convenient spot in the key for you. It might be this page or the start of some subgroup if you are already sure of the subgroup to which a plant belongs. Read the description of characteristics of that subgroup to check the theoretical against what you are looking at. If the brief description matches the plant, proceed to the key that starts with "A" and some described characteristic. If the plant does not have that characteristic then drop down to "AA" (which may be a considerable distance) and read the described characteristics which are closely related to but excluding those in A.

After selecting either A or AA as the best description of the unknown plant, continue to B under the respective choice. Again, select between B and BB, choosing that which best describes the plant, and proceed to C under whichever you select. If neither choice seem to fit try returning to one of the previous steps to make sure a wrong choice was not made earlier. Even if all your previous choices have been correct, the possibility exists that the plant in front of you was not considered when the key was made and either represents a new species or, more likely, an atypical form of one of the species listed. You may also find that the part of the plant necessary for a choice, e.g., a seed or flower, is not present on your specimen. For those reasons it is not always possible to unequivocally identify a plant.

Continue the keying procedure until given a direction at the end of the line you select to continue to a smaller subgroup. This smaller subgroup will generally have a separate key beginning with A again. Work this new key just as you did the first one. Ultimately you will come to a line which will have a Latin name such as *T. californica* at the end of it. If you have made the correct choices, this label is the term (scientific or Latin name) plant taxonomists have given a collection of closely related plants to which your plant belongs. The name represents a distinct plant species. The individuals in a species are usually able to interbreed among themselves but often cannot cross with plants outside the species.

Descriptions of each species follow the species key in each genus. When you have been directed to a species name, compare that species description to your plant. The description, though abbreviated and concise, should fit your plant very closely. If there are marked discrepancies between the description and your plant, an incorrect choice was probably made during the 'keying' process.

Enough for the theoretical. If you are ready to identify a specimen of a fern, conifer or flowering plant (more primitive plant forms such as mosses, fungi, lichens and liverworts are not covered in this handbook) please proceed to the following key:

A Plant not flowering; reproductive parts found usu in cone-like structures or underneath lvs, occasionally grouped elsewhere
  B Plants ± herbaceous; lvs usu much divided — Ferns and Related Plants p. 7
  BB Trees or occasionally shrubs; lvs usu scale- or needle-like — Conifers p. 15
AA Plants producing flowers
  B Fl-parts mly in 4's or 5's; lvs usu pinnately or palmately veined
    Dicotyledons--Use either Family Key p.19 or Genera Key p.23
  BB Fl-parts mly in 3's; lvs mly parallel veined from base
    Monocotyledons p. 208

With very little practice this key becomes extremely easy, and you will generally begin at one of the subgroup keys without even thinking of using this first one. In selecting between A and AA, the only plants you may get confused are some of the fern relatives. They form such a small group, though, that you can easily resolve your suspicions by a brief glance through the drawings of this group. Selection between B and BB under A is easy, for although some ferns can get fairly large, their branches are never woody, and the large ferns all have the typical lacy leaf. Selection between B and BB under AA requires a bit more practice. Plants with grass-like leaves or leaves striped with longitudinal unbranched veins should be suspected as monocots and all others as dicots until you become more familiar with the other characteristics of each group.

KEY TO THE FERN AND RELATED GENERA

A Reproductive structures not borne on underside of lvs
 B Lvs scale-like or linear, without expanded blade
  C Stems erect, jointed; lvs and branches in whorls — *Equisetum* p. 11
  CC Stems prostrate or inconspicuous; lvs and branches not whorled
   D Lvs scale-like; plants terrestrial — *Selaginella* p. 14
   DD Lvs grass-like; plants aquatic — *Isoetes* p. 11
 BB Lvs with conspicuous blade
  C Lvs pinnately lobed or divided; spore cases in clusters — *Botrychium* p. 9
  CC Lvs palmately 2-4-foliolate; spores enclosed in a bean-shaped structure — *Marsilea* p. 12
AA Reproductive structures borne on underside of lvs (true ferns)
 B Fronds of two kinds; sterile fronds with broad lobes or lfts; fertile fronds with narrower ± linear lobes
  C Fronds widest near middle — *Blechnum* p. 9
  CC Fronds widest at base, triangular in outline — *Cryptogramma* p. 10
 BB Fronds all similar
  C Sori located at edge of lfts and ± covered by curled-under margins
   D Fronds 6-12 dm L — *Pteridium* p. 14
   DD Fronds shorter
    E Blades densely scaly or hairy beneath — *Cheilanthes* p. 9
    EE Blades glabrous or nearly so beneath
     F Margins partly curled under; lfts on stalks — *Adiantum* p. 7
     FF Margins completely curled under; lfts usu sessile
      G Blade about as wide as long; margins of lf-segments only partly curled under — *Aspidotis* p. 8
      GG Blade much longer than wide, or if not lf-segments only partly curled under — *Pellaea* p. 12
  CC Margins of lf-segments never curled under and covering sori
   D Blades yellow- or white-powdery beneath — *Pentagramma* p. 13
   DD Blades greenish beneath
    E Fronds 1-pinnate; lfts entire, dentate or occasionally lobed at base
     F Blades of fronds less than 2 cm W; rare — *Asplenium* p. 8
     FF Blades of fronds over 2 cm W
      G Frond pinnately lobed, the lobes attached to midrib by a broad base — *Polypodium* p. 13
      GG Lfts distinct, attached to midrib by a narrow stalk
       H Fronds very large, usu over 1 m L; sori oblong — *Woodwardia* p. 14
       HH Fronds usu less than 1 m L; sori round — *Polystichum* p. 13
    EE Fronds 2-3-pinnate, or if 1-pinnate the lfts deeply lobed
     F Sori covered by a curved indusium; fronds mly more than 4 dm L
      G Indusium slightly curved, elongate — *Athyrium filix-femina* p. 8
  GG Indusium definitely horseshoe-shaped, circular in outline
   H Ultimate lf-segments dentate to deeply lobed — *Dryopteris* p. 10
   HH Ultimate-lf-segments entire or subentire — *Lastrea* p. 11
 FF Sori not covered or indusium scale-like
  G Indusium present and covering sori; petioles smooth — *Cystopteris* p. 10
  GG Indusium none or inconspicuous; petioles scaly at base
   H Blades 2-9 dm L; 6000-11,000' — *Athyrium alpestre* p. 8
   HH Blades mly less than 2 dm L; below 9000' — *Woodsia* p. 14

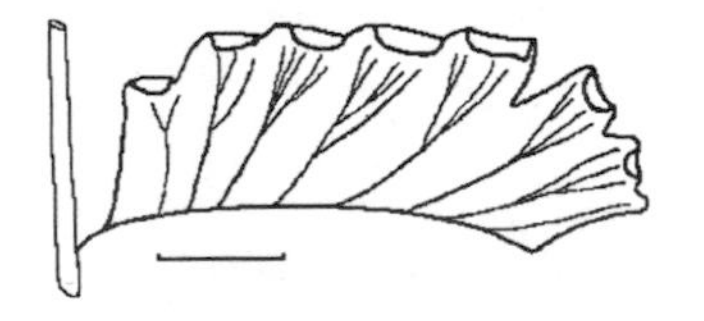

A. pedatum leaflet (1 cm)

Adiantum

Maiden-hair

A. capillus-veneris (1 cm)

Stems and leaf-stalk delicate, brown or black, often shining. Sori found under the reflexed tips of the lfts. Veins on the lfts conspicuous, simple to several times forked, never in a net-like pattern. Herbage bitter and causes increased secretion of mucus.

Lfts roundish above wedge shaped base; rare — *A. capillus-veneris*
Lfts oblong above base; common — *A. aleuticum*

A. capillus-veneris, Common Maiden-hair: Fronds several to many, ascending to pendant, 2-7 dm L; petioles fragile, about as long as blades; blades mly bipinnate or occasionally tripinnate at base, pinnate above; main rachis continuous to tip; ultimate lfts light green, 5-30 mm L. Infrequent, on moist canyon walls below 4000', mly southern SN.

A. aleuticum, Five Finger Fern: Fronds erect, 2-8 dm H; petiole thickish, longer than blade; blades 10-50 cm W; the main rachis usu dividing near base; ultimate lfts oblong with one long edge entire and the other deeply incised between the sori. Common in moist, shady habitats below 10,000', SN (all).

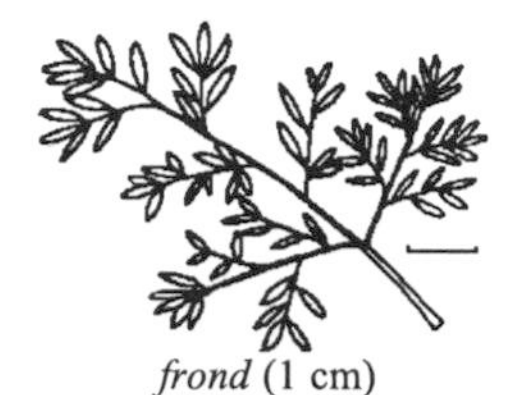

frond (1 cm)

Aspidotis

Indian's Dream

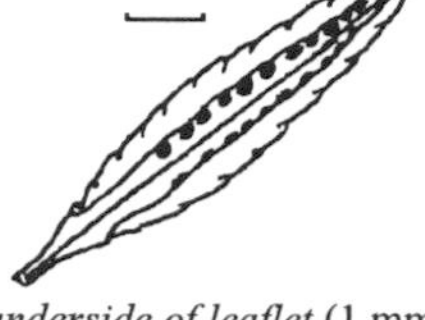

underside of leaflet (1 mm)

A. densum: Fronds mly fertile in dense clusters, 10-15 cm H; stipes thin, reddish-brown, longer than blades; blades 2-6 cm L, 2-3-pinnate; ultimate segments 4-8 mm L, linear to lanceolate. Occasional in exposed rocky habitats, 5000-8900', Tulare to Nevada Co. Similar sp is said to prevent baldness!

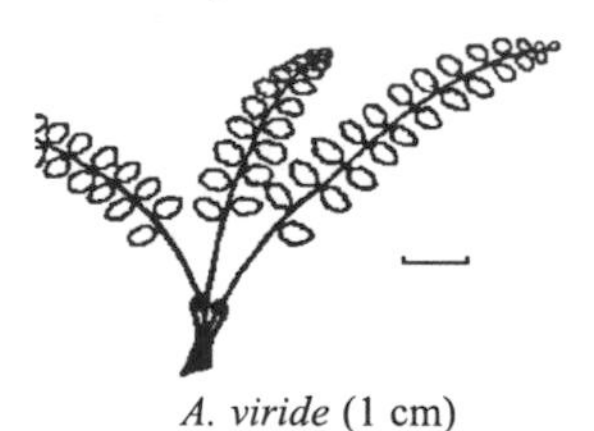

A. viride (1 cm)

Asplenium

Spleenwort

leaflet and sori (1 mm)

Small ferns with dark petioles and rachises. Sori ± oblong. Indusium attached along one long edge. Edibility unknown.

Lfts 2-3, linear; Tulare and Shasta cos. — *A. septentrionale*
Lfts 10-50, roundish; Sierra Buttes — *A. trichomanes-ramosum*

A. septentrionale, Northern Spleenwort: Fronds grass-like, the petiole much longer than the blade, the latter with only 2-3 alternate lfts. Sawtooth Pass and Lassen National Park.

A. trichomanes-ramosum, Green Spleenwort: Fronds 3-20 cm L; petiole red-brown below, green above; lfts ± paired, the margins shallowly toothed or lobed. Only known location in SN is on a north-facing cliff of South Butte, Sierra Co., at 7500'.

A. filix-femina base of blade (1 cm)

Athyrium

Lady Fern

A. filix-femina leaflet and sporangia (1 cm)

Erect medium sized ferns; rhizomes stout; fronds clustered; petiole somewhat scaly; the blade usu 2-3-pinnate, rarely pinnate or simple. Fronds of *A. filix-femina* have been used as a cure for hiccoughs, but neither species is known to be edible.

Indusium absent — *A. alpestre*
Indusium present, curved — *A. filix-femina*

A. alpestre, Alpine Lady Fern: Fronds dying annually, 2-10 dm L; petiole short; blades lacy, oblong, 15-60 cm L, 4-25 cm W; ultimate segments oblong, 1-3 mm L; sori round. Moist habitats, 6000-11,000', SN (all).

A. filix-femina var. *cyclosorum,* Lady Fern: Fronds 6-20 dm L; petiole shorter than blade; blade narrowed at base and apex, 10-40 cm W; ultimate lfts 5-20 mm L, toothed to pinnately incised; sori oblong to horseshoe-shaped. Moist, cool habitats, 4000-8000', SN (all).

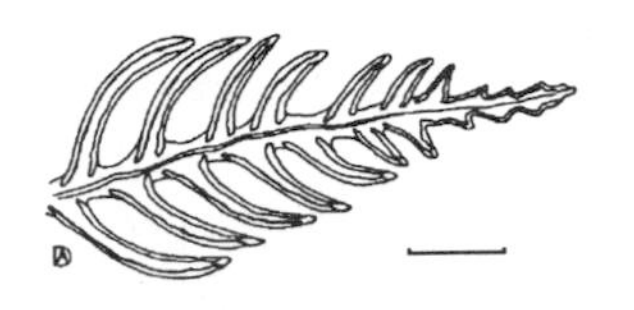

Blechnum

Deer Fern

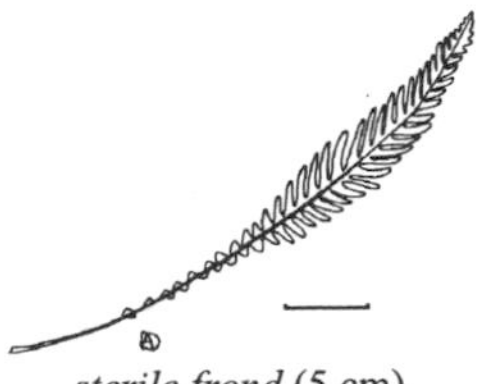

fertile frond (1 cm) *sterile frond* (5 cm)

B. spicant: Fronds of two kinds, the sterile deeply pinnately lobed to 1-pinnate, 5-10 dm L, 2-10 cm W with lfts 5-8 mm W, the fertile appearing later, larger, with lfts 2 mm W. Below 4000', n. SN.

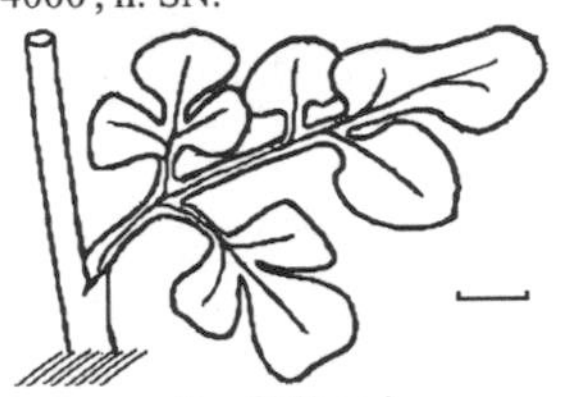

Botrychium

Moonwort

"leaf" (1 cm) *B. multifidum spore cluster* (1 cm)

Fronds consisting of an erect, long-stalked cluster of spores and a divergent sterile "leafy" structure which is lobed or divided. "Lvs" of a related species edible.

A Leafy structure 1-pinnate, usu divergent from about middle of stem; rare
B Lower segments of leafy structure fan shaped, angle at base less than 100° 3) *B. minganense*
BB Lower segments of leafy structure broadly fan shaped angle at base more than 120°
C Leafy segments usu scalloped, well separated; below 8000' 1) *B. crenulatum*
CC Leafy segments usu entire, touching or overlapping; above 9000' 2) *B. lunaria*
AA Leafy structure often 2-3-pinnate, branching off near base of stem
B Lvs 10-40 cm L, 2-3-pinnate 4) *B. multifidum*
BB Lvs 1-10 cm L, simple to pinnate; plant uncommon 5) *B. simplex*

1) *B. crenulatum,* Scalloped Moonwort: Plant to 15 cm H, slender; leafy structure usu less than 6 cm L x 2 cm W; segments usu 3-5 pairs. Rare; in marshes and meadows, 4500-8000', SN (all).

2) *B. lunaria,* Moonwort: Similar to *B. crenulatum;* margins of leaf segments usu 4-9 pairs. Rare; in dryish fields and meadows. Central SN.

3) *B. minganense,* Mingan Moonwort: Plants 5-20 cm H; leafy structure 1-3 cm W; segments less than 10 pairs. Rare; in forests along streams, below 6000', s. SN.

4) *B. multifidum,* Leathery Grape Fern: Plants stout, fleshy, 1-4 dm H; lf-blade 3-20 cm L, 7-30 cm W; ultimate segments oblong, 0.5-2 dm L. Moist, open habitats, 3000-10,000', SN (all).

5) *B. simplex,* Little Grape Fern: Plants 3-15 cm H; leaf-stalk 0.5-2 cm L; blade 1-5 cm L, 0.5-2.5 cm W, the segments usu obovate; fr cluster to 5 cm L, usu unbranched. Moist open habitats, 5000-11,200', SN (all).

Cheilanthes

Lip Fern

C. gracillima leaflet cross section (1 mm) *C. intertexta frond* (1 cm)

Small ferns of usu dry, rocky habitats. The petioles about as long as blades, both usu scaly or hairy. Ultimate lfts small, leathery. The three spp below hybridize, forming plants with intermediate morphology. Similar sp poisonous.

Underside of lfts densely hairy; lf-blade linear, oblong *C. gracillima*
Underside of lfts scaly, the scales sometimes hairy; lf-blades wider
Upper surface of lfts glabrous *C. covillei*
Upper surface of at least some lfts with dissected scales *C. intertexta*

C. covillei, Coville's Bead Fern: Fronds 8-20 cm L, dark green, 3-4-pinnate. lower surface of lfts concave, obscured by scales that extend beyond recurved margin. Rocky places below 8000', SN (all).

C. gracillima, Lace Fern: Fronds 5-20 cm L; petioles reddish brown; blades mly bipinnate, ± glabrous above; ultimate segments 2-4 mm L, ± oblong, yellow-green. Dry, rocky places below 9000', Tulare Co. n.

C. intertexta, Bead Fern: Plants to 10 cm H; fronds 10-30 cm L; blades oblong to narrowly triangular; petiole purplish, wiry; blade 5-15 cm L, 3-4-pinnate; ultimate segments roundish, 1-3 mm L, glabrous above. Dry, rocky habitats below 6500'.

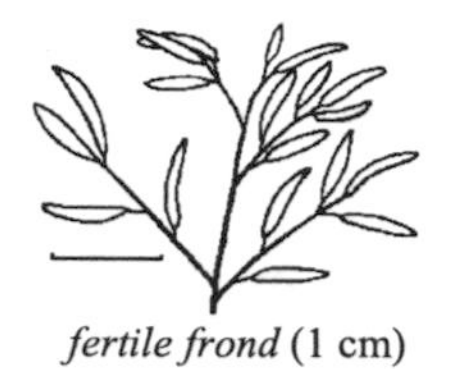

fertile frond (1 cm)

Cryptogramma

Rock-brake Parsley Fern

sterile frond (1 cm)

Plants with stout rhizomes and tufted fronds. Fronds of two kinds, fertile and sterile, the former being more erect with longer petioles and narrower segments. Blades 2-4-pinnate. Plants of rocky habitats. Edibility unknown.

Sterile lvs leathery, the petiole bases persistent *C. acrostichoides*
Sterile lvs thin, the petiole bases shriveling or deciduous *C. cascadensis*

C. acrostichoides, American Parsley Fern: Fronds many, 1-2 dm H; sterile blades 3-12 cm L; ultimate lfts crowded, ± ovate, often with branched hairs; fertile blades more open; ultimate lfts 6-12 mm L, 2 mm W, linear because of strongly revolute margins. Rocky ledges and crevices, 6000-11,000', SN (all).

C. cascadensis, Cascade Rock-brake: Similar to *C. acrostichoides;* ultimate lfts glabrous. Talus slopes and crevices, 6000-11,000', SN (all).

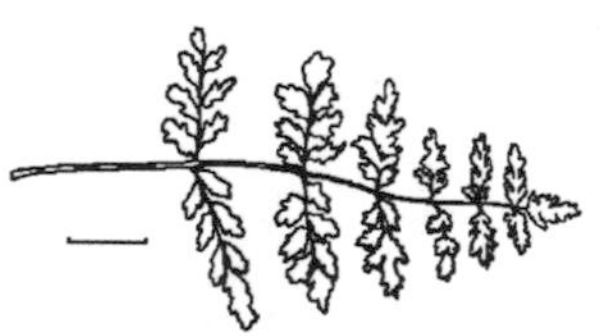

frond (1 cm)

Cystopteris

Brittle Fern

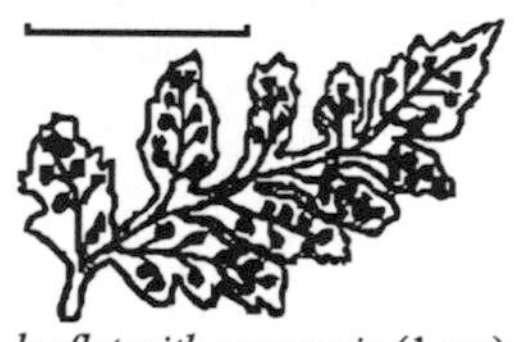

leaflet with sporangia (1 cm)

C. fragilis: Rhizome short, slender; fronds ± clustered, dying annually; petioles slender, mly shorter than blade, fragile, straw-colored or greenish, darker near base; blades lanceolate, thin, 10-25 cm L, 2-7 cm W; indusium hood-like, often toothed at apex. Usu cool, moist habitats below 12,000', SN (all). Edibility unknown.

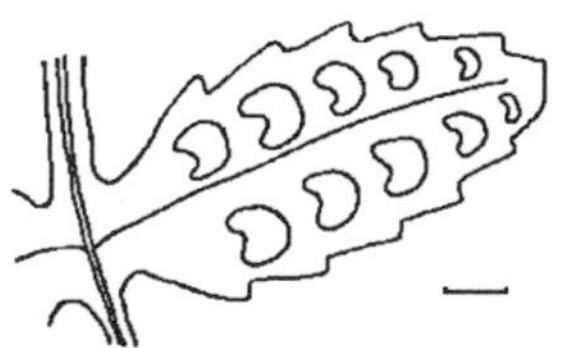

leaflet and indusia (1 mm)

Dryopteris

California Wood Fern

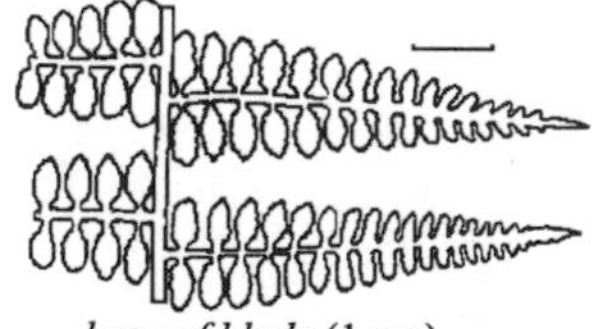

base of blade (1 cm)

D. arguta: Rhizome stout, woody; fronds tufted, evergreen, erect, 3-8 dm L; petioles thick, straw-colored to greenish except at base, scaly, shorter than blades; blades mly 2-pinnate, ± glabrous above, glandular beneath; ultimate segments often spinose-dentate; indusium distinctively kidney or horseshoe-shaped. Moist, cool habitats below 6000', SN (all). Edibility uncertain. Similar spp are purported to be edible, poisonous and of medicinal value.

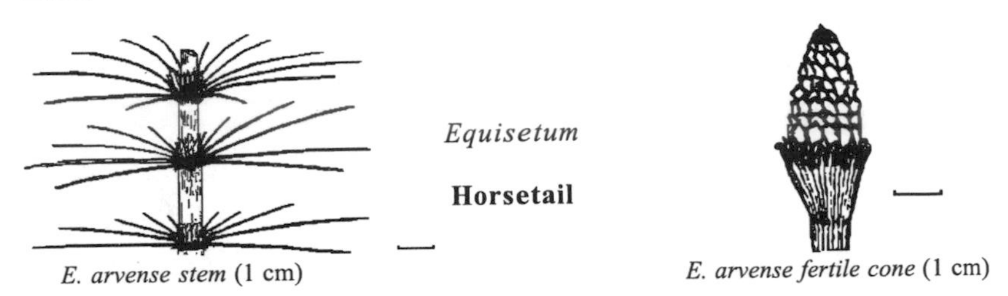

Equisetum

Horsetail

E. arvense stem (1 cm)

E. arvense fertile cone (1 cm)

Perennials from creeping rhizomes. Stems jointed, grooved, usu hollow except at joints. Joints with sheaths representing a whorl of small, fused lvs. Fr a cone-like structure (strobilus) found at the end of the erect stems. Inner pulp of all spp edible in small amounts, toxic in large quantities. The strobili of at least *E. arvense* are edible after boiling.

Stems dimorphous, the sterile with whorls of linear branches, the fertile unbranched with terminal strobilus — *E. arvense*
Stems all the same, fertile
  Sheaths about as wide as long, cylindric; strobili pointed — *E. hyemale*
  Sheaths elongate, ± funnelform; strobili blunt — *E. laevigatum*

E. arvense, Common Horsetail: Sterile stems 1-6 dm H, slender, green with several to many ridges; branches many, jointed; fertile stems 0.5-2.5 dm H, light colored, dying early, with 3-5 joints; strobili 2-4 cm L. Common in wet soil below 9500', SN (all).

E. hyemale vars., Common Scouring Rush: Stems stout, stiff, evergreen, 6-12 dm H, rarely branched; sheaths cylindrical; cones 1-4 cm L, yellow or black. Along stream banks and moist habitats below 8000', SN (all).

E. laevigatum, Smooth Scouring Rush: Stems erect, 3-8 dm H, mly smooth, rarely branched, pale green; strobili 1-2 cm L. Wet habitats below 6500', SN (all).

Isoetes

Quillwort

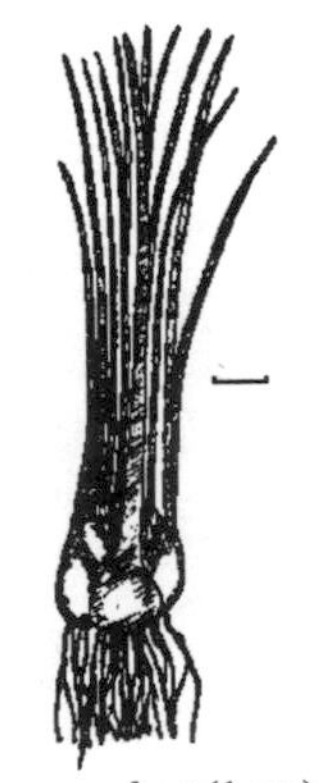

plant (1 cm)

Lvs basal from a fleshy corm; upper portion of lvs sterile with 4 longitudinal air channels with or without peripheral strands. Sporangia located in lf-axils. Widespread in mountain lakes but difficult to key to species. Plants rich in starch and oil; edible raw or cooked.

Plants always submerged; mly above 5000' — *I. occidentalis* complex
Plants amphibious or terrestrial; mly below 5000' — *I. nuttallii*

No description of the spp is given because the major differences observable in the field have been given in the key.

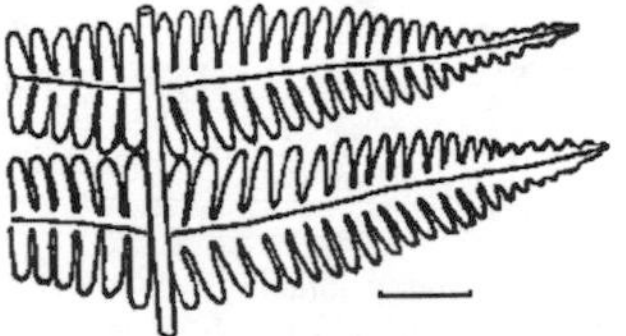

E. arvense stem (1 cm)

Lastrea

Sierra Water Fern

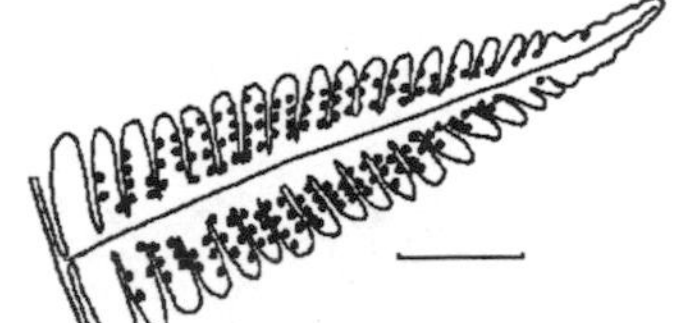

E. arvense fertile cone (1 cm)

L. oregana: Rhizome slender; fronds few, clustered, 5-9 dm L; petiole straw-colored, much shorter than blade, often scaly; blades 4-6 dm L, 1-1.5 dm W, narrowed at both ends, mly pinnate and these lfts deeply pinnately lobed; lobed ± entire, oblong, 5-10 mm L, 1-3 mm W. Wet stream banks below 5000', Madera to Plumas Co. Edibility uncertain; see *Dryopteris.*

Marsilea

Pepperwort

plant (1 cm)

M. vestita: Rhizome long; lvs few to many; petioles 2-20 cm L; blades 1-3 cm W, palmately cmpd; lfts broadly wedge-shaped, 5-15 mm L, nearly as wide, entire; peduncles short, arising from base of petiole; sporocarps solitary, 4-8 mm L, densely hairy when immature. Muddy, wet habitats below 6500', SN (all). A second sp., *M. oligospora,* is occasionally found in the Lassen region between 5000 to 7000'. Related spp edible when dried and ground into a powder.

Pellaea

Cliff-brake

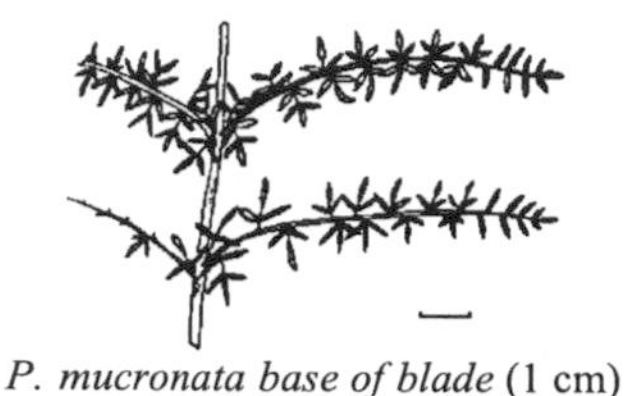

P. mucronata base of blade (1 cm)

P. bridgesii frond (1 cm)

Small ferns mly from dry rocky habitats. Fronds evergreen, glabrous, ± tufted. Sori covered by revolute margins of lfts. A tea can be prepared from the dried fronds of *P. mucronata,* which is purported to stop nosebleeds.

A Blades once pinnate; lfts ± oval, over 3 mm W	
B Lfts entire, thick	4) *P. bridgesii*
BB Lfts mitten-shaped, thin	3) *P. breweri*
AA Blades further divided; lfts usu less than 3 mm W	
B Lfts mly over 3 mm W, not strongly revolute; below 4000'	1) *P. andromedifolia*
BB Lfts mly under 3 mm W, usu revolute to midrib and linear	
C Blade 3-pinnate at base, 2-pinnate above	7) *P. mucronata*
CC Blade 2-pinnate at base	
D Lfts 5-11 per pinna	
E Plumas Co. n.	2) *P. brachyptera*
EE Central SN	6) *P. xglaciogena*
DD Lfts 11-15 per pinna; Placer Co. s.	5) *P. compacta*

1) *P. andromedifolia,* Coffee Fern: Rhizome slender; fronds 1.5-7 dm L; petioles light brown, 0.5-4 dm L; blades 2-4-pinnate, the ultimate segments not crowded, oval to oblong, ± obcordate, 5-10 mm L, 4-8 mm W. SN foothills, Tulare to Butte Co.

2) *P. brachyptera,* Sierra Cliff-brake: Rhizome thick, woody; fronds stiff, 1.5-5 dm L; petioles about as long as blades, purplish brown; blades 1-3 dm L; ultimate segments gray-green, 2-6 mm L, wrinkled; ultimate rachis mly shorter than lfts. Exposed, rocky habitats, 3500-8000'.

3) *P. breweri,* Brewer's Cliff-brake: Rhizome short; fronds clustered, 0.5-20 dm H; petioles 3-10 cm L, red-brown, fragile, with brown scales at base; blades linear-oblong, 5-15 cm L, 1.5-3.5 cm W, light green; lower lfts usu 2-lobed or 2-parted with the upper lobe larger. Rocky habitats, 7000-11,000', Tulare Co. n.

4) *P. bridgesii*, Bridge's Cliff-brake: Rhizome thick; fronds 0.5-2 dm H; petioles 4-10 cm L, brown, glossy; blades oblong, grayish, 4-10 dm L, 1-2.5 cm W; the lfts paired, ± overlapping, oval, sessile, leathery. Dry habitats, especially around rocks, 5000-11,000', Tulare Co. n.

5) *P. compacta*, Desert Cliff-brake: Rhizome thick, woody; fronds many, clustered, 2-3.5 dm L; petioles brown, about 10 cm L; blades gray-green, oblong; ultimate rachis 2-3 times as long as the 3-4 mm L lfts. Dry rocky habitats, 4500-8500'.

6) *P. xglaciogena*. Hybrid between *P. compacta* and *P. mucronata*.

7) *P. mucronata,* Bird's-foot Fern: Rhizome thick, woody; fronds 1.5-4 dm L, stiff; petioles 4-15 cm L, brown to black, fragile; blades bluish gray; lower lfts usu in clusters of 3's much like a bird's foot. Common, rocky habitats below 5000', SN (all).

var. *californica*: Lfts never in 3's. Rare, mly above 6000'.

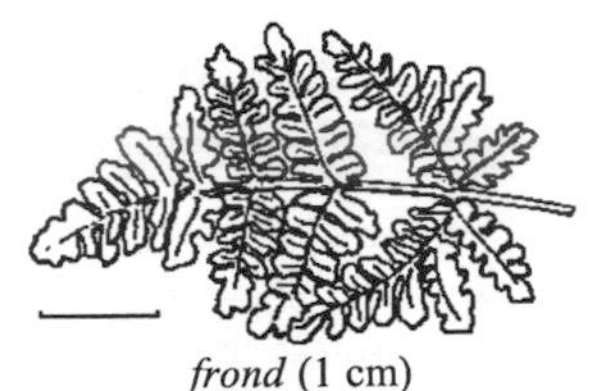
frond (1 cm)

Pentagramma

Goldback Fern

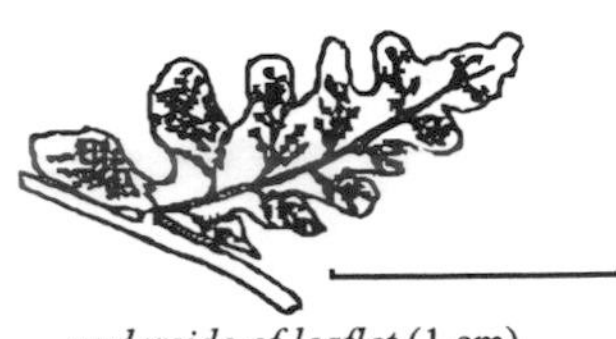
underside of leaflet (1 cm)

P. triangularis: Rhizome thick; fronds many, 1-4 dm L; petioles brittle, red-brown or darker in age, shiny, longer than blades; blades triangular, 1-2-pinnate, leathery; ultimate lfts dark green and glabrous above, waxy yellow beneath. Common in moist habitats below 4500', SN (all). Edibility unknown.

var. *pallida:* Lfts waxy white beneath. Occasional, with sp.

underside of frond (1 cm)

Polypodium

Golden Polypody

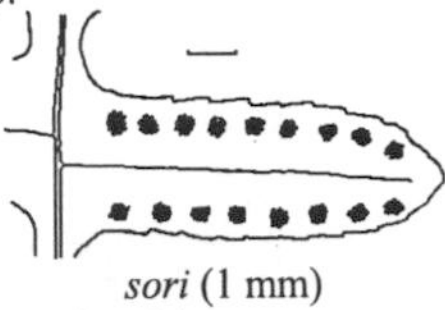
sori (1 mm)

P. hesperium: Rhizome slender; fronds to 25 cm L; 1-pinnate, the ultimate segments ± 2 cm L. Sori ovate, 1-2.5 mm W. Rocky habitats 4500-9000'. SN (all). Rhizome of *P. glycyrrhiza* used as licorice substitute; however, rhizome of *P. hesperium* has an acrid taste.

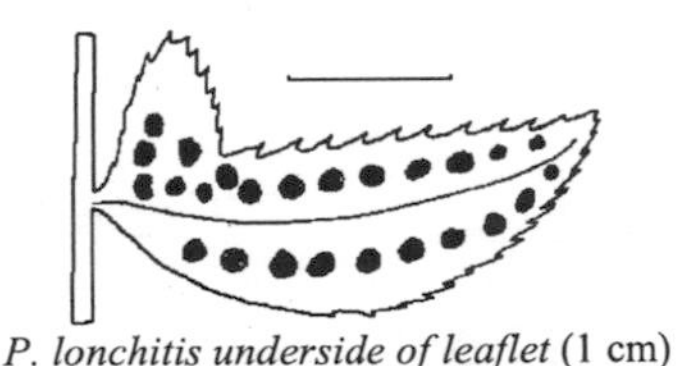
P. lonchitis underside of leaflet (1 cm)

Polystichum

Polystichum Sword Fern

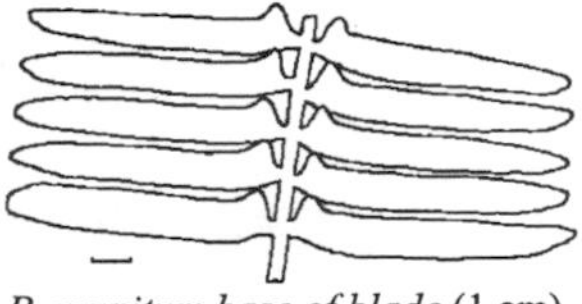
P. munitum base of blade (1 cm)

Evergreen, medium to large ferns. Rhizomes mly stout and woody. Fronds stiff and leathery. The lfts often with sharp spinulose teeth. Sori round, in 2 rows on each lft and covered by an umbrella-like indusium. Rhizome of oriental sp starchy and edible. Roots of the Sierran spp may be edible after roasting. Hybridization common between some species.

A At least lfts near base of frond pinnately lobed (sometimes not so in *P. kruckebergii*)
  B Lft teeth lacking spinose tips; on serpentine; n. SN — 3) *P. lemmonii*
  BB Lft teeth with spinose tips; on serpentine or not
    C Lowest primary lfts deltate, less than 1 cm L — 2) *P. kruckebergii*
    CC Lowest primary lfts lanceolate, 1.5-3 cm L — 6) *P. scopulinum*
AA All lfts entire to serrate, not lobed
  B Basal lfts ovoid-deltoid, shorter than middle lfts — 4) *P. lonchitis*
  BB Basal lfts lanceolate, longer than middle or upper lfts
    C Scales on petiole ovate, persistent above first lfts — 5) *P. munitum*
    CC Scales on petiole lanceolate, those above first lfts deciduous — 1) *P. imbricans*

1) *P. imbricans:* Petiole less than 1/2 blade length, the base scales 2-3 mm W; blade narrow-lanceolate to -ovate. Rocky habitats below 8000', SN (all).

2) *P. kruckebergii,* Kruckeberg's Sword Fern: Petiole usu 1/10 blade length; blade linear to narrow-ovate. Rocky slopes and cliffs, 7000-10,000', SN (all). Probably a hybrid between *P. lemmonii* and *P. lonchitis*.

3) *P. lemmonii,* Lemmon's Sword Fern: Frond 15-40 cm L; blade narrow-lanceolate, 2-3 pinnate. Rocky habitats below 7000'.

4) *P. lonchitis,* Holly Fern: Petiole thick, 1-6 cm L, with brownish scales; blades 10-50 cm L, 2-7 cm W, widest at or above middle. Rare, cool, rocky habitats, 6500-8500', Placer and Plumas cos.

5) *P. munitum,* Sword Fern: Petiole slender to thick, often scaly, shorter than blades; blades ± lanceolate, to 12 dm L, 25 cm W; lfts dark green and glossy above, paler beneath. Open and shaded moist habitats below 7000', SN (all).

6) *P. scopulinum,* Eaton's Shield Fern: Fronds few, 1.5-4 dm L; petioles stout, grooved, densely scaly, 3-15 cm L; blades 1-3 dm L, 2.5-6 cm W, widest near the middle, scaly and fibrous beneath. Dry, rocky habitats, 5000-10,000', n. SN.

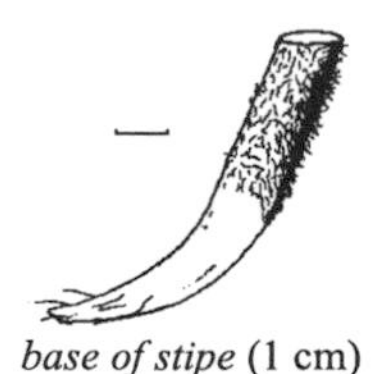

Pteridium

Bracken Fern

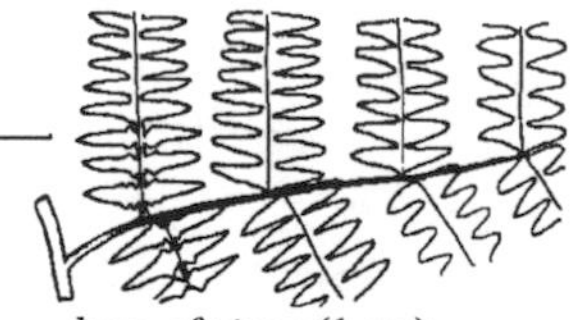

base of stipe (1 cm) *base of pinna* (1 cm)

P. aquilinum var. *pubescens:* Rhizome black, woody; fronds 6-12 dm H; petioles light-colored, stout, densely hairy at base, shorter than blades; blades 1.5-15 dm L, ± triangular, 2-3-pinnate; ultimate segments lanceolate, ± glabrous above, pubescent beneath; sori linear, covered by revolute margin. Common in wooded areas below 10,000', SN (all). Young fronds edible boiled; old fronds may be toxic in large amounts. Rhizome edible roasted or boiled but usu tough.

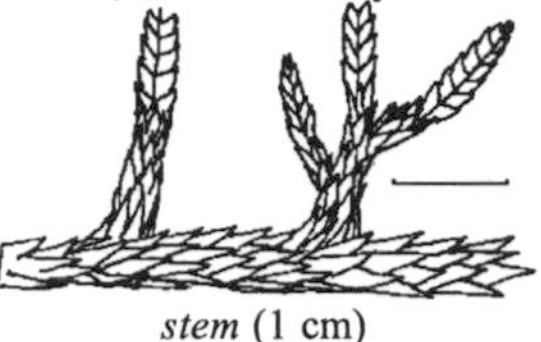

Selaginella

Spike-moss

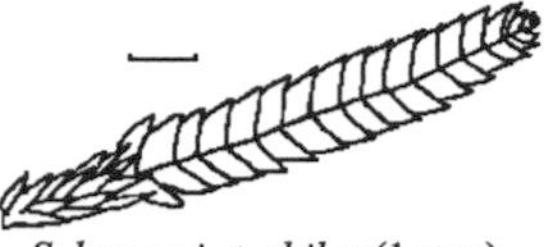

stem (1 cm) *S. hanseni strobilus* (1 mm)

Low, often prostrate, moss-like plants with slender, branched stems and scale-like, overlapping lvs. The lvs short -awned, about 2 mm L. Sporangia found in cone-like strobili at the tips of the branches. Edibility unknown.

Strobili under 10 mm L; moist places below 6000' *S. hanseni*
Strobili over 10 mm L; dry places, mly 7500-13,000' *S. watsoni*

S. hanseni, Hansen's Selaginella: Stems 5-25 cm L; lvs glaucous, usu with many hairs along margins; strobili 5-9 mm L, triangular in cross-section. Forming mats on rocky outcrops, SN (all).

S. watsoni, Alpine Selaginella: Stems 2-20 cm L; branches few and short; lvs crowded, thick, with few or no hairs on margin; strobili 1-2.5 cm L, 4-sided. Rocky habitats, Nevada Co. s.

Woodsia

Woodsia

Small ferns with slender rhizomes. Fronds clustered; petioles yellow-green with dark scaly base. Sori near margin, with cup-like indusium underneath. Indusia with ascending lobes. Edibility unknown.

Blades ± glabrous, bright green *W. oregana*
Blades glandular-puberulent, dull green *W. scopulina*

W. oregana, Oregon Woodsia: Fronds 0.5-2 dm H; petioles 2-10 cm L; blades 5-20 cm L, 1.5-5 cm W, delicate, pinnate and then deeply pinnately lobed; indusia with hair-like segments. Infrequent, dry, rocky habitats mly below 5000', SN (all).

plant (1 cm)

W. scopulina, Rocky Mountain Woodsia: Fronds clustered, 10-25 cm H; petioles 3-15 cm L, dark below, lighter above; blades 1-2-pinnate; indusia with lanceolate lobes. Occasional, rocky habitats, 4000-9000', Eldorado Co. s.

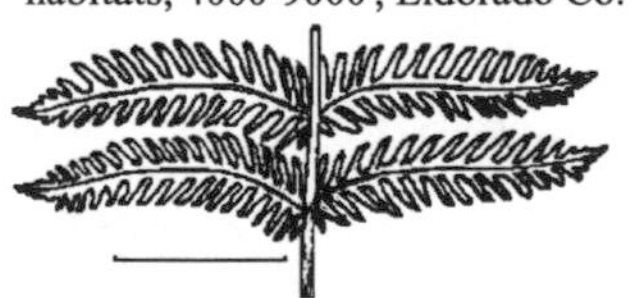

Woodwardia

Giant Chain Fern

section of blade (1 cm) *underside of pinnule* (1 cm)

W. fimbriata: Rhizome large, woody; fronds many, suberect, 10-25 dm H; petioles short, straw-colored except at base; blades nearly bipinnate, 2-5 dm W, ± oblong; ultimate segments lanceolate. Wet seeps, mly below 5000', SN (all). Edibility unknown.

KEY TO THE CONIFER GENERA

A Lvs needle-like, usu more than 1 cm L	
B Needles in clusters of 2 or more	*Pinus* p. 16
BB Needles single, never grouped in clusters	
C Fr ± fleshy, or if cone-like, disintegrating before falling	
D Shrub less than 1 m H; fr a berry	*Juniperus communis* p. 16
DD Trees mly over 5 m H	
E Fr cone-like; common forest trees	*Abies* p. 15
EE Fr nut-like, at least partially covered by a fleshy pericarp	
F Needles 1-2 cm L, ± blunt; Calaveras Co. n.	*Taxus* p. 18
FF Needles 2.5-6 cm L, sharp-tipped; Plumas Co. s.	*Torreya* p. 18
CC Plants producing ± woody cones which remain intact after falling	
D Cones ± globose; the scales thick	*Pinus monophylla* p.17
DD Cones oblong; the scales thin and delicate	
E Cones with notched bracts between scales	*Pseudotsuga* p. 17
EE Cones without bracts between scales	*Tsuga* p. 18
AA Lvs scale-like, usu less than 1 cm L	
B Lvs less than 5 mm L, ± triangular	
C Fr fleshy, berry-like; bark fibrous; widespread	*Juniperus* p. 16
CC Fr woody, cone-like; bark scaly; rare, Plumas Co.	*Cupressus* p. 16
BB Lvs mly 5-10 mm L, ± oblong to lanceolate	
C Lvs in flat sprays; cones splitting lengthwise	*Calocedrus* p. 15
CC Branchlets not flattened; cones opening into many scales	*Sequoiadendron* p. 18

branch with cone (1 dm)

Abies

Fir

branchlet (1 cm)

Large evergreen trees mly with a conical shape. Branches relatively short, often in whorls along the trunk and ending in flat sprays of solitary needles. Needles mly less than 5 cm L, slightly expanded at base and, on falling, leaving an oval or round scar flush with the surface of the branchlet. Male flowers are inconspicuous cylindrical, scaly structures hanging below the upper branches. Cones sit erect on top of upper branches, not particularly woody and usu disintegrating before falling. Seeds eaten by animals but are very bitter and unpalatable.

Needles flat, green on upper surface; older bark ashy-gray — *A. concolor*
Needles ± 4-sided with whitish longitudinal lines on upper surfaces; older bark reddish — *A. magnifica*

A. concolor, White Fir: Trunk 20-70 m H; older bark with wide furrows; younger bark grayish white, smooth; needles 3-6 cm L, blunt or sharp-pointed, the older and lower ones straight, greenish above, whitish below, the younger ones curved upward; female cones greenish or purplish, 7-12 cm L; bracts shorter than scales. A major sp below 7000' and occasional up to 10,000', SN (all).

A. magnifica, Red Fir: Large tree 40-60 m H except at high elevations; older bark with irregular ridges and diagonal furrows; younger bark smooth, grayish white; needles 1.5-3.5 cm L, usu whitish on all sides, the oldest slightly flattened; female cones dark purplish brown, 15-20 cm L; bracts about 2/3 as long as scales. Dominant on many of the mid-elevation slopes, 5000-9000', SN (all).

var. *shastensis:* Bracts exceeding scales. With sp.

branchlet (1 cm)

Calocedrus

Incense-cedar

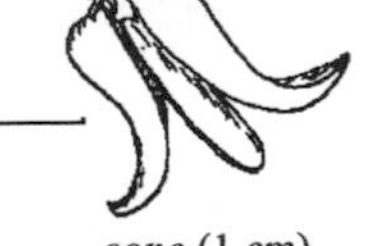

cone (1 cm)

C. decurrens: Evergreen tree mly 20-35 m H, with thick, fibrous, furrowed, cinnamon-brown bark; crowns irregular to conical; younger trees bearing branches to base of trunk; lvs 3-

20 mm L, imbricate, forming champagne glass-like patterns along the branchlets; cones oblong, about 2 cm L with 2 woody scales splitting from a central wall to expose the winged seeds when ripe. Common in canyons and lower slopes, mly below 7000', SN (all).

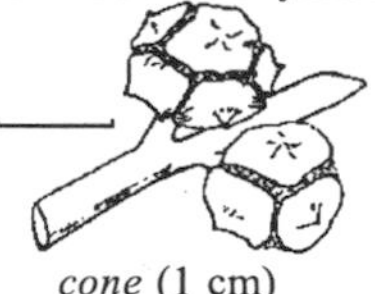

cone (1 cm)

Cupressus

Baker Cypress

branchlet (1 mm)

C. baker: Tree 10-25 m H; older bark grayish or reddish, curling off in irregular layers; branches horizontal or inclined; branchlets slender; lvs mly green, about 2 mm L, pointed, with resin pit on back. Rare, known from only two locations in Plumas Co., about 1.5 miles w. of Wheeler Peak and 1-2 miles ne. of Eisenhammer Peak. Too rare to consider edibility. A second sp., *C. arizonica* is rare but to be looked for at lower elevations in Kern and Tulare cos.

J. communis branchlet (2 cm)

Juniperus

Juniper

J. occidentalis branchlet and fruit (1 cm)

Large shrub or tree with aromatic foliage and fibrous bark. Branches stout and often contorted. Fruit a fleshy berry. The fruit is edible raw, but is better dried, ground and made into cakes. The oil in the frs of *J. californica* is said to contain a diuretic and should not be eaten in quantity. Herbage has unpleasant taste and may be poisonous.

Lvs linear, 6-12 mm L — *J. communis*
Lvs scale-like, 3-4 mm L
  Large shrub, 1-4 m H; bark grayish — *J. californica*
  Tree 5-20 m H; bark cinnamon in hue — *J. occidentalis*

J. californica, California Juniper: Foliage yellow-green; lvs 3-4 mm L, pitted on back; fr 10-20 mm L, red-brown when ripe. Dry, rocky slopes and flats below 5000', Kern River drainage to Trout Mdw and occasional on the lower western slopes of the southern SN.

J. communis var. *saxatilis,* Dwarf Juniper: Low shrub to 1 m H; lvs sharp-pointed, dark green; berries 7-9 mm in diameter, blue with a white powder. Stony or wooded slopes, 6400-11,000', Mono Pass n.

J. occidentalis, Western Juniper: Crown usu dense, roundish; the branches often greatly contorted on exposed windy sites; foliage light green; lvs pitted on back and usu covered with a whitish resin; fr mly less than 10 mm L, blue-black with a whitish powder when ripe. Dry habitats, 3000-10,500', SN (all).

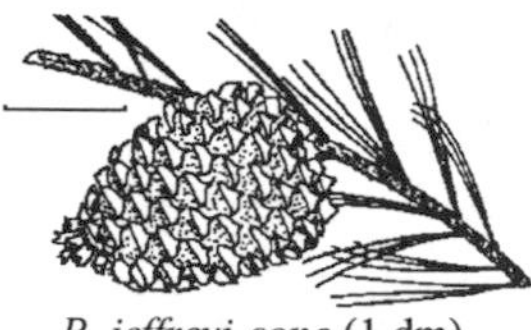

P. jeffreyi cone (1 dm)

Pinus

Pine

P. albicaulis branchlet (1 mm)

Evergreen trees or, at high elevations, shrubs. Lvs mly in clusters with a membranous cylindrical sheath at the base of each cluster. Fr a winged nut located on the upper surface of each scale of the "pine cone." All Sierran spp have edible nutlets but those of *P. monophylla* are largest and best eating. The nutlets are most easily gathered by heating the green cone until it opens. The nuts are shelled and usu ground or roasted. The inner bark of all pine may be used as an emergency food.

A Needles in clusters of 4-5
  B Cones mly less than 8 cm L with thick scales bearing prickles
    C Cones remaining closed, disintegrating at maturity; scales with blunt tips — 1) *P. albicaulis*
    CC Cones opening at maturity; scales tipped with small curved prickle — 2) *P. balfouriana*
  BB Cones mly 8-40 cm L; scales usu thin and flexible, lacking prickles

C Cones with short (less than 2 cm) stalk; needles 3-7 cm L — 4) *P. flexilis*
CC Cones with long (greater than 2 cm) stalk; needles 5-40 cm L
D Cones 10-20 cm L; 6000-9000' — 8) *P. monticola*
DD Cones 25-40 cm L; mly below 7000' — 6) *P. lambertiana*
AA Needles in clusters of 3 or less
B Needles in clusters of 3
C Foliage grayish; bark furrowed; below 4500' — 10) *P. sabiniana*
CC Foliage green; bark with puzzle-shaped scales
D Bark lacking odor; cones with projecting prickles — 9) *P. ponderosa*
DD Bark with vanilla-like odor; cones with incurved prickles — 5) *P. jeffreyi*
BB Needles in clusters of 1 or 2
C Needles solitary, 2.5-3.5 cm L — 7) *P. monophylla*
CC Needles clustered in 2's, 3-6 cm L — 3) *P. contorta*

1) *P. albicaulis,* Whitebark Pine: Low, sprawling shrub to medium sized tree; branches long and often contorted; bark thin, broken into thin whitish scales on the surface, covering the reddish brown deeper bark; needles 3-5 cm L; cones globose, usu torn apart by animals after falling. Rocky habitats near timberline, 7000-12,000', Tulare Co. n.

2) *P. balfouriana* ssp. *austrina,* Foxtail Pine: Small tree, 6-15 m H; older bark cinnamon-brown, broken into squarish sections; upper bark smooth and whitish; branches short; needles thick, 2-4 cm L, clustered at the ends of branchlets; cones 5-14 cm L; tip of scale blunt, the raised portion with prickle actually on the dorsal surface of scale. Rocky slopes, 9000-11,500', vicinity of Onion Valley, Inyo C. s., to Sirretta Peak, Tulare Co.

3) *P. contorta* var. *murrayana,* Lodgepole Pine: Small to medium sized tree to 40 m H; bark gray, scaly; cones ± spherical, 2.5-4 cm in diameter, scales with small incurved prickles. Dry to moist areas, often the first tree to invade moist meadows, 5000-11,000', SN (all).

4) *P. flexilis,* Limber Pine: Tree 10-20 m H; old bark brown or black and scaly; younger bark whitish and smooth; lvs thick, stiff, often curved, dark green; cones 10-25 mm L, oblong with thick scales. Dry slopes, 8000-12,000', Mono Pass s., mly e. SN.

5) *P. jeffreyi,* Jeffrey Pine: Tree 20-60 m H; bark reddish brown, deeply furrowed and broken into relatively narrow plates; needles 15-30 cm L, dull blue-green, glaucous, sharp-pointed; cones oval, 15-35 cm L. Hybridizes with *P. ponderosa.* Dry ground, 6000-9500', SN (all), especially on e. slope.

6) *P. lambertiana,* Sugar Pine: Large tree to 75 m H; bark dark reddish or purplish, narrowly furrowed and with irregular scales; branches widespreading, almost perpendicular to main trunk; cones hanging from the tips of the long branches on stalks 5-7 cm L; scales thin, easily broken. Common on dry slopes in pine forests below 9000', SN (all).

7) *P. monophylla,* Piñon Pine: Small tree 5-15 m H, usu with branched trunk; old bark irregularly furrowed with small reddish brown scales; needles gray-green, usu solitary, sharp-pointed; cones ± spherical, 3-5 cm L with thick scales, the scales knobbed at apex. Dry, rocky place s; below 9000', mly e. SN from Mono Co. s.

8) *P. monticola,* Western White Pine: Tree to 50 m H; older bark brownish, mly divided into small squares, the younger smooth and light gray; needles blunt. In dry habitats, 5000-10,000', SN (all).

9) *P. ponderosa,* Yellow Pine: Large tree to 70 m H; bark thick, yellowish, broken into large rectangular plates to 2 m L and 20 cm W, the furrows shallow; needles 10-25 cm L, deep yellowish green, not glaucous, sharp-pointed; cones oval, the scales flat with the lower surface darker than the upper. Hybridizes with *P. jeffreyi.* Common, often the dominant sp in lower elevation forest, dry ground, 3000-7500', SN (all). A similar species, *P. washoensis,* which has shorter needles and cones 10-12 cm L, is found infrequently along the eastern edge of the SN from Lake Tahoe n.

10) *P. sabiniana,* Foothill Pine: Tree to 25 m H with sparse grayish foliage; trunk often branched near base; bark dark brown with wide furrows and scaly ridges; needles 20-30 cm L; cones ± spherical, 15-30 cm L, borne on long, drooping stalks; scales flat, tipped by a stout outwardly curving prickle. Dry rocky habitats, SN (all).

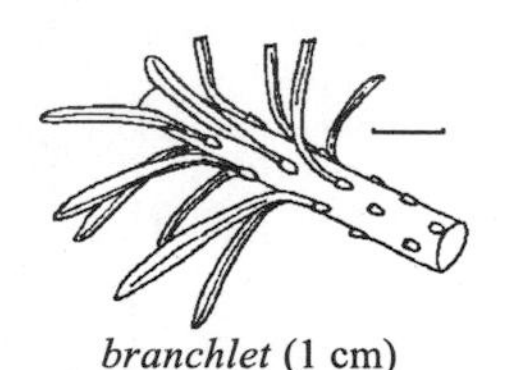

branchlet (1 cm)

Pseudotsuga

Douglas-fir

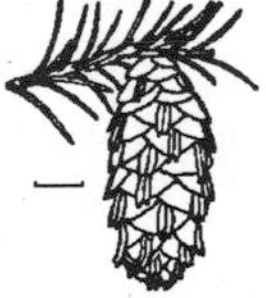

cone (1 cm)

P. menziesii: Tree to 50 m H; older bark dark with deep furrows and broad ridges; younger bark smooth and grayish; branchlets drooping; needles 2-3 cm L, usu flattened, growing from all sides of branchlets; cones hanging at ends of branches, 5-8 cm L, with conspicuous bracts between the scales. Moist habitats below 5000', Fresno Co. n. A delicious tea can be made from the needles.

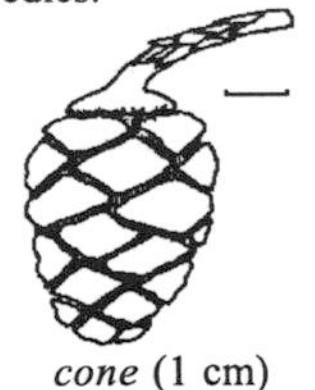
cone (1 cm)

Sequoiadendron

Big Tree
Sequoia

branchlet (1 cm)

S. giganteum: Large tree; bark thick, fibrous, cinnamon-red, often with burn scars; lowest branches usu far up the trunk, thick, short, and often bent upwards; lvs scale-like, 3-10 mm L, appressed to branchlets or the tips spreading; female cones oblong-ovoid, about 5 cm L, reddish. Moist ground along w. slope at 4600-8400', Tulare to Placer Co. Plant inedible.

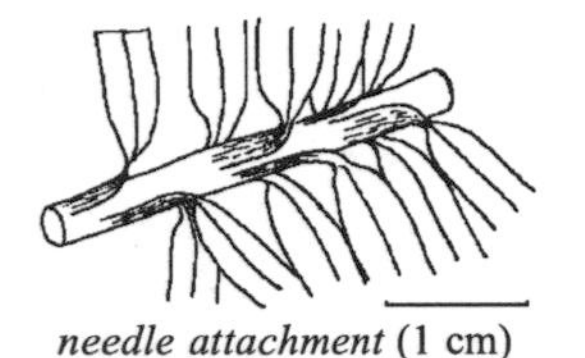
needle attachment (1 cm)

Taxus

Pacific Yew

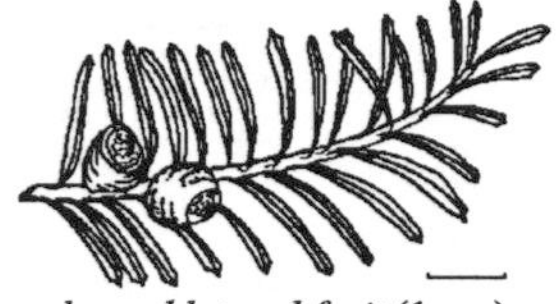
branchlet and fruit (1 cm)

T. brevifolia: Evergreen tree to 25 m H; bark thin, reddish with small scales; branchlets often drooping; needles 1-2 cm L, dark green above, pale below, in flat sprays along branchlets, leaving a short stem on branchlet upon falling; fr reddish, about 1 cm L, seed exposed on distal end. Damp, cool habitats below 7000'. Fleshy seed cover edible, seed and herbage toxic.

branchlet (1 cm)

Torreya

California-nutmeg

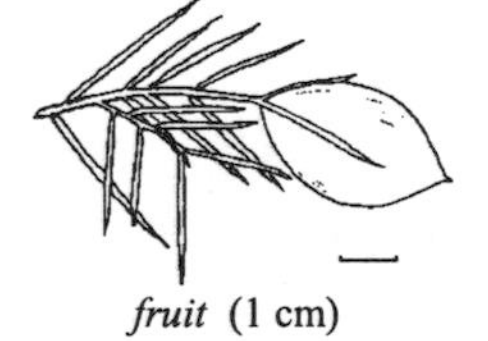
fruit (1 cm)

T. californica: Evergreen tree to 30 m H; bark thin, gray-brown, irregularly broken and scaly; branches whorled or subopposite; needles dark green, very sharp-pointed; seed with leathery, greenish or purplish covering, 2.5-3.5 cm L. Moist shaded slopes below 7000'. White inner part of seed edible, reddish outer part bitter.

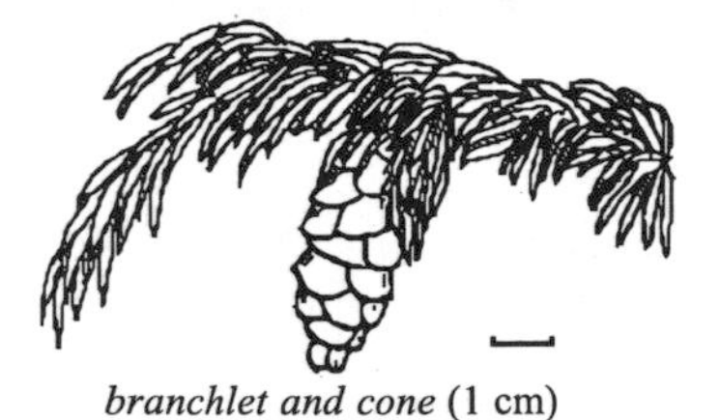
branchlet and cone (1 cm)

Tsuga

Mountain
Hemlock

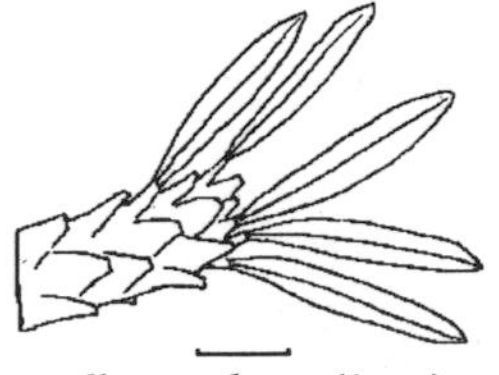
needle attachment(1 cm)

T. mertensiana: Evergreen tree to 45 m H, with slender conical outline and drooping top, bark scaly and deeply furrowed, dark; needles 1-2 cm L, blue-green; cones oblong, 4-9 cm L. Moist, often north-facing slopes, 6000-11,000', Fresno Co. n. Tea can be made from the needles, and in emergencies the inner bark is edible.

KEY TO THE DICOTYLEDON FAMILIES

A Plants aquatic, usu immersed or floating in water
B Lvs opp or whorled along floating or submerged stem; fls small
C Lvs whorled
D Lvs simple, lanceolate HIPPURIDACEAE p. 126
DD Lvs pinnatifid into linear segments HALORAGACEAE p. 126
CC Lvs opposite
D Stipules none; submerged lvs usu linear CALLITRICACEAE p. 84
DD Stipules present, scarious; stem lvs ± 4-ranked ELATINACEAE p. 102
BB Lvs alt or basal, often floating; fls usu showy
C Lvs dissected, submersed; fls yellow or white
D Lvs with urn-shaped structures for capturing insects; corolla 2-lipped LENTIBULARIACEAE p. 137
DD Lvs lacking such structures; fls radially symmetric RANUNCULACEAE p. 165
CC Lvs simple
D Lvs mly less than 2 cm L, in basal tufts; fls small, mly white
E Fls solitary on naked scape SCROPHULARIACEAE (*Limosella*)p. 195
EE Fls usu several per scape BRASSICACEAE (*Subularia*) p. 83
DD Lvs over 2 cm L, floating; fls yellow or reddish
E Lvs peltate, floating; fls 3-merous, reddish purple CABOMBACEAE p. 84
EE Lvs not peltate; fls 5 to many-merous
F Fls bowl-shaped in outline; the yellow sepals about 5 cm L NYMPHACEAE p. 142
FF Fls rotate, smaller; the sepals green MENYANTHACEAE (*Nymphoides*)p. 141
AA Plants terrestrial or if growing in wet areas, the lvs usu well emergent
B Plants woody throughout (not just at base)
C Lvs opposite
D Lvs cmpd
E Lvs palmately cmpd HIPPOCASTANACEAE p. 126
EE Lvs pinnately cmpd or trifoliolate
F Vines RANUNCULACEAE (*Clematis*) p. 167
FF Shrub or tree
G Fr a capsule or drupe, without membranous wings
H Lfts 3; fr a dry capsule STAPHYLEACEAE p. 203
HH Lfts 5-7; fr a fleshy drupe CAPRIFOLIACEAE (*Sambucus*) p. 88
G Fr dry with a membranous wing
H Lvs 3-9 dm L; lfts 13 to 25 SIMAROUBACEAE p. 202
HH Lvs 0.5-2 dm L; lfts 3 to 7
I Lfts usu 3, the margins usu 1-3-toothed ACERACEAE p. 33
II Lfts 3 to 7, the margins serrate OLEACEAE p. 142
DD Lvs simple
E Petals ± united
F Lvs narrow-elliptic, margins revolute; moist places ERICACEAE (*Kalmia*) p. 105
FF Lvs broader; margins not strongly revolute
G Fls irregular; lvs usu sessile; fr a capsule SCROPHULARIACEAE p. 191
GG Fls usu regular; lvs petioled; fr usu fleshy CAPRIFOLIACEAE p. 86
EE Petals separate or none
F Stipules with thick persistent bases RHAMNACEAE (*Ceanothus*) p. 171
FF Lvs without stipules
G Lvs serrate (except sometimes in PHILADELPHACEAE)
H Lvs palmately lobed ACERACEAE p. 33
HH Lvs oblong to roundish, without lobes
I Fls 4-merous; petals 1 mm L CELASTRACEAE p. 95
II Fls usu 5-7-merous; petals more than 6 mm L PHILADELPHACEAE p. 150
G Lvs entire
H Fls in catkins, lacking showy petalloid parts GARRYACEAE p. 121
HH Fls not in catkins, with showy petals or bracts
I Lvs thin; fls white CORNACEAE p. 98
II Lvs firm,; perianth reddish CALYCANTHACEAE p. 84
CC Lvs alternate, whorled, bunched, or basal
D Male and female fls in separate structures, at least male fls in catkins
E Fr an acorn or bur FAGACEAE p. 119

EE Fr a winged nutlet, smooth nut, or capsule
  F Male and female catkins on same plant — BETULACEAE p. 71
  FF Male and female catkins on separate plants
    G Fr a capsule containing many minute seeds; lvs various — SALICACEAE p. 185
    GG Fr 1-seeded; lvs toothed on distal half — MYRICACEAE p. 141
DD Fls not in catkins
  E Lvs cmpd or lobed
    F Vines with tendrils opp the lvs — VITACEAE p. 206
    FF Erect shrubs or vines lacking tendrils
      G Lvs densely yellowish stellate pubescent beneath — STERCULIACEAE p. 203
      GG Pubescence not yellow and rarely stellate if present
        H Fls 6-merous; lvs pinnate; lft-margin prickly — BERBERIDACEAE p. 70
        HH Fls 5-merous; lvs not as above
          I Lvs palmately divided into 5 or more linear to oblong segments
            J Low shrubs; the linear lobes sharp-pointed — POLEMONIACEAE (*Leptodactylon*) p. 154
            JJ Erect shrubs; lfts not sharp-pointed — FABACEAE (*Lupinus*) p. 113
          II Lvs shallowly palmately lobed or pinnately lobed or divided
            J Lvs pinnately 3-foliolate; lfts ± oval, shiny above — ANACARDIACEAE p. 34
            JJ Lvs various, but never pinnately 3-foliolate
              K Lvs palmately lobed; fr fleshy, often prickly — GROSSULARIACEAE p. 124
              KK Lvs mly pinnate, if palmate the fr dry — ROSACEAE p. 173
  EE Lvs entire to serrate, never lobed
    F Petals united at least at base
      G Fls in compact heads — ASTERACEAE p. 44
      GG Fls solitary or in loose clusters
        H Corolla urn-shaped (except *Rhododendron*) — ERICACEAE p. 102
        HH Corolla funnelform — HYDROPHYLLACEAE (*Eriodictyon*) p. 127
    FF Petals distinct to base
      G Lvs aromatic with bay leaf' odor — LAURACEAE p. 137
      GG Lvs not distinctly aromatic
        H Lvs round, cordate, entire — FABACEAE (*Cercis*) p. 110
        HH Lvs elliptic to oblong, never cordate
          I Petals clawed or absent; lvs often 3-veined from base — RHAMNACEAE p. 171
          II Petals present, without extended claw; lvs pinnately veined — ROSACEAE p. 173
BB Plants herbaceous, at least the branches
  C Plants not green
    D Plants parasitic, growing on upper parts of other plants
      E Stems jointed or scale-like; plants on branches of trees — VISCACEAE p. 206
      EE Stems twining, vine-like; on herbs or shrubs — CUSCUTACEAE p. 101
    DD Plants growing from ground (root parasites or saprophytes)
      E Stamens more than 5; plants white, red or brownish — ERICACEAE p. 102
      EE Stamens 4; plant purplish to pale yellow — OROBANCHACEAE p. 147
  CC Plants green
    D Petals or petaloid part of perianth united at base
      E Perianth lacking green sepals or these minute
        F Fls densely clustered in heads
          G Plant matted; lvs opp; fls reddish; rare, Tulare Co — NYCTAGINACEAE p. 142
          GG Plant and fls various; lvs usu alt; common — ASTERACEAE p. 44
        FF Fls in loose clusters or solitary
          G Lvs alt or basal
            H Lvs simple with tendrils opp lvs; dry habitats — CUCURBITACEAE p. 100
            HH Lvs trifoliate, without tendrils; wet habitats — MENYANTHACEAE p. 141
          GG Lvs opp or whorled
            H Stem lvs pinnate or pinnately lobed — VALERIANACEAE p. 204
            HH Stem lvs simple, often whorled — RUBIACEAE p. 184
      EE Perianth with green sepals below petals
        F Ovary 4-lobed; fr 4 nutlets
          G Stem 4-angled; herbage often aromatic — LAMIACEAE p. 133
          GG Stems round; herbage not strongly aromatic
        H Lvs mly alt; otherwise fls white, small — BORAGINACEAE p. 72
        HH Lvs opp; fls mly purple — VERBENACEAE p. 204

FF Ovary not lobed; fr various, never 4 nutlets
G Ovary inferior (below insertion point of petals and sepals)
H Lvs opp, elliptical; stem trailing on ground CAPRIFOLIACEAE (*Linnaea*) p. 86
HH Lvs alt; stem erect CAMPANULACEAE p. 84
GG Ovary superior (positioned above perianth attachment to axis of flower)
H Plants with milky white sap
I Sepals and petals strongly reflexed; fls in umbels ASCLEPIADACEAE p. 43
II Sepals and petals not or slightly reflexed; fls in cymes or pairs APOCYNACEAE p. 41
HH Plants with clear sap
I Corolla irregular
J Lvs mly opp; stamens 5 or less SCROPHULARIACEAE p. 191
JJ Lvs alt or basal; stamens more than 5
K Petals 4, in 2 unlike pairs PAPAVERACEAE p. 148
KK Petals 3, appearing to be 5 due to 2 petaloid sepals POLYGALACEAE p. 157
II Corolla regular
J Stems trailing or low subshrubs
K Matted subshrubs; corolla urn-shaped ERICACEAE (*Gaultheria*) p. 104
KK Stems usu trailing, vine-like; corolla funnelform CONVOLVULACEAE p. 98
JJ Stems ± erect, not vine-like
K Style cleft; calyx mly deeply cleft; corolla with definite tube; fls 5-merous
L Style 2-cleft HYDROPHYLLACEAE p. 126
LL Style 3-cleft or fls in spiny heads POLEMONIACEAE p. 151
KK Style not cleft; calyx often merely lobed; fls often 4-merous
L Lvs alt, petioled SOLANACEAE p. 202
LL Lvs opp or basal, sessile or subsessile
M Stamens placed opp corolla lobes PRIMULACEAE p. 163
MM Stamens inserted alternate to corolla lobes GENTIANACEAE p. 121
DD Petals none or distinct to base
E Plants with milky white sap EUPHORBIACEAE p. 108
EE Plants with colorless or rarely yellow sap
F Ovary ± inferior or perigynous
G Fls lacking petals
H Plants lacking stem or plant attached to branches of trees
I Plants without stem; fls usu on ground ARISTOLOCHIACEAE p. 43
II Plants parasites in branches of trees VISCACEAE p. 206
HH Plants with stem; fls not on ground
I Stem stout; lvs pinnately irregularly incised DATISCACEAE p. 101
II Stems slender; lvs entire SANTALACEAE p. 187
GG Fls with petals
H Infl an umbel or cmpd umbel (rarely capitate)
I Lvs 3-15 dm L; infl umbellate; fr fleshy ARALIACEAE p. 42
II Lvs smaller, or if not infl a cmpd umbel; fr dry; APIACEAE p. 34
HH Infl paniculate or fls solitary
I Stamens perigynous; fls mly 5-merous
J Pistils usu as many as sepals; lvs often pinnate and/or stipulate ROSACEAE p. 173
JJ Pistils mly fewer than sepals; lvs often palmately lobed, never pinnate, usu lacking stipules SAXIFRAGACEAE p. 187
II Ovary definitely inferior; fls often 4-merous
J Fls 2-4-merous; lvs often opp ONAGRACEAE p. 142
JJ Fls 5-merous; lvs alt LOASACEAE p. 139
FF Ovary superior
G Plants insectivorous or with stinging hairs
H Plants with stinging hairs URTICACEAE p. 203
HH Plants with leaves modified for catching insects
I Lvs with gland-tipped hairs DROSERACEAE p. 101
II Lvs cobra-shaped with globose upper section SARRACENIACEAE p. 187
GG Plants without highly modified hairs or lvs
H Lvs opp or whorled, simple, usu entire
I Sepals 2 PORTULACACEAE p. 160
II Sepals 4-5, sometimes united into a 5-lobed tube
J Petals white to red CARYOPHYLLACEAE p. 89

JJ Petals yellow to salmon
K Stamens 5; styles 2 — LINACEAE (*Sclerolinon*) p. 139
KK Stamens many; styles 3 — HYPERICACEAE p. 132
HH Lvs alt or mly basal, sometimes whorled
I Fls 4-merous; stamens 6 — BRASSICACEAE p. 75
II Fls usu 5-merous; stamens rarely 6
J Fr pea pod-like; stamens enclosed in lower petals — FABACEAE p. 109
JJ Fr a capsule, akene or fleshy; stamens not enclosed in petals
K Sepals 2; lvs pinnately lobed to dissected — PAPAVERACEAE p. 148
KK Sepals usu more than 2, if not lvs entire
L Petals absent or if present less than 3 mm L; sepals sometimes petaloid
M Infl a dense terminal spike; lvs simple, over 1 cm W — PLANTAGINACEAE p. 150
MM Infl not a dense spike, or if so the lvs less than 1 cm W
N Lvs with membranous sheath-like stipules extending above node or if not the infl subtended by a whorl of bracts — POLYGONACEAE p. 157
NN Lvs lacking stipules; infl not subtended by bracts
O Infl inconspicuous; lvs simple, usu less than 5 cm L — CHENOPODIACEAE p. 96
OO Infl usu conspicuous; lvs often large or cmpd — RANUNCULACEAE p. 165
LL Petals present, usu longer than 3 mm
M Fls irregular or with spurs
N Stipules present; lower petal spurred — VIOLACEAE p. 204
NN Stipules lacking; one or more sepals may be spurred — RANUNCULACEAE p. 165
MM Fls radially symmetric, lacking spurs
N Stamens more than 10
O Stamens united into tube; lvs mly stellate-pubescent — MALVACEAE p. 140
OO Stamens distinct from base
P Lvs biternate; fls solitary, sepals usu purplish — PAEONIACEAE p. 148
PP Lvs simple to lobed; fls in clusters; petals orange to yellow
Q Subshrub; pubescent usu stellate; fls yellow — CISTACEAE p. 98
QQ Glabrous herbs; petals shiny — RANUNCULACEAE (*Ranunculus*) p. 169
NN Stamens 10 or less
O Lvs deeply lobed or divided
P Fls solitary, axillary; lvs pinnately cmpd — LIMNANTHACEAE p. 137
PP Fls usu clustered, never solitary in axils — GERANIACEAE p. 123
OO Lvs simple, not deeply lobed or divided
P Style one — ERICACEAE p. 102
PP Styles more than one; pistils often more than one
O Pistils more than one; plants fleshy — CRASSULACEAE p. 99
OO Pistil one
P Sepals usu 2; styles 2-3 (to 8 in *Lewisia*) — PORTULACACEAE p. 160
PP Sepals 5; styles 5 — LINACEAE p. 138

KEY TO THE DICOTYLEDON GENERA

A Plants white to purple or brown, not green; saprophytic or parasitic species
  B Plants growing on upper parts of other plants or trees
    C Plants thin yellow vines, on herbs — *Cuscuta* p. 101
    CC Plants thicker with jointed stems, on trees — *Arceuthobium* p. 206
  BB Plants ± erect, growing on ground
    C Plants white to red, sometimes. turning purple or brown with age; fls not 2-lipped — ERICACEAE p. 102
    CC Plants usu brown or purple, fls 2-lipped — OROBANCHACEAE p. 147
AA Plants green, usu not saprophytic or parasitic
  B Plants definitely shrubs or trees; with evident branches bearing lvs
    C Lvs compound
      D Lvs palmately cmpd; lfts 4 or more
        E Plant mly more than 2 m H; lfts 5-10 cm L — *Aesculus* p. 126
        EE Plant about 1 m H; lfts 1-3 cm L — *Lupinus albifrons* p. 113
      D Lvs 3-foliolate or pinnately cmpd
        E Lvs many times divided into small segments, fern-like
          F Lvs mly 3-pinnate; fr an akene — *Chamaebatia* p. 175
          FF Lvs mly 2-pinnate; fr a follicle — *Chamaebatiaria* p. 176
        EE Lvs once divided into several conspicuous lfts
          F Lvs opposite
            G Lfts mly 3
              H Lfts glabrous; margin finely serrate — *Staphylea* p. 203
              HH Lfts pubescent; margin coarsely serrate — *Acer negundo* p. 33
            GG Lfts mly 5 or more
              H Twigs with large spongy pith; fr a berry — *Sambucus* p. 88
              HH Twigs woody throughout; fr dry, winged — *Fraxinus* p. 142
          FF Lvs alternate
            G Stems or lvs with prickles
              H Stems with prickles — *Rosa* p. 182
              HH Lf-margins spiny — *Berberis* p. 70
            GG Plants not prickly
              H Lfts entire, revolute — *Potentilla fruticosa* p. 179
              HH Lfts serrate
                I Lfts 3 — *Toxicodendron* p. 34
                II Lfts 7-25
                  J Lfts 13-25; fr an akene — *Ailanthus* p. 202
                  JJ Lfts 7-13; fr fleshy — *Sorbus* p. 183
    CC Lvs simple
      D Lvs scale-like or narrow (less than 5 mm W), the margins sometimes turned under
        E Lvs ± triangular, about 0.5 cm L; above 7000' — *Cassiope* p. 104
        EE Lvs linear, 0.5-2 cm L
          F Lvs tufted on spur-like branchlets; dry habitats below — *Adenostoma* p. 174
          FF Lvs not tufted, margins usu curled under; moist habitats above 6000'
            G Lvs opp; fls bowl-shaped — *Kalmia* p. 105
            GG Lvs alt; fls urn-shaped — *Phyllodoce* p. 106
      DD Lvs with broader blade
        E Lvs opposite
          F Lvs leathery, evergreen, mly below 7000' on dry slopes
            G Lvs 2-7 cm L — *Garrya* p. 121
            GG Lvs usu 0.5-2 cm L, often on spur-like branchlets
              H Lvs lacking stipules; Mariposa Co. N. — *Paxistima* p. 95
              HH Lvs with stipules, these usu with thick bases — *Ceanothus* p. 171
          FF Lvs ± thin, deciduous; often moist habitats
            G Lvs palmately lobed into 3 or more divisions — *Acer* p. 33
            GG Lvs usu without lobes

H Fls irregular, the petals fused together at least at base
  I Fls in pairs on conspicuous peduncles; fr fleshy — *Lonicera* p. 87
  II Fls on individual stalks (pedicels); fr dry
    J Fls red — *Penstemon* p. 199
    JJ Fls white — *Keckellia* p. 195
HH Fls with 4 or more identical lobes or distinct petals
  I Large bush or tree over 1 m H
    J Lvs aromatic; fls solitary, reddish — *Calycanthus* p. 84
    JJ Lvs not aromatic; fls clustered, white or with white bracts
      K Petals 5 mm L, sometimes subtended by white bracts 40-60 mm L; moist habitats — *Cornus* p. 98
      KK Petals 10-15 mm L; dry slopes — *Philadelphus* p. 150
  II Low shrub, less than 1 m H
    J Lvs 3- to 5-veined from base; below 4500' — *Viburnum* p. 89
    JJ Lvs 1-veined from base; widespread
      K Lvs strongly serrate; fr a dry capsule — *Jamesia* p. 150
      KK Lvs often mitten-shaped, not prominently serrate; fr a berry — *Symphoricarpos* p. 88

EE Lvs alternate
  F Lvs palmately lobed (sometimes. only 3-toothed in *Purshia* and *Artemisia*)
    G Lobes broad, more than 1 cm W at base
      H Lvs stellate-pubescent beneath
        I Fls small, in terminal clusters — *Physocarpus* p. 179
        II Fls large, ± solitary; fls and lvs usu on short branchlets — *Fremontodendron* p. 203
      HH Lvs glabrous or with simple hairs
        I Lvs 10-15 cm W; fr raspberry-like — *Rubus parviflorus* p. 182
        II Lvs less than 8 cm W; fr often spiny — *Ribes* p. 124
    GG Lobes less than 1 cm W at base
      H Lobes linear to needle-like, usu more than 3 — *Leptodactylon* p. 154
      HH Lobes broader, 3, sometimes mere teeth at leaf apex
        I Lvs with sage-like odor, margins plane — *Artemisia* p. 50
        II Lvs ± odorless, margins curled under — *Purshia* p. 181
  FF Lvs not lobed or pinnately so
    G Bark read to red-brown and usu peeling off in thin layers
      H Plant a tree; lvs over 4 cm L — *Arbutus* p. 103
      HH Plant a shrub; lvs less than 3 cm L — *Arctostaphylos* p. 103
    GG Bark not reddish, or if so then not peeling off in thin layers
      H Lf-margins conspicuously cut, lobed or toothed
        I Lf-blades toothed only on distal half
          J Lvs 4-8 cm L, oblanceolate; fls lacking petals — *Myrica* p. 141
          JJ Lvs mly less than 4 cm L, mly rounder; fls mly with petals
            K Fls 1-8 per cluster
              L Fls lacking petals; fr dry with a 2-4 cm L plume — *Cercocarpus* p. 175
              LL Petals white, conspicuous; fr fleshy — *Amelanchier* p. 175
            KK Fls many per cluster
              L Fls white; lvs less than 2 cm L — *Holodiscus* p. 177
              LL Fls red to violet; lvs mly more than 2 cm L — *Spiraea* p. 184
        II Lf-blades cut, lobed, or toothed to base
          J At least male fls in dense cylindrical clusters (catkins); fr dry
            K Fr nut-like or a woody cone, not disintegrating before falling
              L Nut with a cap, acorn-like; lf margins various
                M Cap of nut covered with thick hairs; lvs prominently pinnately veined — *Lithocarpus* p. 119
                MM Cap of nut with smooth scales; lf-veins usu — *Quercus* p. 120
              LL Nut lacking cap or fr cone-like; lvs doubly serrate
                M Fr a woody cone about 2 cm L; often trees — *Alnus* p. 71
                MM Fr a nut about 1.5 cm L; shrubs — *Corylus* p. 72
            KK Fr a capsule or non-woody cone

L Tree with smooth, dark bronze, shiny bark *Betula* p. 71
LL Shrub or tree with lighter bark, usu furrowed in age SALICACEAE p. 185
JJ Fls solitary or in ± open clusters; petals mly present; fr usu fleshy
K Plant a low subshrub less than 1 m H
L Fls on naked stems 4-10 cm L; lvs spatulate *Primula* p. 165
LL Fls on peduncles less than 2 cm L; lvs ± ovate
M Corolla urn-shaped; wet mdws, mly above 5000' *Vaccinium* p. 108
MM Corolla open, the petals separate, dry flats below 6000' *Ceanothus diversifolius* p. 172
KK Plant an erect shrub or tree, more than 1 m H
L Fr dry; lvs usu 3-veined from base *Ceanothus* p. 171
LL Fr fleshy; lvs pinnately veined
M Fr 1-seeded, infl not umbellate *Prunus* p. 181
MM Fr 2-3-seeded; infl mly umbellate *Rhamnus* p. 173
HH Lf-margins ± entire
I Plants of moist habitats; petals at least partially fused or petals absent or tiny
J Petals inconspicuous or none; buds covered by 1 scale
K Infl a 1-3-fld umbel; Placer Co. n. *Rhamnus alnifolia* p. 173
KK Fls in catkins; widespread *Salix* p. 185
JJ Fls with conspicuous petals, these usu fused into an urn-shaped corolla ERICACEAE p. 103
II Plants of dry habitats; petals usu present
J Lvs roundish, over 3 cm W, cordate at base; fr a legume *Cercis* p. 110
JJ Lvs narrower, never cordate at base; fr not a legume
K Lvs usu 3-veined from base; fls white or light blue *Ceanothus* p. 171
KK Lvs with 1 primary vein from base; fls various
L Lvs linear to wedge-shaped, usu less than 5 mm W
M Lvs with margins curled under; fls 1-3, in axils *Cercocarpus* p. 175
MM Lvs flat; fls yellow, in heads
N Herbage glabrous *Ericameria* p. 56
NN Herbage pubescent *Chrysothamnus* p. 54
LL Lvs oblong to elliptic, usu over 1 cm W
M Center of branches hollow, divided by many crosswalls *Oemleria* p. 179
MM Center of branches solid
N Lvs often golden-yellow, pubescent, on petioles 10-15 mm L; fr a spiny bur *Castanopsis* p. 119
NN Lvs ± green, often tufted on spur-like branchlets petioles shorter or none; fr fleshy
O Petals round, 7-8 mm L; fr yellowish, spherical *Peraphyllum* p. 179
OO Petals oblong, 1-3 mm L; fr black, shallowly 2-3-lobed *Rhamnus* p. 173

BB Plants herbaceous or only slightly woody or woody vines
C Lvs cmpd into definite lfts or dissected and fern-like (if lf-blade continuous between segments go to CC)
D Lvs palmately cmpd into more than 3 lfts
E Lft-margins entire
F Lvs opp, appearing like tufts of linear lvs *Linanthus* p. 154
FF Lvs alt or basal
G Fls irregular, pea-like; common *Lupinus* p. 113
GG Fls regular; Placer Co. n. *Cardamine pachystigma* p. 78
EE Lft-margins serrate or lobed
F Lfts less than 1.5 cm L, obovate *Trifolium lemmonii* p. 117
FF Lfts more than 2 cm L, oblong *Potentilla* p. 179
DD Lvs trifoliolate or ternately or pinnately cmpd or dissected
E Stems woody at base or throughout; often vines
F Plants vine-like

G Stems with prickles — *Rubus* p. 182
GG Stems smooth
H Petioles clasping or twining; lfts 3-7 — *Clematis* p. 167
HH Petioles straight; lfts 3 — *Toxicodendron* p. 34
FF Plants erect, usu branched above
G Stems with prickles — *Rosa* p. 182
GG Stems lacking prickles
H Lvs large and fern-like, aromatic; low bush — *Chamaebatia* p. 175
HH Lvs divided into 3-17 distinct lfts — *Sambucus* p. 88
EE Stems herbaceous
F Lvs twice (or more) cmpd
G Fls grouped in umbels, petioles mly sheathing stem at their base
H Infl a raceme of simple umbels; below 5000' — *Aralia* p. 42
HH Infl an umbel or cmpd umbel — APIACEAE p. 34
GG Fls not grouped in umbels; petioles rarely sheathing at base
H Plants submersed
I Lvs bearing small bladder-like traps; fls two-lipped — *Utricularia* p. 137
II Lvs without traps; fls radially symmetric — *Ranunculus* p. 169
HH Plants terrestrial
I Fls solitary or in open clusters, not in heads
J Petals or sepals saccate or spurred
K Fls 4-merous, both outer petals saccate or spurred
L Outer petals both saccate or spurred at base — *Dicentra* p. 149
LL One outer petal spurred, the other not — *Corydalis* p. 149
KK Fls 5-merous
L All sepals with 1-3 cm L spurs — *Aquilegia* p. 167
LL One petal spurred; spur 5-10 mm L — *Viola* p. 204
JJ Petals flat, not saccate or spurred
K Stem lvs opp or usu whorled — *Anemone* p. 166
KK Stem lvs alt
L Sepals 2, falling early; petals golden-yellow — *Eschscholzia* p. 150
LL Sepals 5, persistent; petals often white or inconspicuous
M Male and female fls on separate plants; petals lacking — *Thalictrum* p. 171
MM Fls with both male and female parts
N Fls 4-merous; petals yellow — *Descurainia pinnata* p. 79
NN Fls 5-merous; petals white — *Actaea* p. 166
II Fls in heads
J Lvs with sage-like odor
K Infl elongate; fls yellow — *Artemisia norvegia* p. 50
KK Infl dome-shaped to flat; fls white — *Achillea* p. 46
JJ Lvs without sage-like odor
K Heads globose with few or no bracts — *Cymopterus* p. 36
KK Head ± cylindric, subtended by many green bracts
L Lvs 2-3 times ternate — *Erigeron compositus* p. 57
LL Lvs bipinnate — *Chaenactis* p. 54
FF Lvs once pinnate or trifoliate; the segments often serrate or lobed
G Lvs always 3
H Stem lvs opp or whorled — *Anemone quinquifolia* p. 166
HH Stem lvs alt
I Infl an umbel or cmpd umbel
J Lfts 15-40 cm L — *Heracleum* p. 37
JJ Lfts less than 5 cm L
K Lfts less than 2 cm L, never lobed; fls asymmetric — *Lotus* p. 112
KK Lfts usu over 2 cm L, lobed; fls regular
L Stem present, leafy; dry, open forest — *Sanicula* p. 40
LL Stem none, lvs all basal; usu moist habitats — *Orogenia* p. 38
II Infl various, never umbellate

J Plants ± aquatic; lfts entire *Menyanthes* p. 141
JJ Plants terrestrial; lfts ± serrate
K Fls irregular, often in many-fld clusters; lfts evenly serrate to entire FABACEAE (*Medicago, Melilotus, Psoralea, Trifolium*) p. 109
KK Fls regular, solitary or in open clusters; lfts usu irregularly serrate to lobed or divided
L Stems none; lvs in tufts connected by stolons *Fragaria* p. 176
LL Leafy stems present; stolons none
M Lfts not further lobed or divided, usu serrate distal half
N Petals white; Plumas Co. n. *Suksdorfia* p. 191
NN Petals yellow; widespread above 6000' *Sibbaldia* p. 183
MM Lfts usu lobed or divided
N Petals white; lfts ternately divided *Isopyrum* p. 168
NN Petals shiny yellow; lvs simple to cmpd *Ranunculus* p. 169
GG Lfts mly more than 3
H Stem lvs opp or whorled
I Plants aquatic; stem lvs whorled *Myriophyllum* p. 126
II Plants terrestrial; stem lvs opp
J Lvs 3-8 cm L; 5000-10,000' *Valeriana* p. 204
JJ Lvs 1-5 cm L; mly below 5000' *Erodium* p. 124
HH Stem lvs alt or lvs all basal
I Petioles dilated at base, winged, or lvs with stipules
J Petioles dilated at base or winged, often sheathing stem; stipules seldom present
K Infl an umbel or cmpd umbel; petiole-bases usu conspicuously sheathing stem APIACEAE (*Berula, Oxypolis, Perideridia, Podistera, Sium)* p. 34
KK Infl never umbellate; dilated petiole-bases usu not sheathing stem
L Petals distinct to base, yellow or tinged purplish; pistils many
M Petals shiny yellow above *Ranunculus* p. 169
MM Petals yellow to purplish, never shiny above *Geum* p. 176
LL Petals conspicuously united at base; blue to purple; pistil 1
M Herbage not aromatic; below 4500' *Pholistoma* p. 132
MM Herbage aromatic, above 10,000' *Polemonium eximium* p. 156
JJ Petioles narrow to base, not winged; stipules present
K Lfts serrate or ± lobed, sometimes very small; fls regular ROSACEAE p. 173 (*Agrimonia, Horkelia, Horkeliella, Ivesia, Potentilla, Sanguisorba*)
KK Lfts finely dentate to entire, distinct; fls irregular FABACEAE p. 109 (*Astragalus, Glycyrrhiza, Hoita, Lathyrus, Lotus, Oxytropis, Vicia*)
II Petioles not dilated at base; lvs without stipules
J Corolla regular; petals 4 or 5, occasionally fused below tips
K Petals white, distinct to base, 1-2 mm L *Floerkea* p. 138
KK Petals colored, irregular, usu conspicuous
L Lfts entire; stamens enclosed in lower petals *Lotus* p. 112
LL Lfts mly further lobed; stamens enclosed in upper lip *Pedicularis* p. 198
JJ Corolla regular; petals 4 or 5, occasionally fused below tips
K Petals 4, distinct to base; fr ± linear, usu conspicuous BRASSICACEAE (*Barbarea, Cardamine, Descurainia, Rorippa*) p. 75
KK Petals 5, usu united at base; fr usu inconspicuous
L Fls solitary, drooping; petals separate to base *Paeonia* p. 148
LL Fls clustered; petals united near base
M Style 3-cleft
N Lfts spinose; infl with spiny bracts *Navarretia* p. 155
NN Lfts not spinose; infl never spiny *Polemonium* p. 156
MM Style 2-cleft
N Lvs, including petioles, 10-20 cm L, mly basal *Hydrophyllum* p. 128
NN Lvs shorter, scattered along stem *Phacelia* p. 130
CC Lvs simple, sometimes deeply lobed or divided, but blade still apparent along midvein
D Lvs, at least at lower nodes, opp or whorled
E Plants with milky sap

F Plants annual; lvs occasionally serrate; petals none *Euphorbia* p. 109
FF Plants perennial; lvs entire; petals present
G Stems erect, 3-15 dm H; petals separate, reflexed *Asclepias* p. 43
GG Stems lax, oft horizontal in upper portion; fls urn-shaped *Apocynum* p. 42
EE Plants with clear sap
F Stems 4-angled (with 4 ridges and 4 flat or concave surfaces); plants terrestrial
G Lvs whorled, mly 4 or more per node *Galium* p. 184
GG Lvs opp, 2 per node LAMIACEAE p. 133
FF Stems round, without ridges
G Fls clustered, the clusters subtended by an involucre of green bracts
H Lvs none, the whorl of 'lvs' below fls actually leafy bracts; fls 1-5
Gymnosteris p. 153
HH Lvs present, basal or scattered; fls in sunflower-like heads ASTERACEAE p. 44
(*Arnica, Balsamorhiza, Brickellia grandiflora, Pericome, Whitneya*)
GG Fls not in clusters closely subtended by green bracts
H Plants aquatic, fls inconspicuous
I Lvs whorled
J Submersed lvs pinnatifid *Myriophyllum* p. 126
JJ Submersed lvs entire *Hippuris* p. 126
II Lvs opposite
J Lvs 4-10 mm L; the lower often linear, *Callitriche* p. 84
JJ Lvs 2-3 mm L; lvs all ± obovate *Elatine* p. 102
HH Plants terrestrial or in trees
I Plant a subshrubby parasite in trees *Phoradendron* p. 206
II Plants terrestrial, occasionally of wet habitats
J Fls without petalloid parts or these minute and inconspicuous
K Lvs toothed and 3-5-veined from base; plant often with stinging hairs
Urtica p. 203
KK Lvs serrulate to entire, occasionally 3-lobed, plant lacking stinging hairs
L Sepals distinct to base; lvs not lobed, infl usu a dicotomously
branched cyme CARYOPHYLLACEAE p. 89
LL Sepals united at base; lvs often lobed; fls 1-3 in axils of bracts
Tonella p. 201
JJ Petaloid parts present, usu conspicuous
K Stem lvs appearing like several tufts of pubescent linear *Linanthus* p. 154
KK Stem lvs not appearing tufted, usu wider
L Petaloid fl-parts separate to base or nearly so
M Sepals 4; petals 2 or 4
N Petals less than 5 mm L or if longer then deeply cleft
O Lvs 1-4 cm W; ± ovate; petals 2 *Circaea* p. 143
OO Lvs less than 1 cm W; petals 4 *Stellaria* p. 94
NN Petals over 5 mm L, never deeply cleft
O Stamens 4; lvs white margined *Swertia* p. 123
OO Stamens 8; lvs green throughout *Epilobium* p. 144
MM Sepals 2, 3 or 5-7; petals 5-7
N Petals yellow to yellow-orange
O Lvs 5-10 cm L *Lysimachia* p. 164
OO Lvs less than 3 cm L
P Stamens many; lvs several-veined from base
Hypericum p. 132
PP Stamens 5; lvs 1-veined from base *Sclerolinon* p. 139
NN Petals white, reddish, greenish, bluish, or sometimes lacking
O Sepals 2, rarely 3 PORTULACACEAE p. 160
OO Sepals 5-7
P Lvs in a single whorl on stem
Q Lvs entire *Trientalis* p. 165
QQ Lvs serrate *Chimaphila umbellata* p. 104
PP Lvs opp, usu in several pairs

Q Petals blue; basal lvs ± petioled *Swertia perennis* p. 123
QQ Petals white to pink; lvs sessile CARYOPHYLLACEAE p. 89
LL Petaloid fl-parts united at base into a conspicuous tube
M Sepals red, purple, or less than 2 mm L and inconspicuous
N Lvs often lobed; fls white
O Lvs usu pinnately parted; stems erect *Valeriana* p. 204
OO Lvs often mitten-shaped; stems lax *Symphoricarpos* p. 88
NN Lvs mly entire; fls usu pinkish to purple
O Lvs ± glabrous
P Lvs lanceolate, less than 1 cm W *Kelloggia* p. 185
PP Lvs roundish, 2-4 cm W *Cycladenia* p. 42
OO Lvs densely hairy, oval to elliptic
P Fls pinkish; rare, Tulare Co. *Abronia* p. 142
PP Fls red; widespread *Epilobium canum* p. 144
MM Sepals green, more than 2 mm L
N Corolla irregular, the lobes not all identical
O Fls in long spikes, purplish; below 4500' *Verbena* p. 204
OO Fls not in spikes; widespread SCROPHULARIACEAE p. 191
NN Corolla regular, all lobes identical
O Herbage glabrous
P Corolla blue or white, never rose GENTIANACEAE p. 121
(*Gentianella, Gentiana, Gentianopsis*)
PP Corolla rose with some yellow or white
Q Lvs thick, with revolute margins *Kalmia* p. 105
QQ Lvs thin, flat *Centaurium* p. 121
OO Herbage pubescent
P Lvs ± serrate or lobed
Q Lvs ± serrate on distal half *Linnaea* p. 86
QQ Lvs mly palmately or pinnately lobed *Nemophila* p. 129
PP Lvs entire
Q Corolla less than 5 mm L, white *Plagiobothrys* p. 75
QQ Corolla over 10 mm L, white, pink or violet
R Lvs ovate; style 2-cleft *Draperia* p. 127
RR Lvs linear to lanceolate; style 3-cleft *Phlox* p. 155
DD Lvs alternate or basal, occasionally the lowermost opposite
E Plants aquatic
F Lvs more than 5 cm L, entire
G Lvs peltate, without deep notch *Brasenia* p. 84
GG Lvs with notch reaching summit of petiole
H Fl over 5 cm W; lvs over 10 cm W *Nuphar* p. 142
HH Fl 3-4 cm W; lvs 3-10 cm W; rare *Nymphoides* p. 141
FF Lvs less than 5 cm L, or if more then with several lobes
G Leafy stem 3-40 cm H; petals mly blue; Sierra Co. n. *Downingia* p. 85
GG Lvs all basal; petals whitish; widespread
H Stolons present; lvs oblong to elliptic *Limosella* p. 195
HH Stolons none; lvs linear; rare *Subularia* p. 83
EE Plants terrestrial
F Plants with spine-tipped lvs or bracts
G Infl with spiny bracts or sepals; lvs various
H Lvs linear or with linear lobes
I Heads white-woolly; sepal tips spinescent *Eriastrum* p. 152
II Heads ± glabrous; bracts with linear spine-tipped lobes *Navarretia* p. 154
HH Lvs or lobes wider; plants thistle-like
I Lvs with decurrent spiny wings extending down stems *Carduus* p. 53
II Lvs lacking such decurrent wings *Circium* p. 55
GG Fls not in spiny heads; lvs spiny or spine-tipped
H Lvs broad with many spines along margins *Argemone* p. 148
HH Lvs or lf-lobes linear, spine-tipped

I Lvs palmately lobed; petals conspicuous *Leptodactylon* p. 154
II Lvs linear; petals none *Salsola* p. 97
FF Plants without spine-tipped lvs or bracts
G Fls in involucrate heads, often sunflower- or dandelion-like ASTERACEAE p. 44
GG Plants not in heads surrounded by many bracts
H Plant with membranous stipules surrounding stem at base of petiole
POLYGONACEAE (*Oxyria, Polygonum, Rumex*) p. 157
HH Plant lacking stipules or these not membranous and surrounding stem
I Infl a dense, cylindrical spike on erect peduncle 1-3 dm H; stem none;
common in disturbed sites *Plantago* p. 150
II Infl not a dense spike; stem usu present, leafy
J Fls with 2 or 4 equal sized and ± distinct petalloid parts
K Sepals 2, membranous, ± petalloid; fls often in dense spherical clusters
Calyptridium p. 160
KK Sepals 4, green
L Petals reflexed, lvs all basal *Dodecatheon* p. 164
LL Petals erect or spreading, not reflexed although sepals may be
M Ovary superior; sepals erect, occasionally saccate at base;
petals not lobed BRASSICACEAE p. 75
MM Ovary inferior, sepals often reflexed; petals often lobed
ONAGRACEAE p. 142
JJ Fls 5-merous (sometimes 3- or 6-merous or more), sometimes irregular
and/or petals united at base or petals lacking
K Lvs modified for catching insects; plants of wet areas
L Lvs covered with long, stout, sticky hairs *Drosera* p. 101
LL Lvs erect, cobra-shaped with globose upper section *Darlingtonia* p. 187
KK Lvs not modified for catching insects
L Lvs palmately lobed or peltate
M Lvs peltate *Darmera* p. 188
MM Lvs not peltate
N Stems with tendrils opp the lvs
O Stems ± woody; fr smooth *Vitis* p. 206
OO Stems herbaceous; fr spiny *Marah* p. 100
NN Stems lacking tendrils
O Fls conspicuously irregular
P Sepals and petals not spurred; stems stout, 5-10 dm H
Aconitum p. 165
PP Fls with a conspicuous spur; plants usu shorter
Q Fls solitary; the lower petal spurred *Viola* p. 204
QQ Fls in racemes; upper sepal spurred *Delphinium* p. 167
OO Fls regular
P Petalloid parts yellow
Q Fls in cmpd umbels; below 4500' *Sanicula crassicaulis* p. 40
QQ Fls usu solitary; widespread in moist habitats *Ranunculus* p. 169
PP Petals white, pink, greenish, purplish, or lacking
Q Lvs small, divided into oblong or linear segments; petals
united at base
R Lvs ± fork-shaped with parallel oblong lobes; Lassen Peak
Collomia larsenii p. 151
RR Lvs various with linear lobes; widespread
Allophyllum p. 151
QQ Lvs usu over 2 cm W, often merely bluntly lobed;
petals distinct to base or none
R Petals usu white and less than 10 mm L; styles 2
SAXIFRAGACEAE p. 187
RR Petals usu pink to red and over 10 mm L; styles 5
S Stamens united into tube at base;
pubescence stellate *Sidalcea* p. 140

SS Stamens separate to base; pubescence unbranched *Geranium* p. 124

LL Lvs not lobed or pinnately so

M Lvs pinnately lobed; if not then stamens hidden in upper corolla lip

N Fls irregular; stamens enclosed within a narrowly folded upper lip of corolla

O Plants annual

P Calyx forming a single piece split almost to base ventrally *Cordylanthus* p. 194

PP Calyx 4-cleft or 2-cleft and each half 2-lobed *Orthocarpus* p. 198

OO Plants perennial

P Lvs pinnatifid into many small segments or occasionally simple and dentate *Pedicularis* p. 198

PP Lvs entire or with 1 to few entire lobes *Castilleja* p. 192

NN Fls regular; stamens free and usu conspicuous

O Herbage glabrous or subglabrous; petals none or less than 5 mm L

P Stem none; lvs all basal; petals present *Podistera* p. 40

PP Stem conspicuous, ± leafy; petals none

Q Stem stout, over 1 m H *Datisca* p. 101

QQ Stem 2-6 dm H *Chenopodium botrys* p. 96

OO Herbage pubescent; petals present and over 5 mm L

P Lvs sessile; ovary inferior *Mentzelia* p. 139

PP Lvs mly petioled; ovary superior

Q Style 2-cleft; lvs mly pinnately lobed; corolla mly less than 10 cm L

R Infl spherical; lvs mly basal, 10-20 cm L *Hydrophyllum* p. 128

RR Infl usu elongate; lvs mly cauline, smaller *Phacelia* p. 130

QQ Style 3-cleft; lvs often bipinnate; corolla mly more than 10 mm L

R Stem lvs usu 3-lobed with middle lobe longest *Allophyllum* p. 151

RR Stem lvs pinnately lobed

S Annuals; corolla mly violet or yellow *Gilia* p. 152

SS Perennials; corolla mly red or white *Ipomopsis* p. 153

MM Lvs serrate or entire, not lobed

N Stems 4-angled CAMPANULACEAE p. 84 (*Campanula, Githopsis, Heterocodon, Triodonis*)

NN Stems round

O Plants with milky sap; petals none

P Plants glabrous *Euphorbia* p. 109

PP Plants with stellate pubescence *Eremocarpus* p. 108

OO Plants with clear sap; petals usu present and conspicuous

P Lvs glabrous or subglabrous or if pubescent then lvs all basal

Q Fls irregular

R Ovary inferior; lvs sessile or the cauline reduced to bracts CAMPANULACEAE (*Downingia, Nemacladus, Porterella*) p. 84

RR Ovary superior; lvs with petioles

S Lower petal spurred; lvs with stipules *Viola* p. 204

SS Petals lacking spurs; lvs lacking stipules *Polygala* p. 157

QQ Fls regular

R Petaloid fl-parts present and united at base

S Lvs all basal

T Lvs less than 2 cm L; corolla 1-2 mm L *Androsace* p. 163

TT Lvs 1-20 cm L; corolla obvious, mly over 5 mm L

U Petals reflexed; fls nodding *Dodecatheon* p. 164

UU Petals spreading; fls erect

V Lvs entire; plant herbaceous throughout *Hesperochiron* p. 128

VV Lvs dentate; plant usu woody at base *Primula* p. 165

SS Stem lvs present; these occasionally subglabrous

T Lvs 1-4 cm L, subsessile

U Lvs linear; fls funnelform *Gilia* p. 152

UU Lvs elliptic; fls urn-shaped

V Lvs with wintergreen aroma *Gaultheria* p. 104

VV Lvs not aromatic *Vaccinium* p. 108

TT Lvs 4-12 cm L, the lower with evident petioles
U Lvs mly basal; fls erect — *Swertia* p. 123
UU Lvs mly cauline; fls drooping — *Mertensiana* p. 74
RR Petaloid parts separate to base or none
S Petaloid parts none or fls minute and inconspicuous
T Lvs broad, heart-shaped; fls on ground beneath leaves — *Asarum* p. 43
TT Lvs usu narrow, not heart-shaped; fls not on ground
U Lvs linear, grass-like — *Myosurus* p. 169
UU Lvs lanceolate to ovate, often white-crusty
V Lvs subsessile, entire; infl terminal — *Comandra* p. 187
VV Lvs mly petioled, often hastate or dentate; infl axillary
W Calyx of 1 sepal; mly below 5000' — *Monolepis* p. 97
WW Calyx 3-5-parted; widespread — *Chenopodium* p. 96
SS Petaloid parts present; fls conspicuous
T Sepals 2, rarely more and petaloid (i.e. green sepals none); lvs entire
U Petaloid segments mly small; fls mly in dense — POLYGONACEAE p. 157 (*Chorizanthe, Eriogonum, Oxytheca*)
UU Petaloid segments 5-15 mm L; fls never densely clustered — PORTULACACEAE (*Calandrinia, Claytonia, Lewisia, Montia*) p. 160
TT Sepals 5, green
U Lvs all basal
V Fls solitary
W Lvs entire — *Parnassia* p. 190
WW Lvs serrate
X Fls nodding; stamens 10 — *Moneses* p. 106
XX Fls erect; stamens many — *Caltha* p. 166
VV Fls variously grouped
W Lvs evergreen, ± leathery; style 1 — *Pyrola* p. 107
WW Lvs dying annually; styles 2 — *Saxifraga* p. 190
UU Stem lvs present
V Stems and lvs fleshy — CRASSULACEAE p. 99
VV Stems and lvs thin
W Petals yellow, glossy; stamens many — *Ranunculus* p. 169
WW Petals blue, white or pink; stamens 5 or 10
X Lvs linear to lanceolate, less than 5 mm W; petals usu blue — *Linum* p. 139
XX Lvs broader; petals white to pink or green
Y Infl a 1-sided raceme; style 1 — *Orthilia* p. 106
YY Infl symmetric; styles 2 — *Saxifraga* p. 190
PP Lvs conspicuously hairy, never all basal
Q Ovary 4-lobed; fr 4 nutlets; lvs entire — BORAGINACEAE p. 72
QQ Ovary not lobed; fr not 4 nutlets; lvs often serrate
R Petals distinct to base or absent
S Matted subshrub; lvs crowded along stem — *Petrophytum* p. 179
SS Stems usu erect; lvs not crowded
T Pubescence with some branched to stellate hairs; lvs often with stipules
U Lvs ± linear — *Helianthemum* p. 98
UU Lvs broad, ovate, sometimes palmately lobed — MALVACEAE p. 140 (*Malvella, Sidalcea, Sphaeralcea*)
TT Pubescence unbranched; lvs without stipules
U Petaloid fl-parts none; fls in dense axiallary clusters — *Atriplex* p. 96
UU Petaloid fl-parts conspicuous; infl usu terminal
V Sepals 5, green; petals usu 5, yellow; mly below 7000' — *Mentzelia* p. 139
VV Sepals 6, petaloid; petals none; widespread — *Eriogonum* p. 157
RR Petaloid segments united at least near base
S Plants erect, usu over 1 m H; lvs over 5 cm L
T Infl dense, 1-3 dm L, 2-4 cm thick; fls yellow — *Verbascum* p. 201
TT Infl open; fls not yellow; lvs ± ill-smelling

U Fls white, tubular; widespread — *Nicotiana* p. 202
UU Fls purple, funnelform; Fresno Co. s. — *Turricula* p. 132
SS Plants trailing to erect; lvs usu less than 5 cm L
T Style entire; herbage glabrous to pubescent
U Lvs hastate; plant trailing; fr a capsule — *Calystegia* p. 98
UU Lvs mly ovate and entire; plant usu erect; fr a berry
V Corolla whitish; petioles winged, 15-30 mm L — *Chamaesaracha* p. 202
VV Corolla deep violet with yellow center; petioles 5-10 mm L — *Solanum* p. 203
TT Style 2- or 3-cleft; herbage often glandular-pubescent
U Style 3-cleft; fls pink to white or purplish — *Collomia* p. 151
UU Style 2-cleft; fls bluish to violet
V Stamens equally inserted, subequal in length — *Phacelia* p. 130
VV Stamens unequally inserted to unequal in length — *Nama* p. 128

ACERACEAE - Maple Family

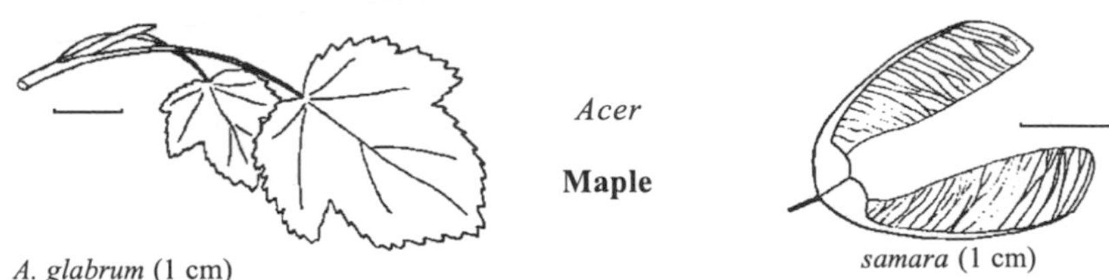

A. glabrum (1 cm) *samara* (1 cm)

Woody plants with opp lvs. Fls regular in several- to many-fld clusters. Fr a 2-winged samara. The sap of the larger maples can be harvested in the same way as the eastern sugar maple, although the product is of inferior quality. The young shoots of mountain maple may be eaten like asparagus. The older lvs may be poisonous. The inner bark of all maples may be eaten in an emergency.

A Lvs pinnately cmpd into 3-5 lfts — 4) *A. negundo*
AA Lvs simple, palmately lobed
B Lvs 2-4 cm W; fr glabrous — 2) *A. glabrum*
BB Lvs usu much larger
C Lvs 5-12 cm W, lobed halfway to center; fr glabrous — 1) *A. circinatum*
CC Lvs 10-25 cm W, deeply lobed; fr hairy — 3) *A. macrophyllum*

1) *A. circinatum,* Vine Maple: Shrub or small tree, 1-6 m H, often vine-like; twigs slender, glabrous; lvs round-cordate in outline, the blades thin, glabrous, palmately 5-11 lobed, 5-12 cm W; petioles 2-5 cm L; samaras 2-3 cm L, widely divergent, reddish when mature. Shaded stream banks below 5000', Butte Co. n., May.

2) *A. glabrum* var. *torreyi,* Mountain Maple: Shrub or small tree, 2-6 m H; twigs usu reddish; lvs usu 3-lobed, sometimes with an additional basal pair, rarely 3-parted, 3-5 cm W, the lobes dentate; samaras 2-3 cm L, usu divergent at an acute angle. Moist to fairly dry slopes and canyons, mly 5000-9000', SN (all). May.

3) *A. macrophyllum,* Big-leaf Maple: Round-topped tree 5-30 m H with glabrous twigs; lvs 1-2.5 dm in diameter, deeply 3-5 parted into coarsely irregularly toothed lobes, paler and more pubescent beneath; petioles usu 5-12 cm L; samaras as variable, the wings 2-4 cm L, divergent at an acute angle. Common on stream banks and in canyons below 6000', SN (all), May.

4) *A. negundo* ssp *californicum,* Box Elder: Round-headed tree 6-15 m H; terminal lft largest, 3-5 lobed, ovate, 5-12 cm L, coarsely serrate, the lateral smaller, all densely pubescent especially beneath and when young; petioles mly 2-8 cm L; samaras red when young, straw color when mature, finely pubescent, 2.5-3 cm L. Along streams and bottom lands below 6000', SN (all), Apr-May.

ANACARDIACEAE - Sumac Family

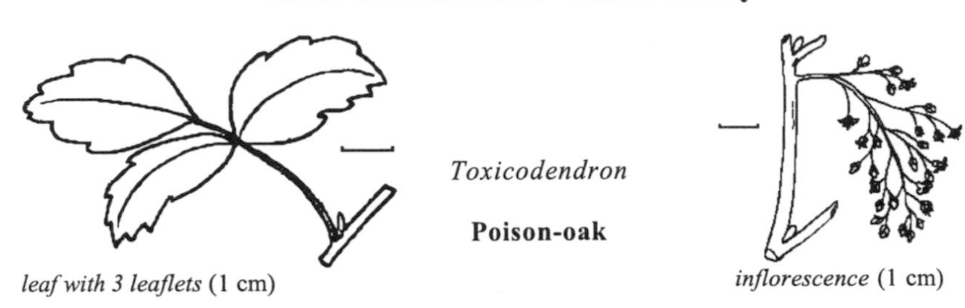

Toxicodendron

Poison-oak

leaf with 3 leaflets (1 cm) *inflorescence* (1 cm)

T. diversilobum: Woody shrub or climbing vine; lvs 3-foliolate; lfts ovate, 2-6 cm L, bright green in the early season, reddish later; terminal lft petioled, the lateral sessile; panicles axillary; fls whitish, 3-5 mm L. Wooded slopes below 5000', SN (all), May. The plant should be avoided for its oil will cause an irritating dermatitis.

APIACEAE - Carrot Family

Aromatic herbs, usu with hollow stems and alt cmpd or simple lvs, the petioles often dilated at base. Fls small, perfect or sometimes unisexual. The umbels usu with an invol. of bracts. Sepals usu 5, sometimes missing. Petals 5. Ovary inferior. Fr dry. Occasionally a difficult family to determine to genus without the fr. Because very poisonous as well as edible plants are found in this family, one should make certain of the identity of a plant before tasting.

A Infl capitate, not umbellate or rays inconspicuous
B Herbage spiny; 4000-6100', Nevada Co. n. — *Eryngium* p. 36
BB Herbage not spiny; above 9000', Placer Co. s.
C Lvs bipinnate, blue-gray, with short hairs — *Cymopterus* p. 36
CC Lvs mly pinnate, yellow-green, $\pm$ glabrous — *Podestera* p. 40
AA Infl a distinct umbel, not capitate
B Ovary and fr bristly or rarely covered with minute knobs
C Plants annual; fls white; below 5000' — *Yabea* p. 41
CC Plants perennial; up to 10,000'
D Fr round, minute; fls yellow — *Sanicula* p. 40
DD Fr torpedo-shaped, 10-20 mm L; fls greenish yellow to white — *Osmorhiza* p. 39
BB Ovary and fr not bristly, occasionally winged
C Ribs of fr not predominantly winged; fr $\pm$ round in cross section
D Fls yellow
E Plants lacking stems; lvs all basal; mly s. SN — *Tauschia* p. 41
EE Plant with leafy stem; Madera Co. n. — *Osmorhiza* p. 39
DD Fls not yellow, although sometimes drying yellow
E Plants with leafy stem, mly rather tall; invol. usu present
F Stems purple-spotted; lvs dissected into small divisions — *Conium* p. 36
FF Stems not spotted; lvs mly pinnately divided into mly larger divisions
G Lvs all once-pinnate; plants of wet areas
H Lfts irregularly serrate or lobed, oblong to ovate — *Berula* p. 35
HH Lfts $\pm$ regularly serrate, linear to lanceolate — *Sium* p. 40
GG Lvs usu further divided, if not then lfts linear and entire
H Lfts entire or serrate and lobed; invol. bracts present — *Perideridia* p. 39
HH Lfts regularly serrate; invol. bracts none — *Cicuta* p. 36
EE Plants lacking stems, mly less than 1 dm H; lvs all basal; invol. bracts none
F Plants glabrous; lfts 10-20 mm L; 4000-6000', — *Orogenia* p. 38
FF Plants gray-pubescent; lfts 1-3 mm L; above 8000' — *Oreonana* p. 38
CC At least 2 of the ribs of the fr with broad wings; fr $\pm$ flattened
D Stems stout, over 5 dm H; lvs mly over 15 cm L
E Lfts 3, 1.5-4 dm L, palmately lobed — *Heracleum* p. 37

EE Lfts many, smaller, usu not palmately lobed
F Umbellets capitate; pedicels minute — *Sphenosciadium* p. 41
FF Infl a distinct cmpd umbel with conspicuous pedicels
G Lvs simply pinnate into usu ovate lfts; wet habitats — *Oxypolis* p. 39
GG Lvs much dissected into oblong to linear divisions; moist to dry habitats
H Fls yellow to purple; lvs mly basal — *Lomatium dissectum* p. 37
HH Fls white to pink; stem lvs several to many — *Angelica* p. 35
DD Stems slender or none, under 5 dm H; lvs less than 5 cm L
E Fls white to pink — *Ligusticum* p. 37
EE Fls yellow to purplish
F Lateral ribs of fr winged, the dorsal ribs filiform — *Lomatium* p. 37
FF All ribs of fr winged — *Cymopterus terebinthinus* p. 36

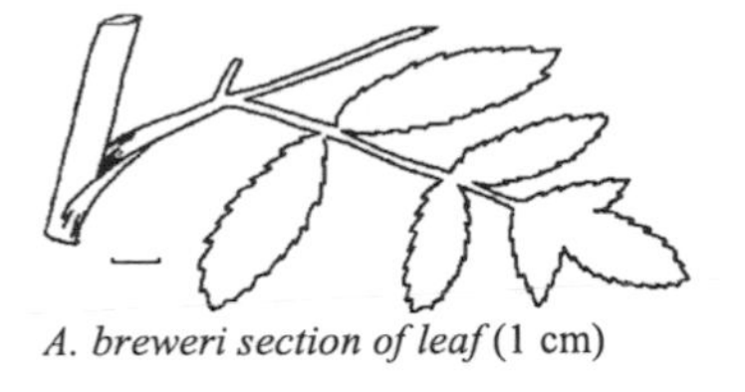
A. breweri section of leaf (1 cm)

Angelica
Angelica

fruit (5 mm) *cross section* (1 mm)

Lvs ternate-pinnately or pinnately cmpd. Petioles sheathing, sheaths sometimes inflated and bladeless. Infl of loose cmpd umbels. Invol. usu none. Involucel of several linear bractlets or lacking. Sepals minute or absent. Petals white or pink. Fr strongly flatted dorsally, the dorsal ribs narrow winged, the lateral broader, about equal to carpel body. Stems of a related sp can be cooked like asparagus and the lvs used for flavoring.

Lvs dissected into linear, entire divisions — *A. lineariloba*
Lvs 1-3-ternate-pinnate into lanceolate to ovate divisions, 80-30 mm W, serrate
Basal lvs 2-3-ternate-pinnate; widespread — *A. breweri*
Basal lvs 1-ternate-pinnate; Kern and Tulare cos. — *A. calli*

A. breweri, Brewer's Angelica: Stems 8-15 dm H; lf-blades 1.5-3.5 dm L, the lfts lanceolate, 4-9 cm L; petioles 2-3 dm L; rays 25-40, unequal, 3-8 cm L; pedicels webbed at base; ovaries densely villous; fr 8-12 mm L, 5-7 mm W. Open woods, 3000-8000', SN (all), Jun-Aug.

A. calli, Call's Angelica: Stems 10-15 dm H; lf-blades 1-4 dm L, ovate, the lfts ovate, 3-12 cm L; rays 25-40, subequal, ascending to reflexed, 3-7 cm L; pedicels webbed at base; ovaries hairy; fr 4-5 mm L. Uncommon, along stream banks, 4000-6500'.

A. lineariloba, Sierra Angelica: Stems 5-15 dm H; lf-blades 1-3 dm L; rays 20-40, subequal, 3-7 cm L; pedicels not webbed at base; ovaries $\pm$ glabrous; fr 10-13 mm L, 5-7 mm W. Gravelly, open slopes, 6000-10,600', Mariposa and Mono cos., Jun-Aug.

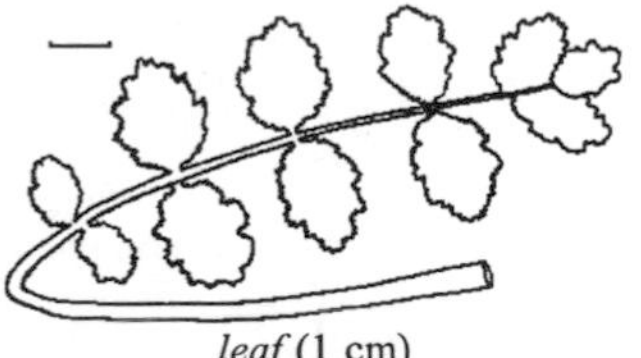
leaf (1 cm)

Berula
Cut-leaved Water-parsnip

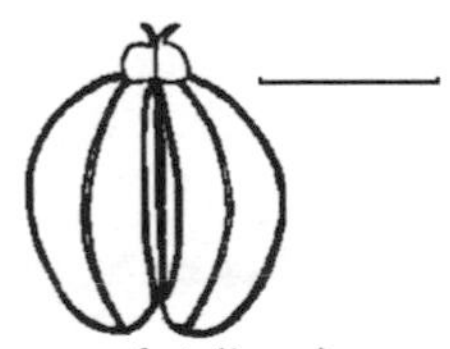
fruit (1 mm)

B. erecta: Perennial; stems erect, branched, glabrous, 2-8 dm H; lvs with sheathing petioles 4-12 cm L, blade 1.5 dm L; lfts in 5-9 pairs, sessile, 2-4 cm L; fls white; infl a loose cmpd umbel; invol. and involucel of conspicuous narrow bracts(lets); rays 6-15; fr 2 mm L, ribs filiform. Marshes and sluggish water, below 5000', SN (all), Jul-Oct. Possibly toxic to livestock.

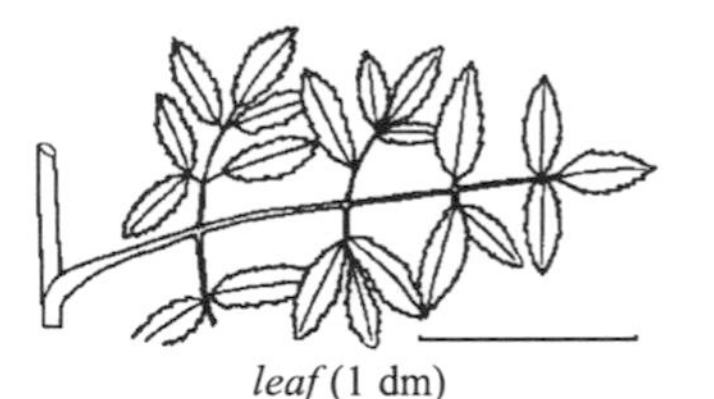

leaf (1 dm)

Cicuta

Water Hemlock

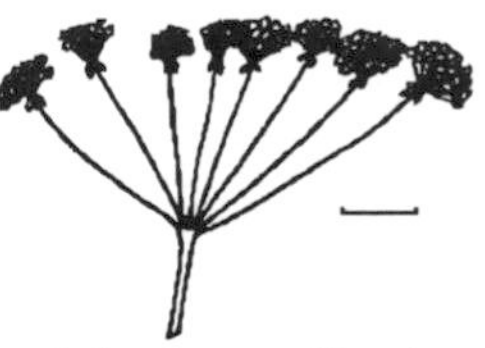

inflorescence (1 cm)

C. douglasii: Perennial; stem stout, branched, glabrous, 0.5-2 m H, from tuberous base; lf-blades 1-3.5 dm L, 1-3-pinnate; lfts linear-lanceolate to ovate, 3-10 cm L; petioles 1-8 cm L; infl of loose cmpd umbels; peduncles 5-15 cm L; involucel of several narrow bractlets, 2-15 mm L; rays 15-30, 2-6 cm L; pedicels 3-8 mm L, slender; fls white or greenish; sepals evident; fr 2-4 mm L; ribs wider than intervals between. Wet places below 8000', SN (all), Jun-Sep. Extremely poisonous and can be lethal if eaten even in small amounts.

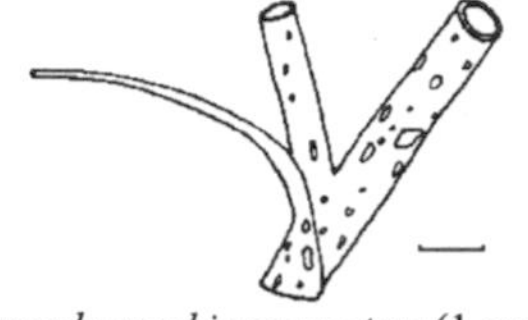

purple markings on stem (1 cm)

Conium

Poison-hemlock

inflorescence (1 cm)

C. maculatum: Stems glabrous, 0.5-3 m H; lower lvs petioled, upper sessile, all finely dissected; lf-blades 15-30 cm L; bracts of invol. many, small; bractlets of involucel many, shorter than the pedicels; rays 10-20, 1-5 cm L; sepals absent; fls white; fr ovoid, glabrous. Common in low waste places below 5000', Apr-Jul. Very poisonous, causing paralysis, coma, and death.

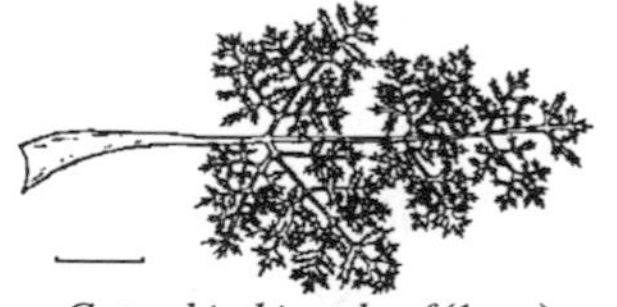

C. terebinthinus leaf (1 cm)

Cymopterus

Cymopterus

sterile frond (1 cm)

Perennials from taproots, usu glabrous; lvs mly basal. Young stem or lvs of *C. cinerarius* edible; however, roots of *C. terebinthinus* contain the cmpd pteryxin, an anticoagulant and spasmodic. Thus, plants in this genus should be sampled with caution.

Rays conspicuous; infl open; fls yellow — *C. terebinthinus*
Rays appearing absent; infl dense, spherical; fls white or purple
  Fls white; bracts conspicuous; fr glabrous; 8000-11,000' — *C. cinerarius*
  Fls usu purple; bracts absent; fr hairy; rare, below 5000', s. SN — *C. ripleyi*

C. cinerarius, Gray's Cymopterus: Lvs 7-8 cm L; blades oblong-ovate in outline, 1.5-2.5 cm L; lfts 2-5 mm L; petioles 3-5 cm L; infl 1-2 cm in diameter. Dry, open slopes, Jun-Jul.

C. ripleyi, Ripley's Cymopterus: Lvs round in outline, ternately divided; lfts wedge-shaped, deeply 3-lobed; peduncles extending beyond lvs; fr 6-7 mm L; marginal ribs winged. Sandy soil.

C. terebinthinus, Rocky Pteryxia: Lvs much divided into ± linear segments 1-8 mm L; blades 3-15 cm L; petioles sheathing, 2-15 cm L; rays unequal, 1-8 cm L; pedicels 1-8 mm L; invol. none; fr 5-10 mm W. Rocky habitats below 11,000', SN (all), May-Jun.

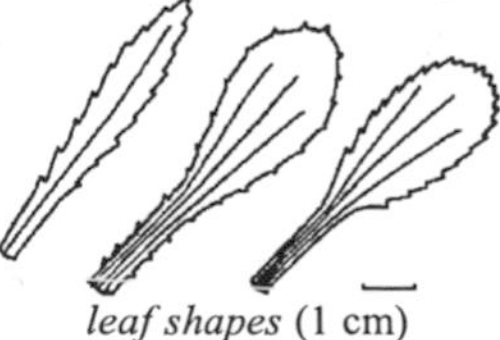

leaf shapes (1 cm)

Eryngium

Coyote-thistle

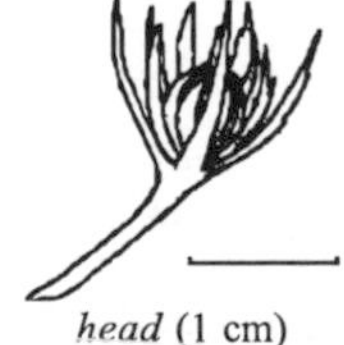

head (1 cm)

E. alismaefolium: Perennial; stems many, glabrous, 0.5-3 dm H; basal lvs lanceolate to ovate, 3-15 cm L, coarsely spinose-serrate to pinnatifid, with short broad petioles; stem lvs reduced; heads many, short-pedunculate, 5-10 mm L; bracts few, rigid, 6-15 mm L, sharp-pointed; bractlets 5-8 mm L. Seasonally moist places, Jul-Aug. Herbage of a related sp inedible.

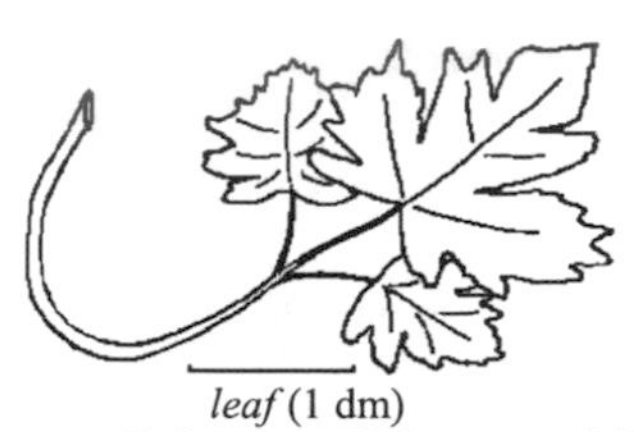

Heracleum

Cow-parsnip

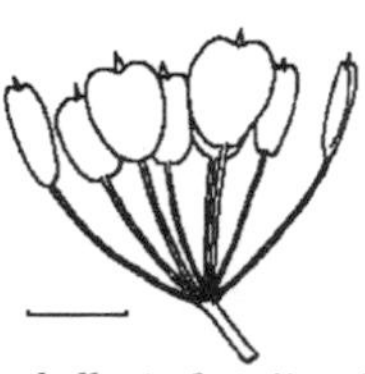

leaf (1 dm) *umbellet in fruit* (1 cm)

H. lanatum: Hairy perennial; lf-blades 2-5 dm L; lfts ovate to roundish, cordate, the middle largest and longest, petiolulate; petioles 1-4 dm L, sheathing; invol. bracts 5-10, 0.5-2 cm L; bractlets similar; rays 15-30; pedicels 8-20 mm L; fr 8-12 mm L, 5-8 mm W. Moist and shaded places below 9000', SN (all), Apr-Jul. The root is edible cooked. The young stems can be peeled and eaten raw or cooked. The hollow basal portion of the plant and the lvs when dried can be used as a salt substitute.

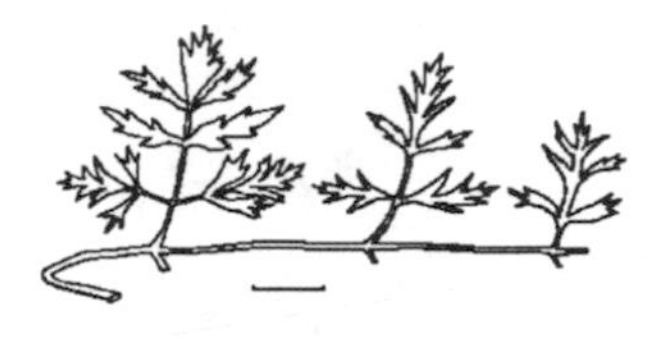

Ligusticum

Gray's Lovage

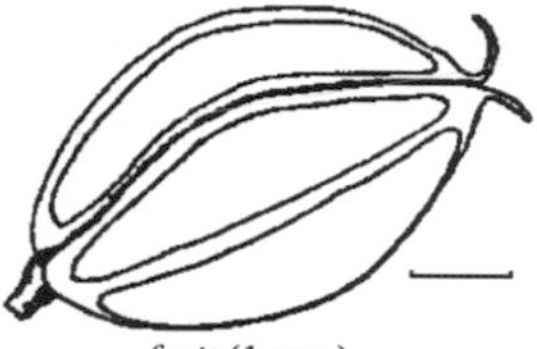

leaf (1 cm) *fruit* (1 mm)

L. grayi: Stems erect, glabrous, slender; lf-blades ovate to oblong in outline, 0.5-2 dm L; the lfts ovate to oblong, 1-2 cm L, pinnatifid; petioles to 1 dm L, sheathing; stem lvs 0-2, reduced; peduncles 0.7-2 dm L; invol. none or inconspicuous; involucel of 4-8 linear bractlets, 2-5 mm L; rays 5-14, 1-3 cm L; pedicels 3-8 mm L, slender; fr 4-6 mm L; ribs narrowly winged. Mdws and slopes, 4000-10,500', SN (all), Jun-Sep. Related spp are known to contain alkaloids, and this sp is best avoided.

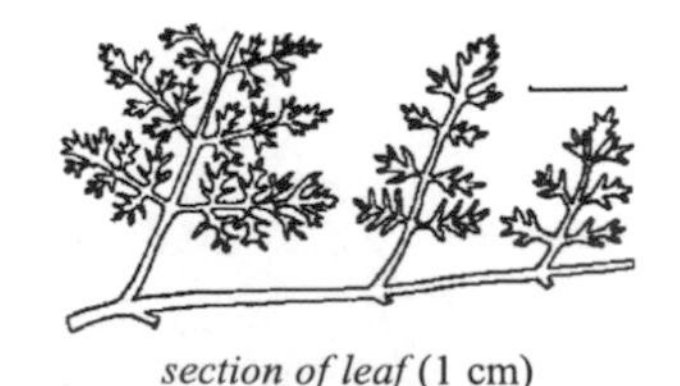

Lomatium

Lomatium

section of leaf (1 cm) *L. torreyi umbellet in fruit* (1 cm)

Plants with or without obvious stem; lf-blades ovate to oblong in outline; fls in cmpd umbels. Invol lacking; involucels present or none. Sepals small. Fr strongly flattened dorsally. The green stems may be eaten in the spring but grow tough later in the season. A tea can be made from the lvs, stems and fls. The stringy roots are edible raw or used for flour. The seeds are also edible raw, roasted, or as a flour.

A Fls white
  B Pedicels 0-2 mm L; below 5000', Sierra Co. n. — 4) *L. piperi*
  BB Pedicels 3-10 mm L; 4000-95000', mly e. SN, Mono Co. n. — 3) *L. nevadense*
AA Fls yellow or purplish
  B Plant with well-developed stem; lvs usu over 15 cm W — 1) *L. dissectum*
  BB Plant without stem or if stem present, the lvs smaller
    C Bractlets none
      D Plants 5-15 cm H; fr 6-9 mm L; rare in central SN — 7) *L. stebbinsii*
      DD Plants 10-30 cm H; fr 10-15 mm L; Tuolumne Co. s. — 8) *L. torreyi*
    CC Bractlets present
      D Plants conspicuously pubescent; fls usu yellow
        E Stem present, short, lvs clustered at base; n. SN — 5) *L. plummerae*
        EE Stem none; e. SN — 2) *L. foeniculaceum*
      DD Plants glabrous throughout; fls purple; Kern Co. — 6) *L. shevocki*

1) *L. dissectum* var. *multifidum,* Fern-leaved Lomatium: Stems usu present, 1-8 dm H, puberulent to almost glabrous; ultimate lf-segments linear-oblong, 5-25 mm L; petioles 5-25 cm L, sheathing at base; stem lvs few, smaller; rays 10-30, 3-10 cm L, subequal; bractlets few, linear; pedicels 1-3 mm L; fr 10-15 mm L, glabrous. Common on rocky slopes, mly below 9500', SN (all), May-Jul.

2) *L. foeniculaceum:* Plant 0.3-10 cm H; petiole wholly sheathing; blade pinnately to ternate-pinnately dissected; segments 1-7 mm L, linear to obovate, pointed; fr 4-12 mm L, usu hairy. Subalpine scrub, below 11,000'.

3) *L. nevadense*, Nevada Lomatium: Stems short or none; plant 1-4 dm H, pubescent; lf-blades 5-6 cm L; petioles sheathing to above the middle, purplish, 4-6 cm L; bractlets usu conspicuous; rays 10-20, spreading, 1-2.5 cm L; fr 6-10 mm L, glabrous or puberulent. Dry slopes 4000-9500', Apr-Jul.

4) *L. piperi,* Piper's Lomatium: Plant usu lacking stem, 1-2 dm H, from a small rounded tuber; lf-blades 3-7 cm L; petioles 3-10 cm L; bractlets few or none, small; rays 3-20, spreading, 1-6 cm L; fr 5-9 mm L. Dry, stony places, Mar-May.

5) *L. plummerae,* Plummer's Lomatium: Stem short, 2-3.5 dm H, usu glabrous; lf-blades 5-10 cm L, the segments 3-7 mm L; petioles 3-6 cm L; bractlets linear lanceolate; rays 10-25, unequal; pedicels 3-8 mm L; fls sometimes purplish; fr 9-13 mm L, glabrous. Sandy slopes and flats, 3000-5000', May-Jun.

6) *L. shevockii,* Owen's Peak Lomatium: Plant 4-12 cm H, very glaucous; lf 1-pinnate with 3-5 pinnately lobed lfts; peduncle 4-12 cm; bractlets 3-6, 1-3.5 mm L; rays 5-9, 1-10 mm L; fr 8-10 mm L, glabrous. Rare, rocky slopes and forest, 7000-8000'.

7) *L. stebbinsii,* Stebbins' Lomatium: Plant from tuber; herbage green and shiny; petiole 2-3 cm L; lf-blade 2-5 cm L; ultimate lfts linear, 2-10 mm L; rays 2-7, 1-10 mm L. Rare, gravelly volcanic soil, yellow pine forest below 5000', Calaveras and Tuolumne cos.

8) *L. torreyi,* Torrey's Lomatium: Plant 1-2.5 dm H, glabrous; lf-blades 2-15 cm L, the segments filiform, 3-8 mm L; petioles 2-5 cm L, wholly sheathing, the sheath with white scarious margin; rays 5-9, unequal, 1-4 cm L; fls yellow; fr 10-15 mm L. Granite rocks and slopes, May-Aug.

base of plant (1 cm)

Oreonana

Mountain-parsley

inflorescence (1 cm)

Tufted perennials, lacking stem; lowest lvs bladeless sheaths; umbels cmpd, dense, spherical, lacking bracts; corolla white, anthers purple. Edibility unknown.

Fls appearing with lf-blades; rays 5-15; calyx lobes yellow — *O. clementis*
Fls appearing before lf-blades; rays 20-35; calyx lobes purple — *O. purpurascens*

O. clementis, Clement's Mountain-parsley: Plants 3-8 cm H; lvs 1-4 cm L on petioles about as long; ultimate segments 1-3 mm L, crowded; infl slightly exserted from tuft of lvs; involucel of about 5 oblanceolate bractlets; rays 2-8 mm L, stout; fr 3-4 mm L, densely hairy. Dry, granitic gravel, mly 8000-13,000', Fresno Co. s., May-Aug.

O. purpurascens, Purple Mountain-parsley: Plants 1-2 cm H, grayish pubescent. Rare, on ridges and open slopes, 8000', s. SN.

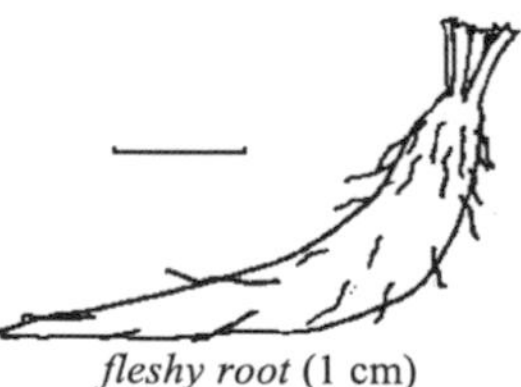
fleshy root (1 cm)

Orogenia

California Orogenia

umbellet (5 mm)

O. fusiformis: Small perennials; lvs ovate in outline, 1-3-ternate, the segments 0.5-6 cm L, linear; bractlets few or none; fls white in cmpd umbels; peduncles 5-10 cm H, slender; rays 2-10; fr 3-4 mm L. Occasional in wet sandy places, Eldorado Co. n., May-Jul. Roots are edible raw or cooked.

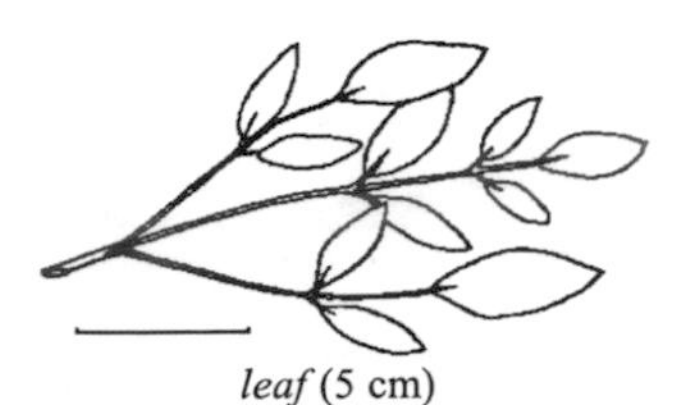

leaf (5 cm)

Osmorhiza

Sweet-cicely

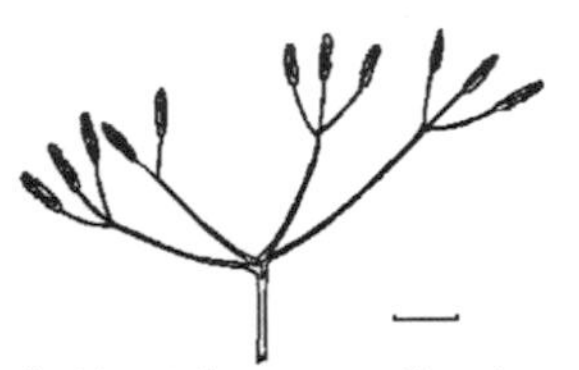

fruiting inflorescence (1 cm)

Slender to stoutish perennials with thick roots. Lvs usu bipinnate; the lfts lanceolate to ovate, petiolulate, coarsely serrate to lobed. Petioles sheathing. Infl of loose cmpd umbels. Sepals absent. Fr cigar-shaped, 10-20 mm L. The roots are sweet and often licorice-flavored and can be used in teas, stew, or soups.

Involucel of several conspicuous bractlets; pedicels 1-3 mm L — *O. brachypoda*
Involucel usu absent; pedicels 3-30 mm L
  Fr densely bristly-hairy — *O. chilensis*
  Fr ± glabrous — *O. occidentalis*

O. brachypoda, California Sweet-cicely: Stems stoutish, 3-8 dm H, pubescent; lf-blades deltoid or ovate, 0.8-2.5 dm L; the lfts ovate, 2-6 cm L; petioles 0.5-2 dm L; involucels 2-10 mm L; rays 2-5; pedicels 1-3 mm L. Shaded woods below 8500', Placer Co. s., Apr-May.

O. chilensis, Mountain Sweet-cicely: Stems slender, 3-10 dm H, usu pubescent; lf-blades roundish; lfts 2-6 cm L; petioles 5-15 cm L; rays 3-6, 2-10 cm L; pedicels 5-30 mm L. Woods, mly below 8000', SN (all), May-Jul.

O. occidentalis, Western Sweet-cicely: Stems stout, 3-12 dm H, villous at nodes, subglabrous to pubescent throughout; lf-blades ovate to oblong, 1-2 dm L; lfts lance-oblong to ovate, 2-10 cm L; petioles 5-30 cm L; rays 5-12; involucel usu none; fls yellow. Wooded slopes, 2500-8700', Mono and Madera cos. n., May-Jul.

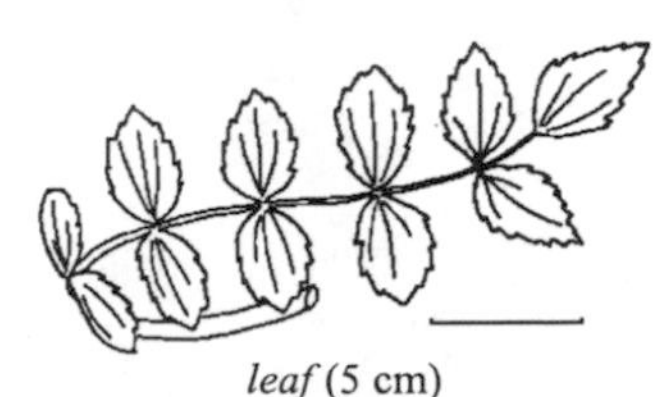

leaf (5 cm)

Oxypolis

Cow-bane

fruit dorsal and lateral aspect (5 mm)

O. occidentalis: Erect glabrous perennial from tuber; stem 6-12 dm H, usu unbranched; lf-blades oblong in outline, 1-3 dm L; lfts 5-13, ± serrate, 3-7 cm L; petioles 1-3 dm L, sheathing; stem lvs few, usu reduced; peduncles 0.5-3 dm L; bracts mly 1-2, 5-25 mm L; pedicels slender; fls white, sepals evident; fr 5-6 mm L. Shallow water and wet places, 4000-8500', SN (all) Jul-Aug. Tubers of a related sp poisonous to cattle.

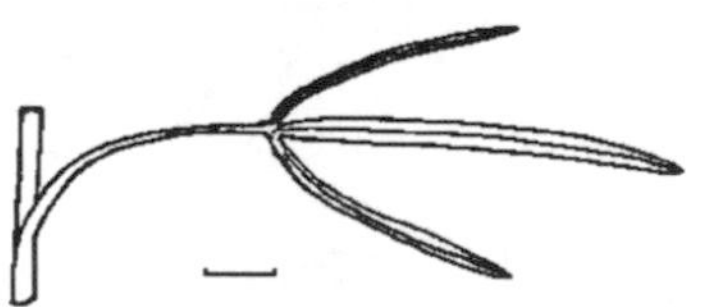

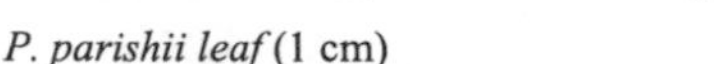

P. parishii leaf (1 cm)

Perideridia

Yampah

inflorescence (1 cm)

Erect branching herbs. Lvs petioled. Infl of loose cmpd umbels. Invol. of few to many entire narrow bracts. Involucel of usu membranous or colored bractlets. Sepals evident. Fls white to pinkish. Fr usu oblong, 3-5 mm L. Plants blooming in summer months. Tuberous roots may be peeled and eaten, or cooked and dried and ground into flour.

A Plants stout; ultimate lfts ± elliptic, irregularly serrate or lobed — 2) *P. howellii*
AA Plants slender; ultimate lfts ± linear, entire
  B Basal lvs with ultimate segments 1-3 cm L — 1) *P. bolanderi*
  BB Basal lvs ternate, ultimate segments 3-9 cm L
    C Rays unequal in fr — 3) *P. lemmonii*
    CC Rays subequal in fr — 4) *P. parishii*

1) *P. bolanderi* ssp. *bolanderi,* Bolander's Yampah: Stems 2-8 dm H; lvs deltoid in outline, 5-15 cm L; ultimate segments 2-8 cm L; petioles 2-8 cm L; peduncles 6-14 cm L; bracts 8-12, 3-12 mm L; bractlets 3-6 mm L. Open forests below 7000', SN (all).

2) *P. howellii,* Howell's Yampah: Stems 6-15 dm H; lvs oblong to oval in outline, 1-3 dm L; 1-2-pinnate, the lfts 2-4 cm L, sometimes entire; peduncles 5-15 cm L, bracts reflexed, narrow, 1-2 cm L; bractlets 3-6 mm L; rays 3-6 cm L; pedicels 4-8 mm L. Mdws below 5000', Mariposa to Nevada Co.

3) *P. lemmonii*, Lemmon's Yampah: Similar to P. parishii except as in key. Moist habitats below 8000', SN (all).

4) *P. parishii* ssp. *latifolia:* Stems 2-9 dm H; lvs lanceolate to ovate in outline; blades 5-15 cm L; lfts 2-10 cm L; petioles 3-7 cm L; peduncles 6-15 cm L; bracts mly absent; bractlets 6-8, 2-4 mm L; rays 10-14, subequal. Moist habitats, 3500-11,200', SN (all).

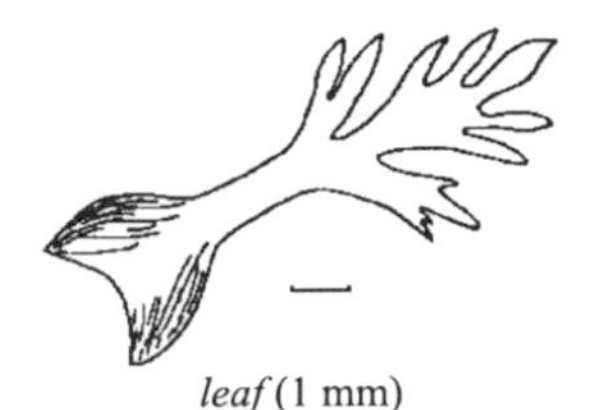
leaf (1 mm)

Podistera

Sierra Podistera

inflorescence (1 mm)

P. nevadensis: Tufted perennial forming compact cushions 2-5 cm H; petiole 3-15 mm L, white membranous on edges; blade 3-10 mm L, with 3-7 lanceolate division, sometimes bipinnate; infl of compact cmpd umbels; peduncles 0.5-3 cm L; fls yellow; fr 1-2 mm L. Rocky areas, mly 10,000-13,000', Mono and Tuolumne to Placer Co., Jul-Sep. Edibility unknown.

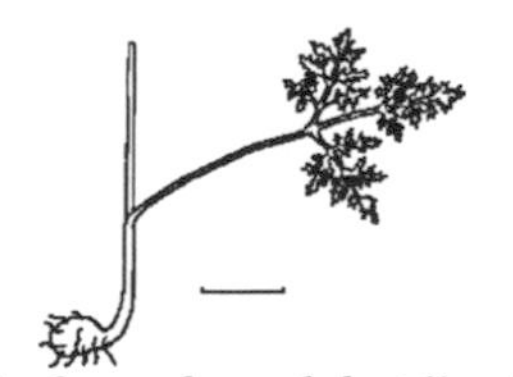
S. tuberosa base of plant (1 cm)

Sanicula

Sanicle

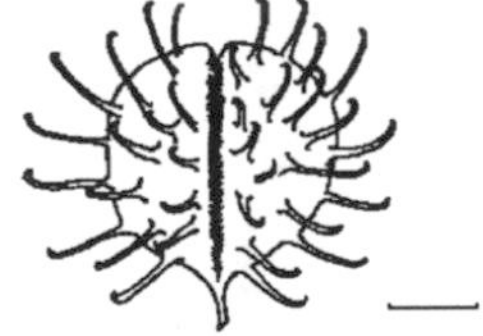
S. crassicaulis fruit (1 mm)

Cauline lvs few or none. Petioles sheathing. Umbels irregularly cmpd, few-rayed, bearing invol. and involucels. Fls yellow, staminate fls often present. Sepals present, persistent. Fr 1-2 mm L. Herbage probably contains alkaloids and should be regarded as inedible.

Stems from globose or irregular tuber; basal lvs 3-12 cm L — *S. tuberosa*
Stems usu from taproot; basal lvs 1.4-4 cm L — *S. graveolens*

S. graveolens, Sierra Sanicle: Stems slender and erect or low and spreading, 0.5-4.5 dm L, branched; lvs ± ovate, the primary divisions oblong-ovate, petiolulate, 3-5-lobed, the segments incised or lobed; stem lvs reduced; invol. bracts 2, pinnatifid, 5-10 mm L; rays 3-5; bractlets 6-10, 1 mm L; bisexual fls 3-5, staminate 7-12. Open forests, 4000-8000', SN (all), Apr-Jun.

S. tuberosa, Tuberous Sanicle: Much like *S. graveolens*; stems 1-8 dm H; umbel rays mly 3. Open or wooded slopes, mly below 8000', SN (all), Apr-Jul.

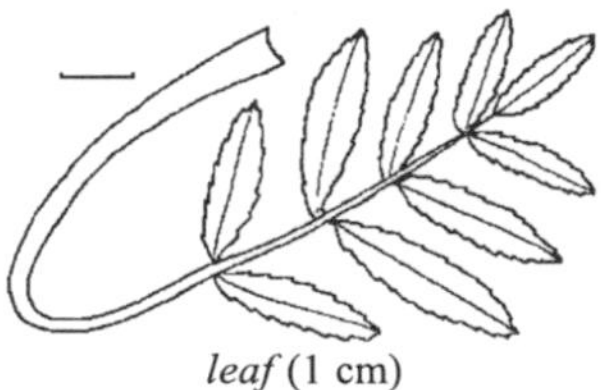
leaf (1 cm)

Sium

Hemlock Water-parsnip

fruit (1 mm)

S. suave: Perennial, stem erect, stout, branched, 6-12 dm H; lower lvs long-petioled, the upper subsessile; submersed lvs with segments like teeth of a comb; emergent lvs with linear to lanceolate divisions 1-9 cm L; infl of loose cmpd umbels; fls white; invol. of ± leafy bracts; involucel of conspicuous narrow bractlets; rays 10-20, 1.5-3 cm L; pedicels 3-6 mm L; fr 2-3 mm L. Wet places below 6500', Feather River n., rare further s., Jul-Aug. Lvs and younger stems edible cooked.

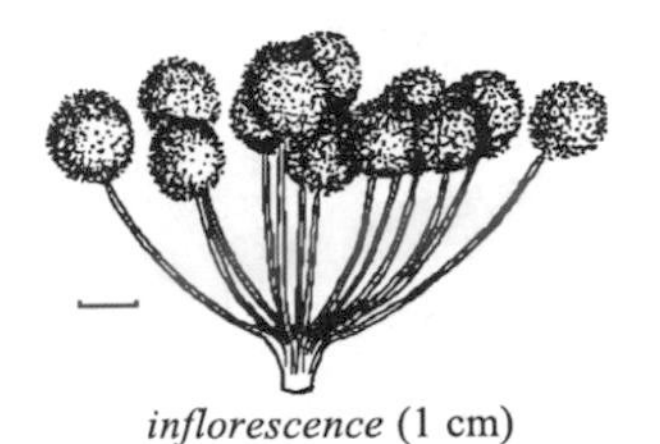

Sphenosciadium

Ranger's Buttons
Swamp White Heads

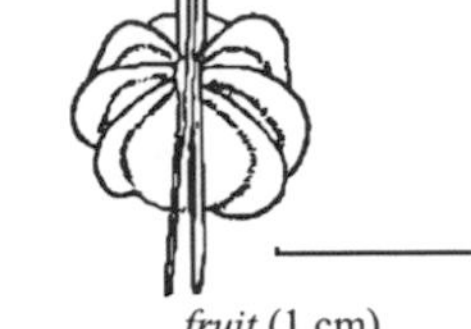

inflorescence (1 cm) *fruit* (1 cm)

S. capitellatum: Stout perennial, glabrous below the pubescent infl; lvs 1-2 pinnate or ternate-pinnate; lf-blades oblong to ovate in outline, 1-4 dm L; lfts irregular in size and shape; petioles 1-4 dm L; invol. none; involucel of many linear pubescent bractlets; rays 4-18 subequal, 2-10 cm L; fls usu white, sometimes purplish; fr 5-8 mm L. Swampy places, 3000-10,400', SN (all), Jul-Aug. Apparently toxic to livestock.

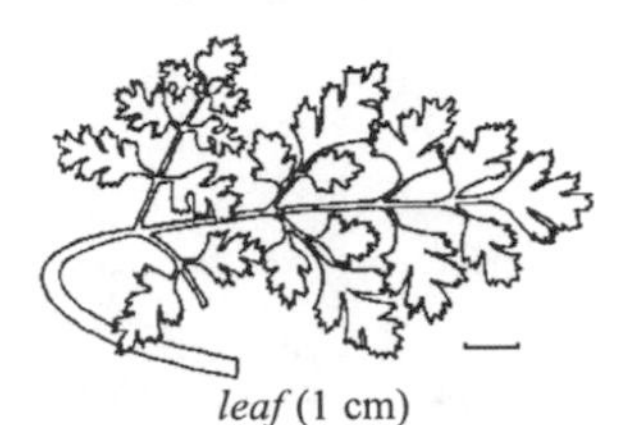

Tauschia

Tauschia

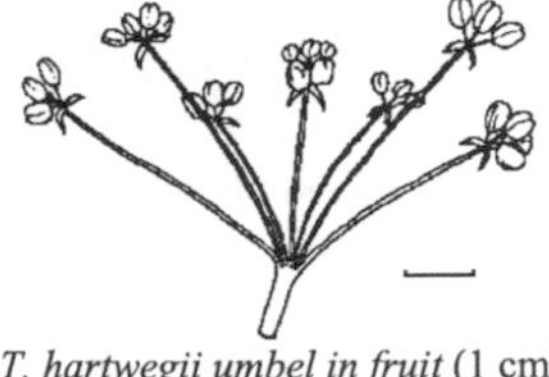

leaf (1 cm) *T. hartwegii umbel in fruit* (1 cm)

Herbage ± glabrous, from taproots or tubers. Lvs petioled, 1-2-ternate-pinnate or bipinnate; lfts serrate. Infl of loose cmpd umbels. Invol. none. Involucel present, usu obvious. Fr slightly flattened laterally, glabrous; ribs filiform. Edibility unknown.

Plants 3-10 dm H; sepals minute; lfts 20-60 mm L; below 5000' *T. hartwegii*
Plants 1-4 dm H; sepals evident; lfts 6-12 mm L; 4000-9000' *T. parishii*

T. hartwegii, Hartweg's Tauschia: Lf-blades 1-2 dm L; lfts oblong to ovate; petioles 5-25 cm L; peduncles 2-8 dm L; bractlets reflexed, 5-12 mm L; rays 2-13 cm L; pedicels 2-7 mm L; fr suborbicular, 4-7 mm L. Occasional on wooded and brushy slopes. Butte Co. s., Apr-May.

T. parishii, Parish's Tauschia: Lf-blades oblong to ovate in outline, 8-15 cm L, the ultimate segments oblong to ovate, 5-8 mm L; petioles 5-15 cm L; peduncles 1-3 dm L; bractlets few, linear, entire, 5-12 mm L; rays subequal, 3-6 cm L; pedicels 2-7 mm L. Common on dry slopes, mly east slope of s. SN, May-Jul.

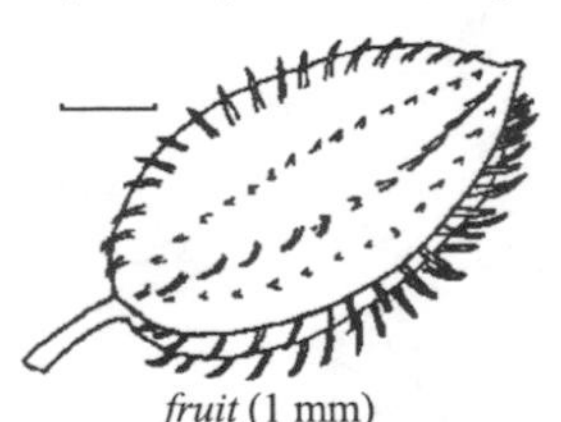

Yabea

Hedge-parsley

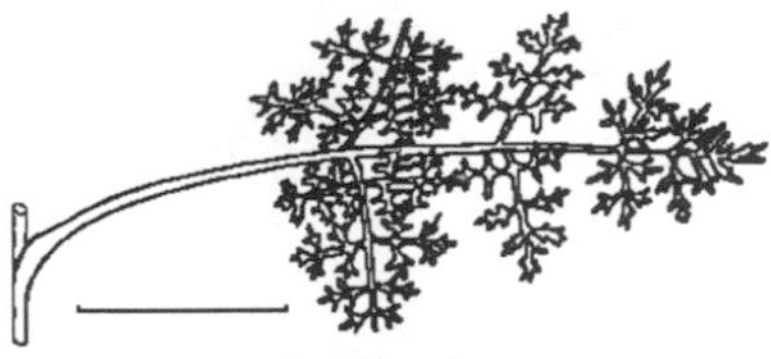

fruit (1 mm) *leaf* (1 cm)

Y. microcarpa: Stems erect, hairy; lvs 2-3-ternate, 2-6 cm L, ultimate lfts linear, 1-3 mm L; rays unequal, 1-9; bracts leafy, pinnate; pedicels 3-7, unequal, sepals evident; fr flattened laterally, oblong, 3-7 mm L. Occasional, open and shaded slopes, SN (all), Apr-Jun. Edibility unknown.

APOCYNACEAE - Dogbane Family

Perennials with entire, opp lvs lacking stipules. Fls clustered, 5-merous. Petals united into a tube at base. Carpels 2, usu distinct. Fr a follicle.

Corolla not over 6 mm L; lf-blade/petiole junction sharp *Apocynum*
Corolla 15-20 mm L; lf-blade usu narrowing smoothly into petiole *Cycladenia*

Apocynum

Dogbane
Indian-hemp

A. pumilum (5 mm) *branch and inflorescence* (1 cm)

Stems erect at base but usu curving to be nearly horizontal in leafy portion. Herbage glabrous. Fls pale, on short pedicels. Corolla-tube short with 5 small sagittate appendages at base opp the lobes. Herbage poisonous to stock and humans but usu avoided due to the bitter, rubbery juice. Seeds edible after being parched and can be ground into meal to make fried cakes. Poisonous constituent a saponin (cymarin) causing hyperthermia, dilation of the pupils and a discoloration in the mouth. The genus is often difficult to key to sp because of hybridization.

Corolla 2-3 mm L, often greenish, scarcely exserted from calyx ... *A. cannabinum*
Corolla 4-6 mm L, often pinkish, well exserted from calyx ... *A. androsaemifolium*

A. androsaemifolium, Bitter Dogbane: Stems 2-4.5 dm H, much-branched; lvs 2-9 cm L, ovate to oblong, usu petioled; corolla white, usu with pink veins; follicles 4-12 cm L. Occasional in open forests, 5000-9500', SN (all), Jul-Aug.

A. cannabinum, Indian Hemp: Stem 3-6 dm H; lvs oblong-lanceolate, 4-10 cm L; corolla greenish to white, cylindric to urn-shaped; follicles 12-20 cm L. Occasional in damp places, SN (all), Jun-Aug.

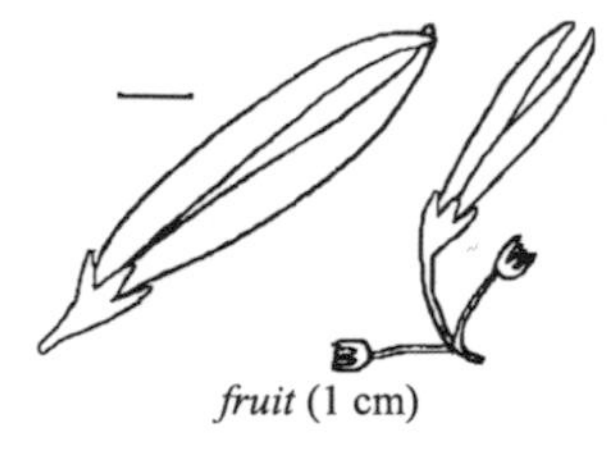

Cycladenia

Cycladenia

fruit (1 cm) *stem and flowers* (1 cm)

C. humilis var. *humilis:* Stems 1 to several, 1-2 dm H, from a fleshy root, glabrous throughout or densely pubescent (var. *tomentosa*); lvs rather thick, in 2-3 pairs, ovate to roundish, 3-7 cm L or the lower smaller, rounded at apex, on petioles 0.5-3 cm L; pedicels 7-12 mm L, corolla rose-purple, about 15 mm L, with small appendages alt with the lobes; follicles 4-6 cm L. Dry slopes and rocky places, 3500-8500', Butte Co. n., May-Jul. Probably poisonous.

ARALIACEAE - Ginseng Family

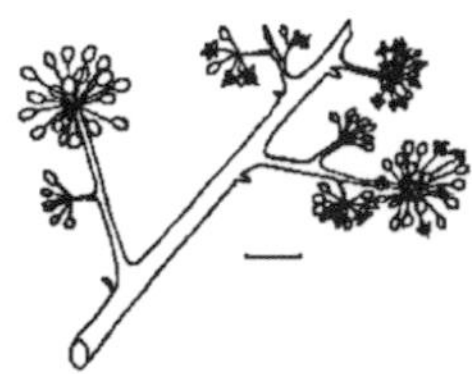

Aralia

Spikenard
Elk-clover

inflorescence (1 cm) *leaf* (1 dm)

A. californica: Perennial; roots large with milky juice; stems 1-3 m H; lvs alt, glabrous, ternate then pinnately 3-5-foliolate; petioles to 3 dm L, sheathing; lfts ovate or oblong, serrate, subcordate at base, 5-25 cm L; infl 3-4 dm L with numerous, glandular-pubescent, many-fld umbels; pedicels 1-2 cm L; sepals minute; petals about 2 mm L. In moist and shady spots below 5000', SN (all), Jun-Aug. Root of a related sp is used in root beer; frs poisonous raw but may be used for jelly.

ARISTOLOCHIACEAE - Birthwort Family

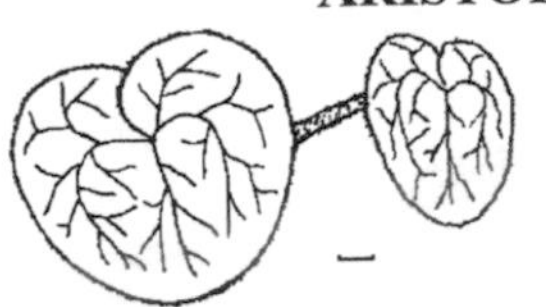

leaf (1 cm)

Asarum

Wild-ginger

A. hartwegii flower (1 cm)

Stemless perennials with basal, long-petioled, cordate or hastate, entire lvs. Fls large, solitary, borne in lower axils. Calyx regular, 3-parted, bell-shaped. Rootstock may be used as a substitute for ginger; it may be dried and kept for later use or candied by tenderizing short pieces in boiling water and then boiling the pieces in a heavy syrup.

Calyx lobes long-attenuate, 25-65 mm L — *A. hartwegii*
Calyx-lobes acute or obtuse, 8-12 mm L — *A. lemmonii*

A. hartwegii, Hartweg's Wild-ginger: Rootstock stoutish; lvs persistent, usu acute at tip, 4-10 cm L, ± mottled above; petioles 5-15 cm L; pedicels 1-2.5 cm L; calyx brownish purple. Shaded places, 2500-7000', Tulare Co. n., May-Jun.

A. lemmonii, Lemmon's Wild-ginger: Much like *A. hartwegii* except rootstock stolon-like; lvs usu rounded at tip, dark green above. Moist places, 3600-6000', Tulare to Plumas Co., May-Jun.

ASCLEPIADACEAE - Milkweed Family

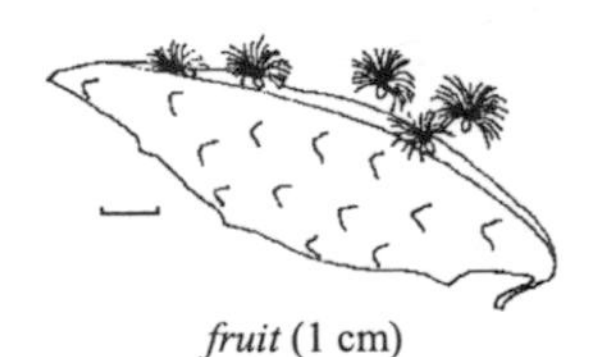

fruit (1 cm)

Asclepias

Milkweed

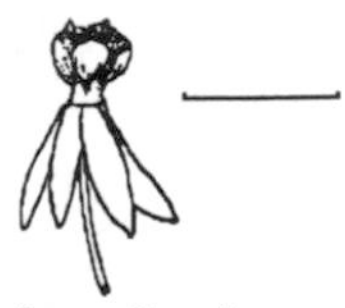

flower (1 cm)

Erect perennials. Lvs mly opp or whorled, entire. Fls 5-merous in axillary or terminal umbels. Petals reflexed at anthesis. Follicles narrow, mly acuminate, 5-10 cm L. The lvs and stems of most broad-leaf milkweeds are poisonous to cattle and humans; symptoms include depression, irregular breathing and changing temperature. The fls of spp 2) and 4) may be eaten raw or boiled although not in large amounts. The buds, young shoots and young lvs of the same spp may be eaten as greens after boiling in 2 waters. In at least *A. speciosa* the seeds and inner wall of the pod may be eaten raw or cooked, and a brown sugar can be boiled down from the fls. The white latex from most spp may be made into chewing gum by stirring and slightly heating until solid and then adding animal fat to give it a more lasting character.

A Pedicels, upper stem and lvs ± glabrous
  B Lvs broad, ± cordate; fls dark red-purple — 1) *A. cordifolia*
  BB Lvs narrow, usu less than 2 cm W, not cordate; fls greenish white — 3) *A. fascicularis*
AA Pedicels, upper stem and often lvs white tomentose; lvs broad
  B Lvs opp; hood 2-3 times stamen and stigma length — 4) *A. speciosa*
  BB Lvs mly whorled; hood about equal to stamens and stigma — 2) *A. eriocarpa*

1) *A. cordifolia,* Purple Milkweed: Stems 3-8 dm H, often tinged purple in upper parts, glabrous to somewhat pubescent; lvs mly opp, ovate and mly acute, 5-15 cm L; corolla-lobes 8-9 mm L. Open or wooded slopes below 6300', SN (all), May-Jul.

2) *A. eriocarpa,* Indian Milkweed: White-hairy throughout; stem unbranched, 4-9 dm H; lvs, at least some, in whorls of 3-4, elongate-oblong, 6-15 cm L, on very short petioles, truncate to subcordate at base; corolla-lobes cream, 4-5 mm L; hoods cream or tinged purplish. Frequent in dry, barren places below 7000', SN (all), Jun-Aug.

3) *A. fascicularis,* Narrow-leaved Milkweed: Stems several, 5-9 dm H; lvs linear to linear-lanceolate, usu in whorls of 3-6, 4-12 cm L, 3-10 mm W, short-petioled, usu folded along midrib; corolla-lobes 4-5 mm L; hoods about as long as stamens. Frequent as colonies in dry places below 7000', SN (all), Jun-Sep.

4) *A. speciosa,* Showy Milkweed: Stems stout, with soft short hairs or sometimes glabrate, 5-12 dm H, leafy to summit; lvs oval to oblong, short-petioled, rounded to cordate at base, 8-15 cm L; corolla-lobes rose-purple, 8-10 mm L, mly below 6000', Fresno Co. n., May-Jul.

ASTERACEAE - Aster Family

Herbs or sometimes shrubs. Infl a head. Fls borne on the enlarged summit of the peduncle and surrounded by the bracts of the involucre. Calyx reduced to a whorl of bristles, awns or scales (the "pappus") or absent. Fr an akene.

A Plants bearing spines or burs; phyllaries sometimes fused into shallow cup
  B Lvs not spiny; fr bur-like; stems often brownish; disturbed habitats — *Ambrosia* p. 47
  BB Lvs with sharp spines; fruit not spiny
    C Stems spiny winged from decurrent lf bases; pappus bristles not plumose — *Carduus* p. 53
    CC Stems not spiny winged; pappus bristles plumose — *Cirsium* p. 55
AA Plants without spinose heads, not thistle-like
  B Disk-fls none; the ligule on ray-fls 5-toothed at apex; stems usu with milky sap
    C Heads solitary, usu on scapes, only 1-2 per plant; plants usu low and ± glabrous
      D Pappus none; lvs grass-like or wider, entire — *Phalacroseris* p. 64
      DD Pappus present; lvs various, often toothed to pinnate
        E Akenes with conspicuous beak; lvs mly lanceolate to oblong; pappus of 50 or more white bristles — *Agoseris* p. 47
        EE Akenes not beaked; lvs filiform or linear, or pinnatifid with rachis and divisions linear; pappus awns or ± plumose scales
          F Pappus 15-20 plumose scales or 5 scales tipped by awns — *Microseris* p. 64
          FF Pappus 30-50 silvery non-plumose awns — *Nothocalais* p. 64
    CC Heads several to many per plant; plants often very hairy; pappus of many bristles
      D Stems ± prostrate, 2-5 cm L; lvs with membranous white margins — *Glyptopleura* p. 59
      DD Stems erect, usu more than 1 dm H; lvs usu green throughout
        E Pappus of plumose bristles
          F Fls white or purplish; akene lacking beak — *Stephanomeria* p. 68
          FF Fls yellow or purple; akene with conspicuous beak — *Tragopogon* p. 68
        EE Pappus bristles not plumose
          F Pappus bristles (at least the outer) deciduous — *Malacothrix* p. 64
          FF Pappus bristles persistent
            G Pappus of many white soft bristles; lvs often lobed — *Crepis* p. 56
            GG Pappus bristles tan to brown and fragile; lvs never lobed — *Hieracium* p. 60
  BB Heads with at least some disk-fls
    C Ray-fls none or inconspicuous; phyllaries sometimes white and ray-like
      D Pappus none or reduced to an inconspicuous crown
        E Lvs broad, over 2 cm W, simple, deltate to elliptic, (see also *Artemesia douglasiana*)
          F Lvs ± elliptic, usu over 10 cm — *Wyethia invenusta* p.70
          FF Lvs + deltate, 3-10 cm L
            G Lvs mly basal, white-woolly beneath — *Adenocaulon* p.47
            GG Lvs cauline, often opp, green on both sides — *Pericome* p. 64
        EE Lvs usu less than 1 cm W or lobed or dissected
          F Heads solitary or in pairs; lvs lacking strong odor
            G Lvs elliptic, 1-2 cm L; head arising from leaf axils — *Iva* p. 61
            GG Lvs dissected into linear segments; heads terminal — *Chamomilla* p. 54
          FF Infl consisting of several to many heads; lvs often 3-toothed or -lobed
            G Phyllaries in 1 row; lvs linear — *Madia glomerulata* p. 63
            GG Phyllaries in 2-3 rows; lvs often with sage odor
              H Heads in racemes or panicles, not flat-topped — *Artemisia* p. 50
              HH Heads in flat-topped clusters — *Sphaeromeria* p. 68
      DD Pappus present, of scales, awns, or bristles
        E Phyllaries scarious (white or slightly colored); plants herbaceous, white-woolly
          F Stems 2-9 dm H; lvs 2-8 cm L; pappus concealed by phyllaries — *Anaphalis* p.48
          FF Stems usu less than 3 dm H; lvs often shorter
            G Plants with basal rosette of lvs; pappus conspicuous, usu exserted beyond phyllaries — *Antennaria* p.48
            GG Plants usu without basal rosette; pappus not exserted — *Gnaphalium* p.59
        EE Phyllaries at least partially herbaceous
          F Fls white, greenish, purplish, brownish, or rarely yellowish white
            G Head (recpt) dark and cone-shaped, 2-6 cm H — *Rudbeckia occidentalis* p. 66

GG Head and recpt ± flat
H Pappus of membranous oblong scales
I Annual; lvs linear, entire *Orochaenactis* p. 64
II Perennial; lvs usu pinnate *Chaenactis* p. 54
HH Pappus of numerous usu conspicuous bristles
I Lvs ± oblanceolate, sessile, alt
J Phyllaries with brown midvein; ray-fls present but small *Conyza* p. 56
JJ Phyllaries lacking brown midvein, often tipped green *Lessingia* p. 62
II Lvs deltoid to ovate, usu petioled, alt or opp
J Phyllaries about 5 mm L, 1 mm W, barely overlapping *Ageratina* p. 47
JJ Phyllaries mly large, strongly imbricate *Brickellia* p. 53
FF Fls yellow to orange
G Plants woody at least at base
H Subshrub; lvs serrate *Hazardia whitneyi* p. 59
HH Woody shrub; lvs entire
I Herbage pubescent (except *C. viscidiflorus*) *Chrysothamnus* p. 54
II Herbage glabrous *Ericameria* p. 56
GG Plants mly herbaceous throughout
H Lvs opposite
I Pappus of many bristles; seed not barbed *Arnica discoidea* p. 49
II Pappus usu of 4 stiff awns; seeds and awns barbed *Bidens* p. 52
HH Lvs alt or basal
I Lvs ± linear (cmpd in *E. compositus*), often all basal
J Pappus bristles plumose
K Stems leafy; rare, Middle Fork Kings River *Raillardiopsis* p. 65
KK Lvs mly basal; widespread *Raillardella* p. 65
JJ Pappus bristles simple
K Ray-fls 25-50, 2-3 mm L, erect, pale violet *Trimorpha* p. 69
KK Ray-fls few or lacking, not as above *Erigeron* p. 57
II Lvs oblong to broader; stems usu leafy
J Phyllaries in 1 row, often black-tipped; lower lvs petioled *Senecio* p. 66
JJ Phyllaries in 2-3 rows, not black-tipped; lvs sessile *Aster* p. 51
CC Ray fls present and usu obvious
D Pappus none or inconspicuous
E Phyllaries usu imbricate or at least in more than 1 row
F Ray-fls usu white
G Heads 3-7 cm across, usu solitary; lvs toothed to pinnate *Leucanthemum* p. 62
GG Heads 0.5-1 cm across, in flat-topped clusters; lvs bipinnate *Achillea* p. 46
FF Ray-fls yellow
G Recpt ± columnar *Rudbeckia* p. 66
GG Recpt flat or slightly convex
H Lf-base hastate to cordate; pappus none *Balsamorhiza* p. 52
HH Lvs gradually tapering to petiole; pappus of two awns *Helianthella* p. 60
EE Phyllaries usu in a single row, not imbricate
F Lvs opposite *Whitneya* p. 69
FF Lvs alt or basal
G Recpt naked; ray-fls conspicuous *Eriophyllum lanatum* p. 58
GG Recpt with ring of bracts between ray- and disk-fls; ray-fls often small
H Ray-fls usu 8-12, occasionally less; widespread *Madia* p. 63
HH Ray-fls 5; mly below 4000' *Lagophylla* p. 62
DD Pappus present
E Rays white to pink, red or purplish (see also *Hulsea heterochroma*)
F Pappus of plumose bristles
G Lvs obovate, the lower usu serrate or lobed; anthers yellow in disk-fls *Layia* p. 62
GG Lvs linear, the margins often curled under; anthers purple in disk-fls *Blepharipappus* p. 53
FF Pappus bristles not plumose; rays various
G Pappus bristles markedly unequal; rays 3-15; disk-fls often purple *Machaeranthera* p. 62
GG Pappus bristles subequal; rays mly more than 15; disk-fls yellow
H Pappus much reduced on ray-fls (see also *Erigeron strigosus*) *Lessingia* p. 62
HH Pappus well developed on ray-fls
I Phyllaries usu in 3-4 rows, the tips often reflexed *Aster* p. 51
II Phyllaries in 1-2 rows, the tips erect *Erigeron* p. 57
EE Rays yellow to orange to brown

F Pappus usu scales or few awns
G Lvs opp; pappus 4-awned — *Bidens cernua* p. 52
GG Lvs alt or basal; pappus of scales
H Lvs with decurrent bases extending down stem — *Helenium* p. 60
HH Lf bases not extending down stem
I Lvs usu over 10 cm L; heads usu over 3 cm W — *Wyethia* p. 70
II Lvs 1-8 cm L; heads various but often smaller
J Phyllaries in 1 row; herbage white-woolly — *Eriophyllum* p. 58
JJ Phyllaries in 2-3 rows; herbage usu green
K Ray-fls deeply and conspicuously 3-lobed, often reddish at base — *Gaillardia* p. 59
KK Ray-fls shallowly lobed, if tinged reddish then withering early
L Lvs entire
M Recpt spherical; ray-fls 1.5-3.5 cm L — *Dugaldia* p. 56
MM Recpt slightly convex; ray-fls shorter, withering early — *Rigiopappus* p. 66
LL Lvs serrate or lobed — *Hulsea* p. 61
FF Pappus of soft capillary bristles (scales may be present)
G Phyllaries many, usu imbricate
H Erect perennials usu from a creeping rhizome — *Solidago* p. 67
HH Shrubs or matted subshrubs or perennials from a large taproot
I Lvs serrate (occasionally entire in *Pyrrocoma*)
J Fr 5-10 mm L, glabrous; phyllaries in 4-5 rows, strongly graduate — *Hazardia whitneyi* p. 57
JJ Fr 2-5 mm L, hairy; phyllaries in 3-4 rows, subequal
K Basal tuft of lvs well developed; stems often red-tinged — *Pyrrocoma* p. 65
KK Basal lvs few; stems usu green — *Tonestus* p. 69
II Lvs entire
J Mat-forming subshrub; lvs usu 3-veined from base — *Stenotus* p. 68
JJ Plants usu upright; lvs 1-veined from base
K Shrubs, usu over 50 cm H; pappus of numerous capillary bristles — *Ericameria* p. 56
KK Perennials; pappus of 10-20 inner bristles and about as many outer scales
L Plant from a thick crown of white, shiny persistent lf-bases — *Erigeron linearis* p. 57
LL Plant without caudex; stems decumbent to erect — *Heterotheca villosa* p. 60
GG Phyllaries fewer or if more at least not strongly imbricate, mly subequal and in 1 row
H Lvs opp except sometimes the uppermost — *Arnica* p. 49
HH Lvs alt or basal
I Plant usu scapose; rays 1-3, short — *Raillardella scaposa* p. 65
II Stems usu leafy; rays mly more numerous — *Senecio* p. 66

inflorescence (1 cm)

Achillea

Yarrow

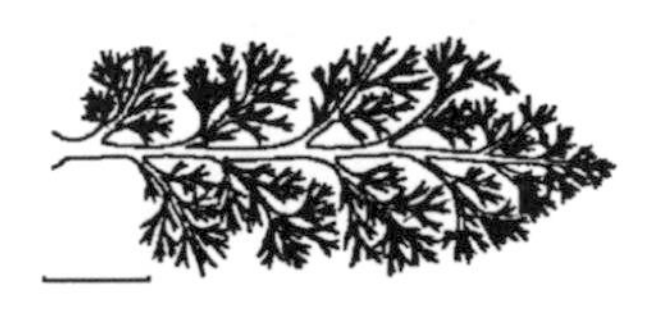

leaf (1 cm)

A. lanulosa: Aromatic perennial; stems to 10 dm H; lvs 5-10 cm L, 10-15 mm W, fern-like, ultimate segments linear, 1-4 mm L; lower lvs petioled, the upper sessile; phyllaries many; heads 5-10 mm W in terminal, flat-topped clusters; rays 4-5, 2-5 mm L, oval. Common in mdws and dampish places, mly below 8000', SN (all), Jun-Aug.

ssp *alpicola*: Lvs less than 10 mm W. Mly 9000-13,000', SN (all), Jun-Aug.

Both forms may contain glucosides and alkaloids that produce severe intestinal disorder if eaten in quantity. A tea can be made from the dried lvs and the lvs can be used as a sage substitute in stews.

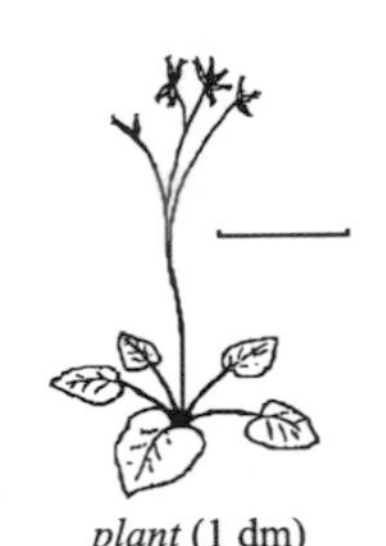

Trail Plant

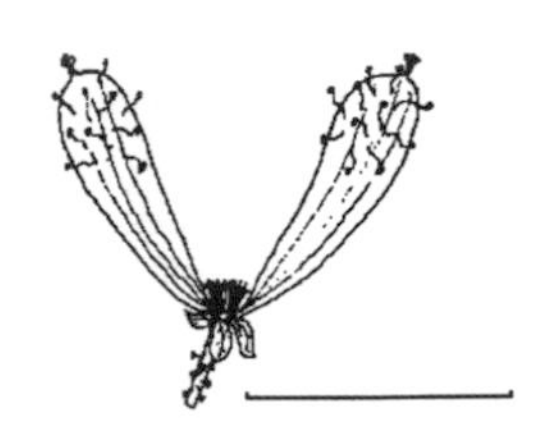

plant (1 dm) *flowers and fruit* (1 cm)

A. bicolor: Stems 5-9 dm H, cobwebby-woolly below, glandular above; lvs deltoid-ovate, sinuate-dentate, 3-10 cm L, mly basal, subglabrous above; petioles mly longer than blades; infl naked, forming upper two thirds or more of plant; phyllaries reflexed in fr, about 1-2 mm L; heads 2-5 mm W, peduncles 5-10 cm L. Shaded woods below 7000', SN(all), Jun-Aug.

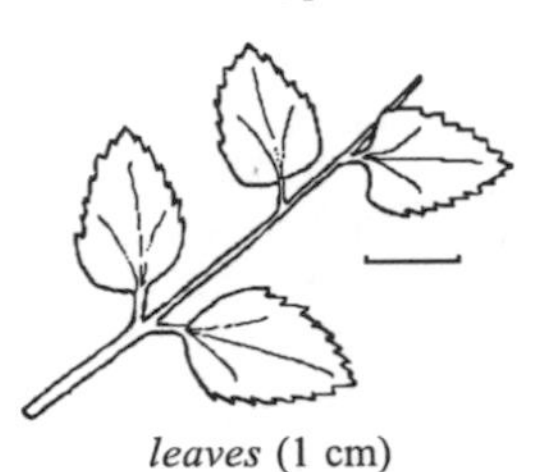

Ageratina

Western Eupatorium

leaves (1 cm) *inflorescence* (1 cm)

A. occidentale: Perennial from a woody base; stems 1.5-7 dm H, greenish or purplish, minutely hairy; lvs mly alt, deltoid-ovate, ± serrate, glabrous, acute, 1.5-3.5 cm L, mly rounded or truncate at base; heads densely clustered at tips of stems and branches, 8-10 mm H; fls pink, red-purple to whitish. Rocky slopes and ridges, 6500-11,000', SN (all), Jul-Sep. A related sp toxic.

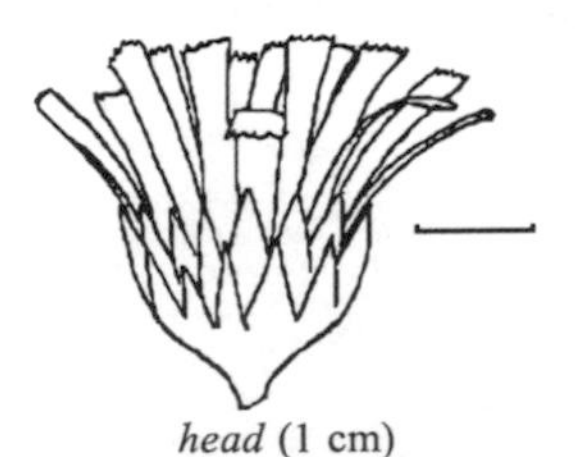

Agoseris

Mountain Dandelion

head (1 cm) *akene, beak and pappus* (1 cm)

Annual or perennial herbs with strong taproot; lvs mly in basal tuft or a few scattered on lower stem. Herbage glabrous or villous. Infl scapose, the head erect; invol. bell-shaped to subcylindric. Phyllaries subequal or imbricate. Fls yellow to orange, often drying darker. Lvs edible but bitter, especially in late season.

A Fls burnt-orange, often drying purplish; above 6000' — 1) *A. aurantiaca*
AA Fls yellow, often drying pinkish
  B Akene-beak less than twice as long as akene-body; above 5000'
    C Akene-beak about half as long as akene-body; scapes 1-3 dm H — 3) *A. glauca*
    CC Akene-beak about as long as akene-body; scapes 3-6 dm H — 2) *A. elata*
  BB Akene-beak 2-4 times as long as akene-body; below 8000'
    C Annual; invol. 5-20 mm H; lvs 5-15 cm L — 5) *A. heterophylla*
    CC Perennial; invol. 25-40 mm H; lvs 10-30 cm L
      D Akene-body truncate at apex; phyllaries in 2 unlike rows — 6) *A. retrorsa*
      DD Akene-body gradually tapered at apex; phyllaries mly in 3 rows — 4) *A. grandiflora*

1) *A. aurantiaca*, Orange-flowered Agoseris: Plants 1-5 dm H; lvs oblong, 5-25 cm L, entire to remotely pinnatifid; invol. about 20 mm H; akene-body 4-8 mm L. Moist mly grassy places, 6000-11,500', SN (all), Jul-Aug.

2) *A. elata*, Tall Agoseris: Scapes stout; lvs entire to pinnatifid, 10-30 cm L; invol. 20-25 mm H; akene-body 8-10 mm L. Uncommon, in moist habitats, 5200-10,500', SN (all), Jul-Sep.

3) *A. glauca* vars., Short-beaked Agoseris: Scapes stout, lvs linear to oblanceolate, 5-20 cm L; invol. 15-25 mm H; akene-body 5-9 mm L. Dryish habitats, 5000-10,500', SN (all), Jul-Aug.

4) *A. grandiflora*, Large-flowered Agoseris: Scapes stout, 1.5-6 dm H; lvs ± lanceolate, entire to pinnately cut; akene-body 4-7 mm L. Dryish habitats below 7400', SN (all), May-Jul.

5) *A. heterophylla*, Annual Agoseris: Plants 0.3-4 dm H; lvs spatulate to linear, entire to sinuate; akene-body 3-5 mm L. Open, grassy habitats below 7500', SN (all), May-Jul.

6) *A. retrorsa*, Spear-leaved Agoseris: Scapes stout, 1.5-5 dm H; lvs + lanceolate, pinnately parted; the segments usu bent backwards, the terminal segments very long; akene-body 5-7 mm L. Common on dry ridges and slopes, SN (all), May-Aug.

Ambrosia

Annual Bur-sage

plant (5 cm)

A. acanthicarpa: Monoecious, gray-green annual, 4-15 dm H, with bristly hairs; lvs often opp below, alt above, petioles winged, blade pinnately divided, to 8 cm L and 7 cm W; infl a raceme with staminate heads above pistillate; pistillate heads 1-fld; fr a bur with body 5-7 mm in diameter and several to many spines.

Anaphalis

Pearly Everlasting

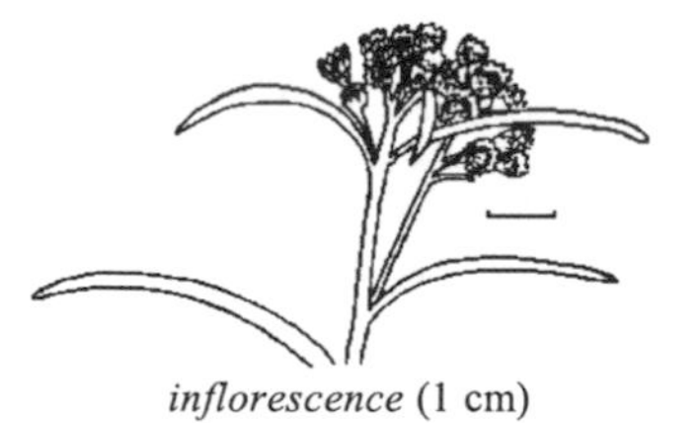

inflorescence (1 cm)

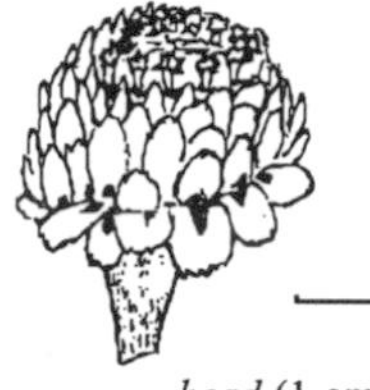

head (1 cm)

A. margaritacea: Erect perennial with leafy stems; lvs alt, lanceolate to linear, sessile ± revolute, greener above than beneath; heads to 1 cm W; invol. about 5-7 mm H; phyllaries white, ovate. Openings in woods, talus, etc., below 8500', SN (all), Jun-Aug. Herbage has been used as a tobacco substitute.

Antennaria

Pussytoes

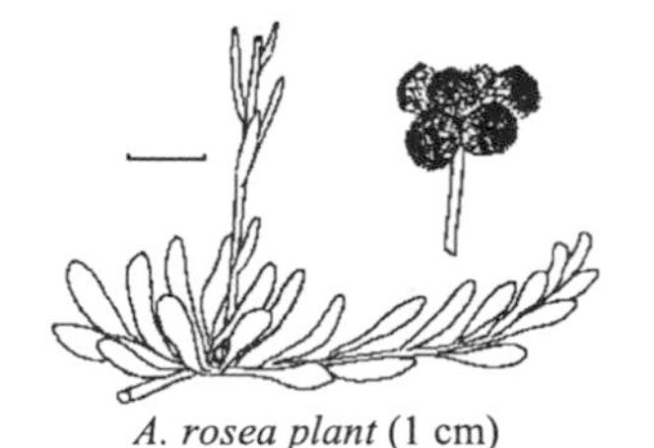

A. rosea plant (1 cm)

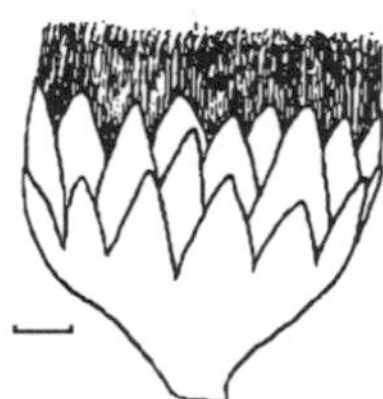

head (1 mm)

Low woolly perennials with simple, entire, alt and basal lvs. Stem lvs usu reduced. Plants dioecious. Staminate fls with scaly pappus of usu club-shaped bristles; pistillate fls with copious pappus. The sap from the stem can be used as a chewing gum. Species difficult to distinguish except for characters in the following key.

A Head solitary; stems 1-4 cm H; lvs 1-3 cm L; 5000-10,000' — *A. dimorpha*
AA Heads in clusters of 3 to many; stems usu taller
  B Plants not mat-forming or stoloniferous; mly below 6000'
    C Basal lvs 3-8 cm L, petioled — *A. luzuloides*
    CC Basal lvs 1-2 cm L, sessile — *A. geyeri*
  BB Plants with stolons, forming dense mats; stems to 3 dm H; mly above 6000'
    C Phyllaries dark brown or blackish green
      D Lowest cauline lf 10-20 mm L; herbage usu not glandular — *A. media*
      DD Lowest cauline lf 6-12 mm L; herbage usu glandular — *A. pulchella*
    CC Phyllaries white to rose or light brown
      D Phyllaries white-tipped with a dark spot at base; Fresno Co. n. — *A. corymbosa*
      DD Phyllaries lacking dark spot at base
        E Phyllaries blunt; Plumas Co. — *A. umbrinella*
        EE Phyllaries acute; SN (all) — *A. rosea*

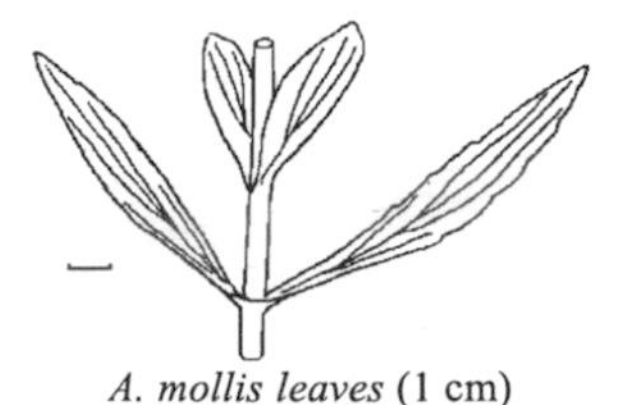

A. mollis leaves (1 cm)

Arnica

Arnica

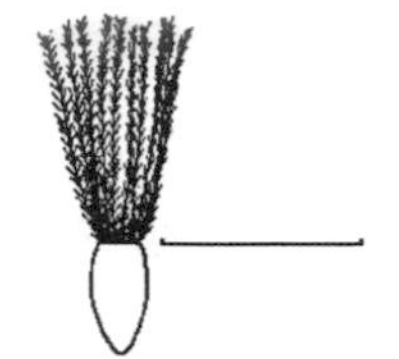

akene and pappus (1 cm)

Perennial herbs, ± glandular or aromatic. Lvs simple, opp or the upper sometimes alt. Heads rather large, turbinate to hemispheric, solitary to many. Fls yellow. Phyllaries green, mly in 2 rows. Recpt flat or convex, naked. Reputed to be poisonous.

A Heads usu without ray-fls — 4) *A. discoidea*
AA Heads usu with ray-fls
  B Stem lvs in mly 5-12 pairs, the upper well developed
    C Phyllaries with tuft of hairs at tip; stems solitary — 2) *A. chamissonis*
    CC Phyllaries without apical tuft of hairs; stems often several from base
      D Stems tufted, often densely so; lvs ± entire — 8) *A. longifolia*
      DD Stems one to several; lvs mly serrate — 1) *A. amplexicaulis*
  BB Stem lvs in 2-4 pairs, the upper often much reduced
    C Lvs with 3 or more principal veins from base, these parallel to each other and rarely branching
      D Lf-blades ± ovate; pappus yellowish, subplumose — 5) *A. diversifolia*
      DD Lf-blades elongate; pappus whitish, barbed but not plumose — 6) *A. fulgens*
    CC Lvs usu with 1 main vein from base; the veination pinnate
      D Rays less than 15 mm L; stem and petioles densely long-hairy at base; young heads nodding — 11) *A. parryi*
      DD Rays 15-25 mm L, or if shorter then plant not long-hairy at base; heads erect
        E Lvs entire to serrulate, the lower rarely cordate; pappus usu brownish, subplumose; widespread above 7000'
          F Stems 2-6 dm H; lvs variable — 9) *A. mollis*
          FF Stems 1-2.5 dm H; lvs ovate — 10) *A. nevadensis*
        EE Lvs mly toothed, the lower often cordate; pappus mly white, barbed but rarely subplumose; usu below 7000'
          F Basal lvs none; stem lvs 6-8; heads several; rare — 12) *A. tomentella*
          FF Basal lvs several; stem lvs 4-8; heads 1-3
            G Basal lvs deeply cordate; lower stem lvs petioled; widespread — 3) *A. cordifolia*
            GG Basal lvs ± deltoid; stem lvs mly sessile; rare — 7) *A. latifolia*

1) *A. amplexicaulis*, Streambank Arnica: Stems 3-7 dm H, ± glandular-pubescent; basal lvs none; stem lvs sessile, ± lanceolate, 4-12 cm L; heads usu several, about 15 mm L; rays 8-14, pale yellow, 10-20 mm L. Moist habitats, 7000-10,000', SN (all), Jul-Aug.

2) *A. chamissonis* ssp *foliosa,* Meadow Arnica: Stem solitary, 3-8 dm H; herbage pubescent, glandular above, stem lvs lanceolate to oblanceolate, entire to denticulate, 4-10 cm L, on petioles to as long; heads 5-15, about 15 mm H; rays pale yellow, 12-18 mm L. Moist habitats, 5000-11,000', SN (all), Jul-Aug.

3) *A. cordifolia,* Heart-leaved Arnica: Stems mly solitary, basal lvs long-petioled; stem-lvs 2-9 cm L, on shorter petioles or the upper sessile; heads mly 1-3, 18-25 mm H; rays 9-14. Dry to moist habitats, 3500-10,000', SN (all), May-Aug.

4) *A. discoidea* var. *alata,* Rayless Arnica: Stems mly solitary, 3-6 dm H, glandular-pubescent, usu branched above; stem lvs in 3 or more pairs, mly crowded near base, on winged petioles; lf-blades ovate to subcordate, 3-8 cm L, dentate; heads several to many. Open woods below 6000', Mariposa Co. n., Jun-Aug.

5) *A. diversifolia,* Lawless Arnica: Stems 1 to several, 1.5-4 dm H, unbranched, subglabrous; stem lvs ovate to elliptic, dentate, acute; the blades 4-8 cm L, usu petioled; heads mly 3-5; rays 12-15, pale yellow. Wet habitats, 7000-11,000', northern Inyo Co. n., Jul-Aug.

6) *A. fulgens,* Hillside Arnica: Rhizomes short, thickly clothed with bases of previous lvs, in the axils of which are dense tufts of long tawny hair; stems 2-6 dm H, ± glandular-pubescent; lvs 3-5-veined from base, the lower petioled, oblanceolate to elliptic, 3-12 cm L; head mly solitary; rays 10-20, 10-15 mm L. Moist, open places, 6000-9500', e. slope, Alpine Co. n., May-Jul.

7) *A. latifolia,* Mountain Arnica: Stems 1 to few, 1-6 dm H; basal lvs long petioled, stem lvs 2-12 cm L, dentate; heads mly 1-3; rays 8-12. Rare, moist habitats, Eldorado and Placer cos., 5500-7000', Jul-Aug.

8) *A. longifolia,* Seep-spring Arnica: Fl-bearing stems mly 3-6 dm H; herbage subglabrous; stem lvs sessile to perfoliate, mly 5-12 cm L; heads several to many; rays 8-13, 10-20 mm L. Wet habitats, 5000-11,000', SN (all), Jul-Aug.

9) *A. mollis,* Cordilleran Arnica: Stems 2-6 dm H, unbranched; stem lvs usu sessile, lanceolate to ovate, entire to denticulate; heads 1 to few; the disks to 3 cm W; rays mly 12-18. Moist places, 6800-11,500', SN (all), Jul-Sep.

10) *A. nevadensis,* Sierra Arnica: Stems solitary or few; herbage ± glandular and puberulent; lvs 3-8 cm L, petioled; heads 1-3. Dry to moist areas, 6600-11,900', SN (all), Jul-Aug.

11) *A. parryi* ssp *sonnei,* Nodding Arnica: Stems solitary, 2-6 dm H, glandular-villous; stem lvs much reduced upwards, the lowermost 5-20 cm L, lanceolate, short-petioled, entire to denticulate; heads usu several. Moist habitats, 6400-11,000', northern Inyo to Nevada Co., Jul-Aug.

12) *A. tomentella,* Recondite Arnica: Stems solitary, 2-5 dm H, glandular-pubescent; lvs grayish hairy, the lower petioled, ovate to ovate-elliptic, 3-7 cm L, mly dentate; heads mly 5-7. Open slopes and woods, about 5000-7000', Tulare Co., Placer Co. n., Jul-Aug.

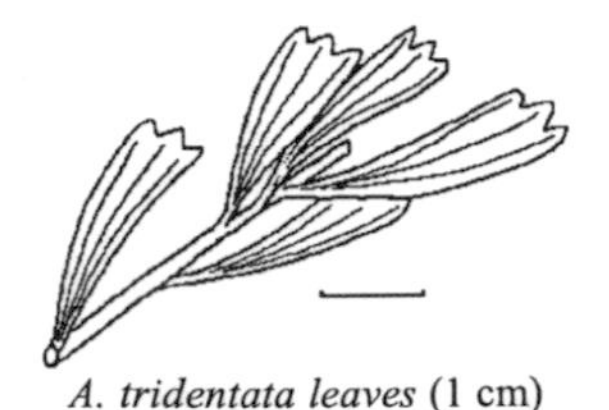

A. tridentata leaves (1 cm)

Artemisia

Sagebrush Wormwood

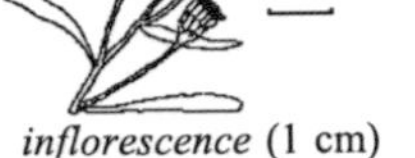

inflorescence (1 cm)

Late summer flowering herbs or shrubs, usu aromatic; lvs alt. Heads small, discoid; recpt flat or hemispheric, naked or with long hairs. Seeds edible raw or as a flour. Herbage may be toxic if eaten in large amounts, but may be used in small quantities to flavor stews, etc.

A. Shrubs
- B Lvs linear mly entire, acute — 2) *A. cana*
- BB Lvs mly wedge-shaped, mly 3-toothed
  - C Plant 1-3 dm H; lvs less than 1 cm L; Alpine Co. n. — 1) *A. arbuscula*
  - CC Plant usu over 3 dm H; lvs over 1 cm L
    - D Heads more than 3 mm in diameter; lvs usu entire on fl-stems; Eldorado Co. s.
      - E Plant sticky resinous; herbage green and very aromatic — 7) *A. rothrockii*
      - EE Plant not sticky; herbage grayish pubescent — 8) *A. spiciformis*
    - DD Heads usu less than 3 mm in diameter; all lvs usu 3-lobed; widespread — 9) *A. tridentata*

AA Herbs
- B Lvs shallowly lobed to entire
  - C Lvs linear, mly entire, 3-8 cm L — 4) *A. dracunculus*
  - CC Lvs lanceolate to ovate, often dentate or lobed, 7-15 cm L — 3) *A. douglasiana*
- BB At least lower lvs parted into linear or lanceolate lobes or dissected
  - C Lvs glabrate above, white tomentose beneath — 5) *A. ludoviciana*
  - CC Lvs ± green on both surfaces — 6) *A. norvegica*

1) *A. arbuscula,* Dwarf Sagebrush: Much like *A. tridentata* except for smaller stature; lvs 0.5-1.5 cm L, almost as wide. Dry habitats, 4000-9500'.

2) *A. cana,* Silver Sagebrush: Evergreen shrub; twigs densely pubescent; lvs mly 2-6 cm L, silvery-pubescent. Dry, rocky habitats, 6000-10,500'.

3) *A. douglasiana,* Douglas' Mugwort: Stems often branched above, stout, 5-15 dm H; lvs ± glabrous and green above, densely gray-tomentose beneath; infl leafy; heads erect or nodding. Usu waste places, mly below 6000', SN (all).

4) *A. dracunculus,* Dragon Sagewort: Stems erect, 5-15 dm H; herbage almost odorless; lower lvs mly deciduous; infl with elongate, leafy, ascending branches; heads many, spreading or nodding. Dry habitats below 9000', s. SN.

5) *A. ludoviciana,* Western Mugwort: Stems 3-10 dm H, unbranched below infl; lower lvs 2-8 cm L, parted into linear or lanceolate lobes, some of these again toothed or lobed; upper lvs less cut to entire. Dry places below 11,600', SN (all).

6) *A. norvegica* var. *saxatilis,* Mountain Sagewort: Stems 2-6 dm H; lvs ovate in outline, the basal with dissected blade 2-10 cm L; upper lvs reduced, sessile; infl loosely racemose. Rocky places, 5000-13,600', SN (all), Jul-Sep.

7) *A. rothrockii,* Timberline Sagebrush: Twigs densely tomentose when young; lvs 0.5-5 cm L, sometimes lanceolate and entire; infl spicate or narrowly paniculate, 0.5-4 dm L. Rocky slopes, mly 65000-11,500', Eldorado Co. s.

8) *A. spiciformis,* Snowfield Sagebrush: Stems widely branching, 3-8 dm H; lvs 2-5 cm L, entire to irregularly toothed, turning yellow and dropping in fall. Common, open slopes and rocky mdws, 6500-12,000', Nevada Co. s. May be a hybrid between *A. cana* and *A. tridentata.*

9) *A. tridentata,* Sagebrush: Trunk short, thick or a few branched from base; herbage silvery throughout; lvs of vegetative shoots mly 3-toothed at tip; lvs of fl-shoots mly entire and linear to oblanceolate; infl many-branched, 1.5-4 dm L. Dry slopes below 10,000', SN (all).

Aster

Aster

head (1 cm) *ray-flower* (1 cm)

Summer or fall flowering usu perennial herbs. Lvs alt, entire or toothed. Heads usu numerous and radiate in panicles or racemes, rarely solitary. Ray-fls usu shades of purple or blue, more rarely white; disk-fls yellow or reddish-purple. Lvs of related sp edible as greens. Some asters will absorb selenium, making them toxic on certain soils.

A Ray-fls none — 3) *A. breweri*
AA Ray-fls present, usu many
  B Rays 5-15, white or pale violet, mly Mariposa Co. n.
    C Lvs oblanceolate, 10-17 mm W, mly entire — 10) *A. oregonensis*
    CC Lvs obovate, 20-50 mm W, mly dentate — 12) *A. radulinus*
  BB Rays mly more than 15; mly blue or purple; lvs ± entire
    C Fl-stems unbranched above basal lvs, supporting one terminal head; lvs linear
      D Stems and invols. glandular, sticky; ray-fls 8-20 — 11) *A. peirsonii*
      DD Stems and invols. not glandular; ray-fls 20-30 — 1) *A. alpigenus*
    CC Fl-stems branched above and supporting several to many heads; lvs usu wider
      D Stem and invols. glandular, sticky
        E Lower lvs obovate, clasping; heads few, 2-5 cm W — 8) *A. integrifolius*
        EE Lower lvs linear to linear-spatulate; heads many, 1.5-2 cm W — 4) *A. campestris*
      DD Stems and invols. not glandular
        E Annual from taproot; wet and marshy habitats — 7) *A. frondosus*
        EE Perennials from rhizomes; dry to moist habitats (*occidentalis* group)
          F Middle stem lvs mly over 1 cm W; mly n. SN
            G Ray-fls 10-20 mm L; 5000-8000', mly w. SN — 6) *A. foliaceus*
            GG Ray-fls 5-10 mm L; mly below 7000', e. SN — 5) *A. eatonii*
          FF Middle stem lvs mly less than 1 cm W; widespread
            G Basal lvs with winged petioles; phyllaries often purple-tipped — 9) *A. occidentalis*
            GG Basal lvs ± sessile; phyllaries green-tipped — 2) *A. ascendens*

1) *A. alpigenus* ssp *andersonii,* Alpine Aster: Stems 5-40 cm H from a fleshy taproot; lvs tufted, 4-15 cm L, 2-10 mm W; stem lvs few, much reduced; invol. 6-10 mm H; rays 7-15 mm L. Moist or boggy mdws, 400-11,500', SN (all).

2) *A. ascendens,* Long-leaved Aster: Stems slender, mly less than 7 dm H; branches erect or ascending, lower lvs narrowly oblanceolate; middle stem lvs clasping; heads few to many in a nearly leafless infl; invol. 4-7 mm H. Moist to dry mdws, e. slope, below 7500'.

3) *A. breweri,* Brewer's Aster: Stems erect, 2-8 dm H, usu much-branched; herbage pubescent; lvs sessile, lanceolate to lance-ovate, 1-3 cm L, ± entire; heads solitary at ends of branches, bell-shaped; invol. 7-10 mm H, 2-3-seriate; pappus of tawny bristles. Open rocky slopes and forests, 4500-10,900', SN(all).

4) *A. campestris* var. *bloomeri,* Western Meadow Aster: Stems slender, erect, 1-3 dm H; lvs firm, 1-5 cm L, 2-7 mm W, ± crowded; heads rarely solitary; invol. 5-6 mm H; phyllaries green-tipped; rays 5-10 mm L. Dry, open slopes, 6000-8000', Lake Tahoe n., mly e. SN.

5) *A. eatonii,* Eaton's Aster: Stems 4-10 dm H, with many erect or ascending branches; lvs lance-linear, numerous, the lower petioled, 5-15 cm L, 4-18 mm W; heads many, in a narrow, leafy panicle; invol. 5-8 mm H; phyllaries green-tipped; rays 20-35. Wet, cool habitats.

6) *A. foliaceus* vars., Leafy Aster: Stems much as in *A. eatonii;* middle stem lvs 1-4 cm W; heads several to many; rays 15-50. Moist habitats.

7) *A. frondosus,* Marsh Aster: Stems decumbent to erect, glabrous, 2-5 dm H; lvs sessile, 2-5 cm L, elliptic; infl narrow; rays many, pinkish, 5-8 mm L. Below 6000', mly s. SN.

8) *A. integrifolius,* Entire-leaved Aster: Stems several, 2-7 dm H; lvs thickish, the basal 4-12 cm L with a slender petiole almost as long; green lvs sessile or clasping; invol. 8-14 mm H; rays 10-15 mm L. Dry habitats, 5500-10,500', Plumas Co. s.

9) *A. occidentalis* and vars., Western Mountain Aster: Stems and lvs various but much as in *A. ascendens;* rays 25-35, 7-10 mm L. Moist to dryish habitats, 4000-10,500', SN (all).

10) *A. oregonensis* ssp *californicus,* Oregon White Aster: Stems erect, 4-12 dm H, leafy; lvs to 9 cm L, sometimes with ± wavy margin; heads many, invol. 7-8 mm H; rays 4-7 mm L. Dry woods, 3500-6500', Eldorado Co. n., rare s. to Tulare Co.

11) *A. peirsonii,* Peirson's Aster: Stems one per rosette of lvs, 2-7 cm H; basal lvs 2-5 cm L, 1-3 mm W, 3-nerved; stem lvs few and reduced; invol. purplish, 7-11 mm H; rays 12-16 mm L. Moist habitats, 11,000-12,250', Tulare, Inyo and s. Fresno cos.

12) *A. radulinus,* Broad-leaf Aster: Stems 1 to several from a woody base, 2-6 dm H; lower stem lvs largest, 4-10 cm L, sessile; heads usu many; invol. 6-9 mm H. Dry forest s below 5000'.

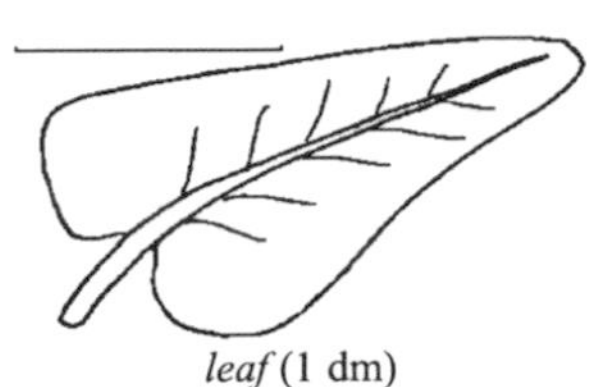

leaf (1 dm)

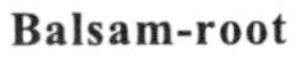

Balsamorhiza

Balsam-root

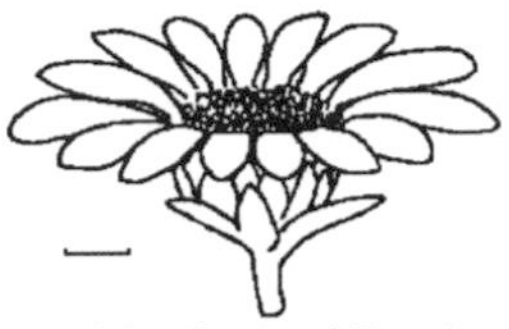

B deltoidea head (1 cm)

Perennial herbs with erect or ascending stems. Lvs almost all in basal rosettes, opp, deltoid to sagittate, entire to dentate, long-petioled. Roots edible raw or cooked. Best peeled and baked. In spring the young stems and lvs may be eaten raw or boiled. As the stems and lvs grow older they remain edible but become tough and fibrous. The seeds may be roasted and eaten or ground into flour.

Herbage sparsely hairy, the lvs green on both surfaces *B. deltoidea*
Herbage densely hairy, the lvs tomentose beneath at least when young *B. sagittata*

B. deltoidea, Deltoid Balsam-root: Stems usu several, 2-8 dm H, usu with several much reduced lvs; blades of basal lvs 10-30 cm L, 5-20 cm W, usu entire or scalloped, sometimes irregularly dentate; heads solitary or few, the summit of peduncle and base of invol. often densely hairy and glandular, but hardly woolly. Deep sandy soil, below 7000', SN (all), Apr-Jun.

B. sagittata, Arrow-leaved Balsam-root: Much like *B. deltoidea;* basal lvs 15-30 cm L, 5-15 cm W, entire; heads solitary, 6-10 cm across the rays; invol. and top of peduncle persistently woolly-tomentose; heads solitary, 6-10 cm across the rays; invol. and top of peduncle persistently woolly-tomentose; phyllaries broadly lanceolate. Deep sandy soils, plains and forest openings, 4300-8300', SN (all), May-Jul.

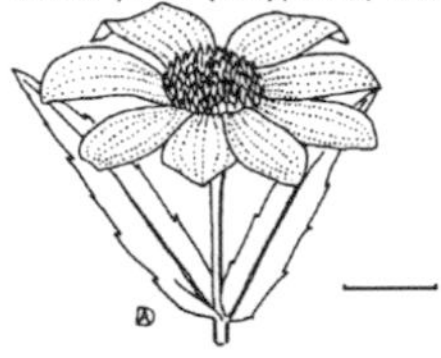

head (1 cm)

Bidens

Bidens

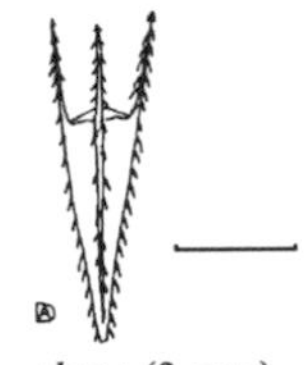

akene (3 mm)

Annuals of wet habitats, with erect stems and opposite, simple lvs. Phyllaries in two rows, the outer leaflike, the inner thinner with membranous margins. Ray-fls few or none, yellow if present. Disk-fls usu many, yellow. Common name comes from the fruit whose sharp awns will penetrate clothing and make the seed difficult to remove. Although small, the seed is edible.

Lvs sessile; Plumas Co., n. *B. cernua*
Lvs petioled; central SN *B. tripartita*

B. cernua, Nodding Bur-marigold: Stems 1-9 dm H, glabrous to short-rough-hairy; lvs with bases fused around stem; blades 4-20 cm L, ± lanceolate, serrate; heads occasionally radiate, nodding in fr; ligules on ray-fls, 8-15 mm L. Below 6000'.

B. tripartita, Sticktight: Stems 4-15 dm H, glabrous; lvs 4-20 cm L, lanceolate, serrate; heads discoid, erect. Below 4000'.

heads (1 cm)

Brickellia

Brickellbush

flower (1 cm)

Late summer and fall-flowering perennial herbs. Lvs simple, veiny. Heads discoid. Edibility unknown

Lvs mly opp, 3-11 cm L; petioles 5-30 mm L — *B. grandiflora*
Lvs alt, 1-4 cm L; petioles short
  Stems 5-10 dm H; heads in small clusters — *B. californica*
  Stems 2-4.5 dm H; heads mly solitary — *B. greenei*

B. californica, California Brickellbush: Stems many, woody at base, ± pubescent; lvs deltoid-ovate, serrate; panicle leafy; heads 12-14 mm H, cylindrical; phyllaries green or purplish, subglabrous. Dry slopes and flats, mly below 5000', SN (all).

B. grandiflora, Large-flowered Brickellbush: Stems 3-7 dm H, usu unbranched up to the infl; lvs deltoid-ovate to lanceolate, often with truncate or cordate base, dentate, acuminate; heads ± nodding; fls greenish or yellowish white; invol. 8-14 mm H, green; phyllaries 5-7-seriate. Dry, rocky slopes, 4500-8000', Mariposa Co. n.

B. greenei, Mountain Brickellbush: Stems many from a woody base, often somewhat branched above, very leafy; lvs ovate, mly serrate; heads terminal and solitary or corymbosely arranged, subtended by leafy bracts; invol. 10-14 mm H; phyllaries linear-lanceolate. Mly dry, open, rocky places, 2700-8000', Placer Co. n.

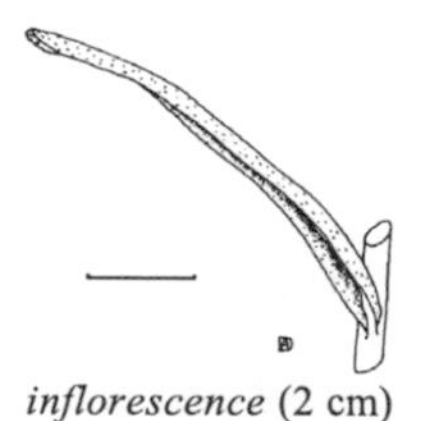

inflorescence (2 cm)

Blepharipappus

Blepharipappus

leaf (3 mm)

B. scaber: Annual, 5-40 cm H, with slender branches; lvs alt, sessile, linear, 10-35 mm L, the margins usu rolled under; heads solitary or in a leafy infl; ray-fls 2-8, white with purple veins, the ligules 2-8 mm L; disk-fls many, white with purple anthers. Open forests, 3500-10,000', Sierra Co. n., May-Aug.

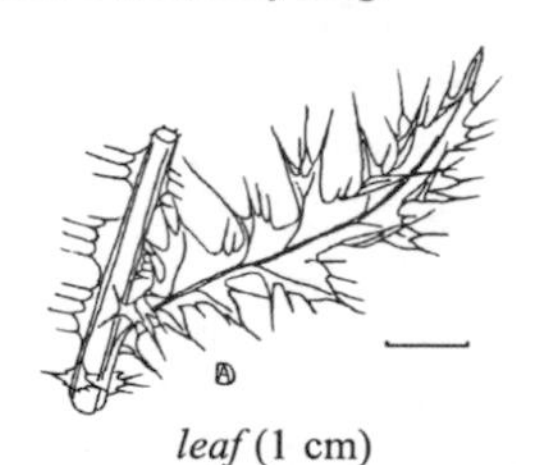

leaf (1 cm)

Carduus

Plumeless Thistle

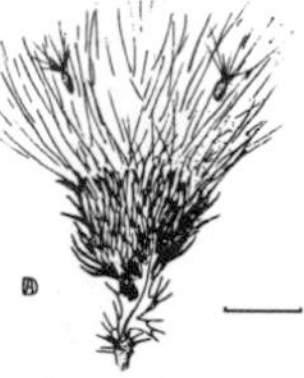

head (1 cm)

Biennials with erect stems and alt lvs. The lf-bases are decurrent down stem as spiny wings. Heads discoid. Plants of waste areas and considered weedy.

Phyllaries usu more than 2 mm W; heads usu solitary, often nodding — *C. nutans*
Phyllaries usu less than 2 mm W; heads often in clusters of 2-5 — *C. acanthoides*

C. acanthoides, Plumeless Thistle: Stems 3-10 dm H, glabrous to woolly, strongly spiny-winged; lvs basal, 10-20 cm L, toothed to lobed; invol. 1-2.5 cm in diameter; phyllaries appressed to spreading. Along roadsides below 3500', Sierra Co. n., May-Jul.

C. nutans, Musk Thistle: Much as *C. acanthoides* except as in key.

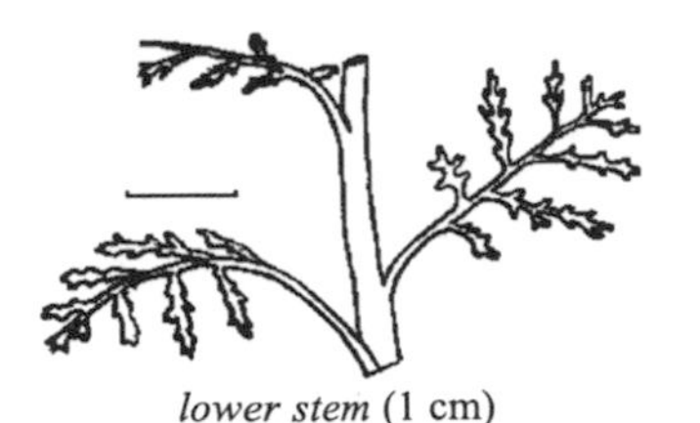

Chaenactis

Pincushion

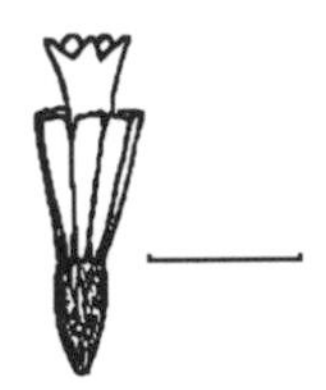

lower stem (1 cm) *flower* (1 cm)

Lvs alt, entire to lobed or dissected. Heads discoid. Herbage probably inedible.

A Stems erect, usu 1-4 dm H
 B Annual; lvs entire to 1-pinnately lobed, glabrate — 4) *C. xantiana*
 BB Perennial; lvs 2-3-pinnatifid, white-woolly — 2) *C. douglasii*
AA Stems mly matted, less than 1 dm H
 B Lvs few-lobed, 3-10 mm W; Tulare to Eldorado Co. — 1) *C. alpigena*
 BB Lvs bipinnate, 10-25 mm W; Sonora Pass to Lassen Peak — 3) *C. nevadensis*

1) *C. alpigena,* Southern Sierra Chaenactis: Stems densely matted, 2-7 cm H; lvs densely yellow- or white-tomentose, 1-3 cm L, palmately to pinnately few-lobed, the petiole about as long as blade; peduncles short; corolla cream-white. Occasional, in sandy or gravelly soils, 8000-12,500', Jul-Aug.

2) *C. douglasii,* Hoary Chaenactis: Stems branched above; herbage thinly cobwebby-woolly; lvs 3-10 cm L, 1-4 cm W; heads about 15 mm H; corolla white to pinkish. Dry, rocky habitats, 4000-7000', SN (all). May-Jul.

var. *alpina:* Stems matted, less than 10 cm L; heads 1-2 on scapes. Alpine fields, 9000-11,000', Alpine and Eldorado cos., Freel Peak and vicinity.

3) *C. nevadensis,* Northern Sierra Chaenactis: Lvs gray-pubescent, 2.5-4.5 cm L, petiole longer than blade; heads solitary; corolla whitish tinged with pink distally. Gravelly soils or talus, 8000-10,900', Jul-Aug.

4) C. xantiana: Stems 1-several; lvs to 7 cm L; the basal rosette withering; phyllaries glabrous, the longest 10-18 mm L; fls white to pinkish, 7-10 mm L.

Chamomilla

Chamomilla

leaf (1 cm) *head* (5 mm)

C. occidentalis, Western Chamomile: Annual, 15-50 cm H; herbage not strongly scented; lvs alternate, sessile, 3-7 cm L, glabrous, irregularly 2-3-pinnately lobed, the segments linear; heads 1-1.5 cm in diameter, solitary or 2-3; phyllaries in 2-3 unequal series; corolla 1-2 mm L. Wet habitats below 7500', SN (all). Used as a substitute for chamomile.

Chrysothamnus

Rabbitbrush

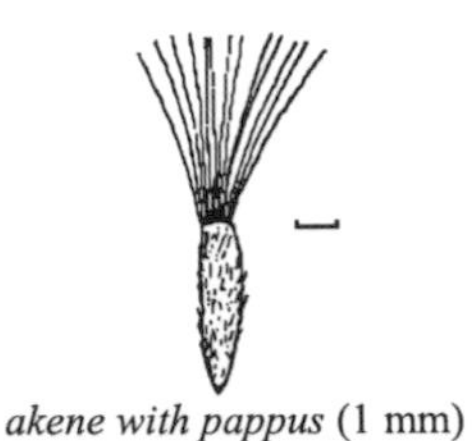

heads (1 cm) *akene with pappus* (1 mm)

Shrubs or subshrubs usu much-branched with erect stems. Lvs alt, entire. Heads numerous, clustered, yellow, narrow, mly 5-fld, blooming in late summer. Herbage may be toxic if consumed in large quantities.

A Twigs white-barked, brittle, not tomentose; e. SN
 B Heads not extending beyond lvs; fls usu 2-3 per head — 1) *C. humilis*
 BB Heads over-reaching lvs; fls usu 5 per head — 4) *C. viscidiflorus*
AA Twigs densely woolly-tomentose

B Infl mly racemose of spicate; phyllaries very long-pointed — 3) *C. parryi*
BB Infl mly cymose; phyllaries obtuse to moderately long-pointed — 2) *C. nauseosus*

1) *C. humilis:* Stems 1-2 dm H, leafy; lvs 1-2 cm L, filamentous; infl dense; invol. 8-10 mm H; fls 2-4 per head, pale yellow. Occasional in sagebrush scrub, above 5000', e. SN, Sierra Co. n.

2) *C. nauseosus,* Rubber Rabbitbrush: Shrub, usu 3-20 dm H; invol. 6-13 mm H; phyllaries usu 3-4-seriate, strongly graduate, in rather definite ranks. Common on e. SN below 10,000'.

3) *C. parryi,* Parry's Rabbitbrush: Shrub up to 5 dm H, the numerous branches erect or spreading, very leafy; lvs narrowly linear to elliptic, 1-8 cm L, 1-8 mm W, 1-3-nerved; heads in short leafy racemes; invol. 9-10 mm H; phyllaries 4-6-seriate, in obscure vertical ranks; pappus brownish white. Mountainsides and flats, 3000-11,500', SN (all).

4) *C. viscidiflorus,* Yellow Rabbitbrush: Stems to 3 dm H; lvs linear or linear-lanceolate, usu densely pubescent, flat or twisted, 2-5 cm L, 2-5 mm W. Dry habitats to 11,000'.

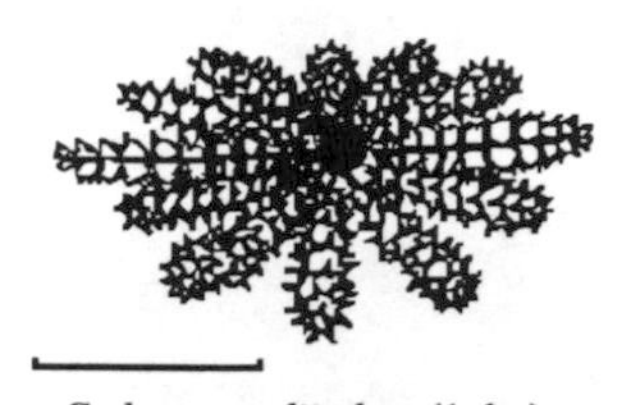

C. drummondii plant (1 dm)

Cirsium

Thistle

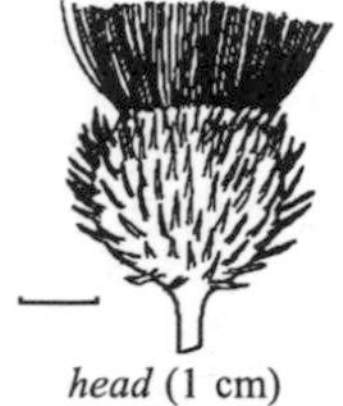

head (1 cm)

Biennial to perennial herbs, spiny. Lvs alt, toothed or more usu pinnatifid. Invol. with several series of phyllaries. All spp have roots that may be eaten raw, boiled or roasted. The peeled stems may be cooked as greens. Young lvs edible raw. Hybridization between spp often makes identification difficult, and descriptions of these tentative species are therefore brief.

A Stem much branched; lvs densely pubescent but not cobwebby; wet habitats — 5) *C. douglasii*
AA Stem branches few; lvs often cobwebby pubescent or glabrous; moist to dry habitats
  B Stem lacking or inconspicuous; fls purple — 7) *C. scariosum*
  BB Stems evident, usu 2-20 dm H; if inconspicuous then fls pale
    C Fls white to faint purple; plants with cobwebby hairs
      D Jointed hairs absent; heads usu in open clusters — 6) *C. occidentalis*
      DD Hairs of two kinds: cobwebby and jointed; heads usu tightly clustered
        E Spines on outer phyllaries 5-10 mm L — 3) *C. canovirens*
        EE Spines on outer phyllaries 2-4 mm L — 4) *C. cymosum*
    CC Fls bright red to purple; hairs usu simple; herbage sometimes glabrate
      D Invol. 1-2 cm in diameter; outer phyllary spines 10-20 mm L — 2) *C. arizonicum*
      DD Invol. 2-4 cm in diameter; outer phyllary spines 2-5 mm L
        E Corollas 30-45 mm L; heads usu on peduncles — 1) *C. andersonii*
        EE Corollas 20-30 mm L; heads sessile, clustered at stem tip — 7) *C. scariosum*

1) *C. andersonii,* Anderson's Thistle: Stem usu 4-7 dm H; lvs usu green above, pubescent beneath; petioles spiny-winged; peduncles to 20 cm L, axillary heads sometimes sessile. Dry habitats, 5000-11,000', SN (all).

2) *C. arizonicum,* Arizona Thistle: Stem angled; lvs 5-15 cm L, linear in outline, pinnatifid with yellow spines; heads 1 to few; invol. 3-3.5 cm H. Dry, stony slopes, Tulare, Alpine and Inyo cos.

3) *C. canovirens,* Gray-green Thistle: Stem usu unbranched below, solitary; heads 1-many; corollas 20-30 mm L. Dry, open habitats, 5000-12,000', mly e. slope, SN (all).

4) *C. cymosum,* Peregrine Thistle: Stem usu 3-6 dm H, usu unbranched below; fls few to many, sessile or on peduncles to 10 cm; corollas 20-30 mm L, dull white. Dry habitats below 7000', SN (all).

5) *C. douglasii,* Swamp Thistle: Herbage gray-pubescent with appressed felt-like hairs; basal lvs 3-10 dm L; heads several to many; peduncles 1-4 cm L. Below 7000', Mariposa Co. n.

6) *C. occidentalis,* Western Thistle: Extremely variable; stems inconspicuous to 30 dm H; corolla pale. Often disturbed habitats, below 12,000', SN (all).

7) *C. scariosum,* Elk Thistle: Stems usu very short but sometimes erect to 10 dm: lvs often basal, glabrous to pubescent above, glabrous to densely pubescent below. Mdws and moist habitats below 11,500', SN (all).

Conyza

Horseweed

plant (1 dm)

C. canadensis: Annual, 1-20 dm H; stem usu much-branched above middle; lvs 1-10 cm L, entire to shallowly few-lobed, glabrous to hairy; heads inconspicuously radiate, many; fresh invol. 2.5-4 cm in diameter; disk-fls 7-13, yellow. Common in waste places below 6500', SN (all), Jun-Aug.

Crepis

Hawksbeard

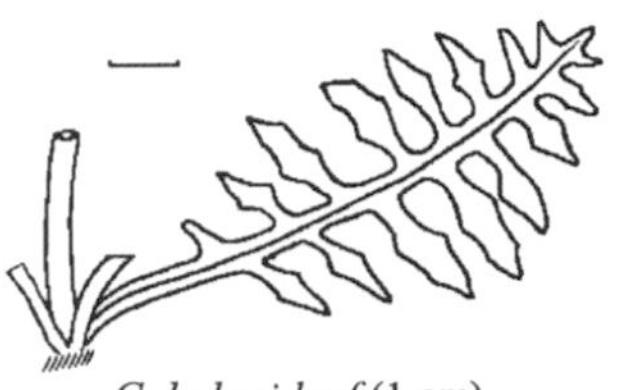
C. bakeri leaf (1 cm)

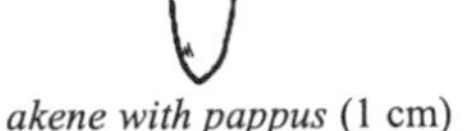
akene with pappus (1 cm)

Perennial herbs, usu with obvious stem. Lvs mly basal, entire to toothed or pinnatifid. Heads in panicles. The inner phyllaries equal, in a single series, with thickened midribs. Recpt naked. Fls yellow. Dry, stony places, SN (all), Jun-Aug. Edibility unknown. The genus is a difficult one to key to sp in the field. Several spp are distinctive, though, so the following key is presented of the Sierran spp. The individual spp will not be described.

A Plants glabrous; lvs entire — *C. nana*
AA Plants pubescent; lvs toothed or pinnatifid
  B Heads mly more than 10 per infl; fls mly less than 10 per head; widespread
    C Stems mly 1-3 dm H; heads 10-20-fld — *C. occidentalis*
    CC Stems 2-6 dm H; heads 4-12-fld
      D Stem lvs much reduced or none; phyllaries white-hairy on margins — *C. pleurocarpa*
      DD Stem lvs 1-3, well developed; phyllaries glabrous or evenly hairy
        E Basal lvs gray-tomentose; infl with 10-60 heads — *C. intermedia*
        EE Basal lvs pubescent; infl with 30-100 heads — *C. acuminata*
  BB Heads mly less than 10 per infl; fls 10-60 per head; n. SN
    C. Herbage not sticky, hairs not glandular — *C. modocensis*
    CC Herbage sticky, glandular hairs present
      D Glandular hairs short, not longer than other pubescence — *C. bakeri*
      DD Glandular hairs long, conspicuous — *C. monticola*

Dugaldia

Sneezeweed

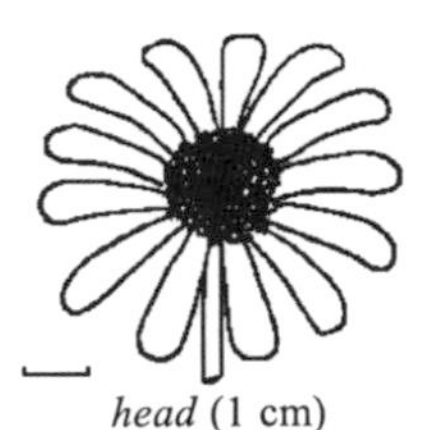
head (1 cm)

leaf (1 cm)

D. hoopesii: Stems stout, erect, 4-10 dm H; herbage yellow-green, hairy when young; basal lvs ± oblanceolate, to 3 dm L, on long, winged petioles; cauline lvs gradually reduced, alt; heads orange or yellow, several per stem; rays 10-20, 15-25 mm L, narrow, toothed but not lobed at apex. Moist places, 7500-11,300', Tulare to Tuolumne Co., Jul-Aug.

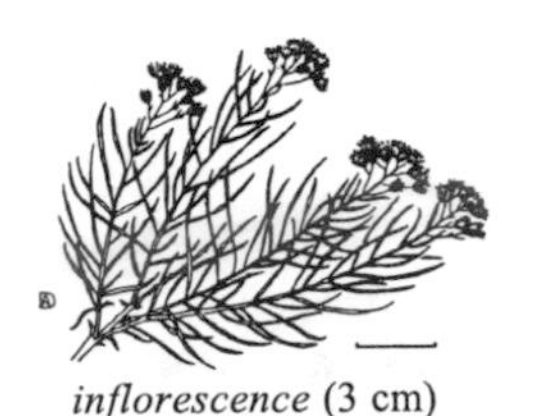
inflorescence (3 cm)

Ericameria

Goldenbush

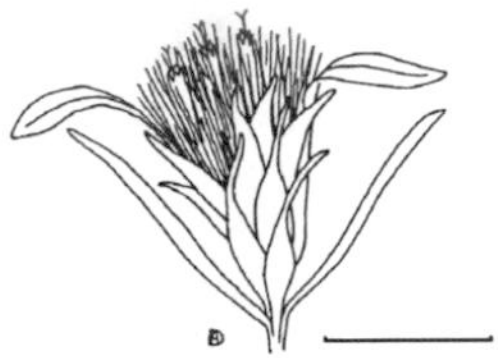
head (1 cm)

Erect, resinous shrubs, blooming in late summer and fall. Lvs alt. Edibility of foliage unknown.

A Plants 1-5 dm H
  B Lvs ± filiform, 2-6 cm L, 1-3 mm W; ray-fls inconspicuous — 2) *E. bloomeri*
  BB Lvs ± oblong, 1-3 cm L, 2-6 mm W
    C Twigs white-woolly; ray-fls none — 4) *E. discoidea*
    CC Twigs glabrous at least below, glandular; ray-fls 3-6
      D Heads 1-few in small cymes — 6) *E. suffruticosus*
      DD Heads many in crowded terminal clusters — 5) *E. greenei*
AA Plant usu over 5 dm H
  B Lvs filiform, 3-6 cm L, 2 mm W — 1) *E. arborescens*
  BB Lvs ± ovate, 0.5-2 cm L, 2-10 mm W — 3) *E. cuneatus*

1) *E. arborescens*, Golden Fleece: Erect shrub, 6-30 dm H with glabrous, resinous branches; lvs thick, crowded; heads discoid, many in rounded terminal panicles; phyllaries in 4 rows. Dry habitats, mly below 4000', but up to 9000' in s. SN, SN (all).

2) *E. bloomeri*, Bloomer's Goldenbush: Compact shrub wider than tall; heads in terminal clusters; invol. 7-12 mm H. Dry habitats, 3500-9500', SN (all).

3) *E. cuneatus*, Wedgeleaf Goldenbush: Much-branched shrub 1-10 dm H; herbage deep green; lvs usu entire; heads in compact clusters; ray-fls 1-5 (n. SN) or usu lacking. Cliffs and slopes below 9000', Plumas Co. s.

4) *E. discoidea*, Whitestem Goldenbush: Lvs ± oblong, sessile, usu entire, green, glandular; heads solitary and terminal or in a small elongate cluster; invol. 10-15 mm L. Rocky, mly open slopes, 9000-12,000', Nevada Co. s.

5) *E. greenei*, Green's Goldenbush: Stems to 2.5 dm H; lvs oblanceolate; invol. 8-12 mm L; rays 7-10 mm L. Rocky openings in forest; n. SN below 6500'.

6) *E. suffruticosus*, Singlehead Goldenbush: Compact subshrub; herbage glandular, aromatic; lvs many on the brittle twigs, linear-oblanceolate to oblong, entire; heads 1-4 at branch tips; invol. 10-14 mm H; ray-fls 3-6, showy. Open rocky slopes and ridges, 7800-12,000', Nevada Co. s.

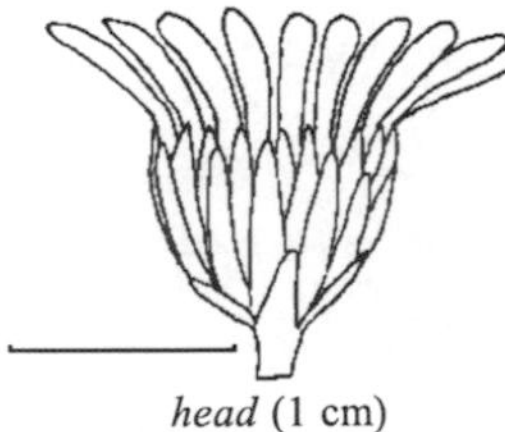
head (1 cm)

Erigeron

Wild Daisy
Fleabane

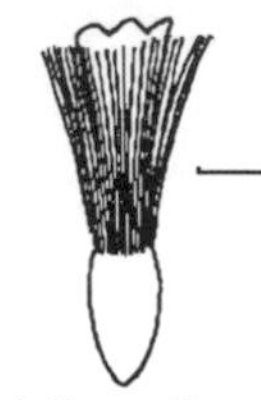
disk-flower (1 mm)

Usu perennials with alt or basal lvs. Heads solitary or in small clusters. Recpt flat, naked. Phyllaries narrow. Common throughout SN. Descriptions of the individual spp will not be given. Lvs of a related sp are edible cooked.

A Lvs 3-lobed or -divided; above 8000'
  B Lvs mly 3-lobed; ray-fls 25-35, 4-7 mm L — *E. vagus*
  BB Lvs 2-3 times ternate; ray-fls often inconspicuous — *E. compositus*
AA Lvs entire to toothed
  B Lvs primarily basal, stem lvs few
    C Ray-fls absent — *E. bloomeri*
    CC Ray-fls present
      D Rays yellow — *E. linearis*
      DD Rays lavender to purple or white
        E Pubescence appressed to stem or absent
          F Ray-fls 75-120 — *E. multiceps*
          FF Ray-fls 15-40

G Rays white; e. SN — *E. eatonii*
GG Rays purple to pink; scattered above 7000'
H Disk 6-12 mm W; rays 4-8 mm L; Fresno Co. n. — *E. tener*
HH Disk 13-18 mm W; rays 7-12 mm L; Fresno Co. n. — *E. barbellulatus*
EE Pubescence spreading, not appressed
F Stems almost leafless; rays 4-7 mm L, not reflexed — *E. pygmaeus*
FF Stems with some lvs; rays 5-13 mm L, coiled or reflexed
G Lvs petioled; stems usu viscid; mly above 10,000' — *E. algidus*
GG Petioles indistinct; stems not viscid except at apex; below 10,000'
H Margins of lower lvs with stiff hairs; below 6000' — *E. pumilus*
HH Margins of lower lvs lacking stiff hairs; Tulare Co. — *E. clokeyi*
BB Stems leafy, the lvs sometimes reduced above
C Ray-fls lacking
D Stem lvs much smaller than basal; heads many — *E. divergens*
DD Stem lvs about as long as basal; heads few
E Lvs threadlike to linear — *E. reductus*
EE Lvs oblanceolate to oblong
F Lvs 5-15 mm L, densely hairy; Donner Pass region — *E. miser*
FF Lvs 10-80 mm L, sparingly pubescent; widespread — *E. inornatus*
CC Ray-fls present, usu over 5 mm L
D Basal lvs present and usu over 5 cm L; rays more than 50 (except *E. lassenianus*); below 6500'
E Rays 10-25; heads usu several to many; Eldorado Co. n. — *E. lassenianus*
EE Rays usu more than 50; below 5000'
F Infl of many heads; stems densely hairy; annual — *E. divergens*
FF Infl 1-4 heads; stems sparsely hairy; perennials
G Rays mly purplish — *E. coulteri*
GG Rays white — *E. peregrinus*
DD Basal lvs usu lacking; cauline lvs less than 5 cm L; rays usu less than 40 (to 60 in *E. foliosus*); widespread
E Lvs 2-5 cm L, threadlike — *E. foliosus*
EE Lvs 0.5-4 cm L, linear to ovate
F Stem densely short-hairy, often with several heads — *E. breweri*
FF Stem glabrous to hairy, with 1 head per stem
G Stem glabrous, not glandular; c. SN — *E. elmeri*
GG Stem loose-hairy, glandular; Tulare Co. — *E. aequifolius*

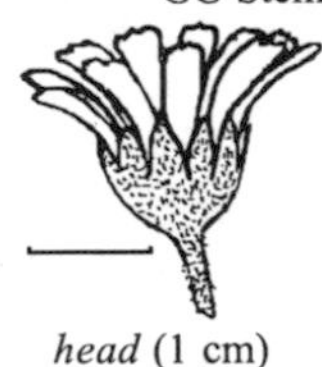

head (1 cm)

Eriophyllum

Woolly Sunflower

akene, pappus, and base of corolla (1 mm)

Annual to perennial, white-woolly herbs or subshrubs. Lvs mly alt, usu toothed or divided. Phyllaries in one row, firm and permanently erect, distinct or somewhat united at base. Rays yellow. Recpt usu naked, strongly convex to flat.

A Small annuals, usu 2-20 cm H; heads solitary; mly rare
B Pappus none or vestigial; Mariposa Co. — 1) *E. ambiguum*
BB Pappus of scales over 0.5 mm L; Kern Co. s.
C Branches open, spreading; ligules 3-5 mm L; below 6000' — 3) *E. congdonii*
CC Branches ascending; ligules less than 2 mm L; above 6000' — 5) *E. nubigenum*
AA Shrubs or subshrubs, heads usu in clusters; common
B Heads many in compact terminal clusters, less than 15 mm in diameter — 2) *E. confertiflorum*
BB Heads solitary or in loose clusters, 15-40 mm in diameter — 4) *E. lanatum*

1) *E. ambiguum:* Stems prostrate to ascending, openly branching; lvs linear to spatulate, 1-4 cm L; rays 5-10, 2-10 mm L. Below 7500', Apr-Jun.

2) *E. confertiflorum,* Golden-yarrow: Stems numerous, 1-3-6 dm H, erect, slender, more leafy toward base; lvs 1-4 cm L, persistently tomentose beneath, usu green and glabrate above, the margins curled under; invol. 3-6 mm W. Common on brush slopes below 9300', Calaveras Co. s., Apr-Aug.

3) *E. congdonii:* Stems 1-3 dm H; lvs 1-4 cm L, entire or lobed near tip; rays 8-10. Iron Mt. and vicinity.

4) *E. lanatum,* Woolly-yarrow: Stems erect or prostrate from a woody base, 1-8 dm L, few to many; lvs drooping, usu 2-4-toothed at tip. Open places below 11,500', SN (all), Apr-Aug.

5) *E. nubigenum,* Yosemite Woolly-sunflower: Stems 5-15 cm H, often branched, leafy; lvs spatulate to oblanceolate, mly entire, 1-2 cm L; invol. 5-6 mm W; rays 4-6. Forest openings, 5000-9000', Jun-Jul.

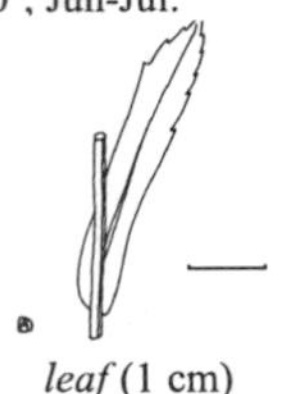

leaf (1 cm)

Gaillardia

Indian-blanket

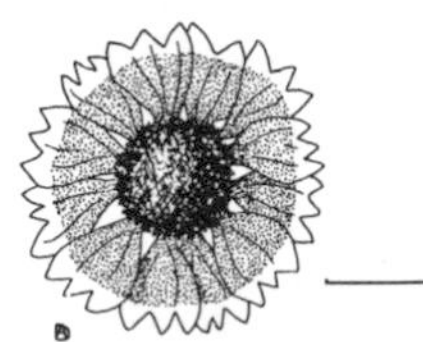

head (1 cm)

G. aristata: Stems erect, 2-7 dm H, usu rough-hairy; lvs alt, entire to pinnately lobed, the lower 5-15 cm L; peduncles 10-20 cm L, heads usu 1-2; disk-fls brown to purple, densely purple-woolly; pappus of 5-10 awn-tipped scales. Open habitats below 6000', SN (all), Jun-Aug.

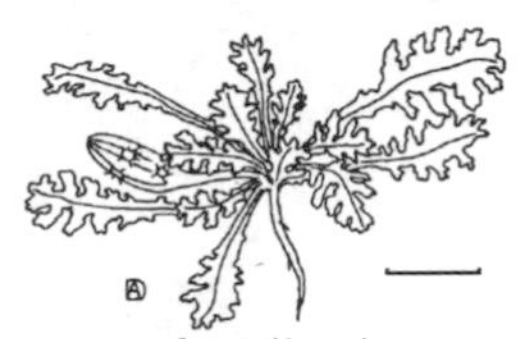

plant (1 cm)

Glyptopleura

Glyptopleura

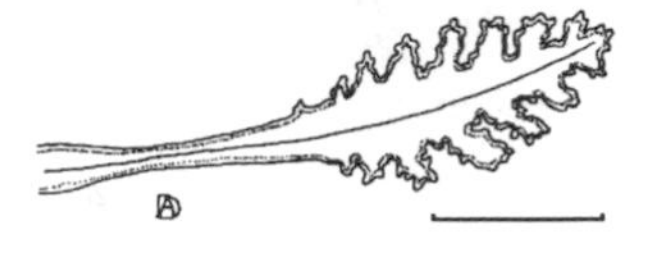

leaf (1 cm)

G. marginata: Tufted annual with milky sap; stems many, prostrate, 2-5 cm L; lvs 2-5 cm L, lobed, the lobes toothed, margins whitish and hard; heads solitary or few; invol. 10-12 mm W; phyllaries in 2 series; fls pale yellow. Sandy flats, below 7000', Kern Co.

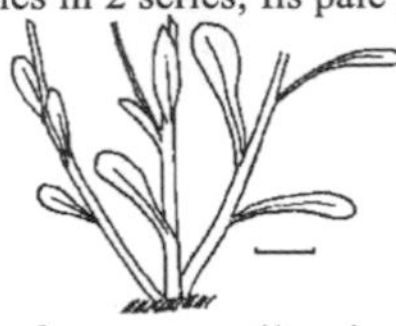

lower stems (1 cm)

Gnaphalium

Cudweed Everlasting

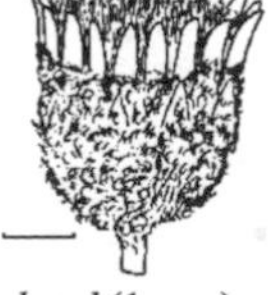

head (1 mm)

Annual or short-lived perennial woolly herbs. Lvs alt, entire, lanceolate to spatulate. Heads discoid, greenish to whitish, clustered at the ends of stems or branches. Pyrrolizidine alkaloids are often present throughout the plant, and Sierran spp. should be regarded as potentially toxic.

Stems 0.5-2 dm H; heads 3-4 mm H — *G. palustre*
Stems over 2 dm H; heads 5-6 mm H
- Stems 2-6 dm H; lvs scarcely decurrent — *G. canescens*
- Stems 5-10 dm H; lvs strongly decurrent — *G. stramineum*

G. canescens sspp: Herbage white-woolly; stems loosely branched; lvs 2-5 cm L. Dry, open habitats below 8700', SN (all), Jul-Oct.

G. palustre, Lowland Cudweed: Stems usu branched at base; lvs 1-3 cm L. Common in moist habitats, below 9500', SN (all), May-Oct.

G. stramineum, Cotton-batting Plant: Stems leafy with greenish yellow pubescence; lvs 2-5 cm L. Moist habitats below 6000, SN (all), Jun-Oct.

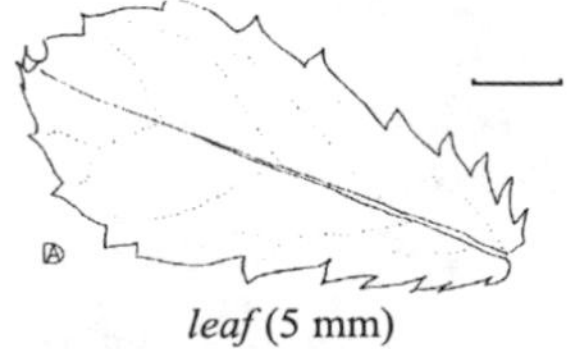

leaf (5 mm)

Hazardia

Hazardia

head (1 cm)

H. whitneyi: Perennial, 2-5 dm H, ± glabrous and resinous; stems several, ascending, unbranched; lvs ± oblong, 2.5-5 cm L, 7-16 mm W, serrate; phyllaries progressing into lvs; ray-fls none or 5-8; disk-fls 15-30. Rocky, open slopes, 4000-10,000', Plumas Co. s., Jul-Sep.

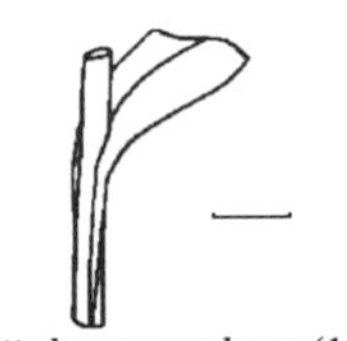
H. bigelovii decurrent base (1 cm)

Helenium

Bigelow's Sneezeweed

H. bigelovii head (1 cm)

H. bigelovii: Stems 4-8 dm H, often few-branched above, white-woolly when young; lower lvs often persisting until anthesis, petiolate, the cauline becoming sessile, oblanceolate to linear-lanceolate, up to 20 cm L and 4 cm W, the upper ones rapidly reduced; heads mly solitary on very long naked peduncles, the disk usu deep yellow like the rays, sometimes red-purple, 1.5-2 cm W; rays 15-30, showy, 8-22 mm L, reflexed. Common in moist places, mly 3000-10,000', SN (all), Jun-Aug.

lower stem (1 dm)

Helianthella

California Helianthella

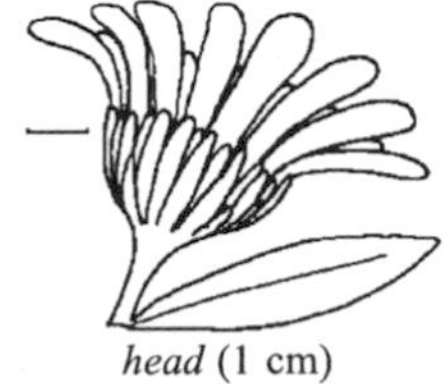
head (1 cm)

H. californica var. *nevadensis:* Stems slender, 2-6 dm H; herbage ± glabrous; lvs lanceolate, 5-25 cm L, 15-30 mm W, often mly basal; heads solitary on long peduncles, the yellow disk 15-22 mm W; rays 2-3 cm L, yellow; phyllaries narrowly lanceolate; pappus of 2 short marginal awns 1-2 mm L. Dry openings or open slopes, below 8000', SN (all), May-Sep. Edibility unknown.

Heterotheca

Hairy Goldenaster

H. villosa: Perennial with decumbent to erect stems 1-12 dm L; herbage hairy, usu glandular above; lvs alt, lanceolate; heads 1-many; ray-fls 5-25, the ligules 5-10 mm L. Rocky habitats, lava flows, below 10,000, SN (all). Edibility unknown..

plant (2 cm)

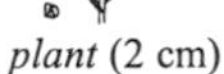

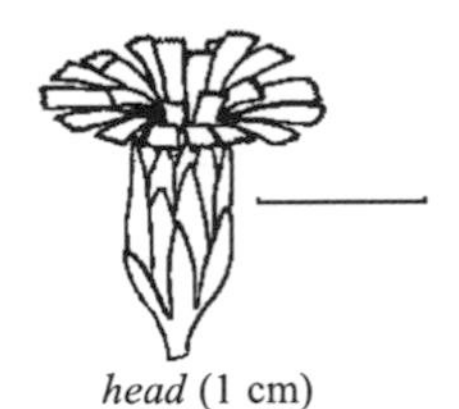
head (1 cm)

Hieracium

Hawkweed

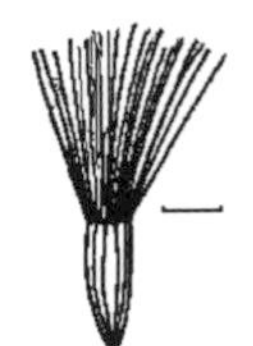
akene with pappus (1 mm)

Perennial herbs, hairy, sometimes glandular. Lvs ± entire, never deeply lobed. Heads variously panicled. Recpt flat, usu naked. Invol. cylindric or bell-shaped. Phyllaries in 1-3 series. Ray-fls white or yellow. The juices of many spp may be dried and used as chewing gum.

A Rays white or pale yellow; stem and invol. glabrous or with long hairs — 1) *H. albiflorum*
AA Rays yellow; stem and invol. pubescent
  B Stems 3-7 dm H; lvs mly basal, 10-20 cm L — 4) *H. scouleri*
  BB Stems often leafy, 1-3.5 dm H; lvs 2-10 cm L
    C Herbage merely pubescent; lvs mly basal — 2) *H. gracile*
    CC Herbage densely long-hairy; stems usu leafy — 3) *H. horridum*

1) *H. albiflorum,* White-flowered Hawkweed: Stems 1 to several, slender, erect, 4-8 dm H; lvs mly basal, 8-15 cm L, 2-4 cm W, the lower with winged petioles; invol. 9-10 mm H; rays 3-4 mm L. Common on dry, open slopes below 9700', SN (all), Jun-Aug.

2) *H. gracile,* Alpine Hawkweed: Stems 1 to several, often branched, slender; lvs spatulate to oblong-oblanceolate, subentire, mly glabrous; invol. 7-8 mm H. Woods and rocky places, 8000-11,000', Tulare to Eldorado Co., Jul-Aug.

3) *H. horridum,* Shaggy Hawkweed: Stems few to several, branched above; lvs spatulate to oblong, entire; invol. narrow, 6-9 mm L; rays about 10 mm L. Common in dry, rocky places, 5000-11,000', SN (all), Jul-Aug.

4) *H. scouleri,* Scouler's Hawkweed: Stems 1 to few, unbranched, rather slender; lvs entire, copiously long-hairy; invol. 8-10 mm H; rays 8-10 mm L. Open woods and rocky places below 6500', n. SN, May-Jul.

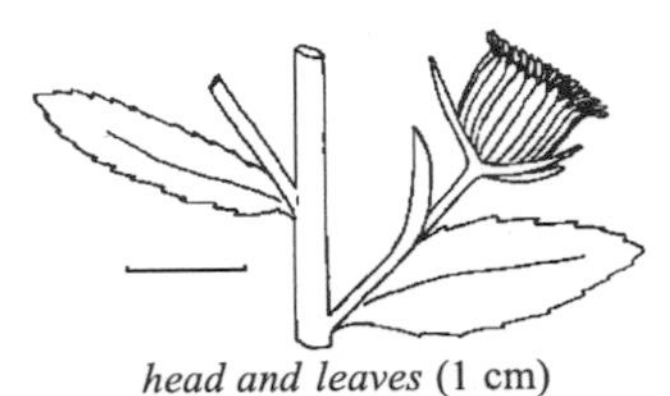

head and leaves (1 cm)

Hulsea

Hulsea

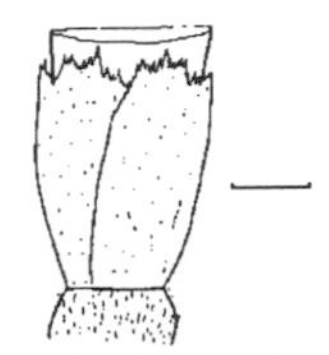

H. vestita pappus (1 mm)

Annual to perennial, rather fleshy, viscid-pubescent aromatic herbs. Lvs alt, usu very numerous, basal ones broadly petiolate, the cauline sessile, entire to pinnatifid. Heads many-fld, yellow or purple. Flowering Jun-Aug. Edibility unknown.

A Lvs not white-woolly, oblong or broader; stems leafy throughout; below 8000'
  B Stems lvs mly 15-35 mm W; rays 25-60, often purplish — 3) *H. heterochroma*
  BB Stem lvs mly 5-10 mm W; rays 10-20, yellow — 2) *H. brevifolia*
AA Lvs often white-woolly, linear or ± pinnatifid if not; stems mly less than 3 dm H, often leafless; above 6000'
  B Lvs densely white-woolly, mly spatulate and entire, 2-5 cm L — 5) *H. vestita*
  BB Lvs glabrous to tomentose, all basal, coarsely toothed or pinnately lobed, mly 5-15 cm L
    C Rays 15-25; peaks of Lassen National Park n. — 4) *H. nana*
    CC Rays 25-50; peaks of Tahoe area s. — 1) *H. algida*

1) *H. algida,* Alpine Hulsea: Stems several, erect, unbranched, 1-4 dm H; herbage glandular and unpleasantly scented; lvs pubescent, petioled; invol. 13-20 mm H, usu woolly. Rocky habitats, 9500-14,000'.

2) *H. brevifolia,* Short-leaved Hulsea: Stems several, usu erectly branched from base, moderately villous, 3-5 dm H; lvs obscurely dentate, 2-4 cm L; heads terminal, solitary; invol. 10-12 mm H. Occasional in forest openings, 6000-8000', Fresno to Tuolumne Co.

3) *H. heterochroma,* Red-rayed Hulsea: Stems several, erect, viscid-villous and heavy scented, 4-12 dm H; lvs oblong, usu sharply dentate, to 10 cm L, sessile or clasping; invol. 12-15 mm H. Infrequent, in forest openings, 3000-9000', Tulare to Eldorado Co.

4) *H. nana,* Dwarf Hulsea: Stems less than 15 cm H, unbranched; lvs mly basal in dense rosettes, glandular and unscented; heads solitary; rays 15-25. In volcanic ash, gravels or talus, 8000-10,500'.

5) *H. vestita*, Pumice Hulsea: Perennial 8-30 cm H; lvs in dense rosettes; heads solitary; invol. 10-13 mm H; rays linear, yellow tinged with reddish or purplish. Sand flats or gravelly soils, 6000-11,000', Mono Co. s.

var. *pygmaea:* Stems less than 6 cm H; lvs sometimes green on upper surface. High peaks, 8000-11,400', Tulare Co. n.

Iva

Poverty Weed

I. axillaris: Stems several, erect from an often decumbent or prostrate base, mly 2-6 dm H; herbage appressed-hairy, red-glandular-punctate; lvs blunt, entire, subsessile, 1-4 cm L, 2-12 mm W, thickish; heads greenish white, discoid, in axils of lvs 5-6 mm W, the phyllaries united into a deeply lobed to subentire cup. Alkaline places below 6700', SN (all), May-Sep.

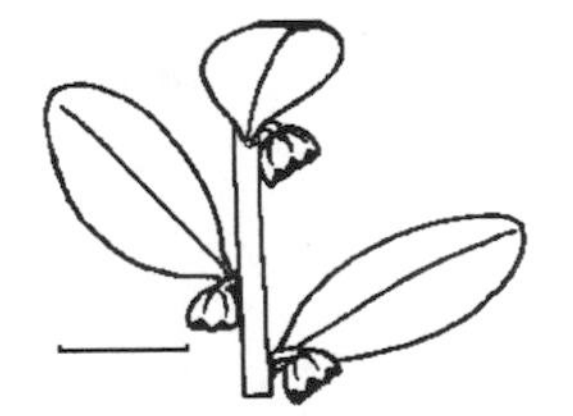

leaves and heads (1 cm)

Lagophylla

Common Hareleaf

L. ramosissima: Annual, stems rigid, erect, 2-10 dm H, usu much branched; herbage grayish or dull green, prominent yellow stipitate glands on upper lvs and heads; lvs linear-oblanceolate, lost before flowering, 3-12 cm L, 5-10 mm W; heads short-peduncled or subsessile; invol. 4-7 mm H; rays inconspicuous. Occasional, in open places below 4000', Fresno Co. n., May-Oct.

upper stem (1 cm)

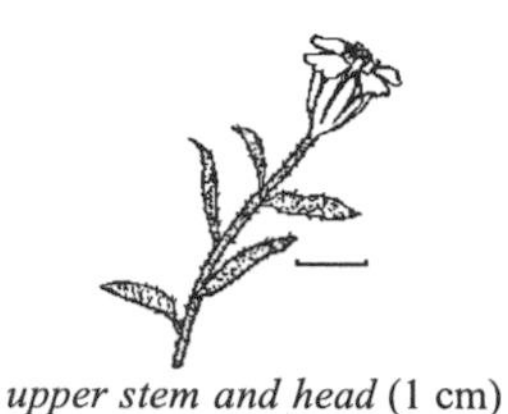

upper stem and head (1 cm)

Layia

Layia

akene, pappus, and corolla base (1 mm)

L. glandulosa: Stem branched, 1-4 dm H, often purplish; lvs alt, often long-hairy above; basal lvs dentate or lobed, the cauline mly entire; rays white, often fading purplish, 6-15 mm L. Common in sandy soil below 7800', SN (all), May-Jun. Edibility unknown.

leaves (1 cm)

Lessingia

Lessingia

head (1 cm)

L. leptoclada: Stems 3-9 dm H, branched, the divergent branches slender, straight; herbage tomentose or becoming glabrate; basal lvs oblanceolate, entire or toothed, 2-5 cm L, upper lvs glandular, much reduced; heads solitary or 2-5 in a dense cluster; invol. 6-10 mm H; corollas lavender to blue-purple. Common in open ground below 6200', Eldorado Co. s., Jul-Oct. Edibility unknown.

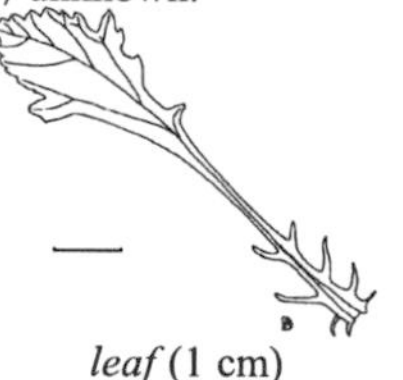

leaf (1 cm)

Leucanthemum

Ox-eye Daisy

head (1 cm)

L. vulgare: Perennial, 2-8 dm H, stout, usu unbranched, usu glabrous; lvs alt, serrate, the lower oblanceolate with winged petioles, the cauline sessile, lanceolate; heads solitary, 3-5 cm in diameter; ray-fls 20-30, the ligules 1-2 cm L. Naturalized in disturbed habitats, Mariposa Co. n. (rare further s.), below 6500'.

upper stem (1 cm)

Machaeranthera

Hoary-aster

phyllaries (5 mm)

M. canescens var. *shastensis:* Stems tufted, usu ascending, mly 5-20 cm H; herbage usu densely pubescent; lvs entire to coarsely toothed; basal lvs spatulate, 2-4 cm L, including petiole, 3-8 mm W; cauline lvs reduced; heads solitary to many; invol 5-10 mm H; phyllaries usu 3-seriate, oblong, often purplish; ray-fls 5-15, violet, 6-8 mm L. Dry wooded or open slopes, 7000-11,500', Tuolumne and Mono cos. n.

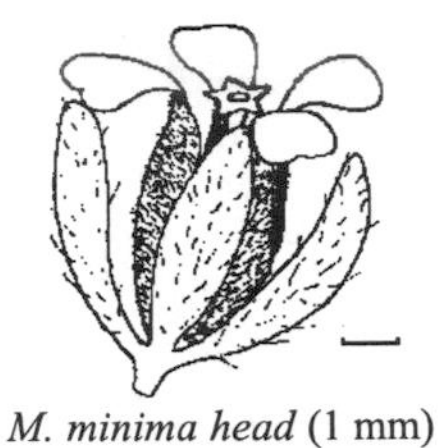

M. minima head (1 mm)

Madia

Tarweed

scales between ray- and disk-flowers (1 mm)

Herbs, usu very glandular and heavy-scented. Lvs linear or oblong, the basal entire to subdentate. Phyllaries enclosing and falling with ray akenes. Pappus lacking or minute scales. Seeds of at least some spp edible raw, cooked, or dried and ground into meal.

A Perennial; lower stem lvs opp
- C Stem 5-12 dm H; phyllaries 10-12 mm L — 1) *M. bolanderi*
- CC Stem 1-7 dm H; phyllaries 5-6 mm L — 5) *M. madioides*

AA Annual; lvs usu alt
- BB Heads small, ray plus disk fls usu less than 12
  - C Stem usu over 15 cm H; lvs 3-9 cm L — 3) *M. glomerata*
  - CC Stem usu less than 15 cm H; lvs 1-3 cm L
    - D Cymes open, usu with several heads; anthers black — 8) *M. yosemitana*
    - DD Heads solitary of few in dense cluster; anthers yellow — 6) *M. minima*
- BB Ray- and disk-fls both present and many
  - C Ray-fls 6-20 mm L, 3-lobed at tip — 2) *M. elegans*
  - CC Ray-fls 4-9 mm L, not lobed
    - D Disk-fls less than 15; pappus none — 4) *M. gracilis*
    - DD Disk-fls more than 15; pappus often minute scales — 7) *M. rammii*

1) *M. bolanderi,* Bolander's Madia: Stem unbranched, from woody root-stock; lvs linear, entire, the basal crowded, forming an erect tuft; lower stem lvs 10-30 cm L; heads few, on peduncles to 25 cm L; ray-fls 8-12, disk-fls many. Moist habitats, 3500-8300', SN (all), Jul-Sep.

2) *M. elegans,* Common Madia: Stems branched above, 2-8 dm H; lvs linear to broadly lanceolate; invol. to 10 mm H; rays 8-16, 6-15 mm L, yellow or with a red blotch at base; disk-fls many, yellow or maroon. Common on dryish slopes, 3000-6000', w. SN , May-Aug.

3) *M. glomerata,* Mountain Tarweed: Stems rigid, leafy, 1.5-8 dm H; lvs linear; invol. 5-9 mm H; rays about 2 mm L, disk-fls 1-10. Openings in woods, 3500-8800', SN (all), Jul-Sep.

4) *M. gracilis,* Slender Tarweed: Stem usu slender, 1-10 dm H, often with shorter branches; lvs not crowded, mly linear, to 10 cm L; invol. 6-9 mm H; rays 8-12. Common on wooded slopes and open areas below 7800', SN (all), May-Aug.

5) *M. madioides*: Stems usu bristly below; lvs 6-10 cm L, opp most of the way up stem, often fused at base; heads few, on long peduncles; rays 8-15, 5-10 mm L. Openings in forests below 4500', Mariposa Co n., May-Jun.

6) *M. minima,* Hemizonella: Stems 2-15 cm H, often much-branched; lvs often much-branched; lvs often clustered at nodes, otherwise scattered, 1-2 cm L; heads small, rays 5 or less. Common, gravelly slopes, 3500-8600', SN (all), May-Jul.

7) *M. rammii,* Ramm's Madia: Stems branching above, slender; lvs scattered, ± entire, linear, long-hairy, 1.5-6(-10) cm L; heads on long, naked peduncles; rays 7-10, 5-9 mm L; disk-fls many. Common in open areas, below 5000', Calaveras to Butte Co., May-Jul.

8) *M. yosemitana,* Yosemite Tarweed: Much like *M. rammii;* heads small; rays 4-7, about 3 mm L; disk-fls 2-7. Rare, in moist habitats, 4000-7500', Tuolumne Co. s., May-Jul.

Malacothrix

Malacothrix

M. floccifera: Erect annual with milky sap; stems 5-40 cm H, glabrous; lvs usu with tufts of white hairs at lobe bases; basal lvs oblanceolate to obovate, evenly 5-15-lobed; heads 2-5 mm in diameter; invol. 5-10 mm H; phyllaries in 2-3 rows, the outer glabrous; corollas usu white. Disturbed areas and loose soil in chaparral below 6500', Mariposa Co.

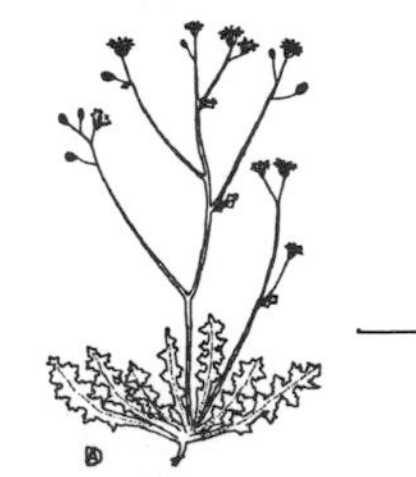

plant (4 cm)

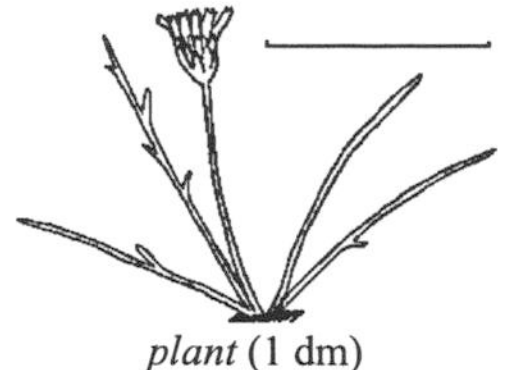

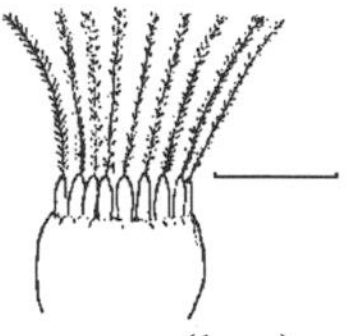

Microseris

Nodding Scorzonella

plant (1 dm) *pappus* (1 cm)

M. nutans: Stems 1 to several, slender, erect, 1-4.5 dm H; herbage ± glabrous; lvs mly basal, those on lower stem lance-linear, 1-2.5 dm L, entire to pinnatifid; heads terminal on leafless branches, usu nodding in bud; invol. 12-14(-20) mm H; pappus of 15-20 narrow scales, each bearing a plumose white bristle. Mly in woods or moist habitats, 4000-10,900', SN (all), Jun-Aug. The slender roots are edible raw.

Nothocalais

Alpine Lake-agoseris

lower stem (1 cm) *pappus* (1 cm)

N. alpestris: Stems usu leafless, 5-20 cm H, ± glabrous; lvs linear to oblanceolate, subentire to pinnately lobed; invol. 10-20 mm H, the inner phyllaries subequal, acuminate; the outer phyllaries wider, covered with fine purple dots. Infrequent, open slopes, Eldorado Co. and perhaps n. Can be confused with *Agrostis glauca*. Lvs edible but bitter.

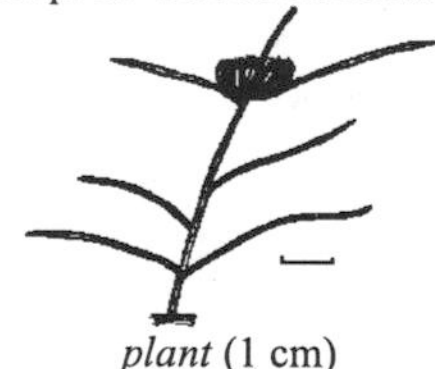

Orochenactis

Orochenactis

plant (1 cm) *akene with pappus* (1 cm)

O. thysanocarpha: Stems ± fleshy, reddish, white-woolly at nodes, branched from base, 5-30 cm H; lvs linear, often white-hairy, entire, alt, or the lowest opp, 1-6 cm L; heads sessile, few-fld, solitary or in clumps. Fls yellow; corolla about 6 mm L. In well drained granitic soils, 7000-11,500', Tulare, Inyo and Kern cos., Jun-Aug. Edibility unknown.

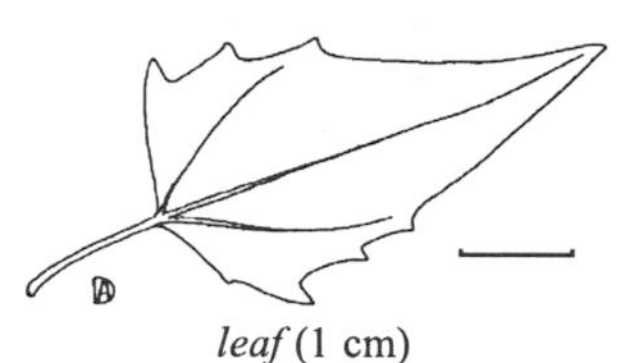

Pericome

Pericome

leaf (1 cm) *upper stem* (1 cm)

P. caudata: Aromatic perennial with many stems from base; stems to 2 m H, much branched; lvs simple, deltoid-lanceolate, usu opp; petioles 1-5 cm L; blades 3-5 cm L; invol. 5-6 mm W; phyllaries 20-25. Dry habitats below 7500', mly e. SN from Mono Co. s., Jul-Oct.

Phalacroseris

Bolander's-dandelion

P. bolanderi: Glabrous perennial with naked 1-headed scapes; lvs slightly fleshy, 6-30 cm L; heads 2-2.5 cm W at anthesis; invol. 8-10 mm H, of 12-16 subequal phyllaries; rays yellow, conspicuous; recpt naked. Wet mdws, 6000-9300', Tulare to Nevada Co., Jun-Aug. Edibility unknown.

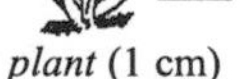

plant (1 cm)

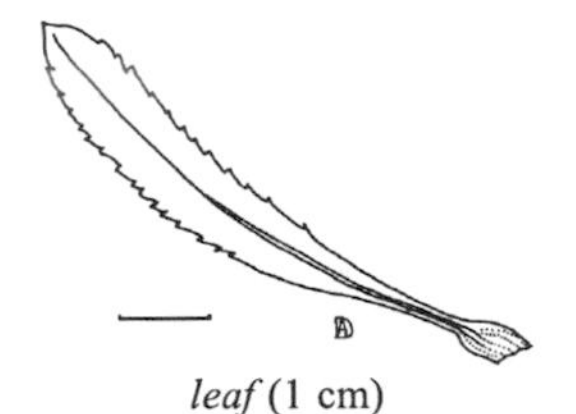

leaf (1 cm)

Pyrrocoma

Pyrrocoma

inflorescence (2 cm)

Usu glabrous perennials from taproots. Stems usu reddish. Lvs alt, the basal petioled, the cauline reduced and often clasping. Phyllaries in 2-6 series. Ray-fls usu present, yellow; disk-fls yellow. Pappus of 15-60 bristles.

A Ray-fls inconspicuous; Plumas Co. n. — 2) *P. carthamoides*
AA Ray-fls evident, much larger than disk-fls; widespread
  B Plants shiny-glandular; heads many, crowded; below 5000' — 3) *P. lucida*
  BB Plants not glandular; heads usu less than 10
    C Heads solitary; above 8000' — 1) *P. apargioides*
    CC Heads 3-15; below 8000' — 4) *P. racemosa*

1) *P. apargioides:* Stems 5-20 cm H; lvs mly basal, 4-10 cm L, oblong, thick, usu coarsely dentate, the cauline reduced, usu entire; ray-fls 10-40, ligules 5-15 mm L; rocky slopes and mdws, SN (all).

2) *P. carthamoides* var. *cusickii:* Stems 5-40 cm H, leafy; lvs 5-20 cm L, oblanceolate, puberulent; heads 1-4 in open cluster, subtended by leafy bracts; ray-fls less than 25, less than 7 mm L, about as long as pappus. Open rocky habitats to 9000'.

3) *P. lucida:* Stems 20-70 cm H; lvs variable, entire to serrate; ray-fls 10-20, the ligules 5-15 mm L. Alkali soils, open areas; Nevada Co. n.

4) *P. racemosa:* Stems to 90 cm H; lvs mly basal, 5-30 cm L, lanceolate to widely elliptic, entire to serrate, the cauline usu serrate; ray-fls 10-30, the ligules 5-10 mm L. Many habitats, SN (all).

Raillardella

Raillardella

plant (1 cm)

Scapose perennials with simple, sessile, ± oblanceolate, entire to subentire lvs. Fls yellow. Invol. 10-15 mm H, the phyllaries subequal. Recpt naked, flat. Hybrids between the two species have been found. Edibility unknown.

Herbage green, glandular; ray-fls sometimes present — *R. scaposa*
Herbage white-silky, not glandular; heads discoid — *R. argentea*

R. argentea, Silky Raillardella: Stems 1-10 cm H, reddish; lvs mly 1-6 cm L. Dry, rocky habitats, 9000-12,000', SN (all), Jul-Aug.

R. scaposa, Green-leaved Raillardella: Herbage pubescent and glandular; lvs mly 3-15 cm L; scape 0.5-4 dm H; phyllaries purple-tipped; disk-fls bright orange. Dry, stony places and edge of mdws. 6500-11,000', SN (all), Jul-Aug.

plant (3 cm)

Raillardiopsis

Muir's-Raillardella

plumose pappus (2 mm)

R. muirii: Perennial with several stems from the branched root-crown; stems glandular, often matted to 2-3 dm H; herbage rough-hairy; lvs ± linear, sessile; heads solitary or few in a loose cluster; invol. about 12 mm H. Open slopes 4000-7000'.

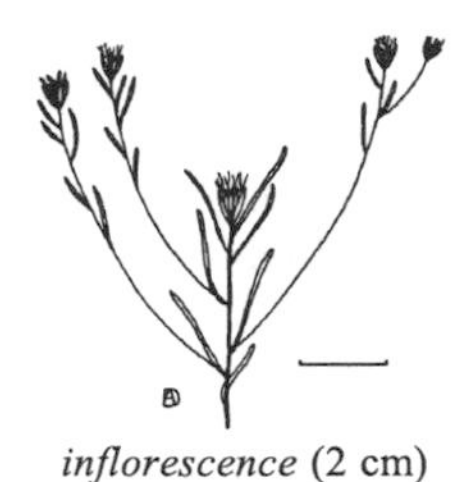

Rigiopappus

Rigiopappus

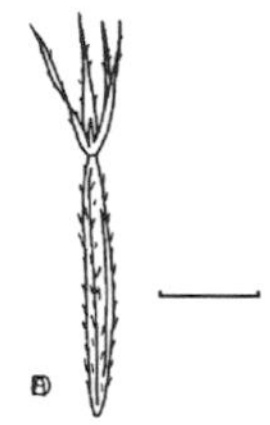

inflorescence (2 cm) *akene and pappus* (2 mm)

R. leptocladus: Erect leafy stems to 3 dm H, the branches usu taller than main stem; lvs filiform, 1-3 cm L; invol. irregularly 2-3-seriate; outer phyllaries often partly enclosing ray-akenes; ray-fls 5-15, barely exceeding disk; pappus of 3-5 rigid scales. Dry, open ground, usu below 4000', SN (all), Apr-Jun.

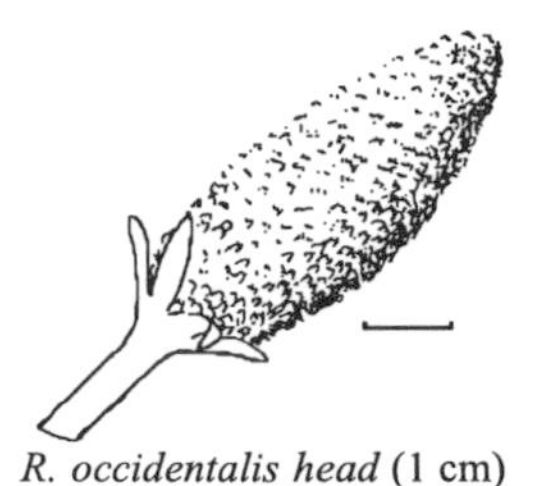

Rudbeckia

Cone-flower

R. occidentalis head (1 cm) *R. californica head* (1 cm)

Tall perennial herbs, flowering mid-summer. Lvs alt, simple, entire to dentate. Heads large. Invol. 2-3-seriate. Rays yellow when present; disk-fls brown or purple. Several spp including *R. occidentalis* and *R. hirta* are suspected of poisoning livestock when eaten in quantity.

Heads without ray-fls — *R. occidentalis*
Heads with ray-fls
  Pappus none; disk globose — *R. hirta*
  Pappus present; disk ± columnar — *R. californica*

R. californica, California Cone-flower: Stems leafy, unbranched, 6-18 dm H, with a single showy head on a long peduncle; lvs broadly lanceolate to elliptic, firm, glabrous on the upper, pubescent on the lower surface, the lower with blades 10-25 cm L, on long slender petioles, entire or irregularly dentate or incised, the upper becoming sessile and entire; rays 8-20, 2.5-6 cm L; disk 3-5 cm H. Occasional in moist mdws, 5500-7800', Eldorado Co. s.

R. hirta var. *pulcherrima,* Black-eyed Susan: Stems usu sparingly branched, 3-8 dm H; herbage rough-hairy throughout; lvs oblanceolate to lanceolate, the lower petioled, entire; heads on long peduncles; disk 1-2 cm W, usu black-purple or brown; rays 8-20, 1.5-3 cm L. Introduced in mid-altitude mdws of the SN, Mariposa to Amador Co.

R. occidentalis, Black Heads: Stems unbranched, 6-15 dm H, with the head surmounting a stout peduncle; lvs broadly ovate to lanceolate, acute, 5-30 cm L, 2-8 cm W, subglabrous or pubescent beneath, entire or irregularly toothed. Wet ground in woods, 4000-6000', Placer to Plumas and Butte cos.

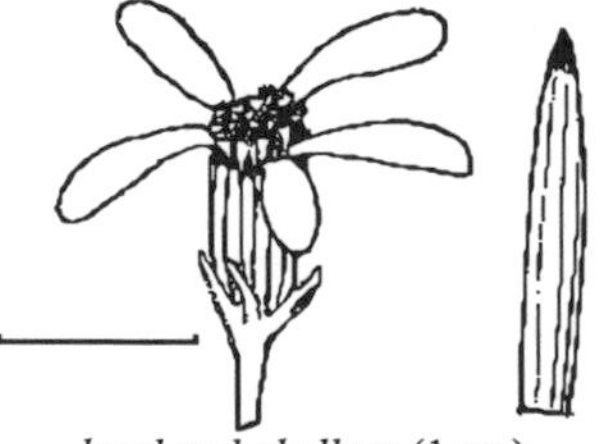

Senecio

Groundsel
Ragwort

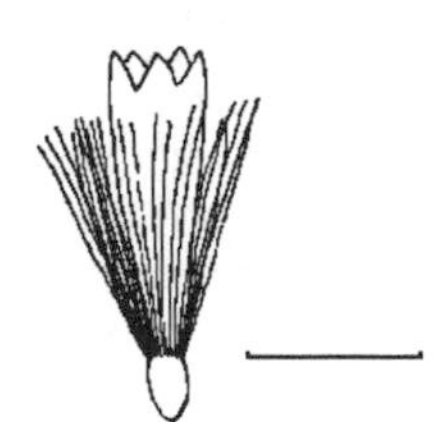

head and phyllary (1 cm) *disk flower* (5 mm)

Annual or perennial herbs. Lvs alt or basal, entire to toothed or pinnatifid. Heads solitary to many. Phyllaries ± equal, usu in 1 row, often with smaller bracteoles at base. Several and perhaps many *Senecio* contain toxic alkaloids and all spp should be avoided. The genus is large and often difficult to identify to sp. The key below may be used to help identify a specimen. Descriptions of the individual spp will not be given.

A Stems mly over 3 dm H; lvs mly over 4 cm L
  B Lvs mly basal; stem lvs rapidly reduced up the stem; stems mly solitary
    C Ray-fls few or none; below 8500'
      D Stems glabrous; wet habitats, Mono Co. n. — *S. hydrophilus*
      DD Stems woolly when young; dry habitats, SN (all) — *S. aronicoides*
    CC Ray-fls mly 5-12, conspicuous; widespread
      D Invol. mly 8-12 mm H; stems 3-9 dm H; 2500-10,600' — *S. integerrimus*
      DD Invol. mly 4-7 mm H; stems 2-5 dm H; 6000-11,500' — *S. scorzonella*
  BB Stems leafy to infl
    C Lvs pinnatifid, oblong; phyllaries not black-tipped; mly 7000-9000', Mariposa Co. s. — *S. clarkianus*
    CC Lvs entire, dentate; phyllaries usu with black tips
      D Lvs linear, entire; stems 2-6 dm H; 7800-10,400', e. SN, Mono co. s. — *S. spartioides*
      DD Lvs mly triangular to rhombic; stems 5-15 dm H; 4000-11,000'
        E Lower lvs with deltoid to subcordate base — *S. triangularis*
        EE Lower lvs tapering at base — *S. serra*
AA Stems mly less than 4 dm H; lvs 1-4 cm L; mly high alpine habitats
  B Stems leafy throughout
    C Lvs ± linear, strongly revolute; 9500-10,600', e. SN, Mono Co. — *S. pattersonianus*
    CC Lvs ± obovate; 8500-12,400', SN (all) — *S. fremontii*
  BB Lvs mly basal, reduced above
    C Herbage white-tomentose at anthesis; plants of dry habitats; lvs ± entire
      D Stems 1-3 dm H; rays mly 5-8; 4200-11,750', SN (all)
      DD Stems 0.5-1.5 dm H; rays 9-16; 10,400-13,000', s. SN — *S. werneriaefolius*
    CC Herbage subglabrous; plants of generally moist habitats; lvs often serrate or pinnatifid
      D Heads solitary, radiate; 6500-10,500' — *S. subnudus*
      DD Heads several to many, or if few then discoid
        E Head discoid or lower lvs ± cordate; 8000-11,000' — *S. pauciflorus*
        EE Heads radiate; lower lvs not cordate; 6000-10,700' — *S. cymbalarioides*

Solidago

Goldenrod

S. californica inflorescence (1 cm) *S. canadensis head* (1 cm)

Perennial herbs with leafy, usu unbranched stems. Lvs alt, entire or toothed. Heads numerous, small, yellow, in showy clusters. Phyllaries usu with obscurely herbaceous tips. Ray-fls small. Pappus copious. Young lvs can be used as potherbs. The dried lvs and fls can be used to make a tea; however, large amounts of the herbage (especially raw) may be toxic.

A Middle stem lvs largest — 2) *S. canadensis*
AA Lower stem lvs largest
  B Infl spherical or flat-topped; heads 10-20; mly above 8000' — 3) *S. multiradiata*
  BB Infl elongate with many heads; mly below 8500'
    C Herbage usu densely hairy; dry habitats — 1) *S. californica*
    CC Herbage + glabrous; wet habitats — 4) *S. spectabilis*

1) *S. californica,* California Goldenrod: Stems 2-12 dm H, minutely but densely pubescent; lower stem lvs 5-12 cm L, 1-3 cm W, upper stem lvs usu reduced; invol. 3-5 mm H. Common in dry or moist clearings below 7500', SN (all), Jul-Oct.

2) *S. canadensis* ssp *elongata,* Meadow Goldenrod: Stems 3-10(-15) dm H; lvs nearly uniform, 5-12 cm L, 1-2 cm W; panicle 5-20 cm L, dense; invol. 3-5 mm H. Mdws and moist openings in woods, SN (all), May-Sep.

3) *S. multiradiata,* Alpine Goldenrod: Stems erect, 0.5-4 dm H, hairy at least above; basal lvs oblanceolate to elliptic, entire to serrulate, mly 2-10 cm L, 5-8 mm W, petioled; stem lvs spatulate to lanceolate, usu sessile and entire; invol. 4-7 mm H; ray-fls usu 13. Sunny, rocky or grassy places, 8000-12,500', SN (all), Jun-Sep.

4) *S. spectabilis*, Showy Goldenrod: Stems to 18 dm H; lvs 10-20 cm L, oblanceolate, entire or serrate distally; invol. 3-4 mm L; ray-fls 5-20, to 3 mm L. Bogs and alkaline mdws, e. SN. Mono Co. n.

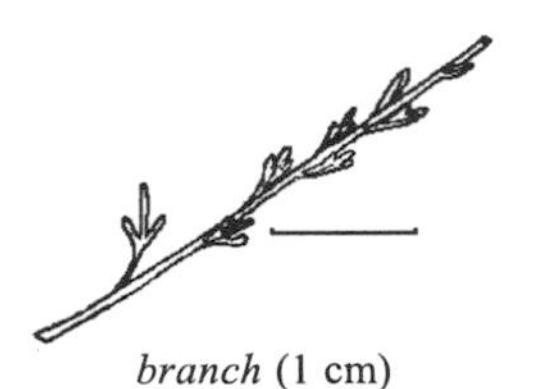

Sphaeromeria

Gray-tansy

branch (1 cm) *upper stem* (1 cm)

S. cana: Aromatic perennial, with several slender stems 1-2 dm H; herbage pubescent; lvs well distributed, alt, subsessile, linear-oblong and entire to narrowly obovate with 3 linear-oblong segments; infl dense, 2-several headed; heads 3-6 mm W; disk-fls lemon-yellow. Dry, rocky places, 9000-12,000', Madera and Mono cos. s., Jul-Aug. Young lvs may be used to flavor stews, but the herbage is probably poisonous if eaten in quantity.

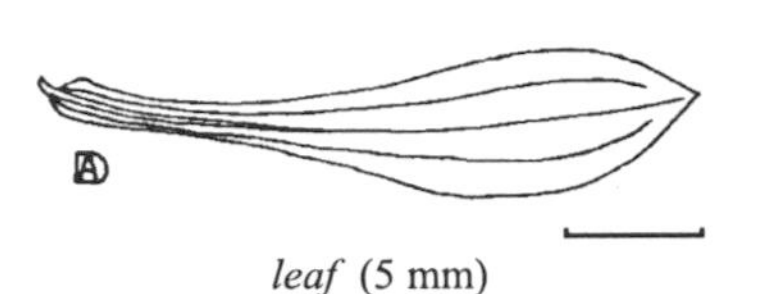

Stenotus

Cushion Stenotus

leaf (5 mm) *plant* (2 cm)

S. acaulis: Stems numerous, densely clustered from a woody base, subscapose, 5-10 cm H, sometimes forming mats to 5 dm W; lvs mly erect, entire, spatulate or narrower, 1-6 cm L; invol. 7-10 mm H; ray-fls 6-10 mm L. Dry, rocky habitats, 5300-10,500', mly e. SN, May-Aug.

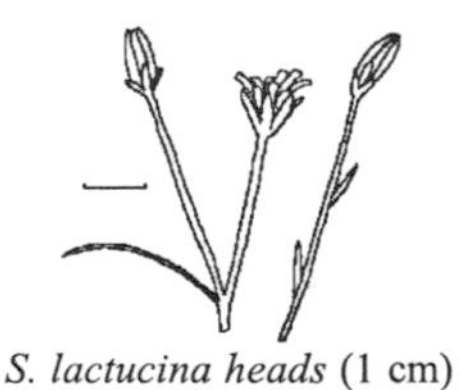

Stephanomeria

Stephanomeria

S. lactucina heads (1 cm) *akene with pappus* (1 mm)

Annual or perennial herbs with slender branched stems. Lvs alt, linear to oblong, entire to toothed, uppermost usu much reduced. Invol. of several equal phyllaries with some small outer bracts. Edibility unknown.

Stems stiff, 5-20 dm H, with long straight branches; heads clustered along branches *S. virgata*
Stems slender, 1-5 dm H; heads terminal, mly solitary
  Lvs filiform, the upper bract-like *S. tenuifolia*
  Lvs wider, the upper not much reduced *S. lactucina*

S. lactucina, Large-flowered Stephanomeria: Stems single, 1-3 dm H, often branched at or near the base; lvs mly linear, 3-10 cm L, the lower with a few teeth; invol. 12-15 mm H; rays bright pink to purple, 8-11 mm L. Dry flats and ridges, 4000-8000', Mariposa Co. n., Jul-Aug.

S. .tenuifolia, Narrow-leaved Stephanomeria: Stems few to several from a woody root-crown, erect, with many slender ascending branches; lvs inconspicuous; heads often only 3-4-fld; invol. 5-8 mm H, the phyllaries 3-5; rays pink. Dry slopes, 4000-11,000', Tulare Co. n., especially e. SN, Jul-Aug.

S. virgata, Tall Stephanomeria: Lower lvs oblong to spatulate, 1-2 dm L, often sinuate, dying before anthesis; upper lvs small, linear, entire; heads 4-15-fld; invol. 6-8 mm H; rays purplish on back, pinkish to white above, 7-9 mm L. Common, late summer annual, disturbed places below 6000', SN (all), Jul-Oct.

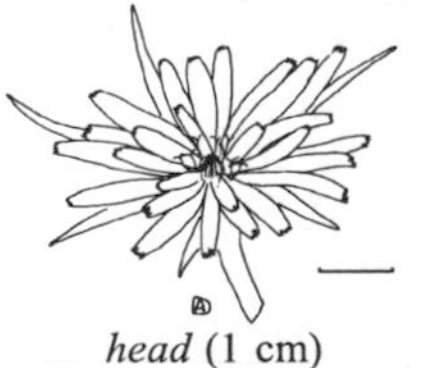

Tragopogon

Salsify
Goat's Beard

head (1 cm) *akene, beak and pappus* (1 cm)

Glabrous biennials from taproot. Sap milky. Stems stout, few-branched. Lvs alt, linear, 2-4 dm L, grass-like with parallel veins. Heads solitary on long, naked peduncles. Phyllaries in one row, conspicuous. Recpt naked. Fruit 2-3 cm L with obvious beak and wide-spreading pappus of plumose bristles. Plants of disturbed habitats and probably inedible.

Fls purple; phyllaries longer than ligules *T. porrifolius*
Fls yellow; phyllaries shorter than ligules *T. pratensis*

T. porrifolius, Salsify, Oyster Plant: Stem to 1 m H; peduncle much wider towards tip; phyllaries 5-10, 2-4 cm L (to 7 cm in fr). Occasional at lower elevations. A similar sp *T. dubius*, with phyllaries much longer than its pale yellow fls may rarely occur in the same range.

T. pratense: Stem 2-8 dm H; peduncle not wider towards tip; phyllaries 6-9, 1.5-3 cm L (to 4.5 cm in fr). Moist areas below 5000'.

Trimorpha

Short-rayed Aster

T. lonchophylla: Stems erect, slender, solitary or several in tuft, 1-6 dm H, mly pubescent; basal lvs oblanceolate, smoothly tapering to long petioles, the blade 3-6 cm L, the petioles about as long; cauline lvs narrower, alt; heads usu several, erect; invol. 5-10 mm H; phyllaries imbricate; rays pale violet to white; disk-fls yellow. 5000-11,000', Madera and Mono cos. s.

lower stem (1 cm)

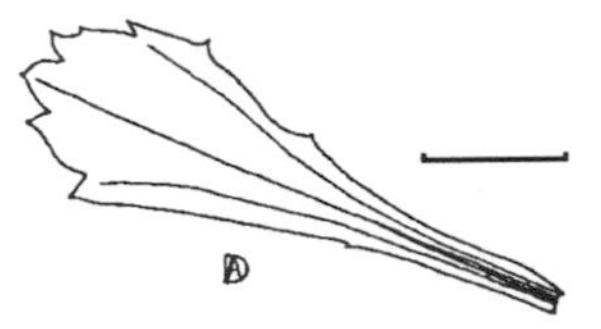

Tonestus eximium leaf (1 cm)

Tonestus

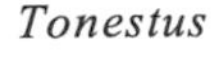

Tonestus

Tonestus peirsonii head (2 cm)

Cespitose, glandular perennials Lvs alt, oblanceolate. Infl open; heads radiate. Phyllaries in 3-4 rows. Ray-fls and disk-fls yellow. Pappus of many white bristles.

Invol. less than 10 mm H, 15 mm W; Alpine, Eldorado and Inyo cos *T. eximius*
Invol. 14-18 mm H, 20-30 mm W; Inyo and Fresno cos. *T. peirsonii*

T. eximius, Tahoe Tonestus: Stems erect, to 15 cm H, lvs dentate above middle, 2-5 cm L, wedge-shaped. Alpine summits, 8600-9600'.

T. peirsonii, Inyo Tonestus: Stems few to several; lvs 3-8 cm L, 10-25 mm W, dentate. Rocky summits near timberline, 9600-12,000', mly e. SN, Jul-Aug.

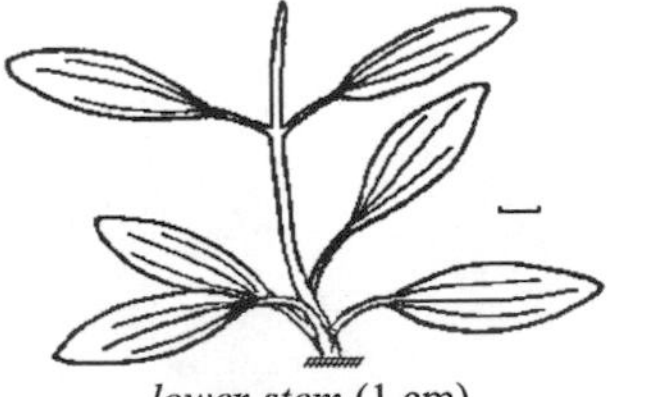

lower stem (1 cm)

Whitneya

Whitneya

inflorescence (1 cm)

W. dealbata: Stems erect, 1.5-3.5 cm H; basal lvs 5-10 cm L including petiole; stem lvs opp, relatively few, narrower and soon reduced, sessile; peduncles 5-15 cm L; heads solitary or in small clusters; invol. 8-11 mm H; rays 7-9, about 2 cm L, spreading, yellow. Uncommon, on open hillsides and forest openings, 4000-7000', Fresno to Shasta Co., Jun-Aug. Edibility unknown.

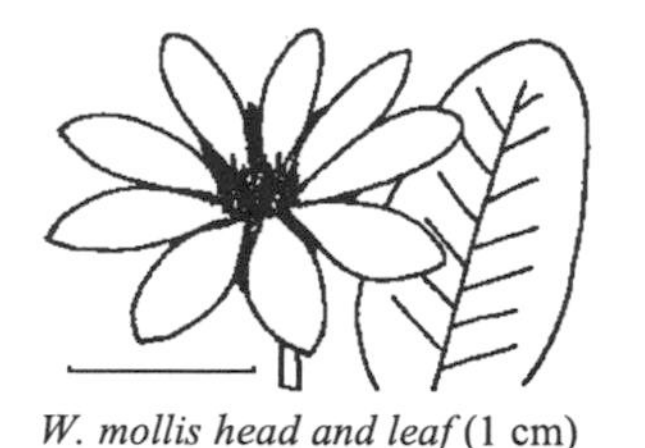

Wyethia

Wyethia
Mule-ears

W. mollis head and leaf (1 cm) *akene with pappus* (5 mm)

Coarse perennial herbs with erect or ascending usu unbranched stems from a branching crown. Lvs cauline and often in basal rosettes, alt, entire or toothed. Heads usu large, yellow. Phyllaries in 2-4 rows. Seeds edible, resembling sunflower seeds in taste.

A Basal lvs similar to or smaller than stem lvs, or absent; rare or local
  B Lvs round to deltoid; invol. bell-shaped — 6) *W. ovata*
  BB Lvs deltoid to elliptic or lanceolate; invol. hemispheric
    C Rays conspicuous, to 5 cm L; pappus a crown of scales — 2) *W. elata*
    CC Rays short or none; pappus none — 4) *W. invenusta*
AA Basal lvs present, usu larger than stem lvs; widespread
  B Basal lvs lanceolate; phyllaries shorter than head; below 5500' — 1) *W. augustifolia*
  BB Basal lvs mly elliptic to ovate; phyllaries slightly to much surpassing head
    C Heads 2-3 cm W; mly above 5000' — 5) *W. mollis*
    CC Heads 3.5-6 cm W; mly below 5000' — 3) *W. helenioides*

1) *W. augustifolia,* Narrow-leaved Mule-ears: Stems 2-6 dm H; pubescent; basal lvs 2-4 dm L, 4-8 cm W, usu with winged petioles, the cauline reduced; head usu 1, 3-4 cm W on a long slender peduncle; phyllaries green, long-ciliate, ± imbricate; akenes 7-9 mm L; pappus of ± united scales, usu prolonged into 1-4 awns. Open, grassy slopes, SN (all), May-Jul.

2) *W. elata,* Hall's Wyethia: Stems 5-10 dm H, leafy throughout, glandular pubescent; largest lvs usu 15-20 cm L, 8-12 cm W, acute or apex, truncate or subcordate at base, densely canescent and glandular beneath, entire to shallowly toothed, short-petiolate; heads 1-4, 2.5-4 cm W; outer phyllaries equaling or exceeding the disk; akenes 6-12 mm L. Dry, open slopes, 3000-4600', Tulare to Mariposa Co., Jun-Aug.

3) *W. helenioides,* Gray Mule-ears: Stems 2-6 dm H; herbage tomentose, becoming glabrate in age; basal lvs about 30-50 cm L, 10-12 cm W, abruptly short-petiolate; stem lvs similar but smaller; outer phyllaries to 8 cm L, exceeding the ray-fls. Open fields and sunny woodland borders, up to 6000', Mariposa to Eldorado Co., May-Jul.

4) *W. invenusta,* Coville's Wyethia: Stems 3-8 dm H, leafy throughout, hairy and becoming densely glandular-puberulent above; lvs as in *W. elata;* heads usu solitary, 2-3 cm W; phyllaries ± equaling the disk; akenes 7-8 mm L. Open woods, 3800-6000', Fresno Co. s., Jul-Aug.

5) *W. mollis,* Mountain Mule-ears: Stems 4-10 dm H; herbage densely tomentose when young, becoming glabrate; basal lvs elliptic to ovate, acute, 2-4 dm L, 6-17 cm W, narrowed to an ample petiole, entire; stem lvs similar but reduced; heads 1-4; outer phyllaries 4-6, white tomentose, slightly exceeding the disk; akenes 8-11 mm L. Dry, wooded slopes and rocky openings, 4500-10,600', Fresno Co. n., May-Aug.

6) *W. ovata,* Southern Wyethia: Stems few, stout, 1-3 dm H, much exceeded by lvs; herbage silky-villous when young; lvs broadly ovate to suborbicular, 7-20 cm L, with prominent veins, entire, petiolate; heads often hidden by lvs, 1.5-2 cm W; phyllaries in 2 rows, the outer phyllaries 4-6, erect, usu exceeding the 5-8 short rays. Grassy, openly wooded hillsides below 6000', Tulare Co., May-Aug.

BERBERIDACEAE - Barberry Family

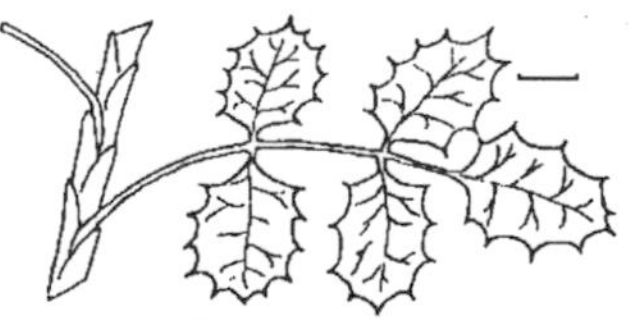

Berberis

Barberry

B. sonnei leaves and fruit (1 cm) *B. pumila leaf* (1 cm)

Evergreen shrubs with inner bark and wood yellow. Lvs alt, odd-pinnate. Lfts with bristle-tipped teeth; petioles ± clasping. Fls yellow, in drooping racemes. Sepals 6, in 2 series, petaloid, falling early, subtended by 3 bractlets. Petals 6, in 2 series. Stamens 6, opposite petals. Ovary superior. Fr a few-seeded berry. Berries of most spp are edible. The juice may be used to make a drink (with lemon, mint and sugar). Pies and jellies may be made from the berries which contain pectin. The berries can be dried for winter use or added to soup to improve the flavor.

Lfts pinnately veined; bud bracts deciduous; mly below 6000' *B. aquifolium*
Lfts ± palmately veined; bud bracts persistent; rare, at 7000' *B. nervosa*

B. aquifolium, Dwarf Barberry: Stems ascending or erect, 1.5-4 dm H; lvs 8-15 cm L, lfts mly 5-9, 3-6 cm L, 2-4.5 cm W, with 3-10 spiny teeth on each side; racemes 2-5 cm L; berries oblong-ovoid, blue-black, 5-8 mm L. Rocky outcrops and clay slopes, Mariposa Co. n., Apr-May.

B. nervosa: Stems 2.5-6 dm H; lvs 10-25 cm L; lfts 5, 4-8 cm L, 2.5-3.5 cm W, with 12-20 bristle-like teeth on each side; raceme 4-7 cm L; berries ovoid, blue-black, about 6 mm L. Rare, rocky banks, central SN, Apr-May.

BETULACEAE - Birch Family

Deciduous trees and shrubs with simple, usu serrate, petioled lvs and deciduous stipules. Male and female fls in separate compact clusters appearing before lvs.

Lvs cordate at base, soft-pubescent on both surfaces; fr a nut *Corylus* p.
Lvs rarely cordate, usu ± glabrous; fr cone-like
  Female "cone" woody, clustered, falling whole; bark not aromatic *Alnus* p.
  Female "cone" solitary, disintegrating at maturity; bark aromatic *Betula* p.

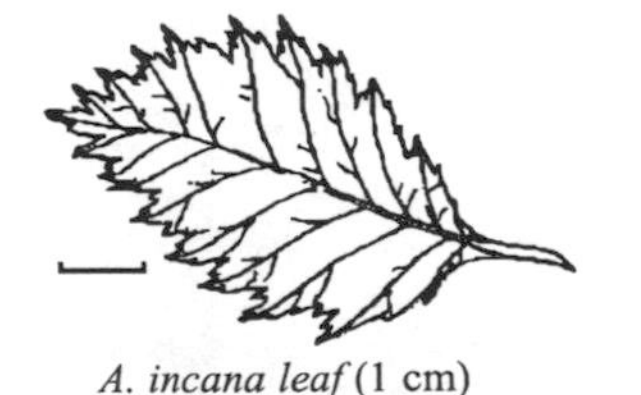

Alnus

Alder

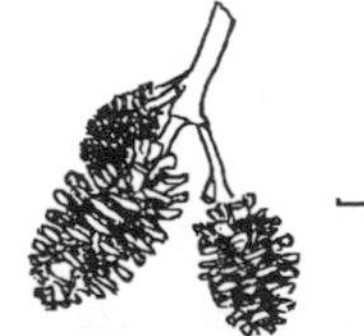

A. incana leaf (1 cm) *cones* (1 cm)

Plants with scaly bark and few-scaled lf-buds. Staminate aments clustered at ends of branchlets, pendulous. Pistillate fls in erect, spike-like catkins. Edibility unknown.

Shrubs 3-5 m H; lvs with about 10 tooth-like lobes on each edge *A. incana*
Trees 10-25 m H; lvs not lobed; mly below 5000' *A. rhombifolia*

A. incana var. *tenuifolia,* Mountain Alder: Shrub or small tree with smooth gray or red-brown bark; lvs rounded to cordate at base, coarsely toothed then finely serrate, 2.5-7 cm L; petioles 1-2 cm L. Moist places, 4500-8000', SN (all), Apr-Jun.

A. rhombifolia, White Alder: Bark whitish to gray-brown; lvs finely or coarsely double serrate, darker above; petioles 1-2 cm L. Along streams SN (all), May-Apr.

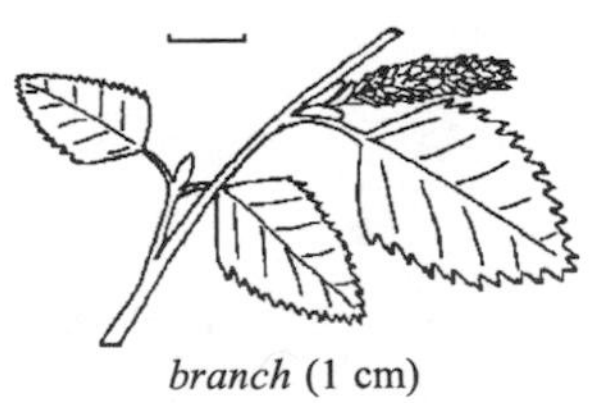

Betula

Birch

branch (1 cm) *cone* (1 cm)

B. occidentalis: Tall shrub or tree to 8 m H; bark smooth, dark bronze, shining, with many long lenticels; twigs rough with large resinous glands; lvs broadly ovate, sharply serrate except near the entire base, gland-dotted when young, darker above; petioles about 1 cm L. Moist places, 2000-8000', SN (all), Apr-May. Sap drinkable in spring. Juice of lvs used for mouthwash.

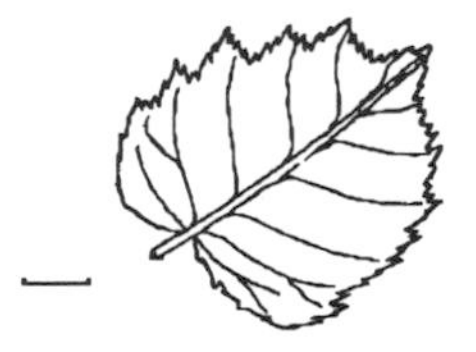

Corylus

Hazelnut

A. **Filbert**

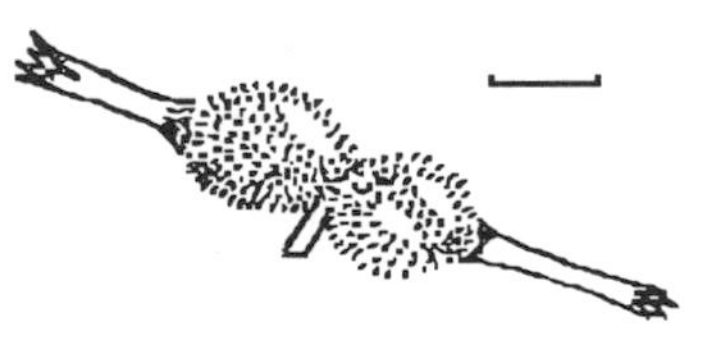

leaf (1 cm) *fruit* (1 cm)

C. cornuta: Open spreading shrub, 2-4 m H, with smooth bark; lvs rounded to obovate, 4-7 cm L, doubly serrate, sometimes ± 3-lobed, becoming glabrous in age. Damp slopes and banks below 7000', Kaweah River, Tulare Co. n., Mar-Apr. Nuts ripening in late summer or fall are excellent food high in fat and protein and may be ground and used in bread.

BORAGINACEAE - Borage Family

Herbage usu rough-hairy. Lvs simple, mly entire, mly alt. Calyx usu 5-parted or -lobed, often slightly irregular. Corolla 5-lobed. Stamens 5, alt with corolla-lobes, inserted mly in corolla-tube.

A Fls small, white; nutlets 1-3 mm L
  B Nutlet attachment scar flush with or indented into nutlet — *Cryptantha* p. 72
  BB Nutlet attachment scar raised above surface of nutlet — *Plagiobothrys* p. 75
AA Fls usu more than 10 mm L, brightly colored; nutlets usu larger
  B Corolla yellow to orange; nutlets smooth
    C Plants from a stout root; infl open, not scorpioid — *Lithospermum* p. 74
    CC Plants from filiform roots; infl a dense scorpioid cyme — *Amsinckia* p. 72
  BB Corolla blue or pinkish, rarely white
    C Plants glabrous; corolla 10-20 mm L; nutlets smooth — *Mertensia* p. 74
    CC Plants hairy; nutlets with prickles
      D Nutlets widely spreading in fr, evenly covered with about 1 mm L prickles — *Cynoglossum* p. 73
      DD Nutlets appressed to each other in fr; prickles longer and often unevenly distributed — *Hackelia* p. 73

Amsinckia

Fiddleneck

A. menziesii var. *intermedia:* Stems 2-8 dm H, sparsely to moderately bristly; lvs linear to lanceolate, rough-hairy, 2-5 cm L, the lower petioled; racemes leafy-bracteate at base, 5-20 cm L; calyx 5-10 mm L in fr; corolla 8-10 mm L. Common in grassy or open places below 5000', SN (all), May-Jun. Herbage and seeds contain toxic pyrrolizidine alkaloids.

inflorescence (1 cm)

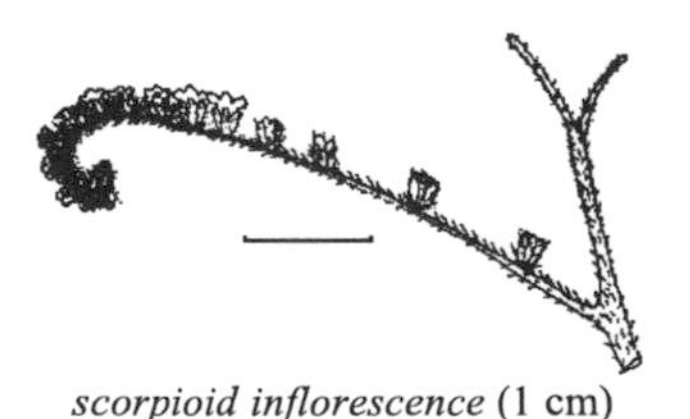

Cryptantha

Popcorn Flower

scorpioid inflorescence (1 cm) *nutlet* (1 mm)

Usu hairy annuals or perennials. Earliest lvs opp, the others alt. Fls in bractless or bractate usu scorpioid spikes or racemes. Nutlets 1-4. The many spp and their wide distribution makes *Cryptantha* a very common genus in the Sierra. The individual spp are very difficult to tell apart, usu requiring careful study of the nutlets and using equipment rarely available in the field. For this reason the spp will not be treated individually. Edibility unknown.

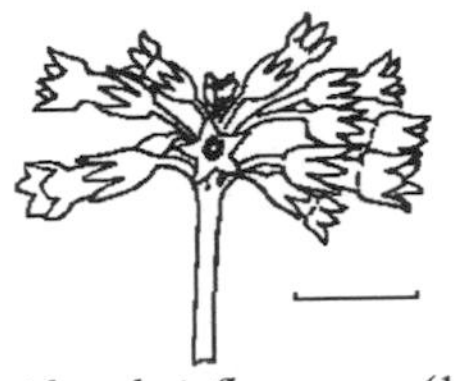

Cynoglossum

Hound's Tongue

C. occidentale inflorescence (1 cm) *nutlets* (1 cm)

Tall perennial herbs from a thick base. Basal lvs long-petioled, the upper sessile or nearly so, pubescent. Infl a bractless (usu) terminal panicle of scorpioid racemes. Calyx deeply 5-parted; corolla-tube closed by 5 obtuse projections at base of corolla-limb. Lvs of related spp contain pyrrolizidine alkaloids which are toxic; however at least one similar sp has been used as a vegetable.

Stems glabrous; lvs about 1-2 times as long as wide *C. grande*
Stems hairy; lvs usu much longer than wide *C. occidentale*

C. grande, Large Hound's Tongue: Stems erect, 3-9 dm H; lvs mly basal or on lower stem, ovate, the blades 8-15 cm L, abruptly narrowed into petioles often as long; peduncle well developed; pedicels usu 1-2.5 cm L; calyx-lobes 5-7 mm L; corolla-tube 4-6 mm L; nutlets about 8 mm L. Dryish openings in woods, 4000-7000', SN (all), May-Jul.

C. occidentale, Western Hound's Tongue: Stems erect, 2-4 dm H, hairy; lower lvs oblanceolate, the blades 5-15 cm L; petioles winged, 5-10 cm L; upper lvs reduced, sessile or cordate-clasping; infl rather compact, long-peduncled; pedicels about 4-8 mm L; calyx 5-7 mm L; corolla-tube 4-6 mm L; nutlets about 8 mm L. Dryish openings in woods, 4000-7000', SN (all), May-Jul.

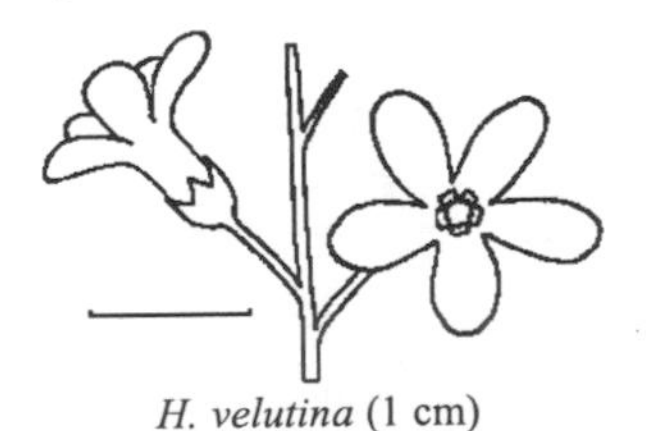

Hackelia

Stickseed

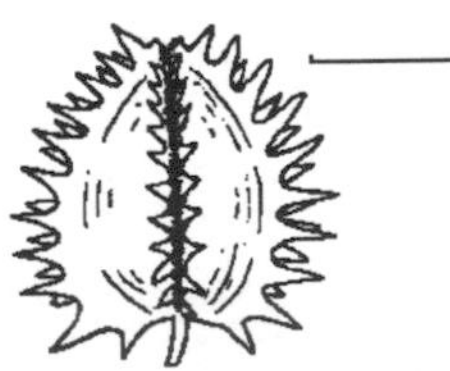

H. velutina (1 cm) *H. setosa nutlets* (5 mm)

Perennials with well developed stems. Lvs alt, pubescent, linear to oblong, the basal long-petioled, the upper ± sessile. Infl paniculate; pedicels recurved in fr. Calyx often reflexed in fr. Nutlets attached below the middle to the broadly pyramidal gynobase, each nutlet 4-6 mm L. Edibility unknown.

A Outer face of each nutlet naked or with few prickles, though rimmed with a conspicuous border of flattened ones
  B Corolla 4-8 mm across
    C Stem 1-3 dm H; corolla with a yellow center; above 10,000' near Mt. Whitney — 8) *H. sharsmithii*
    CC Stem 3-11 dm H; widespread below 10,000
      D Facial prickles 4-10; corolla pale blue with white center — 4) *H. micrantha*
      DD Nutlet facial prickles 1-3; infl usu narrow — 3) *H. floribunda*
  BB Corolla 10-18 mm broad; below 6000'; Sierra Co. n. — 7) *H. setosa*
AA Outer face of nutlet with more than 10 prickles
  B Corolla-tube inconspicuous
    C Fls 6-12 mm W; Alpine Co. n.
      D Fls white to pinkish; facial prickles ± as long as marginal — 2) *H. californica*
      DD Fls blue to pinkish; facial prickles shorter than marginal — 1) *H. amethystina*
    CC Fls 10-18 mm W; Tuolumne Co. s. — 5) *H. mundula*
  BB Corolla-tube present, exceeding the sepals
    C Fls 6-8 mm W — 6) *H. nervosa*
    CC Fls 12-20 mm W — 9) *H. velutina*

1) *H. amethystina,* Amethyst Stickseed: Stem stout, erect, 4-8 dm H; basal lvs narrow-elliptic, the blades 8-12 cm L on petioles almost as long; mid-cauline lvs subclasping. Uncommon, moist habitats and clearings, 5000-6500', Plumas Co., Jun-Jul.

2) *H. californica,* California Stickseed: Stems erect or ascending, leafy, 4-8 dm H, densely villous; basal lf-blades oblong-oblanceolate, 6-15 cm L; upper lvs lanceolate to narrow-oblong; panicles widely branched in age; pedicels 6-8 mm L in age. Gravelly slopes and open, dry woods, 4000-8000', June-Aug.

3) *H. micrantha:* Stems few, erect or ascending, usu densely hairy; basal lvs few, oblanceolate, the blades 7-15 cm L; upper lvs well distributed along the stems, gradually reduced upward; panicle open; pedicels 5-10 mm L; calyx 2-3 mm L; corolla-tube not exserted beyond calyx. Common in moist places 4500-11,000', Tulare Co. n., Jul-Aug.

4) *H. mundula,* Pink Stickseed: Much like *H. californica,* with softer pubescence; 8-14 mm L in fr; corolla usu pink with whitish center, fading blue; appendages conspicuous. Open woods, 7000-9500', Mariposa Co. s., Jun-Jul.

5) *H. nervosa,* Sierra Stickseed: Stems 1 to few, erect, 2-5 dm H; basal lf-blades ± oblong, 4-15 cm L; stem lvs shorter, the upper lance-ovate; infl open; moist places, 5000-10,000', Fresno to Plumas Co., Jul-Aug.

6) *H. setosa,* Bristly Stickseed: Stems 1 to several, suberect, bristly-hairy; lower lvs linear-oblanceolate, 5-10 cm L; upper lvs reduced; infl few-branched; pedicels 5-10 mm L; corolla blue; dorsal surface of nutlet with several short prickles, the margins with long and short alternating prickles. Open wooded ridges or slopes, Jun-Jul.

7) *H. sharsmithii,* Sharsmith's Stickseed: Stems several, erect or ascending, slender; lvs bright green, elliptic to ovate, 5-10 cm L; the cauline lanceolate to somewhat narrow-oblong, gradually reduced upward; infl of several few-fld branches; pedicels 5-10 mm L in fr; corolla blue to pink, the limb 8-12 mm W; appendages spreading, broad, conspicuously exserted. Dry, wooded slopes, 5000-10,000', Jun-Aug.

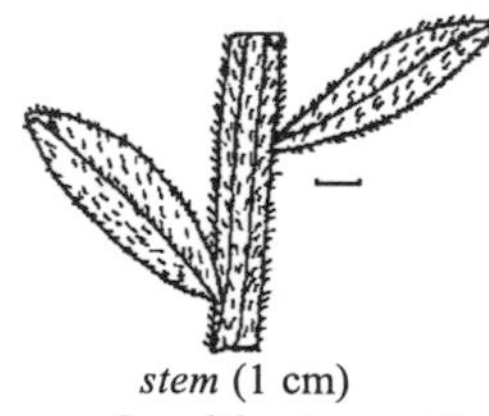

stem (1 cm)

Lithospermum

Stoneseed
California Gromwell
Shasta Puccoon

flower (1 cm)

L. californicum: Stems few to several, erect or ascending, often ribbed, 1.5-4 dm H, hairy; lower lvs lance-linear to lanceolate, 4-10 cm L, 5-10 mm W, the upper lance-oblong to -ovate, 10-20 mm W; infl rather congested; calyx 6-9 mm L; corolla 12-18 mm L; nutlets white, shining, about 6 mm L. Dry slopes and ridges below 5000', Placer Co. n., May-Jun. Roots contain a red or violet dye; those of a similar sp are edible after cooking.

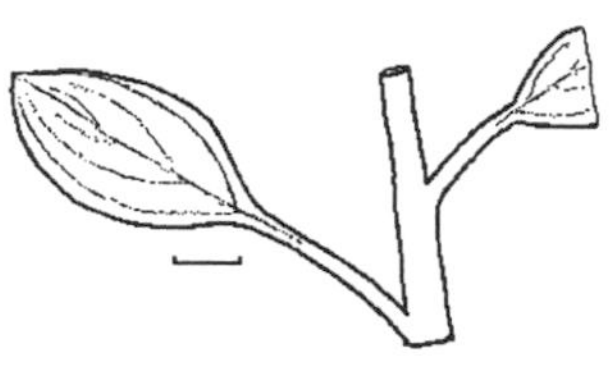

leaf (1 cm)

Mertensia

Lungwort
Bluebell

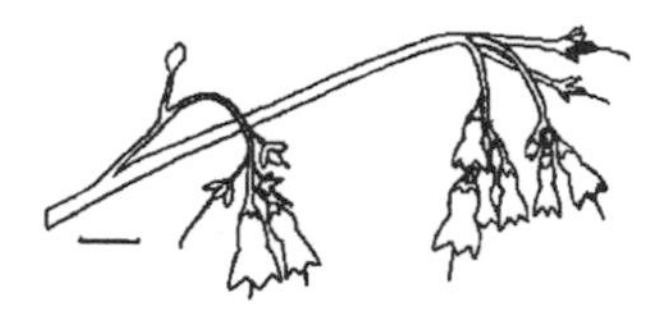

inflorescence (1 cm)

M. ciliata var. *stomatechoides:* Stems erect or ascending, mly 5-10 dm H, usu several; lvs alt, glaucous; lf-blades lanceolate to ovate or oblong, 5-12 cm L, 1-4 cm W, ciliate, subentire, the lower long-petioled, the upper sessile; infl a ± drooping panicle; pedicels 3-12 mm L; calyx-lobes 2-6 mm L; corolla pink at first, maturing lilac to blue, usu constructed just above calyx; corolla-tube 6-8 mm L, the limb 5-10 mm W; style exserted. Moist mdws, 5000-10,000', SN (all), May-Aug. Rootstock of similar sp edible.

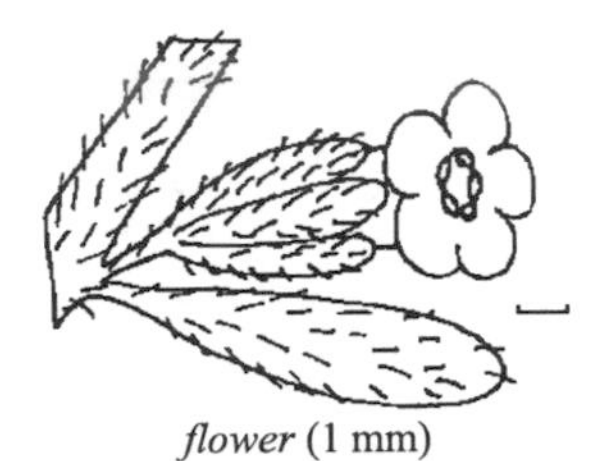

flower (1 mm)

Plagiobothrys

Popcorn Flower

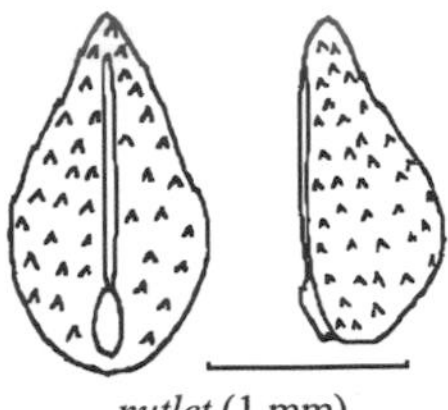

nutlet (1 mm)

Annuals usu with weak slender appressed hairs. Lower lvs opp or in a rosette and crowded. Calyx divided to middle or lower, usu persistent, often tawny or brown. Corolla-tube usu included in calyx. Difficult genus to key. The most common Sierran spp are *P. torreyi* with stem lvs all alt, and *P. cognatus* and *P. hispidulus* with lower stem lvs opp. The seven or more spp found in the SN are almost impossible to differentiate without considerable magnification of the nutlets. Therefore the spp will not be treated individually here. Edibility unknown.

BRASSICACEAE - Mustard Family

Herbs with pungent watery juice. Lvs alt, without stipules. Fls in terminal racemes or corymbs, rarely solitary, terminal. Sepals 4, deciduous; petals 4, forming a cross-shaped pattern, hence the former name of the family: Cruciferae. Stamens 6 (rarely 2 or 4), 2 of which are shorter than the other 4. Ovary and fr usu split into 2 cells by a thin partition.

A Fr elongate, 3 or more time longer than wide or thick
  B Fr flattened, linear or wider
    C Plants pubescent, at least some hairs forked
      D Stems ± glabrous; lvs densely stellate pubescent; pubescence felt-like, the individual hairs difficult to resolve — *Anelsonia* p. 76
      DD Stems pubescent at least below; individual hairs easily distinguished
        E Fr usu very much longer than wide; valves usu 1-nerved — *Arabis* p. 76
        EE Fr mly 3-6 times longer than wide; valves nerveless — *Draba* p. 79
    CC Plants essentially glabrous; pubescence simple if any
      D Calyx flask-shaped, often colored — *Streptanthus* p. 82
      DD Calyx erect or ascending, not flask-shaped
        E Valves of fr not nerved; lvs often pinnate — *Cardamine* p. 78
        EE Valves of fr 1-nerved; lvs simple — *Arabis* p. 76
  BB Fr round or 4-angled in cross section
    C Lvs entire to dentate, never lobed
      D Petals 15-25 mm L; fr 4-angled
        E Fls white to purple — *Hesperis* p. 81
        EE Fls yellow to orange — *Erysimum* p. 80
      DD Petals 5-7 mm L; fr ± round — *Arabis glabra* p. 76
    CC Lvs pinnately lobed to pinnately cmpd
      D Stems and herbage ± pubescent; plants of dry to moist habitats — *Descurainia* p. 79
      DD Stems and herbage glabrous; plants of wet habitats
        E Stems reclining; frs 6-20 mm L — *Rorippa* p. 82
        EE Stems erect; frs 20-40 mm L — *Barbarea* p. 78
AA Fr 1-3 time longer than thick or wide
  B Fr turgid or inflated, not flattened
    C Lvs pinnatifid or pinnately lobed
      D Fls white, 4-5 mm L; rocky ridges on Lassen Peak — *Smelowskia* p. 82
      DD Fls yellow, 1-3 mm L; wet habitats — *Rorippa* p. 82
    CC Lvs simple, entire or nearly so
      D Plant aquatic; rare, central SN — *Subularia* p. 83
      DD Plant terrestrial
        E Lvs 2-5 cm L; fr stellate-pubescent; below 8000' — *Lesquerella* p. 81
        EE Lvs usu less than 1 cm L; fr with mly simple hairs; usu alpine — *Draba* p. 79
  BB Fr evidently flattened
    C Fr with a plain, smooth broad surface
      D Fls solitary at the summit of a leafless stalk — *Idahoa* p. 81
      DD Fls in clusters; stems usu leafy
        E Lvs 2-5 cm L; fr 1-seeded — *Thysanocarpus* p. 83
        EE Lvs less than 2 cm L; fr 2 to many seeded — *Draba* p. 79

CC Fr with a conspicuous seam or vein extending up the broad surface
D Stems less than 1 dm H; above 10,000' — *Anelsonia* p. 76
DD Stems 2-6 dm H; below 8000'
E Fr heart-shaped, broader distally — *Capsella* p. 78
EE Fr roundish — *Lepidium* p. 81

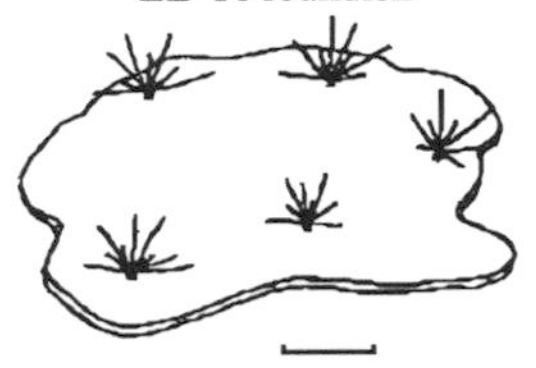

stellate hairs (1 mm)

Anelsonia

Phoenicaulis

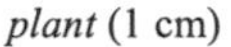

plant (1 cm)

A. eurycarpa: Perennial, 2-5 cm H; lvs in basal cluster and reduced upward, oblanceolate, 1-1.5 cm L; fls few; pedicels 5-9 mm; petals yellow, wider towards tip; fr 2-2.5 cm L, less than 10 cm W, glabrous, the valves 1-nerved. Open slopes and flats 10,500-14,000', Tuolumne Co. s., Jul-Aug. Edibility unknown.

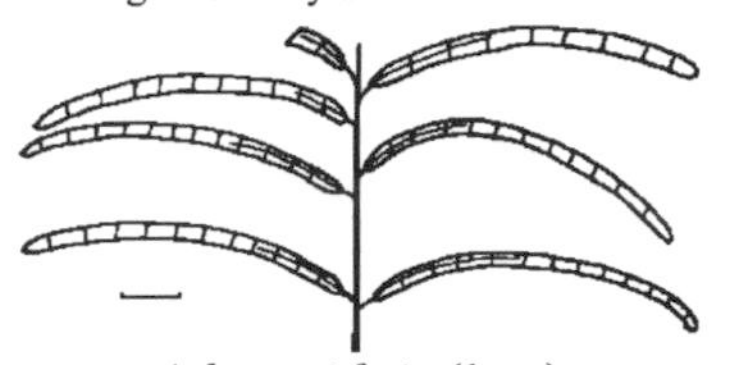

A. breweri fruits (1 cm)

Arabis

Rock-cress

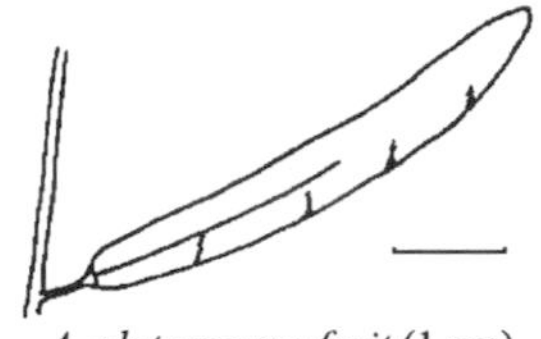

A. platysperma fruit (1 cm)

Biennial or perennial. Stems leafy, often branched above. Lvs dentate to entire, the basal petioled, the cauline usu sessile. Infl racemose, bractless, elongating. Sepals erect, oblong to ovate, one pair sometimes saccate. Related sp has edible lvs.

A Fr 3-8 mm W; seeds (including wings) 2.5-5 mm L
B Fr and pedicels reflexed — 18) *A. suffrutescens*
BB Fr and pedicels erect or ascending
C Basal lvs 10-30 mm W, margins usu undulating and toothed — 16) *A repanda*
CC Basal lvs 1-8 mm W, margins entire
D Basal lvs oblanceolate to spatulate, ± glabrous — 12) *A. platysperma*
DD Basal lvs linear, less than 2 mm W, with prominent hairs — 14) *A. pygmaea*
AA Fr usu less than 3 mm W; seeds less than 2 mm L
B Lvs mly green, ± glabrous; frs erect to divaricate
C Lvs rounded at tip, over 1 cm W; sepals often saccate at base
D Fr ± round in cross-section; petals yellowish white — 5) *A. glabra*
DD Fr flattened; petals white to pink
E Stems 5-15 cm H; fr 1.5-2 mm W — 2) *A. davidsonii*
EE Stems 20-70 cm H; fr 1 mm W — 6) *A. hirsuta*
CC Lvs usu acute, often less than 1 cm W; sepals not saccate at base
D Stems 4-25 cm H; lvs usu less than 5 mm W
E Fls white; lvs from thick caudex; rare, above 10,000, n. SN — 19) *A. tiehmii*
EE Fls pale rose to purple; caudex not particularly thick; widespread
F Cauline lvs with lobed base ± clasping stem; rare — 11) *A. microphylla*
FF Cauline lvs tapered at base, not clasping; common — 10) *A. lyalli*
DD Stems 30-90 cm H; lvs usu 7-15 mm W
E Frs erect, rather crowded; pedicels erect — 4) *A. drummondii*
EE Frs ± divaricate; pedicels divaricate to descending — 3) *A. divaricarpa*
BB Lvs usu grayish with appressed hairs; frs wide-spreading to reflexed
C Petals 4-6 mm L
D Frs strictly reflexed, appressed to stem; stem 2-8 dm H — 15) *A. rectissima*
DD Frs horizontal or slightly descending; stems 0.6-2 dm H — 9) *A. lemmonii*
CC Petals 6-14 mm L
D Frs reflexed; sometimes appressed to stem
E Petals white to pinkish; stem lvs revolute — 7) *A. holboellii*
EE Petals mly rose to purple; stem lvs not revolute — 13) *A. puberula*
DD Frs wide-spreading, sometimes curved
E Plants 0.6-2 dm H; basal lvs wide-spatulate with 3-forked hairs — 1) *A. breweri*
EE Plants 2-9 dm H; hairs not 3-forked

F Basal lvs often over 3 cm L; fr 6-12 cm L — 17) *A. sparsiflora*
FF Basal lvs 2-3 cm L; fr 4-6 cm L — 8) *A. inyoensis*

1) *A. breweri*, Brewer's Rock-cress: Stems several, unbranched, densely hairy below with mly simple hairs, often glabrous above; basal lvs mly entire, short-petioled, 4-6 mm W; stem lvs sessile, usu less than 2 cm L; pedicels 5-10 mm L; petals red-purple to pink, 6-9 mm L; fr usu curved, 3-7 cm L. Dry, rocky slopes and summits below 7400', Yuba Co. n., May-Jul.

2) *A. davidsonii*, Davidson's Rock-cress: Stems 1 to several, slender, glabrous, unbranched; lower lvs ± oblong, entire, few, sessile; pedicels 8-15 mm L; petals white to pinkish, 8-10 mm L; frs divaricate, straight to somewhat curved, 3-5 cm L. Rocky places, 5000-11,500, Tulare to Plumas Co., Jul-Aug.

3) *A. divaricarpa*, Bent-pod Rock-cress: Stems 1 to few, glabrous above, glabrous to sparsely pubescent below; lvs oblanceolate to spatulate, dentate to subentire; pedicels 6-12 mm L; petals pink to purplish, 6-10 mm L; fr mly straight, 2-8 cm L. Dry slopes in open woods, 7000-11,000', SN (all), Jul-Aug.

4) *A. drummondii*, Drummond's Rock-cress: Stems 1 to few, ± glabrous; lower lvs oblanceolate, entire to dentate, 2-7 cm L; stem lvs usu clasping, crowded toward base; pedicels 1-2 cm L; petals white to violet, 7-10 mm L; fr 4-10 cm L. Dryish or dampish benches and slopes, 5500-10,900', SN (all), Jun-Jul.

5) *A. glabra*, Tower Mustard: Stems 1 to few, rarely branched above, usu hairy below, glabrous above, 4-12 dm H; lower lvs oblanceolate to oblong, usu dentate, with coarse hairs, 3-12 cm L, 1-3 cm W; stem lvs lanceolate to ovate, mly entire and glabrous, often with revolute margins; racemes long, erect; white or rarely purplish; fr 4-10 cm L. Shaded canyons and mountains, up to 9300', SN (all), May-Jul.

6) *A. hirsuta* vars., Hairy Rock-cress: Stems erect, 1 to few, hairy below, glabrous above; basal lvs obovate to oblanceolate, 3-7 cm L, 1-2.5 cm W; pedicels 5-15 mm L; petals white, 5-9 mm L; fr 3-6 cm L. Moist places 4000-8200', SN (all), May-Jul.

7) *A. holboellii*, vars., Holboell's Rock-cress: Stems 1 to several, erect, 1-9 dm H, mly pubescent, sometimes glabrous above; basal lvs oblanceolate to spatulate, 1-5 cm L, 2-6 mm W; frs 2.5-8 cm L. Dry, rocky places, 6000-11,000', SN (all), May-Jul.

8) *A. inyoensis*, Inyo Rock-cress: Stems several, erect, densely pubescent below, glabrate above; basal lvs many, narrowly oblanceolate to spatulate, entire, 2-3 cm L, 2-5 mm W; stem lvs sessile, oblong; petals pink to purplish, 7-9 mm L; fr 4-6 cm L. Dry, rocky places, 5000-12,000', Tulare Co., Inyo to Mono Co., May-Jul.

9) *A. lemmonii*, Lemmon's Rock-cress: Stems several, slender, pubescent throughout or usu glabrous above; basal lvs ± spatulate, entire to few-toothed, 1-2 cm L; stem lvs sessile to clasping; pedicels 2-5 mm L; petals pink to purple, 4-6 mm L; fr 2-4 cm L. Dry, rocky places, 8000-14,000', Tulare Co., to Lassen Peak, Jun-Aug.

10) *A. lyalli*, Lyall's Rock-cress: Stems usu more than 1, glabrous; lower lvs oblanceolate, entire, 1-3 cm L, 1-6 mm W; stem lvs lanceolate, remote, sessile; pedicels 4-10 mm L; sepals about 4 mm L; petals white to purplish, 6-10 mm L; fr 3-5 cm L, 2-3 mm W; seeds winged. Rocky places 8000-12,000', SN (all), Jul-Aug.

11) *A. microphylla* var. *microphylla*, Small-leaved Rock-cress: Stems several, usu unbranched, hairy below; lvs mly basal, 5-30 mm L, ± linear, entire; petals spatulate, pink to purple; fr spreading, 2-5 mm L, glabrous. Rocky outcrops, 4000-8000', n. SN, May-Jul.

12) *A. platysperma* and var., Flat-seeded Rock-cress: Stems glabrous, erect to decumbent, 0.5-4 dm L; basal lvs many, 2-5 cm L; stem lvs few but present, sessile; pedicels 5-15 mm L; petals pink to white, 4-7 mm L. Dry benches and slopes, 5500-12,750', SN (all), Jun-Aug.

13) *A. puberula*, Blue Mountain Rock-cress: Stems 1 to few, mly unbranched, densely pubescent, 1.5-5 dm H; basal lvs oblanceolate, entire or few-toothed, 1-2.5 cm L, 3-6 mm W; stem lvs crowded, densely pubescent, lanceolate to oblong, entire to irregularly toothed; pedicels 4-8 mm L, pubescent; fr 3-6 cm L. Dry, stony places, 4000-10,300', e. SN, Jun-Jul.

14) *A. pygmaea*, Dwarf Rock-cress: Stems several, slender, erect to somewhat decumbent, unbranched, 5-10 cm H, basal lvs tufted; stem lvs few, sessile, linear; pedicels ascending, 5-8 mm L; petals white, about 4 mm L; fr 2-4 cm L. Dry flats of volcanic sand or gravel, 8500-11,000', Tulare Co. from Rock Creek to Templeton Mdws, Jun-Jul.

15) *A. rectissima*, Bristly-leaved Rock-cress: Stems 1 to several, glabrous to sparsely hairy; basal lvs spatulate to oblanceolate, entire, 1-3 cm L, 4-10 mm W; stem lvs crowded below, oblong, few above; pedicels glabrous, 4-12 mm L; petals white or pinkish; fr 5-8 cm L. Dry slopes and benches, 4000-9000', SN (all), Jun-Jul.

16) *A. repanda* and var., Repand Rock-cress: Stems ascending, mly 1 or 2, densely pubescent below, subglabrous above, 2-8 dm H; basal lvs obovate, toothed, 3-8 cm L; stem lvs few, oblanceolate; pedicels stout, 3-8 mm L; petals white to pinkish, narrow, 4-6 mm L; fr 4-10 cm L. Dry slopes, 4600-11,600', Tulare to Nevada Co., Jun-Aug.

17) *A. sparsiflora* var. *arcuata*, Elegant Rock-rose: Stems pubescent below; basal lvs linear-lanceolate, long-petioled, entire, coarsely hairy, 3-10 cm L, 3-8 mm W; stem lvs linear-oblong;

pedicels 5-15 mm L; petals pink to purple, 8-14 mm L; fr 6-12 cm L. Dry slopes, 2500-6000', w. SN (all), and below 9000', e. SN, Mono Co. n., Apr-Jul.

18) *A. suffrutescens,* Woody Rock-cress: Stems somewhat woody, several, glabrous, 2-5 dm H; basal lvs linear to spatulate, 1-4 cm L, 2-6 mm W; stem lvs few, sessile; pedicels 4-10 mm L; petals rose to purplish, 6-8 mm L; fr 3-6 mm W. Dry places, 5500-9000', Fresno Co. n., SN, Jun-Jul.

19) *A. tiehmii*, Tiehm's Rock-cress: Slender stems 8-18 cm H, glabrous, ascending; basal lvs erect, 1.5-2.5 cm L, acute; cauline lvs 3-5, sessile, about 1 cm L; pedicels slender, 3-4 mm L. Rare on rocky formations and decomposed granite, 10,000-11,000', n. SN, Jul-Aug.

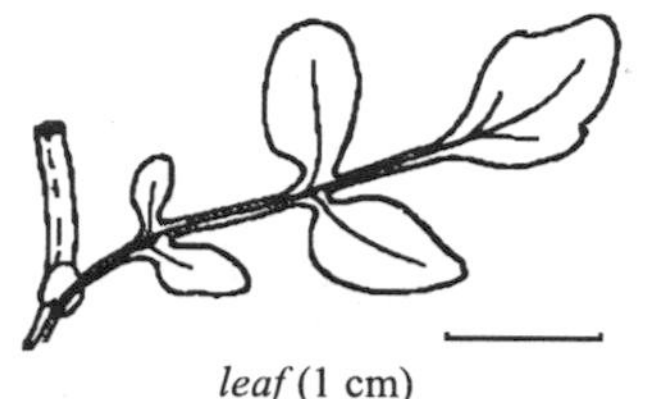

leaf (1 cm)

Barbarea

Winter-cress

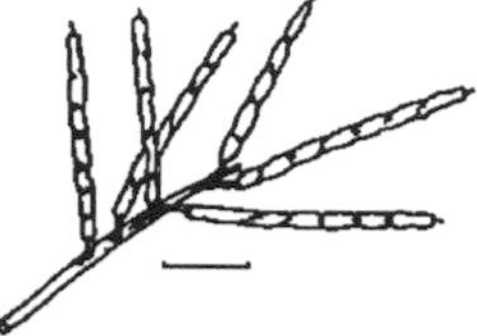

fruits (1 cm)

B. orthoceras: Perennial with stout, erect, angled stems, 2-4 dm H, usu few-branched above; basal lvs 3-10 cm L, petioled, elliptic, simple or with 2-4 small lfts and a large terminal one; middle and upper stem lvs pinnatifid, sessile, clasping; pedicels thick, ascending, 3-6 mm L; sepals yellow to yellow-green, the outer 2 saccate at base; petals yellow, 4-6 mm L; fr linear, erect and appressed to spreading. Damp to wet habitats, below 11,000', SN (all), May-Sep. Lvs edible raw but may leave a bitter aftertaste; boiling removes bitterness.

inflorescence (1 cm)

Capsella

Shepherd's-purse

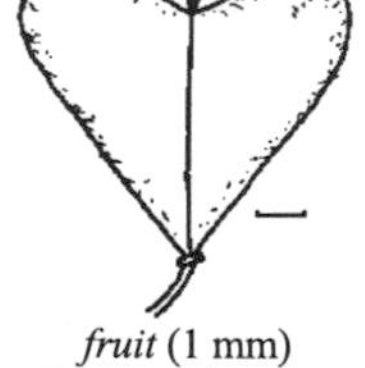

fruit (1 mm)

C. bursa-pastoris: Stems mly branched; fls white, about 2 mm L. Common in disturbed sites below 7000', SN (all), May-Sep. Young plant edible raw, but better cooked and is high in ascorbic acid; seeds can be collected just before maturity, parched and eaten, or ground into flour.

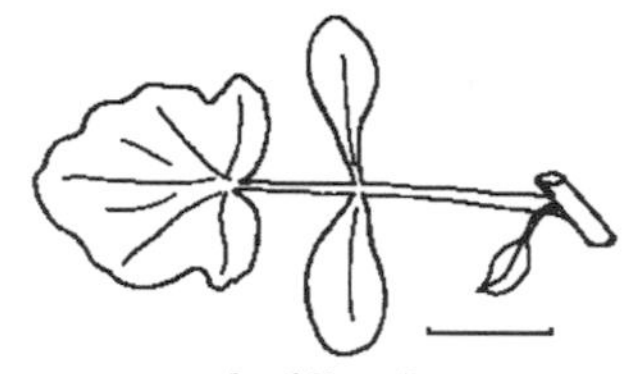

leaf (1 cm)

Cardamine

Bitter-cress

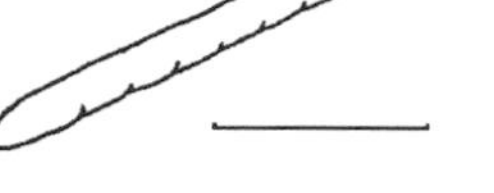

fruit (1 cm)

Perennials, mly glabrous, with leafy stems. Fls white. Sepals equal at base. Fr linear, narrow, many-seeded, opening elastically from base. Plants may be eaten raw in salads but are better cooked.

A Lvs all simple, not divided although sometimes with small lateral lfts
- B Flowering stalk ± leafless; the basal lvs ovate to elliptic, ± entire — 1) *C. bellidifolia*
- BB Flowering stalk leafy; lvs serrate or lobed
  - C Lvs ovate, usu dentate towards tip — 5) *C. pachystigma*
  - CC Lvs cordate or reniform with sinuate margins — 4) *C. cordifolia*

AA Lvs, at least those on the stem, cmpd
- B Stem lvs 3-5-foliolate; lfts of upper lvs broad
  - C Petals about 5 mm L; wet habitats — 2) *C. breweri*
  - CC Petals 10-15 mm L; moist, shady slopes — 3) *C. californica*
- BB Stem lvs 5-9-foliolate; lfts of upper lvs linear — 6) *C. pensylvanica*

1) *C. bellidifolia* var. *pachyphylla,* Alpine Bitter-cress: Stems 3-10 cm H; lvs fleshy, long-petioled, sometimes 3-lobed, 4-12 mm L; infl subumbellate, about 10-fld; fruiting pedicels 5-10 mm L; petals 3-5 mm L; fr erect, 2-3 cm L. At 7000-8000', Lassen Co. n., Jun-Jul.

2) *C. breweri,* Brewer's Bitter-cress: Stems often procumbent at base and rooting at nodes, 2-6 dm H; lvs mly cauline; terminal lfts ovate, sometimes 3-lobed or cordate, 1-4 cm L; fruiting pedicels 7-15 mm L; petals about 5 mm L; frs ascending to erect, 1.5-2.5 cm L. Along streams, 4000-10,200', SN (all), May-Jul.

3) *C. californica,* California Milkmaids: Herbage glabrous; rhizomal lvs mly 3-foliolate, the lfts broadly ovate, 2-5 cm W, sinuate to dentate; stems slender, 1-4 dm H; stems lvs few, the lfts lanceolate to ovate, toothed to entire; racemes many-fld; fruiting pedicels usu 1-2.5 cm L; petals pale rose to white, mly 10-14 mm L; fr 2-5 cm L. Shady slopes below 5000', May-Jun.

4) *C. cordifolia* var. *lyallii,* Lyall's Bitter-cress: Stems erect, 2-5 dm H; lvs 5-12, mly cauline, 1-7 cm W, petioled; pedicels 8-15 mm L in fr; petals 7-9 mm L; frs spreading, 2-3 cm L. Along streams, 5500-7300', Placer Co. n., Jun-Jul.

5) *C. pensylvanica,* Pennsylvania Bitter-cress: Stems erect or somewhat procumbent, 2-10 dm L; lvs 2-10 cm L; lfts of lower lvs ± ovate, 5-15 mm L; lfts of upper lvs oblanceolate to linear, 6-25 mm L; fruiting pedicels ascending 8-10 mm L; petals 2-5 mm L; frs erect, 2-2.5 cm L. Infrequent, in moist places, 3000-6800', Amador and Nevada cos., May-Jun.

6) *C. pachystigma,* Stout-beaked Toothwort: Rhizomal lvs simple, ovate, cordate at base, 4-5 cm L, often coarsely 5-9-toothed in upper half; stems glabrous, racemes short, dense; fruiting pedicels 5-25 mm L; petals usu pink, 7-10 mm L; fr 3-5 cm L. Wooded slopes, 5000-9500, SN(all), May-Jun.

Descurainia

Tansy-mustard

leaf (1 cm)　　　　*inflorescence* (1 cm)

Stems erect, branched especially above. Fls racemose, yellow, small. Fr linear-cylindric, the valves opening from below, 1-nerved. Herbage bitter, but may be eaten as greens; the seeds may be parched, ground and used as flour or cereal. *D. pinnata* is poisonous in large quantities, causing blindness and paralysis of the tongue.

Style minute; fr over 1 mm W, ± oblong　　　　*D. pinnata*
Style prominent; fr less than 1 mm W, linear
　　Fr 9-15 mm L　　　　*D. incana*
　　Fr 3-7 mm L　　　　*D. californica*

D. californica, Sierra Tansy-mustard: Stems 3-8 dm H; lvs 2-6 cm L, the lower with 2-4 pairs of lanceolate, entire to incised pinnae; sepals about 1 mm L, yellow or greenish; petals slightly longer; fr ± erect. Dry slopes, 7000-11,000', especially e. SN, Nevada and Mono cos. s., May-Aug.

D. incana:, Mountain Tansy-mustard: Stems 3-12 dm H, usu branched above; lvs 2-10 cm L, the lower occasionally bipinnate, the upper pinnate; pedicels divaricate; sepals 1-3 mm L, yellow; petals 1-4 mm L. Dry habitats, 4000-11,000', SN (all), Jun-Aug.

D. pinnata vars., Western Tansy-mustard: Pubescent annual, 1-6 dm H; often branched; lower lvs 3-9 cm L; upper lvs usu pinnate; pedicels ascending to wide-spreading, 8-15 mm L; sepals 1-2 mm L; petals to as long. Dry, sandy places below 8000', mly s. and e. SN, May-Jun.

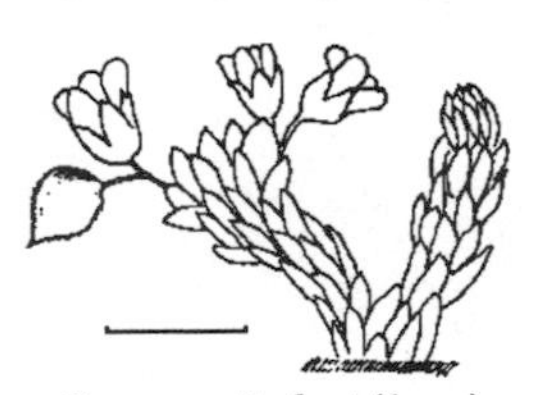

Draba

Draba

D. paysonii plant (1 cm)　　　　*fruit* (1 mm)

Annuals or perennials, mly low tufted, alpine plants. Fls racemose. Sepals equal at base. Plants mly inedible. The genus is somewhat difficult to key and the spp are very localized. The following is a key to the Sierran spp, but no descriptions of the individual spp will be presented.

A Fr elongate, the main body at least 3 times as long as wide, shorter frs sometimes present
B Petals white; fr definitely elongate; mly above 10,000'
C Stems leafless or stem lvs much reduced; Convict Lake Basin, Mono Co. *D. nivalis*
CC Stems with 1 to few lvs, these similar to basal; in scattered localities, Dana Plateau s. *D. praealta*
BB Petals yellow
C Lvs 5-12 mm L; petals 5-6 mm L, above 9000', Tulare Co. and Lake Tahoe region *D. cruciata*
CC Lvs mly 10-40 mm L; petals 1-4 mm L
D Annual; lvs ± ovate, 4-8 mm W; mly damp, shaded habitats, 7000-12,000', SN (all) *D. crassifolia*
AA Fr definitely less than 3 times as long as wide
B Lvs usu 1-2 cm L (low forms of *D. lemmonii* may have smaller lvs)
C Petals 2-3 mm L, white; annual; below 6000', Mariposa Co. N. *D. verna*
CC Petals 4-6 mm L, yellow; perennial; 8500-14,200'
D Stem lvs present, evident; Lassen Peak *D. aureola*
DD Stem leafless; rocky habitats; Eldorado Co. s. *D. lemmonii*
BB Lvs usu less than 12 mm L
C Petals white; above 8500'
D Basal lvs 2-4 mm W, fr ± elongate, stellate pubescent; SN (all) *D. breweri*
DD Basal lvs 1-2 mm W, fr ± oval, glabrous or with simple hairs: Tulare Co. *D. monoensis*
CC Petals yellow
D Lvs obovate, 2-7 mm W; fl-stems 3-8 cm H; 8000-10,200', Eldorado and Tuolumne cos. *D. asterophora*
DD Lvs ± linear, usu less than 2 mm W, often densely overlapping; plants often compact, low
E Lf surfaces ± glabrous, margins long-hairy; 8500-13,000', Fresno Co. n. *D. densifolia*
EE Lf surface hairy, often densely so; usu e. SN
F Stems 1-3 cm H; lvs 2-6 mm L; fr stellate-pubescent; 11,000-12,500', Inyo Co. *D. sierrae*
FF Stem and lvs often larger; fr with mly simple hairs
G Underside of lvs with comb-like appressed hairs; 8700-14,200', Inyo to Eldorado Co. *D. oligosperma*
GG Underside of lvs without such hairs; above 9000', Mono and Nevada cos. *D. paysonii*

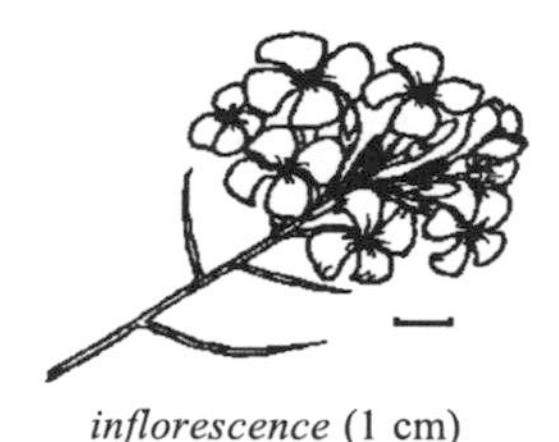

inflorescence (1 cm)

Erysimum

Wallflower

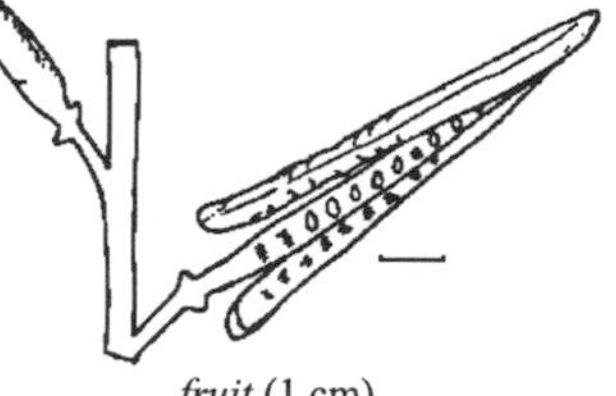

fruit (1 cm)

Annual to perennial; stems leafy, stout, with appressed 2-3-forked hairs. Lvs narrow. Fls in dense terminal racemes. Sepals erect, narrow, the 2 outer usu saccate at base. Related spp contain small amounts of toxic alkaloids.

Perennial; petals 15-20 mm L, yellow to orange *E. capitatum*
Annual; petals less than 10 mm L, light yellow *E. cheiranthoides*

E. capitatum, Western Wallflower: Stems to 10 dm H, usu unbranched, hairy; lower lvs usu dentate, 4-15 cm L, ± oblanceolate; upper lvs lance-linear, 2-5 cm L; pedicels 5-10 mm L in fr; fr 5-10 cm L. Common on dry slopes below 12,000', SN (all), May-Aug.
E. cheiranthoides, Wormseed Wallflower: Stem 1-5 dm H; lvs mly basal, 2-8 cm L, lanceolate to oblanceolate, shallowly dentate to entire; petals mly 3-6 mm L; fr 1-3 cm L, round in cross-section. Uncommon, disturbed sites; below 8000', SN (all). Jun-Aug.

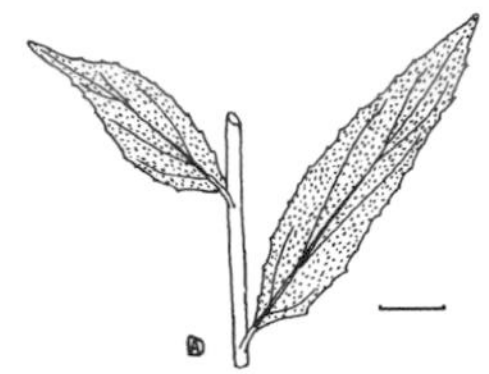

Hesperis

Dame's Rocket

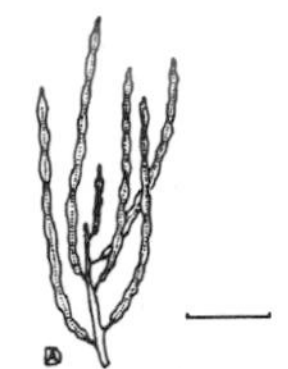

leaves (1 cm) *fruit* (3 cm)

H. matronalis: Stems simple to few-branched, to 10 dm H, with simple and stellate hairs; lvs dentate, pubescent, 5-20 cm L, the lower petioled; fls many, in racemes; sepals erect, hairy; petals white to purple, 2-2.5 cm L; fr 4-10 cm L, round in cross-section. Occasional in n. SN below 5000', Apr-Jun.

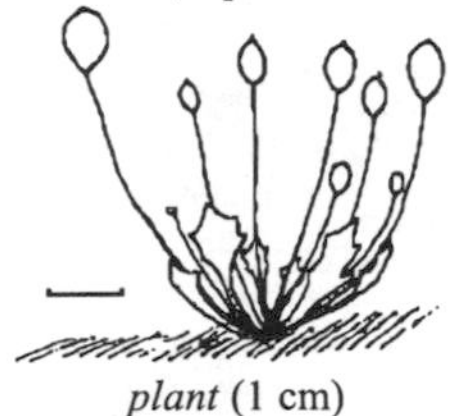

Idahoa

Flat-pod

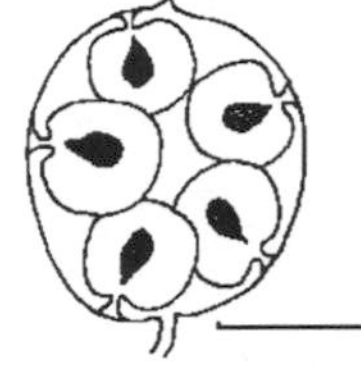

plant (1 cm) *fruit pod* (5 mm)

I. scapigera: Low glabrous annual with slender scape, 2-15 cm H; lvs in basal rosette, ovate, petioled, usu pinnatifid, 2-3 cm L; fls solitary, white, about 2 mm L; fr 6-10 mm in diameter. Moist places, 2000-6000', Sierra Co. n., Apr-May. Edibility unknown.

Lepidium

Pepper-grass

Fls minute in racemes, these elongating into unique cylindrical-connate clusters of fr. Young stems and lvs may be eaten raw or dried for future use; the seeds may be used as pepper to flavor meat, soups, etc. Plants high in vitamin C.

Plants mly glabrous; upper lvs perfoliate — *L. perfoliatum*
Plants mly pubescent; upper lvs not perfoliate
  Petals mly lacking; Yosemite Valley — *L. densiflorum*
  Petals present and as long as sepals; widespread — *L. virginicum*

inflorescence (1 cm)

L. densiflorum, Common Pepper-grass: Stems 3-5 dm H; lvs mly oblanceolate, the basal 4-8 cm L, coarsely toothed, the divisions also toothed, the cauline entire to somewhat toothed; racemes many, 6-15 cm L; fr about 2.5 mm L.

L. perfoliatum, Shield-cress: Stems erect, branched, 2-5 dm H; lower lvs bipinnatifid into linear segments, the middle stem lvs entire; pedicels 4-8 mm L; fr about 4 mm L, minutely notched. Occasional, below 7000', SN (all), Apr-Jun.

L. virginicum var. *pubescens,* Wild Pepper-grass: Stems 2-6 dm H, stout, minutely pubescent, usu branched above; lvs ± oblong, 1-3 cm L, 4-8 mm W, clasping at base, serrate or denticulate; fr slightly notched at tip, about 7 mm L. Usu disturbed areas, mly below 8000', SN (all), May-Aug.

Lesquerella

Western Bladder-pod

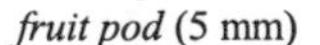

plant (1 cm) *fruit pod* (5 mm)

L. occidentalis: Perennial with stems to 2 dm L; herbage stellate pubescent; basal lvs elliptic, subentire, 2-7 cm L on petioles about half as long; stem lvs oblanceolate, reduced; sepals

5 mm L; petals 10 mm L; pedicels S-shaped in fr, 8-15 mm L; fr 5 mm L, slightly flattened at apex and on margins, with persistent style 5 mm L. Dry slopes below 8000'. Placer Co. n., May-Jul.

inflorescence (1 cm)

Rorippa

Yellow-cress

R. obtusa fruits (1 cm)

Stems often decumbent; lvs usu glabrous. Fls in short racemes. Herbage edible if water is not polluted.

A Plants perennial; stems decumbent; petals over 3 mm L
  B Plants terrestrial; pinnately lobed; fr 3-5 mm L; Lake Tahoe area — 5) *R. subumbellata*
  BB Plants aquatic; pinnately cmpd; fr-7-18 mm L; widespread — 2) *R. nasturtium-aquaticum*
AA Plants annual or biennial; petals 1-3 mm L
  B Stems erect, branched above; pedicels 5-6 mm L — 3) *R. palustris*
  BB Stems diffuse, branched from base; pedicels 1-4 mm L
    C Fr strongly curved; lf-segments linear to oblong, usu acute — 1) *R. curvipes*
    CC Fr not curved, lf-segments obovate to rounded or lf lobed — 4) *R. sphaerocarpa*

1) *R. curvipes* vars.: Stems 1-4 dm L; lower lf-blade 2-8 cm L, entire to deeply pinnately lobed, usu on shorter petioles; upper lvs reduced, clasping stem. Occasional in wet places, often with roots immersed, below 7500', SN (all).

2) *R. nasturtium-aquaticum,* Water-cress: Stems prostrate or ascending, often rooting at nodes, 1-6 dm L; lvs 1-10 cm L, with 3-11 ovate lfts, 5-25 mm L; pedicels 6-15 mm L; fls in dense terminal cluster. Common, below 8000', May-Oct.

3) *R. palustris*, Marsh Yellow-cress: Stems 3-8 dm H; lower lvs pinnatifid with large terminal lobe, 8-14 cm L; petioles winged; upper lvs subsessile, dentate or somewhat lobed; racemes long; fr 6-12 mm L. Occasional, wet places, often with roots immersed, below 6000', SN (all), Apr-Jul.

4) *R. sphaerocarpa*, Yellow-cress: Stems 1-3 dm L; lvs pinnatifid, the lower lvs less divided, with blades 2-5 cm L, petioled, the upper 1-3 cm L, subsessile; fr subglobose to oblong, 3-8 mm L, 2-3 mm W. Wet places, mly 5000-8000', SN (all), Jun-Sep.

5) *R. subumbellata,* Umbellate Yellow-cress: Stems 5-18 cm L, branched, hairy; lvs short-petioled to sessile, subpinnatifid, 1-3 cm L, 3-10 mm W; infl subumbellate to somewhat elongate; pedicels erect to divaricate; fr broadly oblong to subglobose, glabrous, 3-5 mm L. Moist places, 6000-7000', Jun-Jul.

lower stem (1 cm)

Smelowskia

Alpine Smelowskia

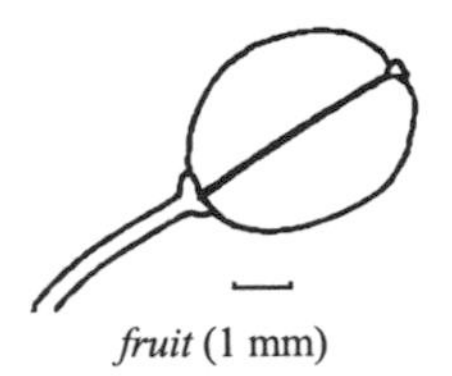

fruit (1 mm)

S. ovalis var. congesta: Low cespitose perennial; stems 5-15 cm L, densely white-woolly with simple hairs; lvs pinnatifid; the basal petioled, 2-6 cm L, the segments obovate, whitish tomentose; stem lvs few, smaller; pedicels 4-8 mm L; petals white or pinkish. Rare on dry, volcanic soils at about 10,000', Lassen Peak, Jul-Aug.

Streptanthus

Jewel Flower

Petals usu narrowed, with crisped or channeled blades. Stamens often in 3 pairs according to length. Frs linear, flattened parallel to partition.

upper stem (1 cm)

A Perennial; e. SN — 1) *S. cordatus*
AA Annual or biennial; w. SN
  B Middle stem lvs oblong to obovate; frs spreading to erect; to 11,000'
    C Frs arcuate-spreading; widespread — 6) *S. tortuosus*
    CC Frs erect; Kings-Kern Divide — 5) *S. gracilis*
  BB Middle stem lvs linear or pinnate; frs deflexed; below 5000'
    C Fls usu yellow, rarely with purplish tinge; lvs entire and linear or with linear lobes — 2) *S. diversifolius*
    CC Fls white or violet; lvs not with linear segments
      D Fls violet or purplish; Fresno and Tulare cos. — 4) *S. fenestratus*
      DD Fls whitish with purple veins; 3000-4000', Madera Co. s. — 3) *S. farnsworthianus*

1) *S. cordatus*, Perennial Jewel Flower: Stems mly unbranched, glabrous, stoutish, 3-8 dm H; basal lvs spatulate-obovate, variously dentate, 2.5-7 cm L, petioled; stem lvs broadly oblong, 2-6 cm L, often clasping, entire; pedicels ascending, 5-10 mm L, stout in fr; sepals greenish or purplish, 5-8 mm L; petals purple, white-margined, recurved, 10-14 mm L; fr ascending or spreading, 5-8 cm L. Dry slopes, 4000-10,000', May-Jul.

2) *S. diversifolius,* Varied-leaved Streptanthus: Stems glabrous, erect, 2-5 dm H, branched above; stem lvs 1-5 cm L; upper lvs and bracts entire, cordate-ovate, clasping; pedicels ascending, 2-5 mm L; sepals yellowish, 5-6 mm L with recurved tips; petals recurved, 8-10 mm L; fr flattened, 4-8 cm L. Dry, rocky slopes, Tulare to Butte Co., Apr-Jul.

3) *S. farnsworthianus,* Farnsworth's Jewel Flower: Basal lvs pinnatifid, middle stem lvs deeply pinnately lobed to subentire; pedicels to 5 mm L; fls 10-15 mm L; fr ascending, straight or curved, 7-9(-12) cm L. Rare.

4) *S. fenestratus,* Tehipite Jewel Flower: Much like *S. diversifolius;* lvs deeply divided, the segments usu oblong or broader, not linear-filiform.

5) *S. gracilis,* Alpine Jewel Flower: Stems slender, glabrous, often branched from near base, 1-3 dm H; lower lvs rounded to oblanceolate, sinuate-dentate to lobed, slender-petioled, 1-4 cm L; stem lvs entire to lobed, 0.5-1.3 cm L, mly sessile and clasping; pedicels erect or ascending, 3-6 mm L; sepals purple, spreading at tip, 4-5 mm L; petals pinkish, 7-8 mm L; fr 3-7 cm L. Dry slopes of disintegrated granite, 10,000-11,000', Jul-Aug.

6) *S. tortuosus,* Mountain Jewel Flower: Form variable; stems 2-10 dm H, occasionally branched above, glabrous; lower lvs spatulate-obovate, short-petioled, entire or toothed, 30-8 cm L; middle stem lvs toothed to subentire, clasping, 2-9 cm L; uppermost lvs in lower part of infl, deeply clasping; racemes mly 5-10 cm L; pedicels 3-10(-15) mm L, ascending; sepals usu purplish, 5-10 mm L, recurved at tips; petals purplish or yellowish white, usu with purple veins, 6-12 mm L; fr 6-12 cm L. Common, dry, rocky slopes below 8500', SN (all), May-Aug.

var. *flavescens:* Stems 1-2 dm H; usu bushy; sepals yellow, 4-6 mm L; petals yellow, 6-8 mm L. At 10,000-10,600', Farewell Gap, Tulare Co., Aug.

var. *orbiculatus:* Stems mly 1-2 dm H, bushy; sepals purple, 5-6 mm L; petals 7-9 mm L. Dry slopes, 7000-11,500', SN (all), Jun-Sep.

Subularia

Awlwort

S. aquatica: Small, stemless aquatic perennials, lvs subulate, in basal tufts; fls few, minute, on naked scapes; scapes 3-10 cm L; pedicels 2-5 mm L; sepals greenish; petals white. Rare, wet banks and shallow water, 7000-10,000', East Lake, Mono Pass, Dana Mdws, Donner Lake, Lake Tahoe, are known locations, Jul-Aug. Too rare to consider edibility.

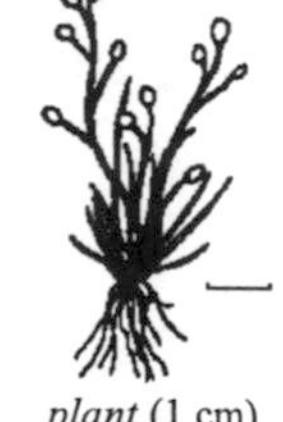
plant (1 cm)

basal leaves (1 cm)

Thysanocarpus

Fringe-pod

fruits (1 cm)

T. curvipes: Stems erect, slender, 2-5 dm H, ± pubescent, branched above; basal lvs in rosette, sinuate-dentate to subentire, 2-5(-10) cm L; stem lvs lanceolate, clasping, the upper entire; pedicels slender, recurved, 5-12 mm L; fls about 1 mm L, white to purplish, in slender racemes. Dry openings, 3000-5000', Tuolumne Co. s., May. Seeds may be parched and eaten or ground and mixed with flour.

CABOMBACEAE - Watershield Family

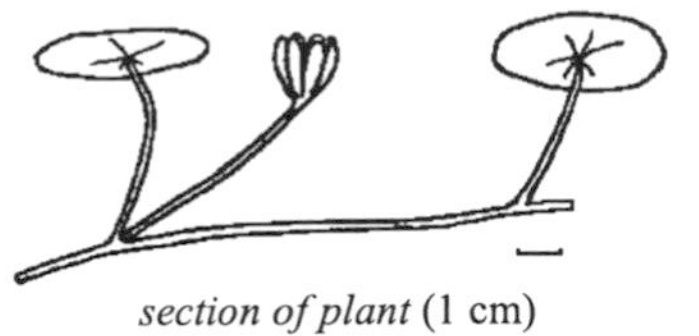

section of plant (1 cm)

Brasenia

Watershield

leaf (1 cm)

B. schreberi: Lvs entire, oval to elliptic, 5-15 cm L, green above, often purple below; fls small, purple; sepals and petals similar, mly 3, persistent, 10-15 mm L; fr leathery, indehiscent. Ponds and slow streams below 7000', SN (all), Jun-Aug. The starchy rootstocks may be peeled, boiled and eaten, or dried and stored or ground into flour. The unexpanded leaves and lf-stems may be eaten in salad, slime and all.

CALLITRICHACEAE - Water-starwort Family

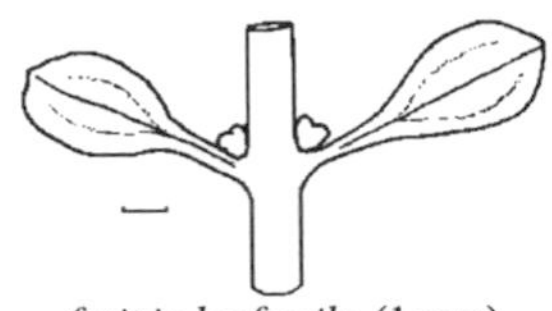

fruit in leaf axils (1 mm)

Callitriche

Water-starwort

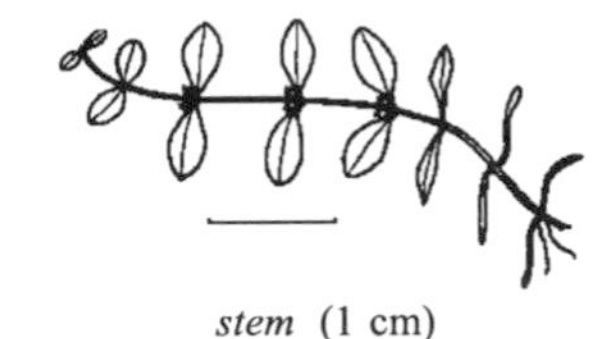

stem (1 cm)

C. verna: Slender perennial, 0.5-2.5 dm L; the lvs opp, variable, the lower submersed, often linear, 1-nerved, to 1 mm W, the upper often dilated, the terminal in a floating rosette, obovate, 4-10 mm L, on petioles about as long; fr nut-like, compressed, 4-lobed, separating into four 1-seeded parts. Shallow water or on mud, to 11,400', SN (all), May-Aug. Edibility unknown.

CALYCANTHACEAE - Calycanthus Family

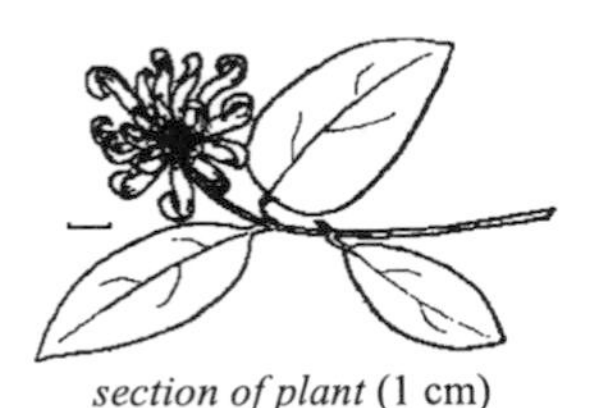

section of plant (1 cm)

Calycanthus

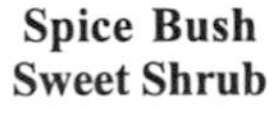

Spice Bush
Sweet Shrub

leaf (1 cm)

C. occidentalis: Erect usu rounded shrub, 1-3 m H; lvs opp, ovate to lance-oblong, 5-12 cm L, 2-7 cm W, pleasantly aromatic when bruised; fls red, solitary, terminal, short-peduncled; sepals and petals linear-spatulate, 2-6 cm L, rather fleshy. Along streams and moist places below 4000', Tulare Co. n., Apr-Aug. Berries and herbage suspected to be poisonous.

CAMPANULACEAE - Bellflower Family

Plants with milky or acrid sap. Lvs without stipules. Fls my perfect, usu 5-merous except carpels (usu 2). Ovary inferior. Little is reported on the edibility of members in the following genera.

A Corolla regular; stamen filaments not fused
  B Lvs ovate, dentate, 5-10 mm L — *Heterocodon*
  BB Lvs linear to lanceolate, usu over 10 mm L — *Campanula*
AA Corolla irregular; anthers and filament fused into a tube
  B Stems purplish; stem lvs reduced to bracts; fls minute — *Nemacladus*
  BB Stems green; stem lvs many; fls showy
    C Fls sessile, the elongate ovary often mistaken for a pedicel; below 5500' — *Downingia*
    CC Fls pedicelled; above 5000' — *Porterella*

Campanula

Bellflower
Harebell

C. prenanthoides inflorescence (1 cm) *C. prenanthoides leaves* (1 cm)

Perennials, lower lvs oblanceolate to ovate, the upper narrower. Fls showy; calyx mly 5-merous; corolla bell-shaped or nearly so. Similar sp has edible root.

Lvs sessile or nearly so; corolla bright blue *C. prenanthoides*
Lvs on long petioles; corolla pale blue to white *C. scouleri*

C. prenanthoides, California Harebell: Stems slender, erect, 2-8 dm H, often angled, sparsely stiff-pubescent; lvs 1.5-4 cm L, coarsely toothed, infl narrow; fls mly in clusters of 2-5; pedicels 2-6 mm L; corolla-lobes 2-3 times as long as tube. Dryish wooded places below 6000', Tulare Co. n., Jun-Sep.

C. scouleri, Scouler's Harebell: Stems slender, 1-3 dm H, glabrous or ± puberulent; lvs serrate, the blades 1-4 cm L, narrowed into a margined petiole at least half as long; infl mly racemose, the few fls single in upper axils; pedicels slender, 0.5-2 cm L; corolla-lobes almost as long as the tube. Wooded places below 5000', Sierra Co. n., Jun-Aug.

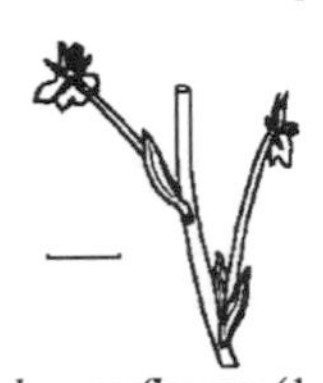

Downingia

Downingia

D. elegans flowers (1 cm) *C. prenanthoides leaves* (1 cm)

Low glabrous soft-stemmed annuals, rather succulent and tender. Lvs sessile, mly entire. Fls solitary, sessile in the axils of bracts, but appearing pedicelled because of the long inferior twisted ovaries. Sepals narrow, ± unequal. Corolla mly blue, blotched on lower lip, the upper 2 lobes smaller than the 3 fused lower.

Corolla densely white-hairy within *D. bicornuta*
Corolla ± glabrous within
  Stem slender, 0.3-1.5 dm L *D. montana*
  Stem stoutish, 0.5-4 dm H *D. bacigalupii*

D. bacigalupii, Bacigalupi's Downingia: Lvs 5-25 mm L, 0.5-4 mm W; infl 5-25 cm L, with bracts 8-25 mm L; sepals 4-10 mm L; capsule 2.5-4.5 cm L. Mud flats, vernal pools, etc., Sierra Co. n., Jun-Sep.

D. bicornuta: Stems 5-25 cm L; lvs 5-15 mm L; infl 4-10 cm L with bracts 5-20 mm L; fls 7-20 mm L, the lower lobes blue-purple, with a central white section containing yellow-green dots. Occasional, below 5500', SN (all).

D. montana, Sierra Downingia: Lvs 5-13 mm L, 0.3-1 mm W; infl 3-10 cm L, the bracts 8-16 mm L, often toothed; sepals 4-8 mm L, corolla 9-12 mm L, the lobes of the lower lip light blue or violet, the central part of lip white, the base dark blue-purple; 2 upper lobes erect, 4-6 mm L; capsule 1.5-3.5 cm L. Open grassy mdws, mly 3000-5500', Tuolumne Co. n., May-Aug.

Heterocodon

Heterocodon

H. rariflorum: Stems very slender, 5-30 cm L, scattered bristly-pubescent; lvs 5-8 mm L, roundish, dentate, cordate-clasping at base, fls axillary, the lower cleistogamous with inconspicuous corolla, the upper with blue corolla. Below 8000', SN (all), May-Jul.

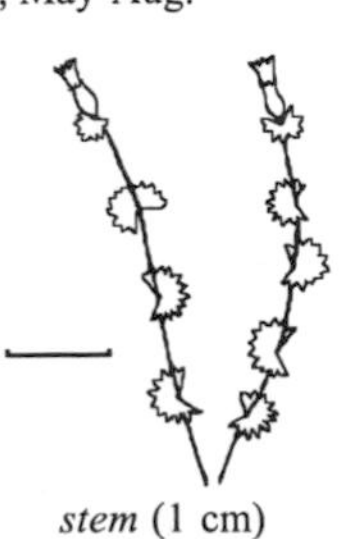

stem (1 cm)

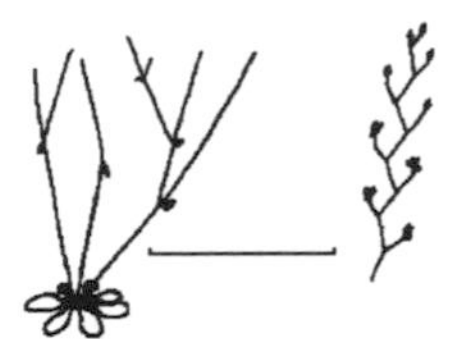

Nemacladus

Nemacladus

N. interior stem (1 dm) *N. interior flower* (1 mm)

Small annuals with capillary diffusely branched brownish or purplish stems. Basal lvs in a compact rosette, cauline mly reduced to bracts. Fls loosely racemose, borne on capillary pedicels from most axils. Fr a capsule.

Rosette lvs few; corolla about 2 mm L — *N. capillaris*
Rosette lvs usu conspicuous, usu 5 or more; corolla about 3 mm L — *N. interior*

N. capillaris, Common Nemacladus: Stems 0.5-1.6 dm H, usu shiny; rosette-lvs ovate, entire or nearly so, 0.3-1.5 cm L; pedicels mly straight, 8-12 mm L. Dry slopes and burns, below 4500', w. base of SN, May-Jul.

N. interior, Sierra Nemacladus: Plants 1.5-3 dm H; rosette-lvs about half as wide as long; corolla pale lilac or whitish with pink tinge. Dryish slopes and disturbed places below 5600', Butte Co. s., May-Jul.

Porterella

Porterella

upper stem (1 cm) *flower* (1 mm)

P. carnosula: Erect annual, 0.5-2 dm H, glabrous; stem lvs soft, sessile, 1-2.5 cm L; infl 3-10 cm L; pedicels 5-25 mm L; bracts 4-18 mm L; sepals 3-9 mm L; corolla 6-9 mm L, blue with yellow or whitish eye, strongly bilabiate, the upper 2 lobes erect. Wet places, 5000-10,000', Tulare to Lassen Co., Jun-Aug.

CAPRIFOLIACEAE - Honeysuckle Family

Lvs opp, mly without stipules. Fls regular or irregular, mly 5-merous, with inferior ovaries.

A Lvs pinnately cmpd — *Sambucus* p. 87
AA Lvs simple
  B Lvs 3-5 veined from base; below 4500' — *Viburnum* p. 89
  BB Lvs with one main vein from base
    C Fr dry; fls 2, on erect peduncles 4-8 cm H — *Linnaea* p. 85
    CC Fr fleshy; peduncles, if present less than 2.5 cm L
      D Berry white; plants often low subshrubs — *Symphoricarpos* p. 88
      DD Berry red or black; plants shrubs or vines — *Lonicera* p. 86

Linnaea

Twin Flower

L. borealis ssp *longiflora*: Stems trailing, slender, to 1 m L, finely pubescent; lvs elliptical, acute at apex, with scattered longish hairs mly on margins, slightly toothed in upper half, mly 1-2 cm L, the petioles 2-4 mm L; peduncles usu 4-8 cm L; pedicels 2 per peduncle, about 1 cm L; calyx-lobes 3-4 mm L; corolla 12-15 mm L, funnelform, the tube exceeding calyx. Dense woods below 8000', Plumas Co. n., Jun-Aug. Edibility unknown.

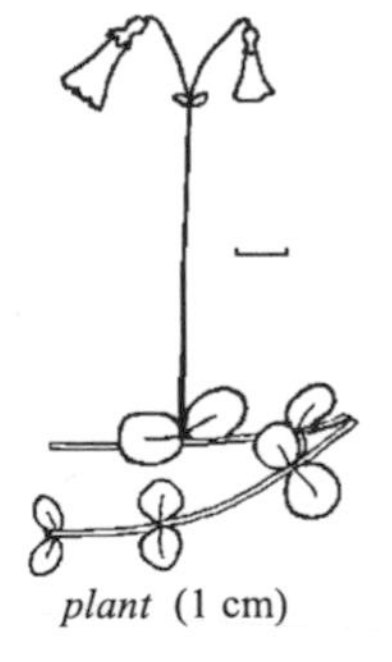

plant (1 cm)

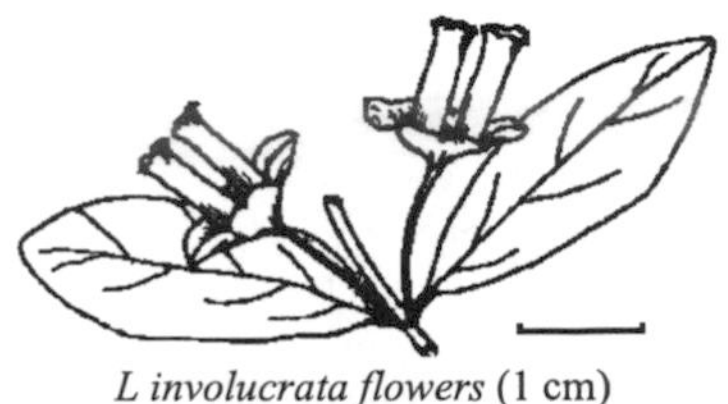

L involucrata flowers (1 cm)

Lonicera

Honeysuckle

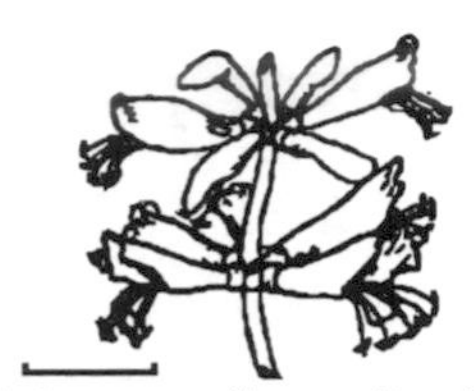

L. interrupta flowers (1 cm)

Erect or vine-like shrubs with entire lvs. Sepals 5 or obsolete. Corolla bell-shaped to tubular, ± gibbous at base. The sweet but seedy berries are edible raw and can be dried for future use.

A Stems usu leaning on other vegetation; fls in sessile whorls
  B Upper lf-pairs not fused around stem; Butte Co. — 7) *L. subspicata*
  BB Upper lf-pairs fused around stem
    C Fls mly in a single whorl; lvs 3-8 cm L — 2) *L. ciliosa*
    CC Fls in several whorls; lvs 1.5-3.5 cm L
      D Stipules green or scale-like; corolla glandular-hairy — 4) *L. hispidula*
      DD Stipules none; corolla glabrous — 5) *L. interrupta*
AA Stems erect; fls in pairs on peduncles
  B Peduncles 2-5 mm L; bractlets fleshy and enclosing fr — 1) *L. cauriana*
  BB Peduncles 12-25 mm L; bractlets not enclosing fr
    C Corolla lavender; fr bright red; bracts inconspicuous — 3) *L. conjugialis*
    CC Corolla yellowish; fr black; bracts about 1 cm L — 6) *L. involucrata*

1) *L. cauriana*, Mountain Fly Honeysuckle: Undershrub, 2-9 dm H, with shreddy light brown bark; lvs membranous, oval to obovate, 1.5-3.5 cm L; bracts at base of ovaries green, oblong-linear, 6-7 mm L; bractlets surrounding the ovaries of the 2 fls and forming a sac-like cup; corolla yellow, funnelform, 9-12 mm L, somewhat 2-lipped; fr blue-black, about 6 mm in diameter. Moist banks, 5000-10,500', Tulare to Nevada Co., May-Jul.

2) *L. ciliosa*, Orange Honeysuckle: Deciduous shrub, stems 1-5 m L; lvs ± oval, green above, glaucous beneath, ciliate on margins. Otherwise glabrous; petioles 3-5 mm L; corolla yellow to red-orange, 2-3 cm L, swollen on one side at base, slightly 2-lipped, the lobes about half as long as tube; fr red, 5-6 mm in diameter. Dry slopes, 2000-5000', Butte Co. n., May-Jun.

3) *L. conjugialis*, Double Honeysuckle: Deciduous, straggling shrub with slender stems, 6-15 dm H; lvs oblong-ovate, 2-6 cm L, acute, thin, light green, subglabrous to soft pubescent above; petioles 2-4 mm L; bracts minute; ovaries of the paired fls ± united; corolla 6-8 mm L, bilabiate; fr about 5-6 mm across. Wooded slopes, 4000-10,200', Tulare Co. n., Jun-Jul.

4) *L. hispidula* var. *vacillans*: Stems sprawling, to 6 m L; lvs 4-8 cm L, oblong to ovate; infl a long, interrupted spike. Canyons and woodlands below 7000', SN (all).

5) *L. interrupta*, Chaparral Honeysuckle: Evergreen bushy shrub with the branches twining or leaning on other vegetation; lvs round to elliptic, green above, glaucous beneath, the uppermost pair usu connate; spikes interrupted, 3-16 cm L, in an open panicle; corolla yellowish, 10-14 mm L; fr red, about 5 mm in diameter. Dry slopes below 6000', SN (all), May-Jul.

6) *L. involucrata* , Twinberry: Upright deciduous shrub, 6-30 dm H; lvs ± oval, acutish, 3-12 cm L, 2-5 cm W; petioles 5-12 mm L; peduncle rather coarse, with 2 bracts at its summit, these ovate to oblong, 1-1.5 cm L, often turning reddish or purplish; bractlets united, resembling the bracts; corolla 12-16 mm L, viscid pubescent; ovaries not united; fr about 8 mm in diameter, almost enclosed in the bractlets. Moist places, 6000-10,800', SN (all), Jun-Aug.

7) *L. subspicata*: Stems to 25 dm; lvs 1-4 cm, oblong to ovate; infl a long, interrupted spike, usu glandular-hairy; fls 10-12 mm L, yellow, 2-lipped; stamens and style exerted. Below 6000', in chaparral.

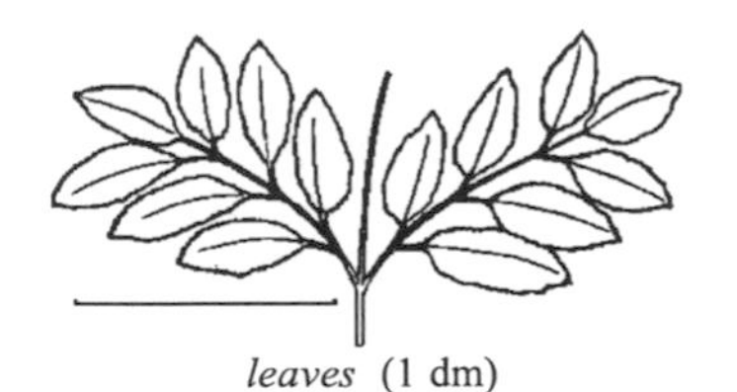

Sambucus

Elderberry

leaves (1 dm) *S. microbotrys fruits* (1 cm)

Lvs odd-pinnate with serrate lfts. Twigs with large spongy center. Fls small, mly whitish, in large clusters. The blue or black berries are edible fresh or in pies, jams or wine; they may also be dried for future use. Some people have experienced nausea from eating the raw fr; cooking the berries will prevent this side effect. The red berries produced by *S. microbotrys* are poisonous in many localities. The young shoots of young shrubs of a similar sp are used as an asparagus substitute; however the inner bark, lvs and roots of all spp are strong purgatives.

A Berry blue or white; infl flat-topped
  B Lfts usu 5-9, 3-15 cm L — 1) *S. caerulea*
  BB Lfts usu 3-5, 1.5-6 cm L; usu below 4500' — 3) *S. mexicana*
A Berry bright red or black; infl dome-shaped
  B Berry red; low shrub with ill-smelling herbage — 4) *S. microbotrys*
  BB Berry black; lfts usu 5; e. SN — 2) *S. melanocarpa*

1) *S. caerulea*, Blue Elderberry: Large shrub or small tree 2-8 m H; lfts usu asymmetrical at the base; fls 5-6 mm W; infl 0.5-2 dm across; berries nearly black but densely glaucous, thus appearing bluish, 5-6 mm in diameter. Open places, up to 10,000', SN (all), Jun-Sep.

2) *S. melanocarpa*, Black Elderberry: Shrub 1-2 m H; lfts 5-7, oval to lance-ovate, coarsely serrate, 4-12 cm L; infl 4-7 cm across; fr 4-5 mm in diameter. Occasional, in moist places, 6000-12,000', Jul-Aug.

3) *S. mexicana*, Desert Elderberry: Much like *S. caerulea*; lfts often falling during dry season; infl mly 0.3-1 dm across; berries either blue or white under the white bloom, often ± dry at maturity. Open flats, valley and canyons, SN (all), May-Sep.

4) *S. racemosa*, Red Elderberry: Stems 0.5-1 m H; lvs thin, glabrous or nearly so; lfts 5-7, oval to elliptic, 3-8 cm L, coarsely serrate; infl mly 3-6 cm across; fr 4-5 mm in diameter. Common in moist places, 6000-11,000', SN (all), Jun-Aug.

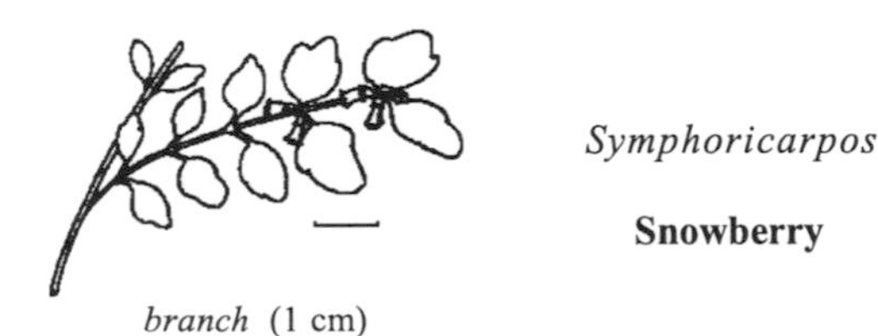

Symphoricarpos

Snowberry

branch (1 cm) *fruits* (1 cm)

Lvs short petioled, mly entire, deciduous. Fls pink, in small terminal or axillary cluster, sometimes solitary, 4-5 merous. Stamens inserted on the corolla. Fr white, with 2 nutlets. The berries are rather tasteless but are edible raw or cooked, although large quantities of the fruit may be harmful. The lvs contain a poisonous saponin.

Stems erect to 1.5 m H; corolla 7-9 mm L — *S. vaccinoides*
Stems spreading to procumbent; corolla mly shorter
  Berry 4-5 mm W; corolla bright pink to reddish; moist habitats — *S. acutus*
  Berry 6-8 mm W; corolla pink; dry habitats — *S. rotundifolius*

S. acutus, Creeping Snowberry: Stem procumbent or trailing; branches 4-8 dm L; lvs oval to ovate, 1-3 cm L, dark green and ± pubescent above, somewhat paler and densely pubescent beneath; fls 1 or 2 in upper axils; sepals less than 1 mm L; corolla 4-5 mm L. Wooded areas, 3500-8000', SN (all), Jun-Aug.

S. rotundifolius, Snowberry: Branches declined, 5-10 dm L, often rooting at tips; lvs glaucous, often lobed on young shoots, oval to narrow-elliptic, usu acutish, 1-2 cm L, grayish green, thickish, short-pubescent to subglabrous; fls 2 to several; sepals about 1 mm L; corolla 6-7 mm L. Rocky slopes, 4000-11,000', SN (all), Jun-Aug.

S. vaccinoides, Mountain Snowberry: Stems branched, with underground runners; lvs dark green above, pale beneath, puberulent, oval, acutish, 1-2 cm L, occasionally revolute; fls 1-2 in uppermost axils; sepals about 1 mm L; corolla pink; berry 6-7 mm W. Rocky slopes, 5000-10,500', Fresno Co. n., Jun-Aug.

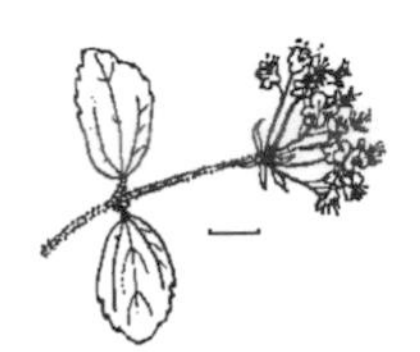

Viburnum

Western Viburnum

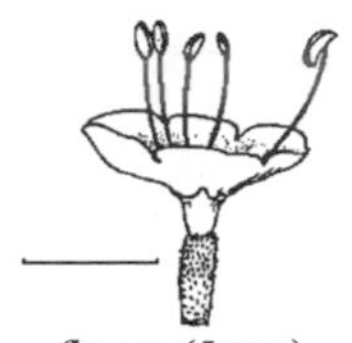

inflorescence (1 cm) *flower* (5 mm)

V. ellipticum : Slender-stemmed shrub, 1-4 m H; lvs elliptical to roundish, coarsely dentate except at base, 2-6 cm L, 3-5 veined from base, glabrous or somewhat pubescent above, paler and more pubescent beneath; petioles 6-12 mm L; fls 6-8 mm W in peduncled umbel-like clusters; fr fleshy, 10-23 mm L. Occasional below 4500', Fresno and Eldorado cos., May-Jun. Several similar spp have edible berries.

CARYOPHYLLACEAE - Pink Family

Annual or perennial herbs. Lvs mly opp, entire, simple. Infl often a dichotomously branching cyme. Fls regular, 4-5-merous. Ovary superior, styles 3-5. Few of the genera mentioned have much to offer in terms of edibility, nor are any particularly toxic.

A Petals none; sepals 1 mm L
  B Stipules absent; hypanthium very hard — *Scleranthus* p. 93
  BB Stipules present, 1 mm L, white; hypanthium herbaceous in fr — *Herniaria* p. 91
AA Petals usu present; sepals 2-15 mm L
  B Sepals fused into a cylinder; petals with long claws
    C Styles 3-5; common throughout SN — *Silene* p. 93
    CC Styles 2; introduced spp in disturbed habitats mly below 5000'
      D Infl subtended by 2 or more bracts
        E Bracts green; lf-base not sheathing — *Dianthus* p. 90
        EE Bracts reddish to brown; lf-base sheathing stem — *Petrorhagia* p. 92
      DD Infl not subtended by bracts
        E Calyx 3-5 mm L; lvs less than 5 mm W — *Gypsophila* p. 91
        EE Calyx 5-20 mm L; lvs over 10 mm W — *Vaccaria* p. 95
  BB Sepals distinct to base
    C Petals lobed or absent
      D Lvs 5-10 cm L — *Pseudostellaria* p. 92
      DD Lvs 1-4 cm L
        E Styles 5; fruit cylindric — *Cerastium* p. 90
        EE Styles 3; fruit ovoid — *Stellaria* p. 94
    CC Petals entire, rarely lacking
      D Stipules present; petals pink — *Spergularia* p. 94
      DD Stipules none; petals usu white
        E Styles 5, alt with sepals — *Sagina* p. 92
        EE Styles opp sepals and usu 3
          F Capsule splitting along 3 values — *Minuartia* p. 91
          FF Capsule splitting along 6 valves
            G Lvs lanceolate to oblanceolate, mly 4-8 mm W — *Moehringia* p. 92
            GG Lvs needle-like, 1-2 mm W, often spine-tipped — *Arenaria* p. 89

Arenaria

Sandwort

A. nuttallii upper stem (1 cm) *A. macrophylla flower* (5 mm)

Low branched perennials; commonly tufted or matted. Lvs sessile, needle-like, 0.5-2 mm W, often spine-tipped. Fls small, white, sometimes rose to cream, 5-merous. Styles 3. Difficult genus to key.

A Infl usu compact, few- to many-fld; pedicels less than 7 mm — 2) *A. congesta*
AA Infl usu open; pedicels mly longer
B Sepals blunt; plant glaucous — 1) *A. aculeata*
BB Sepals pointed; plant green
C Lvs 1-2 cm L; petals 4-7 mm L — 3) *A. kingii*
CC Lvs 2-6 cm L; petals 6-10 mm L — 4) *A. macradenia*

1) *A. aculeata*: Mat-forming, glaucous, glandular-hairy; lvs 1-3 cm L; sepals 3-4 mm L; petals 5-10 mm L. Rocky slopes and volcanic soils, Nevada Co. n., May-Jul.

2) *A. congesta*, Capitate Sandwort: Stems tufted, 10-40 cm L; herbage green; lvs 1-8 cm L; sepals 3-6 mm L; petals 5-8 mm L. Dry ridges and slopes, SN (all).

3) *A. kingii* var. *glabrescens*, King's Sandwort: Stems slender, 1-2 dm H, from woody base; lvs mly basal, linear, ± fleshy, erect or ascending, 1-2 cm L; cymes open, few- to several-fld; pedicels mly 1-2 cm L; sepals 3-5 mm L; petals occasionally slightly notched at apex, 4-7 mm L. Frequent on dry rocky slopes, 6000-11,000', Inyo to Nevada Co., Jun-Aug.

4) *A. macradenia*, Desert Sandwort: Plants tufted, stems 40-60 cm; lvs blunt to sharp-pointed. Highly variable in numerous habitats, mly c. and s. SN below 8000'.

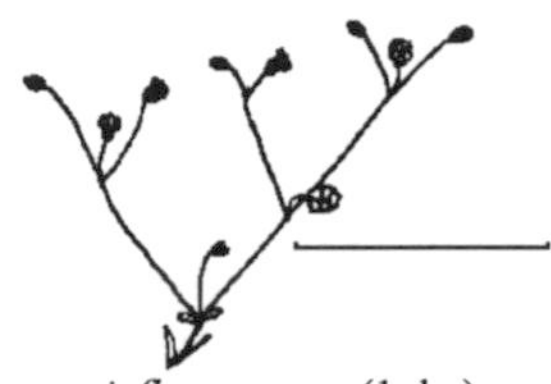

inflorescence (1 dm)

Cerastium

Mouse-ear Chickweed

flower (5 mm)

Pubescent perennials. Fls in terminal dichotomous cymes, white. Rarely 4-merous. Similar sp edible as a pot herb when collected in early season.

Bracts of infl green or only upper with narrow margin; above 8500' — *C. beeringianum*
Bracts of infl with broad membranous margins, 5000-8000'
Petals 2-3 times as long as sepals — *C. arvense*
Petals about as long as sepals — *C. fontanum*

C. arvense, Meadow Cerastium: Basal branches prostrate or creeping; the fl-stems tufted, erect or ascending, usu glandular, 1-3 dm L; main lvs lance-oblong, 1.5-3.5 cm L, 8-12 mm W, hairy; fls loosely clustered; sepals 5-7 mm L; petals broadly lobed. Moist, rocky or grassy banks, SN (all), May-Aug.

C. beeringianum, Alpine Cerastium: Plant matted, the stems spreading, glandular, 4-10 cm L; lvs mly 2-7 pairs, oblong, 7-20 mm L; infl 1- to few-fld; pedicels 5-20 mm L; sepals 4-8 mm L; petals 4-8 mm L, bluntly 2-lobed. Occasional, near snowbanks, 8500-12,200', Tuolumne, Mono and Alpine cos., Jul-Aug.

C. vulgatum, Common Mouse-ear Chickweed: Basal branches matted; fl-stems ascending or decumbent, 1-4 dm L, glandular-pubescent; lvs ± oblong, 1-2.5 cm L; fls loosely clustered; sepals 4-6 cm L. Occasional, in mdws, SN (all), May-Aug.

D. armeria inflorescence (1 cm)

Dianthus

Pink

D. deltoides flower (1 cm)

Annual or perennial. Basal lf-blade oblanceolate, the cauline ± linear. Infl terminal cymes, ± open. Fls pink. Introduced genus.

Fls few-several; pedicels less than 3 mm L; c. SN — *D. armeria*
Fls 1-few on pedicels over 5 mm L — *D deltoides*

D. armeria ssp *armeria*, Grass Pink: Annual or biennial with stems 15-60 cm H; bracts linear to lanceolate; calyx 15-20 mm L, hairy; petals 4-5 mm L. Disturbed habitats below 4500'.

D. deltoides ssp *deltoides*, Meadow Pink: Perennial with stems 20-40 cm H; bracts ovate; calyx 12-15 mm L, ± glabrous; petals 5-10 mm L. Wet, disturbed habitats below 8000', SN (all).

plant (5 cm)

Gypsophila

Baby's-breath

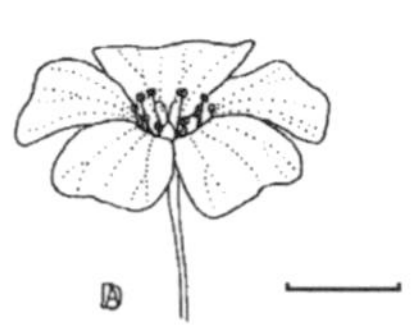

flower (1 cm)

G. elegans var *elegans*: Erect annual or biennial to 50 cm H; lvs linear-lanceolate; infl terminal, glabrous; fls few-many; calyx 3-5 mm L; petals 5-20 mm L, white to pink with purple veins; styles 2. Open forest about 6500'.

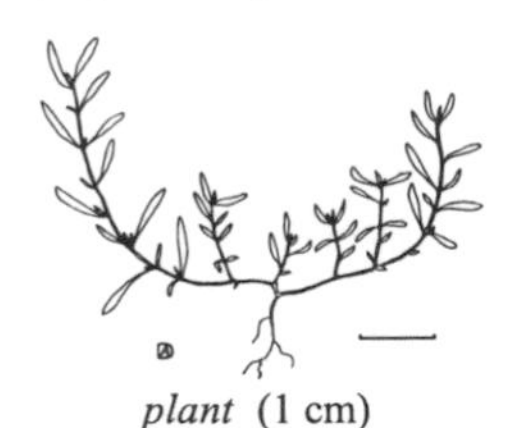

plant (1 cm)

Herniaria

Herniaria

inflorescence (5 mm)

H. hirsuta: Annual to 20 cm H; lvs opp below, alt above, oblanceolate; stipules white, ovate, 1 mm L; fls 3-8, lacking petals, ± sessile; sepals 5, 1 mm L; fr an utricle. C. SN below 6000'.

Minuartia

Sandwort

Annual to perennial. Lvs awl-shaped to thread-like or oblong. Fls 5-merous, petals occasionally absent; styles 3. Difficult genus to key to species.

M. rubella plant (1 cm)

A Plants annual; mly below 7000'
  B Lvs thread-like, 5-15 mm L; plant 5-30 cm H — 1) *M. douglasii*
  BB Lvs lanceolate, 2-5 mm L; plant 2-5 cm H — 4) *M. pusilla*
AA Plants perennials; mly above 10,000
  B Stem glabrous; lvs linear, fleshy; petals none — 6) *M. stricta*
  BB Stem usu glandular-pubescent; petals present
    C Sepal tip rounded with margins incurved and hood-like — 3) *M. obtusiloba*
    CC Sepal tip pointed, margins flat
      D Plants densely tufted — 5) *M. rubella*
      DD Plants mat-forming — 2) *M. nuttallii*

1) *M. douglasii*, Douglas' Sandwort: Stems slender, freely branched, erect, 0.5-2 dm H, subglabrous to sparsely glandular-pubescent; pedicels filiform, 6-25 mm L; sepals ovate, 2-3 mm L; petals conspicuous, 3-6 mm L. Common in dry, barren places below 7000', SN (all), May-Jun.

2) *M. nuttallii* sspp, Nuttall's Sandwort: Stems loosely matted, prostrate, 5-15 cm L, densely leafy; lvs erect to slightly recurved, black-tipped, 5-8 mm L; fls rather few in open cymes; pedicels 4-10 mm L; sepals 4-6 mm L; petals shorter. Dry granitic gravel, mly 10,000-12,000', Eldorado Co. s., Jul-Aug.

3) *M. obtusiloba*, Alpine Sandwort: Stems slender, decumbent, densely clustered, from a woody base, leafy below; lvs awl-shaped, 4-8 mm L; fls 1-2; sepals 4 mm L; petals 6-7 mm L. Dry, rocky places 10,500-12,500', Fresno and Inyo cos., Jul-Aug.

4) *M. pusilla*, Dwarf Sandwort: Stems slender, glabrous, usu few-branched; stem lvs in 1-3 pairs; pedicels capillary, 2-10 mm L; sepals 2-3 mm L; petals shorter or none. Occasional, dry woods and open slopes below 6800', Tuolumne Co. n., May-Jul.

5) *M. rubella*, Red Sandwort: Stems 2-5 cm H, tufted; lvs crowded at base of stems, 3-10 mm L; fls 2 to few; sepals 2.5-3 mm L; petals somewhat shorter. Dry, rocky and gravelly places, 10,000-12,000', s. SN, Jul-Aug.

6) *M. stricta*: Stems densely tufted, 1-5 cm H; lvs 5-7 mm L, sometimes pubescent; fls solitary to 2 or 3; pedicels 5-10 mm L; sepals 2-2.5 mm L, weakly 3-nerved; petals scarcely as long. Local on high plateaus, 12,000-13,200', Mono Co. s., Aug.

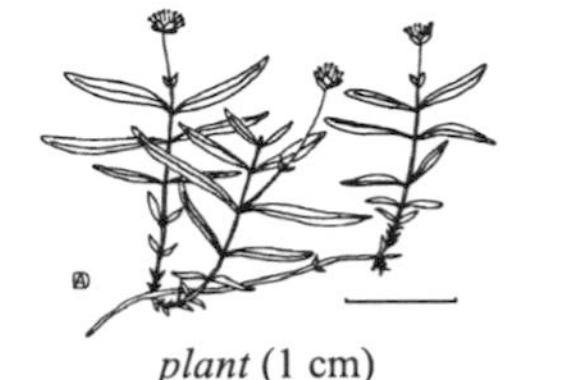

plant (1 cm)

Moehringia

Large-leaved Sandwort

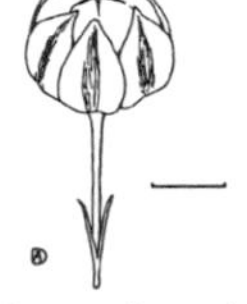

flower (2 mm)

M. macrophylla: Stems slender, ascending to suberect, mly 0.5-1.5 dm H; herbage pubescent throughout; lvs bright green, 1.5-6 cm L, ± sessile; fls 1-5 in short cymes; pedicels capillary, mly 5-20 mm L; sepals about 3-4 mm L; petals 5-8 mm L. Occasional on shaded slopes below 8800', SN (all), Apr-Jun.

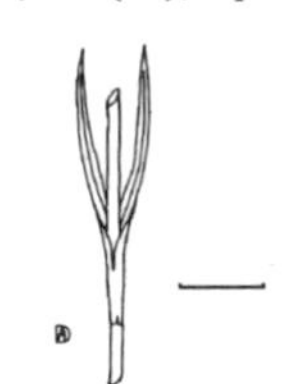

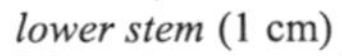

lower stem (1 cm)

Petrorhagia

Petrorhagia

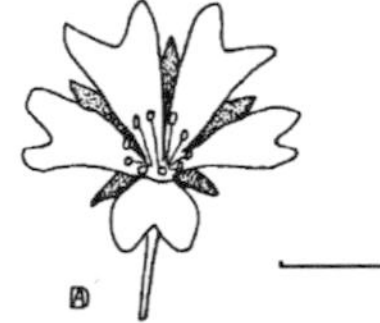

flower (1 cm)

P. dubia: Erect annual to 40 cm H; lvs ± linear, 1-6 cm L, sheathing stem at base, 3-veined from base; infl head-like; sepals fused into a cylindrical tube 8-12 mm L; petals 10-15 mm L, clawed, 2-lobed, pink. Disturbed habitats below 5000', Mariposa Co. n.

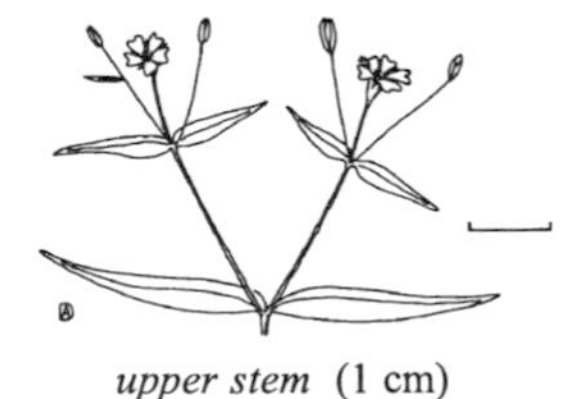

upper stem (1 cm)

Pseudostellaria

Sticky Starwort

flower (5 mm)

P. jamesiana: Diffusely branched perennial to 3 dm H; herbage glandular-pubescent; lvs narrow-lanceolate, 5-10 cm L, to 10 mm W, sessile; cymes terminal and axillary; pedicels 5-20 mm L; bracts green; sepals 3-5 mm L; petals white, 6-10 mm L and cleft. About mdws and damp places, 4000-8500', SN (all), May-Jul.

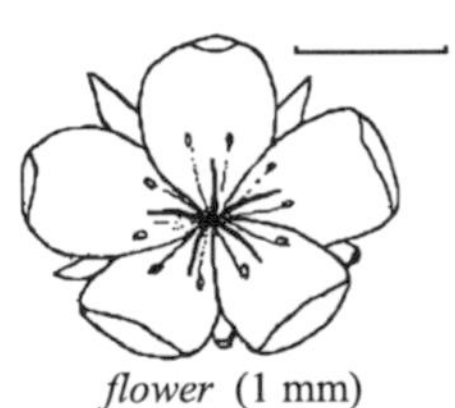

flower (1 mm)

Sagina

Pearlwort

S. saginoides plant (1 cm)

Fls whitish, small, terminal on stems or branches, 5-merous. Stamens usu 10, Seeds minute.

Plants annual, minute, without sterile basal rosette	*S. apetala*
Plants perennial, low, matted; stems many with basal rosette	*S. saginoides*

S. apetala, Western Pearlwort: Plant glabrous except for the calyx and pedicels, 2-8 cm H; lvs filiform, 6-10 mm L; pedicels 6-12 mm L; sepals ovate, blunt, 2 mm L; petals almost as long.

Uncommon or perhaps overlooked, in wooded and brushy or grassy places below 8000', SN (all), May-Jun.

S. saginoides, Pearlwort: Glabrous, matted or tufted, stems 2-9 cm L; lvs thickish, linear, 5-10 mm L; pedicels 5-12 mm L, often curved at tip; sepals ovate, blunt, 2 mm L; petals about 1 mm L. Common on moist banks, 4000-12,000', SN (all), May-Sep.

Scleranthus

Knawel

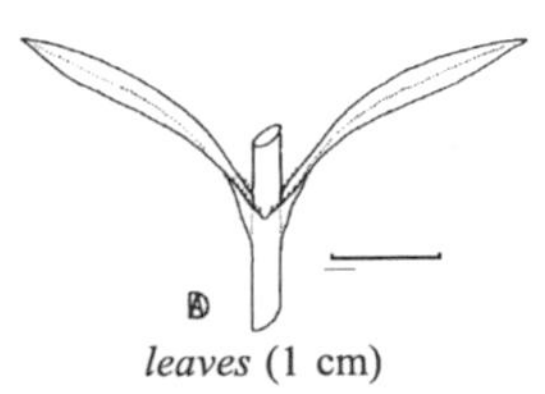
leaves (1 cm)

S. annuus: Much-branched annual with rigid, prostrate to erect stems 5-15 cm L; lvs needle-like, 5-10 mm L; sharp-pointed; fls 2-5, axillary; hypanthium 3-4 mm W, very hard; petals none; sepals 5, about 1 mm L. Disturbed habitats below 4000', Tuolumne Co. n. Introduced species.

S. californica flower (1 cm)

Silene

Catchfly Campion

S. douglasii flower (1 cm)

Perennial, usu somewhat viscid. Calyx cylindric to ovoid or bell-shaped, 5-toothed, 10- to many-nerved. Petals usu with scale-like appendages at base of blade. Stamens 10. Ovary on a well developed stipe. Shoots of several similar spp edible after cooking.

A Petals 2-lobed, or entire
  B Lvs mly over 5 mm W; fls nodding or stems decumbent
    C Fls nodding at anthesis; stems erect — 2) *S. bridgesii*
    CC Fls erect; stems usu decumbent — 7) *S. menziesii*
  BB Lvs mly less than 5 mm W; fls and stems erect
    C Basal lvs 5-9 cm L; Tulare and Fresno cos. — 12) *S. verecunda*
    CC Basal lvs 2-5 cm L; Alpine Co. n.
      D Petal blades ± entire; appendages lacking — 5) *S. invisa*
      DD Petal blades 2-lobed; appendages 2 — 11) *S. suksdorfii*
AA Petals 4-lobed; the lobes sometimes shallow or unequal
  B Calyx cleft half its length; Tulare Co. — 1) *S. aperta*
  BB Calyx shallowly cleft; widespread
    C Petals over 5 mm L; all petal-lobes similar
      D Corolla bright red; stem lvs usu many — 3) *S. californica*
      DD Corolla flesh to rose; stem lvs 4-10
        E Petals deeply divided into linear lobes — 6) *S. lemmonii*
        EE Petals cleft less than halfway into lanceolate lobes — 9) *S. occidentalis*
    CC Petals mly less than 5 mm L; outer petal-lobes often reduced
      D Lvs linear, 1-2 mm W; outer lobes much reduced — 10) *S. sargentii*
      DD Lvs oblanceolate, 2-6 mm W; lobes usu subequal — 8) *S. montana*

1) *S. aperta*, Naked Campion: Stems slender, 1.5-3.5 dm H, unbranched; lvs 3-10 mm L, 1-3.5 mm W; pedicels 5-30 mm L; calyx 10-nerved, 8-10 mm L, the lobes lanceolate; petals 1-2 cm L, shallowly to unequally 4-lobed, without appendages. Dry benches in pines, 8000-9000', Jul-Aug.

2) *S. bridgesii*, Bridge's Campion: Stems several, 3-8 dm H, puberulent below, glandular-pubescent above; stem lvs ± elliptic; infl long, open, the pedicels slender, 5-10 mm L; calyx 8-15 mm L, viscid-puberulent, the lobes 2-3 mm L; petals white or tinged purple, the blades 7-12 mm L; appendages 1-2.5 mm L. Dry, open slopes, 2600-8000', Tulare Co. n., Jun-Jul.

3) *S. californica*, California Indian Pink: Stems several, leafy, suberect to decumbent, 1.5-4 dm L; lvs ovate to oblanceolate, 3-8 cm L, 1-3 cm W; fls solitary to few in cluster; pedicels mly 2-3 cm L; calyx 1.5-2.5 cm L, the lobes 2.5-5 mm L; petals much exserted, 2-3 cm L, the blades usu 4-lobed with appendages 1-2 mm L. Open, habitats below 5000', SN (all), Apr-Aug.

4) *S. douglasii*, Douglas' Campion: Stems several, 1-4 dm H; lvs numerous in lower part, mly oblanceolate, 2-7 mm W, sometimes linear (var. *monantha*); infl 1-6-fld, not glandular; pedicels 5-30 mm L; calyx 12-15 mm L, the lobes 1-3 mm L; petals usu creamy-white; appendages linear to oblong, 1 mm L. Dry flats or slopes, 5000-9500', Jun-Aug.

5) *S. invisa*, Hidden-petal Campion: Stems few, 1-4 dm H; lvs mly basal, oblanceolate to spatulate, 2-4 cm L, 2-6 mm W, the cauline in 1-2 pairs; fls few; pedicels 5-25 mm L; calyx 10-nerved, about 9 mm L; petals pale lavender, equal to or slightly exceeding the calyx. Open forests, 7000-8600, Alpine Co. to Lassen Peak, Jul.

6) *S. lemmonii*, Lemmon's Campion: Stems slender, erect or decumbent, 1.5-4.5 dm L; lvs lance-elliptic to oblanceolate, 2-3 cm L, 5-10 mm W, the cauline mly in 2 pairs; infl open; pedicels slender, 1-2 cm L; calyx 8-10 mm L, with prominent green ribs; petals yellowish white to pinkish, the blades 4-8 mm L, appendages 1 mm L. Open woods, 3500-8000', SN (all), Jun-Aug.

7) *S. menziesii* ssp *dorii*, Menzie's Campion: Stems 0.5-2 dm L; lvs many, 2-6 cm L, 3-20 mm W, lanceolate to elliptic or oblanceolate; infl leafy; calyx 5-8 mm L; petals white, the blades 2-3 mm L; appendages very short. Moist woods and near small streams, 6000-10,800', c. SN, Jun-Jul.

8) *S. montana*, Mountain Campion: Stems many, ± woody at base, 1.5-4.5 dm H; lvs 2-6 cm L, erect, the cauline in 2-4 pairs; infl mly several-fld, erect; petioles 1-2 cm L; calyx 13-16 mm L, the ribs prominent, greenish or purplish; petal-blades 4-6 mm L, whitish, cleft into mly 4 linear subequal lobes; the appendages 1-3 mm L. Widespread, rocky habitats, 4500-10,400', Tulare Co. n., Jun-Aug.

9) *S. occidentalis*, Western Campion: Stems few, stout, 3-6 dm H; lower lvs 5-8 cm L, the cauline smaller, mly in 3 pairs; infl rather open, many-fld; pedicels short; calyx 15-25 mm L, densely glandular-pubescent; petal-blades 7-12 mm L, almost equally 4-lobed; appendages 2-4 mm L, linear. Dry, open places, 4000-7000', Alpine Co. n., Jun-Jul.

10) *S. sargentii*, Sargent's Campion: Stems mly 1-1.5 dm H; basal lvs linear-oblanceolate, 1.5-2.5 cm L, the cauline usu 1-2 pairs; fls 1 to few; calyx with prominent, somewhat purplish ribs; petals white to rose-purple, the blades 2-4 mm L; appendages oval, 1-2 mm L. About rocks and talus, 6500-12,000', Plumas Co. s., Jul-Aug.

11) *S. suksdorfii*, Suksdorf's Campion. Much like *S. sargentii* except that petals lack lateral teeth. A species of the Cascade Range, reaching its southern limit on the volcanic peaks in the Lassen area. Open, rocky habitats, Jul-Aug.

12) *S. verecunda* ssp *platyota*, Cuyamaca Campion: Stems several, erect, 3-7 dm H; basal lvs linear, 2-6 mm W; calyx glandular-pubescent, usu greenish; petals white to greenish or rose, the appendages 1-2 mm L. Dry forest slopes, 5000-11,000', Jun-Aug.

stem and leaves (1 cm)

Spergularia

Purple Sand-spurrey

inflorescence (1 cm)

S. rubra : Stems ± matted, slender, 6-25 cm L, glabrous except in infl; lvs linear, 6-12 mm L, densely clustered and apparently whorled; stipules 4-5 mm L, silvery; infl many-fld, leafy; sepals 4-5 mm L; petals pink, 3-4 mm L. Waste places and roadsides below 7000', SN (all), Jun-Sep. Edibility unknown.

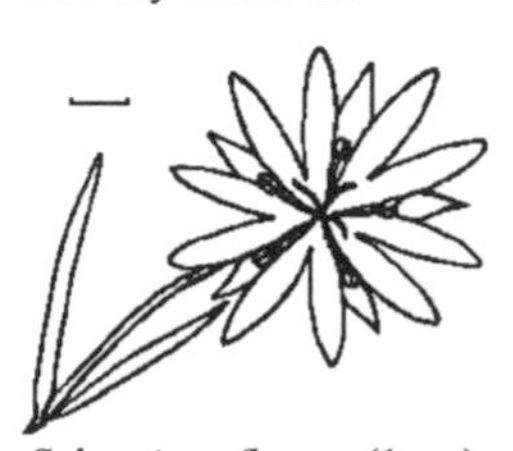

S. longipes flower (1 cm)

Stellaria

Chickweed
Starwort

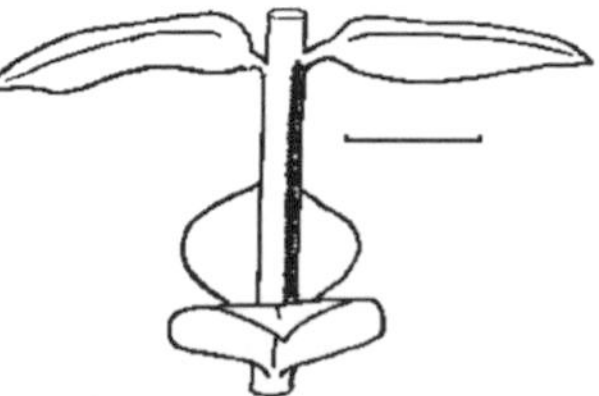

S. media stem with line of hairs (1 cm)

Low diffuse herbs. Fls white, styles mly 3, opp as many sepals. The young or upper stems of *S. media* are edible either raw as salad greens or boiled. The young growing tips are best since the plant becomes stringy with maturity.

A Petals equal to or longer than sepals — 4) *S. longipes*
AA Petals much shorter than sepals or none
  B Lvs lanceolate, much longer than wide; petals usu present
    C Mature sepals 2-3 mm L; fruiting pedicels ascending — 2) *S. calycantha*
    CC Mature sepals 4-5 mm L; fruiting pedicels reflexed — 1) *S. borealis*
  BB Lvs ovate to oblong; petals none

C Annual; internodes with longitudinal line of hairs — 5) *S. media*
CC Perennial; stems glabrous
D Sepals usu 4; lf-margins ± flat — 6) *S. obtusa*
DD Sepals usu 5
E Lf-margins crisped; infl with leafy bracts — 3) *S. crispa*
EE Lf-margins plane; infl with small scarious bracts — 7) *S. umbellata*

1) *S. borealis*, Northern Starwort: Stems 4-angled, usu unbranched, 2-4 dm L, somewhat coarse; lvs 2.5-4 cm L, the upper not much reduced; fls few, terminal or axillary, on recurved pedicels; cymes leafy-bracted; sepals sharply acute; petals minute or none. Wet places below 6000', Mariposa Co. n., Jun-Jul.

2) *S. calycantha* ssp *interior*, Interior Starwort: Stems strongly branched, 1-3 dm L; lvs somewhat serrulate, about 1 cm L; fls in terminal cymes with leafy bracts; pedicels ascending; petals minute or none. Infrequent, moist places, 4500-6000', Tuolumne Co. n., Jun-Aug.

3) *S. crispa*, Chamisso's Starwort: Stems glabrous, very slender, ± unbranched, 1-4 dm L; lvs sessile or nearly so, 0.8-2 cm L; pedicels axillary, 6-10 mm L, sometimes deflexed in fr; sepals lanceolate, acute, scarious-margined. Moist banks and mdws below 11,000', SN (all), May-Aug.

4) *S. longipes*, Long-stalked Starwort: Stems ± tufted, glabrous, unbranched, erect or ascending, 1-2.5 dm H; lvs rigid, lanceolate, about 1.5 cm L, erect or ascending, mly green; fls solitary or few, terminal on slender suberect pedicels; pedicels to 4 cm L; bracts scarious, small; sepals 3-5 mm L; petals about as long as sepals (or longer). Common in moist places, 4500-11,000', SN (all), May-Aug.

5) *S. media*, Common Chickweed: Stems procumbent or ascending, 1-4 dm L; lvs short-petioled or the upper sessile; cymes leafy; sepals 4-5 mm L; petals slightly shorter, deeply cleft, or lacking. Common weed in shady places, mly below 8000', SN (all), May-Sep.

6) *S. obtusa*, Obtuse Stellaria: Stems usu prostrate, glabrous, 4-20 cm L; lf-blade 5-10 mm L, ± ovate; fls solitary in axils; pedicels ascending, reflexed in fr; sepals 2-4 mm L, ovate, glabrous. Uncommon in moist habitats, below 7000', Mariposa Co. n.

7) *S. umbellata*, Umbellate Chickweed: Stems slender, branched, 1-2 dm L; lvs acute, 1-2 cm L, 4-6 mm W, thin; fls in upper axils and in terminal umbel-like cymes with scarious bracts, 2-5 mm L; pedicels filiform, recurved at tip; sepals 1.5-2 mm L, acute; petals minute or none. Infrequent in damp, shaded places, 6000-11,500', SN (all), Jul-Aug.

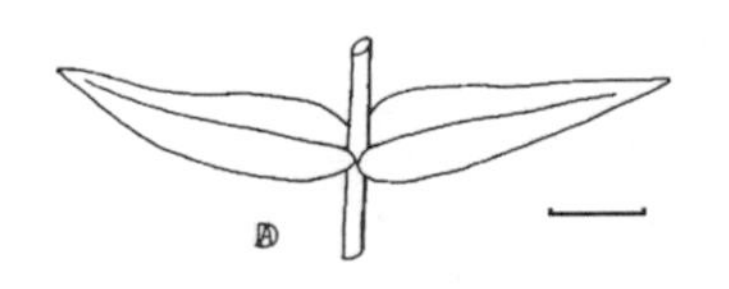

leaves (1 cm)

Vaccaria

Cow-herb
Cockle

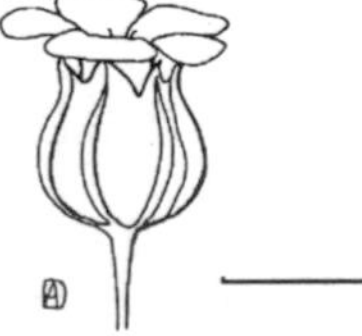

flower (1 cm)

V. hispanica: Glabrous annual to 1 m H; lvs 2-12 cm L, lanceolate to ovate, 3-7 veins from base; infl a terminal, many-fld cyme; bracts leafy; pedicels 5-40 mm; sepals fused into a tube, 5-20 mm L; petals 5, 15-25 mm L, clawed, pink to red. Below 7500', Tuolumne Co. n.

CELASTRACEAE - Staff-tree Family

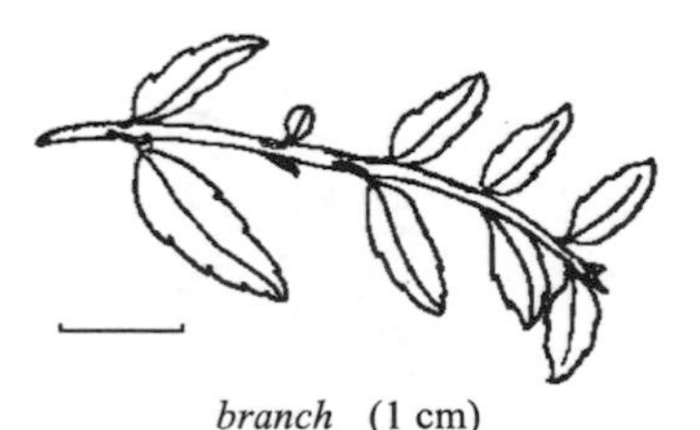

branch (1 cm)

Paxistima

Mountain Lover
Oregon-boxwood

flower (1 mm)

P. myrsinites: Low evergreen, very leafy, densely branched shrub, 3-10 dm H, sometimes spreading and almost prostrate, with corky squarish branchlets; lvs opp, 1-2.5 cm L, oval to oblong, serrulate on distal two-thirds, dark glassy green above, paler beneath, on short petioles; lf-margins thickened or slightly revolute; fls 1-3 in lf-axils; petals reddish brown. Scattered in ± shady places, 2000-6000', Mariposa Co. n., May-Jul. Plant inedible.

CHENOPODIACEAE - Goosefoot Family

Annual or perennial herbs usu with a "mealy" or scurfy surface and occasionally succulent. Lvs alt, entire to lobed, without stipules, sometimes reduced to scales. Fls small, greenish, usu in small cymose glomerules. Stamens as many as sepals or fewer. Sepals 5 or fewer. Styles 2-3. Species in this family are not likely to catch ones attention very often in the Sierra because they are mly inconspicuous plants with small fls and are found in habitats (alkali flats, roadsides, waste areas) not commonly frequented by hikers and campers. In addition, the small fls and relatively few conspicuous and unique morphological features of the species make identification of certain taxa difficult and frustrating. For these reasons, detailed descriptions of individual species will not be given. The family is much more common and prolific in the Great Basin, and many of the taxa found in the Sierra represent the fringes of a Great Basin distribution.

A Lvs linear, 1-5 cm L, spine-tipped — *Salsola*
AA Lvs flattened, not spine-tipped
  B Calyx of 1 persistent sepal — *Monolepis*
  BB Calyx 3-5-parted
    C Fls imperfect; lvs ± entire, e. SN — *Atriplex*
    CC Fls mly perfect; lvs mly serrate or lobed, widespread — *Chenopodium*

Atriplex

Saltbush

stem and leaves (1 cm) *staminate flower* (1 mm)

Annuals, often scaly or powdery; herbage usu white to gray. Usu associated with Great Basin biotic zone and rarely found within the region covered by this text. The lvs and young shoots are salty and can be used as greens in stews. Seeds may be ground and used as flour

A Lvs greenish above, often coarsely toothed: lf-base usu tapering
  B Plant mat-like; stems decumbent to ascending; below 6500', s. SN — 3) *A. serenana*
  BB Plant erect; stems ascending; disturbed habitats below 7500' — 2) *A. rosea*
AA Lvs white or gray on both surfaces, never coarsely toothed; lf-base truncate to hastate
  B Lvs all alt, the upper subsessile; e. SN below 8000' — 4) *A. truncata*
  BB Lowest lvs opp; all lvs petioled; n. SN below 5600' — 1) *A. argentea*

1) *A. argentea*, Silverscale: Stems decumbent to erect, much-branched, 2-8 dm L, scurfy-gray; lf-blades 2-5 cm L, triangular to roundish, entire to wavy margined; fls in axillary glomerules and terminal spikes. Alkaline places below 6000', mly e. SN.

2) *A. rosea*, Tumbling Oracle: Stem erect, 4-10 dm H; branches ascending; lvs stiff, the blades 1-6 cm L, greenish above, becoming red, coarsely wavy-toothed. Common in disturbed places.

3) *A. serenana*, Bractscale: Stems 3-10 dm L; lvs subsessile, the blade 1-4 cm L, elliptic to lanceolate, dentate. Alkaline places.

4) *A. truncata*, Wedgescale: Stems erect to decumbent, whitish, the branches 2-5 dm L; lvs ± ovate, 1.5-4 cm L, sessile or the lower petioled. Mly below 8000'.

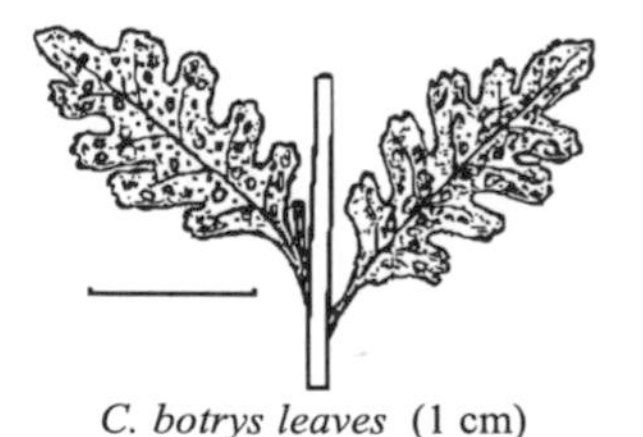

Chenopodium

Goosefoot

C. botrys leaves (1 cm) *C. foliosum* (1 cm)

Stamens mly 5, styles 2. All spp are ± edible. The young shoots and lvs may be collected through most of the year; the lvs may be eaten raw or both stems and lvs cooked as greens. The seeds ripen in late summer and may be eaten raw or added to flour whole or after grinding. Plants contain oil of chenopodium which may be poisonous in large amounts.

A Plants usu glandular-pubescent; herbage aromatic; lower lvs sinuate-pinnatifid (resembling an oak leaf); disturbed habitats
  B Axillary infl ± open, elongate, many-fld; lvs to 6 cm L — *C. botrys*
  BB Axillary infl compact, spherical, 5-10-fld; lvs less than 2 cm L — *C. pumilio*
AA Plants mly mealy, glabrous; lvs various but not pinnately lobed and oak-like
  B Lvs bright green, not scurfy, usu lobed, truncate at base; uncommon — *C. simplex*
  BB Lvs dull green to grayish, the margins entire to toothed, sometimes lobed at base
    C Lvs relatively narrow, lanceolate, oblong, or elliptic, not broader at base
      D All lvs 1-veined from base, entire — *C. leptophyllum*
      DD Lower lvs 3-veined from base, entire or toothed, not lobed
        E Stems prostrate to ascending; lvs often toothed; e. SN below 7000' — *C. glaucum*
        EE Stems erect; lvs entire; below 9000', SN (all)
          F Stems diffusely branched from base; uncommon — *C. desiccatum*
          FF Stems usu few branched; open habitats — *C. hians*
    CC Lvs broad in outline (ovate, widely elliptic or deltoid), often broader at base
      D Plants prostrate; lvs entire; uncommon, disturbed places, s. SN — *C. vulvaria*
      DD Plants erect
        E Fr wall easily detached from seed
          F Lf blades definitely longer than wide; to 11,000' — *C. atrovirens*
          FF Lf blades about as wide as long;
            G Stems usu less than 25 cm H; lvs thick; e. SN below 7500' — *C. incanum*
            GG Stems usu 25-40 cm H; lvs thin; to 10,000' — *C. fremontii*
        EE Fr wall adhering to seed, not easily removed intact
          F Lower lvs usu dentate above base
            G Lower lvs usu powdery below, deltate but not hastate; below 6000' — *C. album*
            GG Lower lvs glabrous on undersurface, often hastate; to 11,000' — *C. foliosum*
          FF Lower lvs entire or ± lobed, not dentate; densely powdery below
            G Plants much branched; sepals ± enclosing fr; e. SN — *C. nevadense*
            GG Plants few branched; sepals not enclosing fr; to 8500' — *C. incognitum*

M. nuttalliana leaves (1 cm)

Monolepis

Monolepis

flowers (1 mm)

Low branched herbs. Stems prostrate or ascending, 0.5-2 dm L. Lvs alt, often mealy when young, glabrous in age, entire or hastate. Fls imperfect, sessile. Stamen 1 or none; styles 2, slender. Disturbed or saline habitats. The whole plant, except perhaps roots, of *M. nuttalliana* may be used as a potherb. The seeds are also edible.

Lvs 10-50 mm L, usu hastate or diamond-shaped; below 5000' — *M. nuttalliana*
Lvs 5-15 mm L, entire; mly 5000-8000'; rare — *M. spathulata*

M. nuttalliana: Nuttall's Monolepis: Stems several from base, ascending, stout, succulent, 1-2 dm L; lvs 1-4 cm L, petioled; fl-clusters sessile, dense, often reddish; sepal about 1 mm L. Rather common, dry or moist often saline places and on burns, Apr-Sep.

M. spathulata, Club-leaved Monolepis: Stems 3-15 cm L, decumbent or ascending, branched from base; lvs narrowly spatulate to oblanceolate, ± fleshy; fl-clusters many-fld, sessile; sepal about 0.5 mm L. Moist subalkaline places, SN (all), Jun-Sep.

upper stem (1 cm)

Salsola

Russian-thistle Tumbleweed

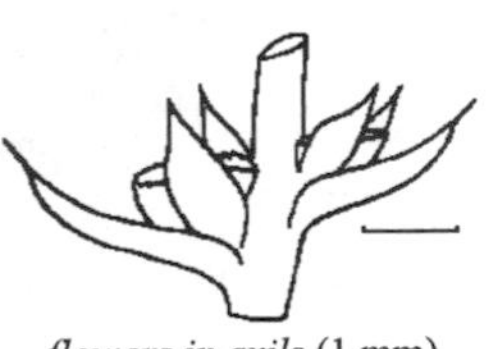
flowers in axils (1 mm)

S. tragus: Stems 3-10 dm H, densely branched and forming a rounded clump, usu glabrous; fls perfect. Common as a tumbleweed in fields and waste places, SN (all), Jul-Oct. At least the young parts may be eaten as potherbs; the older parts contain significant quantities of nitrates and oxalates and may be toxic if eaten in quantity.

CISTACEAE - Rock Rose Family

Helianthemum

Peak Rush-rose

base of stem (1 cm) *upper stem* (1 cm)

H. scoparium: Subshrub to 3 dm H; stems ascending; lvs alt, linear, 5-30 mm L, early deciduous. stellate pubescent; infl a narrow terminal panicle, leafy, few-fld; sepals 5, the outer two usu narrower and shorter; petals yellow, 5-7 mm L; fr a capsule. Below 4000', Mariposa Co. to Eldorado Co, Apr-May. Edibility unknown.

CONVOLVULACEAE - Morning-glory Family

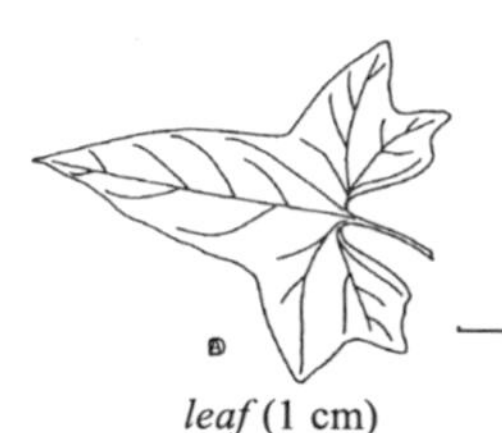

Calystegia

Morning-glory

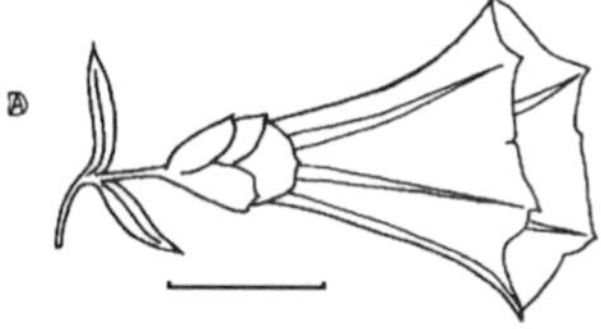

leaf (1 cm) *C. occidentalis flower* (1 cm)

Pubescent perennials with trailing to ascending stems. Lvs petioled, usu triangular-hastate. Peduncles axillary, with a pair of bracts just below calyx. Sepals often unequal. Corolla pleated, about 3 cm L. Stamens included, inserted on corolla tube. Style filiform, stigmas 2; fr a capsule. The two spp listed below probably intergrade. Herbage of some spp contains a purgative which may cause stomach distress.

Bractlets on peduncle longer than and concealing calyx — *C. malacophylla*
Bractlets on peduncle shorter than or distant from calyx — *C. occidentalis*

C. malacophylla, Sierra Morning-glory: Plant densely gray-hairy; lvs triangular-hastate, 2-4 cm L; peduncles 1-fld, 6-9 cm L (3-5 cm L in ssp *malacophylla*); bractlets 5-20 mm L, entire, ± ovate; corolla white, 2-5 cm L. Dry slopes in chaparral, to 10,000'.

C. occidentalis, Western Morning-glory: Plant pubescent but not densely so; lvs 1-4 cm L; peduncles 1-several fld, about as long as subtending lf; bractlets 5-10 mm L, linear (to triangular in ssp *fulcrata*); corolla white to cream. Dry slopes to 9000'.

CORNACEAE - Dogwood Family

Cornus

Dogwood

C. nuttallii inflorescence (1 cm) *C. stolonifera upper stem* (1 cm)

Lvs opp, entire. Fls small, 4-merous; ovary inferior; style 1. Fr a drupe. The fr of at least *C. nuttallii* is edible raw or cooked.

A Fls in compact heads subtended by petaloid bracts; fr-clusters dense — 2) *C. nuttallii*
AA Fls in open cymes or umbels; each drupe pedicelled
  B Lvs pinnately 8-14-veined — 4) *C. sericea*
  BB Lvs pinnately 6-8 veined; below 5000'
    C Drupes black when mature; infl an umbel subtended by bracts — 3) *C. sessilis*
    CC Drupes whitish when mature; infl a bractless cyme — 1) *C. glabrata*

1) *C. glabrata,* Brown Dogwood: Shrub to small tree, 1.5-6 m H, often forming thickets by means of underground shoots; twigs brownish to reddish purple, slender; lvs lanceolate to elliptic, acute, ± glabrous, mly 2-5 cm L, 1.5-2.5 cm W, on petioles 3-8 mm L; infl 2.5-4.5 cm across; pedicels 2-3 mm L; petals 4.5-5 mm L; drupes white to bluish, 9 mm in diameter, the stone 5-6 mm W, almost smooth. Moist places, SN (all), May-Jun.

2) *C. nuttallii*, Mountain Dogwood: Stem 4-25 mm H; twigs at first green, later dark red to almost black; lvs usu 6-12 cm L, 3-7 cm W, elliptic to obovate; petioles 5-10 mm L; infl appearing in autumn, subtended by 2 lvs and 2 bracts that persist until spring; at anthesis the head of fls subtended by 4-7 white, yellowish or pinkish bracts 4-6 cm L, 3-6 cm W; petals 4 mm L. Mountain woods below 6000', SN (all), Apr-Jul.

3) *C. sessilis*, Blackfruit Dogwood: Shrub or small tree, 1.5-4 m H; twigs pale; lvs usu 4-9 cm L, 2-3.5 cm W, elliptic, acute, ± tomentose beneath on vein-axils; petioles 5-10 mm L; infl subtended by 2 pairs of deciduous bracts 1 cm L, these brown or with yellow edges; pedicels about 1 cm L, white-villous; petals 3 mm L; drupes at first whitish, turning yellow and red to very dark, 1-1.5 cm L, ellipsoid. Stream banks, Calaveras Co. n., Apr-Jun.

4) *C. sericea*, American Dogwood: Spreading shrub 2-5 m H; twigs bright red-purple, lvs usu 5-10 cm L, 2-6 cm W, lanceolate to ovate, acute, often pubescent beneath; petioles 5-8(-25) mm L; infl 3-6 cm across, ± dome-shaped; pedicels puberulent, mly 4-8 mm L; drupes white, 7-9 mm L. Moist places below 9000', SN(all), May-Jul.

CRASSULACEAE - Stonecrop Family

Lvs fleshy, glabrous, without stipules, entire. fls clustered, regular, 4-5-merous. Stamens as many or twice as many as petals. Carpels as many as calyx-segments, distinct or united below.

Plants annual, lvs 3-5 mm L — *Parvisedum*
Plants perennial, lvs 5-30 mm L
  Petals 10-20 mm L; fl-stems axillary — *Dudleya*
  Petals less than 10 mm L; fl-stems terminal — *Sedum*

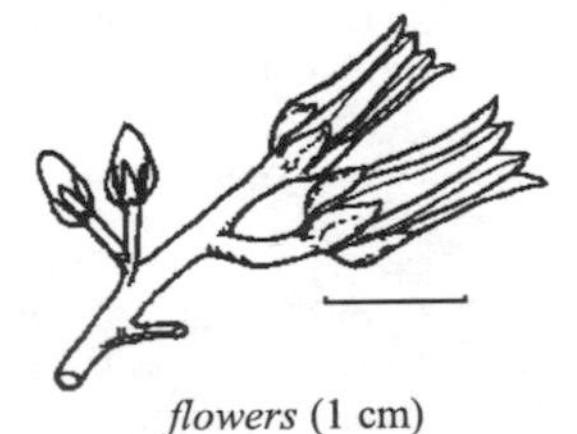

flowers (1 cm)

Dudleya

Hen-and-chickens

basal leaves (1 cm)

D. cymosa : Plants ± glaucous; rosette lvs mly oblong, acute, 5-10 cm L, 1.5-4 cm W, in a dense rosette; fl-stems reddish, 1-2.5 dm H, arising laterally from axils of rosette lvs; stem lvs lanceolate, 1.5-3.5 cm L, rounded at base, acute; infl dense; pedicels rather slender, 5-12 mm L; petals mly bright yellow to red. Rocky habitats, up to 9000', Butte Co. s., Apr-Jun. Stem and lvs edible at least after cooking. Another sp, *D. calcicola* with pale yellow petals is to be looked for on limestone outcrops below 7000' in Kern Co.

Parvisedum

Congdon's Sedella

P. congdonii: Stems 3-9 cm H, usu branched; fls scattered, petals bright yellow with median reddish line, 2 mm L. Rocky places below 5000', Eldorado Co. s., Apr-May. Plants too small to be of much food value.

inflorescence (1 cm)

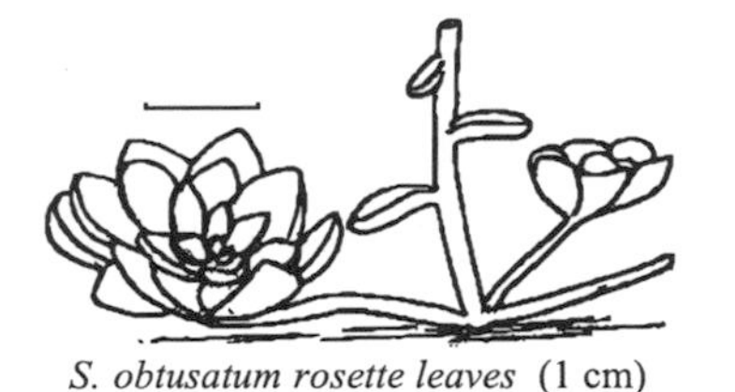

Sedum

Stonecrop

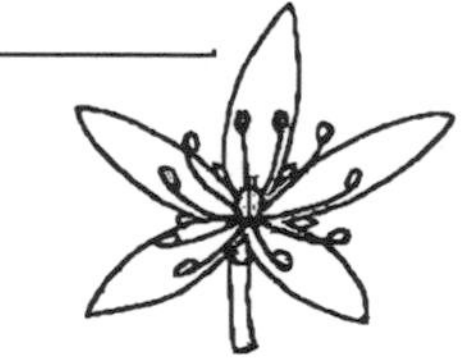

S. obtusatum rosette leaves (1 cm) *flower* (1 cm)

Usu glabrous herbs. Lvs mly alt, often small and imbricated. Plant can be used in salads or as a potherb. Young stems and lvs have the best flavor, and some spp are better tasting than others.

A Petals purple; fls 4-merous; lvs scattered, not forming rosettes 3) *S. rosea*
AA Petals yellow; fls 5-merous; lvs usu in rosettes and often also scattered
B Rosette lvs spatulate, larger and quite different from stem lvs
C Petals separate ± to base; below 7500', Eldorado Co. n 4) *S. spathulifolium*
CC Petals united at least 1/4 their length; 5000-13,000' 2) *S. obtusatum*
BB Rosette lvs not much different than stem lvs
C Lvs linear, ± round; 6000-12,000' 1) *S. lanceolatum*
CC Lvs lanceolate to elliptic, flattened; below 10,000' 5) *S. stenopetalum*

1) *S. lanceolatum*, Narrow-petaled Stonecrop: Stems tufted, the fl-stems 7-20 cm L; lvs crowded, sessile, 5-15 mm L, glaucous or dull green; infl elongate with short branches; petals narrow-lanceolate, 6-7 mm L; follicles 4 mm L. Rocky places 6000-12,000', Tulare to Alpine Co., Jun-Aug.

2) *S. obtusatum*, Sierra Stonecrop: Fl-stems erect, 3-16 cm H, often reddish; rosette lvs glaucous, usu obtuse, 5-25 mm L, usu with some red; stem lvs oblong-spatulate; infl 2-10 cm L, 2-5 cm across; petals fading to buff or pink, 6-9 mm L; follicles erect, 6-7 mm L, red in maturity. Rocky ridges and slopes, Tulare to Plumas Co., Jun-Jul.

3) *S. rosea* ssp *integrifolium*, Rosy Sedum: Stems several, 7-15 cm H; lvs flat, rather thin, sessile, obovate, 10-15 mm L, acute, entire or slightly dentate above the middle; infl terminal, congested; petals 3 mm L; follicles erect, dark purple, 3-5 mm L. Moist, rocky places, 7500-12,500', Tulare to Eldorado Co., May-Jul.

4) *S. spathulifolium*, Pacific Stonecrop: Sterile stems 1-8 cm L; fl-stems erect or decumbent, 5-30 cm H; rosette lvs obtuse, glaucous, 5-30 mm L; stem lvs spatulate to elliptic-oblong, 6-20 mm L; infl 10-50-fld; petals rarely orange or white, lanceolate, 5-8 mm L; follicles yellow-green, erect or divergent. Rocky habitats, May-Jul.

5) *S. stenopetalum*, Star-fruited Stonecrop: Stems 5-10 cm H; lvs narrow-lanceolate, with scarious-sheathing dilated base, ± flattened, 1-3 cm L; infl racemose, sparsely branched, with scattered fls; petals narrow-lanceolate, 6-10 mm L; follicles widely spreading, 5-6 mm L. Rocky places 4500-5500', Lassen Co.

ssp. radiatum: Lvs elliptic-oblong, 0.3-1 cm L. Below 8000', Tulare to Eldorado Co., Jun-Aug.

CUCURBITACEAE - Gourd Family

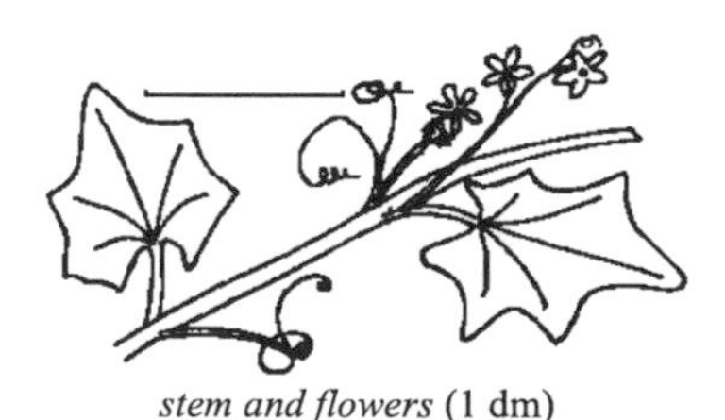

Marah

California Man-root

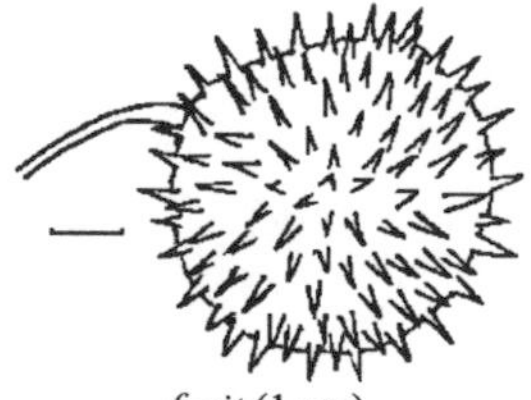

stem and flowers (1 dm) *fruit* (1 cm)

M. fabaceus: Stems vine-like, 3-7 m L; lvs ± round in outline, 5-7-lobed, 5-10 cm W; petioles 3-6 cm L; staminate fls 8-20 in an axillary raceme; pistillate fl solitary, from same axil as staminate fls; fr globose, 4-5 cm in diameter, covered with soft spines less than 5 mm L. Plants ± inedible.

CUSCUTACEAE - Dodder Family

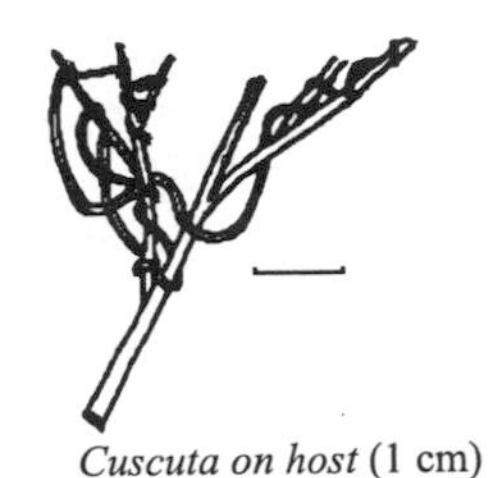

Cuscuta

Dodder

Cuscuta on host (1 cm) *flower* (1 mm)

C. californica: Parasitic plants without chlorophyll; the stems slender, twining, yellow to orange and fastened to their hosts (herbs or shrubs) by knobs (haustoria). Lvs reduced to minute scales. Fls small, perfect, mly waxy-white, 4-5-merous. Calyx and corolla of the same color, both with parts united at least part-way. Plants may cause digestive upset if eaten.

DATISCACEAE - Datisca Family

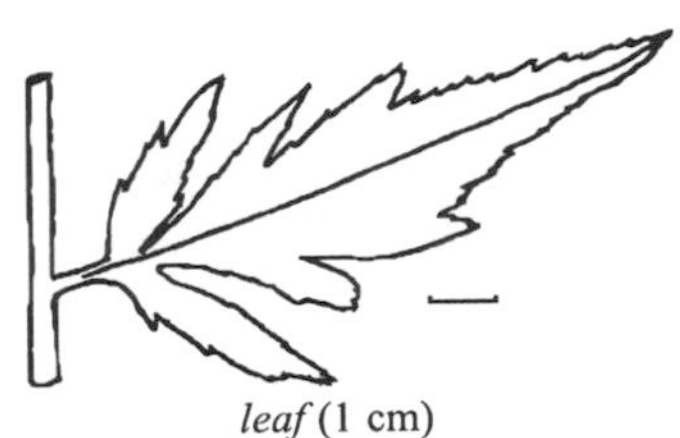

Datisca

Durango Root

leaf (1 cm) *axillary flowers* (1 cm)

D. glomerata: Stout glabrous perennial, 1 m or more H; lvs alt, pinnately irregularly incised, ovate to lanceolate in outline, 1-2 dm L; petioles 2-3 cm L; fls sessile, clustered in each axil of a leafy raceme. Dry stream beds and washes, occasional in SN below 5500', May-Jul. Plant distasteful and believed to be toxic.

DROSERACEAE - Sundew Family

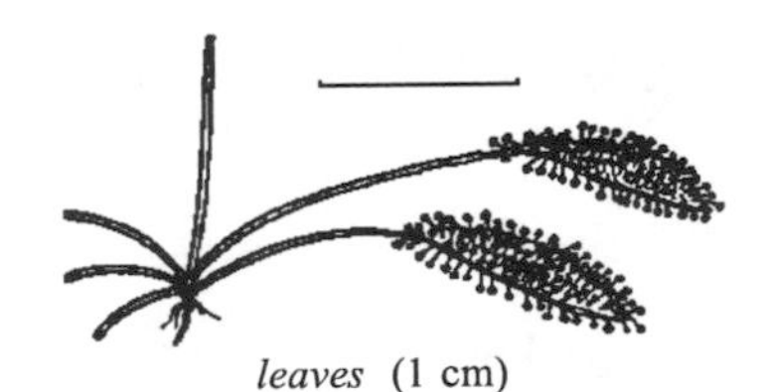

Drosera

Sundew

leaves (1 cm) *inflorescence* (1 cm)

Perennials in cold swamps below 8000'. Lvs usu in a basal rosette, mly bearing gland-tipped sensitive hairs. Fls perfect, 5-merous, rarely 4- or 8-merous. Hybrids of the following two spp have been found. Bitter juice of the stem and lvs makes plants unpalatable.

Lf-blades elongate-spatulate *D. anglica*
Lf-blades ± orbicular, broader than long *D. rotundifolia*

D. anglica, English Sundew: Lvs erect in the rosette, the petiole 3-7 cm L, the blade 15-35 mm L, 3-5 mm W, gradually narrowed to the petiole; scape 1-8-fld; otherwise this sp much like *D. rotundifolia.* Infrequent, below 6000', Nevada Co. n., Jul-Aug.

D. rotundifolia, Round-leaved Sundew: Lvs in a spreading basal rosette, the petiole 1.5-5 cm L, flat, the blade 4-10 mm L, abruptly tapering to the petiole, the upper surface covered with reddish hairs that are longest on the margin; scape glabrous, rarely forked, 5-25 cm H, 2-15 fld. Intermittent, SN (all), Jul-Aug.

ELATINACEAE - Waterwort Family

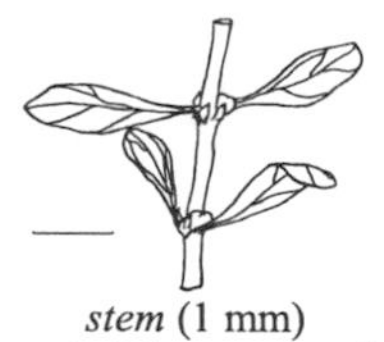

stem (1 mm)

Elatine

Waterwort

axillary flower (1 mm)

Dwarf glabrous annuals or subperennials, often rooting at nodes. Lvs opp. Fls 2-4-merous, usu 1 per node. Capsule membranous. Edibility unknown.

Stamens 3, opposite sepals — *E. chilensis*
Stamens 6 (rarely 1) or if 3, these opposite petals — *E. heterandra*

E. chilensis: Stems trailing to erect, to 10 cm L; lvs 2-4 mm L, about equal to internodes, oblong to elliptic, on short petioles; stipules present; suborbicular; sepals 2, with a third reduced or lacking; petals 3. Shallow water or mud banks below 6500', SN (all), May-Sept.
E. heterandra:: Stems slender, 2-5 cm L; similar to *E. chilensis* except as in key.

ERICACEAE - Heath Family

Trees, shrubs, green perennials or white to brown saprophytes. Lvs ± entire, often thick and leathery, without stipules. Fls perfect, regular or slightly irregular, mly 5-merous.

A Lvs scale-like, not green; stems usu fleshy
  B Stems striped red and white; sepals absent; petals 5 — *Allotropa* p. 103
  BB Stems not striped red and white; sepals and petals both present
    C Stems and fls reddish; petals united more than half way to tip
      D Stem glabrous; fls usu erect in cluster on top of plant — *Hemitomes* p. 105
      DD Stem glandular; fls arising laterally on side of main axis
        E Corolla 7-8 mm L; stems slender, usu 3-10 dm H — *Pterospora* p. 107
        EE Corolla 12-18 mm L; stems stout, usu 1-3 dm H — *Sarcodes* p. 107
    CC Stems white to brownish; petals distinct
      D Corolla glabrous; 4000-8500', widespread — *Pleuricospora* p. 106
      DD Corolla pubescent within; below 5000', Fresno Co. — *Pityopus* p. 106
AA Lvs green; plants often woody
  B Lvs usu basal or on a short simple stem; plants small subshrubs; petals distinct to base
    C Fls solitary of few in umbel-like raceme
      D Lvs thin, ovate, ± basal — *Moneses* p. 106
      DD Lvs leathery, alt or whorled on the stem — *Chimaphila* p. 104
    CC Fls few to many in an elongate raceme
      D Raceme 1-sided; petals greenish yellow — *Orthilia* p. 106
      DD Raceme symmetric; petals often white to red — *Pyrola* p. 107
  BB Lvs cauline; stem usu woody and branched; petals usu fused into an urn-shaped corolla
    C Lvs less than 1 cm W and ± linear or thin and less than 3 cm L, occasionally opp
      D Lvs thin, ovate to oblong, 1-3 cm L — *Vaccinium* p. 108
      DD Lvs thicker, ± linear or scale-like
        E Lvs opp, the margins usu strongly curled under — *Kalmia* p. 105
        EE Lvs ± alt, needle- or scale-like
          F Lvs needle-like, 6-15 mm L; corolla rose-purple — *Phyllodoce* p. 106
          FF Lvs scale-like, 3-6 mm L; corolla white to pinkish — *Cassiope* p. 104
    CC Lvs mly over 1 cm W, mly thick or over 3 cm L, alt
      D Fr fleshy, usu conspicuous in later season; often dry habitats
        E Plants scarcely woody, matted; lvs aromatic — *Gaultheria* p. 104
        EE Plants definitely woody, usu with smooth red stems
          F Plant a tree; lvs 5-12 cm L — *Arbutus* p. 103
          FF Plant shrubby; lvs 1-4 cm L — *Arctostaphylos* p. 103
      DD Fr dry, often hidden by persistent sepals or petals; mly moist habitats
        E Corolla funnelform, 3.5-5 cm L — *Rhododendron* p. 107
        EE Corolla urn-shaped, about 1 cm L
          F Infl elongate, about 10 cm L — *Leucothoe* p. 105
          FF Infl flat or dome-shaped, about as long as wide — *Ledum* p. 105

Allotropa

Sugar Stick

A. virgata: Stem erect, unbranched, glabrous, 1-5 dm H, densely clothed at base with elongate, scale-like lvs; lvs 2-3 cm L; raceme many-fld, forming the upper 1/3 to 1/2 of plant; petals 5-6 mm L, whitish. Occasional, in thick humus, 2000-10,000', SN (all), Jun-Aug.

upper stem (1 cm)

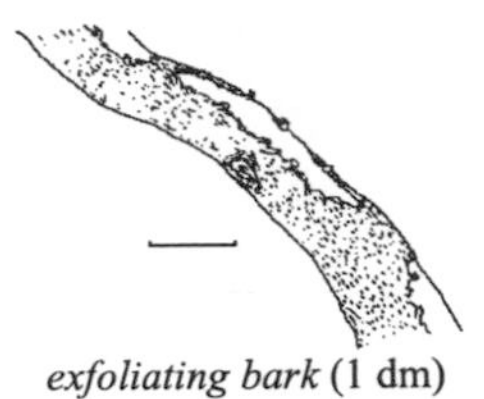

Arbutus

Pacific Madrone

exfoliating bark (1 dm) *leaf* (1 cm)

A. menziesii:: Bark freely exfoliating leaving a polished reddish surface; lvs entire to serrulate; berry red to orange, roundish, 8-10 mm in diameter. Wooded slopes and canyons below 5000', Mariposa Co. n., Mar-May. The berries may be eaten raw or may be boiled or steamed. After boiling they may be dried for future use.

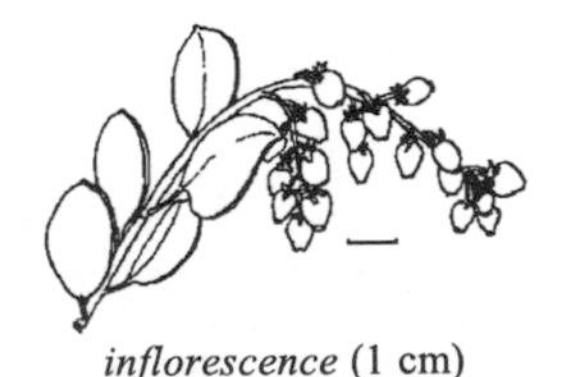

Arctostaphylos

Manzanita

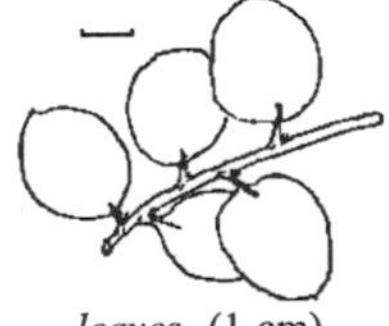

inflorescence (1 cm) *leaves* (1 cm)

Branches usu crooked, usu smooth with thin red to brown bark that exfoliates. Calyx 4-5-lobed, the lobes broad. Fr berry-like, with rather copious granular pulp or with thin pericarp and dry. The berry may be eaten raw when ripe, although not in large quantities. Cooking the berries renders them perfectly edible. A cider can be made by crushing the berries, ripe or green, scalding with an equal volume of water and allowing the solids to settle. The berries may also be dried. The seeds may be ground in a flour or meal for cereal.

A Branches prostrate or procumbent, usu less than 6 dm H, rooting on contact with ground
  B Hairs of branchlets and infl gland-tipped; corolla pink — 3) *A. patula*
  BB Hairs of branchlets and infl not gland-tipped; corolla usu white
    C Stomates on both surfaces of lvs; widespread — 2) *A. nevadensis*
    CC Stomates on lower surface of lvs; rare — 4) *A. uva-ursi*
AA Branches erect or ascending, taller, not usu rooting at tips
  B Pedicels glabrous; stem from an enlarged basal burl
    C Lvs pale gray-green; branchlets glabrous or nearly so — 1) *A. mewukka*
    CC Lvs bright green; branchlets usu glandular-hairy — 3) *A. patula*
  BB Pedicels ± glandular-pubescent; ovary usu glandular-pubescent; lvs pale green; stems without basal burl — 5) *A. viscida*

1) *A. mewukka*, Indian Manzanita: Stems 1-2.5 m H with smooth deep red to purplish bark; lvs oblong to obovate, 2.5-5 cm L, glabrous, on stout petioles 7-9 mm L; infl mly branched, open, glaucous, the peduncle and rachises usu dark red; pedicels 4-5 mm L; corolla white or pinkish, 6-7 mm L; fr depressed-globose, dark red to red-brown, 12-14 mm in diameter. Dry slopes, mly 2500-6000', Butte Co. s., Mar-Apr.

2) *A. nevadensis*, Pinemat Manzanita: Stems intricately branched, without basal burl, rooting freely and forming erect branchlets 3-6 dm H; bark brownish to deep red; lvs bright green, lanceolate to elliptic or oblanceolate, 2-2.5 cm L; petioles puberulent, 3-7 mm L; infl compact, many-fld, erect; pedicels 2-5 mm L; corolla 6-7 mm L; ovary glabrous; fr depressed-globose. Moist places to dry slopes in woods, mly 5000-10,000', SN (all), May-Jul

3) *A. patula*, Greenleaf Manzanita: Stems several, spreading, 1-2 m H (a dwarf from less than 0.5 m also exists); bark bright red-brown; lvs broadly ovate to almost round, ± glabrous, 2.5-4 cm L; petioles 6-10 mm L; infl various; pedicels 5-7 mm L; corolla 5-8 mm L; ovary glabrous; fr ± globose, 7-10 mm W. Open forest, 2000-11,000', SN (all), Apr-Jun.

4) *A. uva-ursi* var. *coactilis,* Bearberry, Sandberry: Stems trailing without basal burl, branchlets erect, 5-15 cm L; bark dark brown or somewhat reddish; lvs oval or obovate, shining, obtuse, 1-2.5 cm L; petioles 1-2 mm L; infl dense, short; corolla pinkish to white, 4-5 mm L; fr bright red, smooth, 6-12 mm W. Sonora Pass, Convict Lake Basin.

5) *A. viscida*, Whiteleaf Manzanita: Stems erect, 1-4 m H, with smooth dark red-brown bark; branchlets pale glaucous-green; lvs white-glaucous, 2-4 cm L, on petioles 8-12 mm L; infl open; pedicels 10-12 mm L; corolla pink to whitish, 6-7 mm L; fr globose, light brown or red, glabrous to sticky, 6-8 mm in diameter. Dry slopes below 5000', SN (all), Mar-Apr.

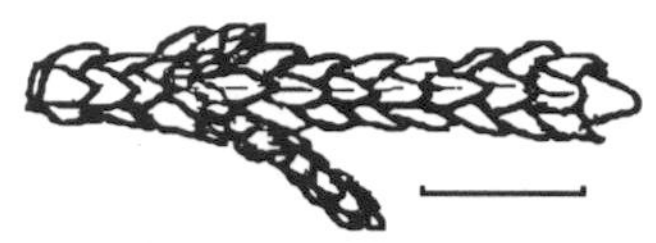

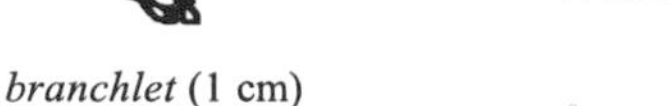

Cassiope

White-heather

branchlet (1 cm) *flowers* (1 cm)

C. mertensiana: Stems creeping with ascending branches, 1-3 dm H; lvs ovate-lanceolate, keeled on back, 3-6 mm L; fls 4-5-merous, solitary in lf-axils near ends of branches, nodding on slender pedicels, 6-20 mm L; corolla 5-6 mm L. Rocky ledges and crevices, 7000-12,000', Fresno Co. n., Jul-Aug. Herbage probably poisonous.

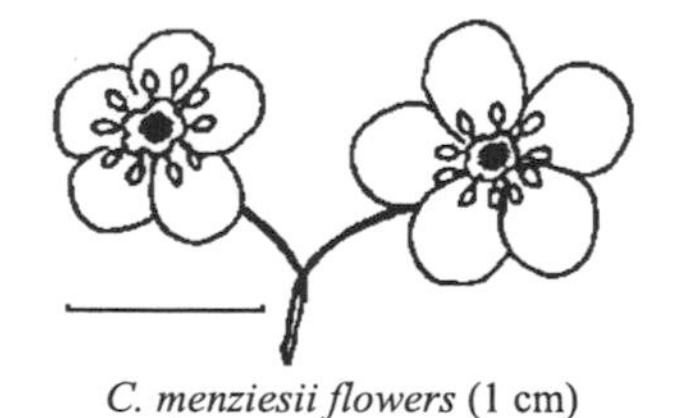

Chimaphila

Pipsissiwa

C. menziesii flowers (1 cm) *C. menziesii leaves* (1 cm)

Plants low, evergreen, with branching stem. Lvs thick, shining, short-petioled. Fls nodding at anthesis. Ovary 5-lobed. The lvs of *C. umbellata* may be eaten in small quantities raw. The roots and lvs may be boiled, and the resulting liquid cooled may be used as a cold drink.

Lvs ovate; fls greenish white, usu 1-3 in the infl	*C. menziesii*
Lvs oblanceolate; fls pink, usu 3-7 in the infl	*C. umbellata*

C. menziesii, Little Prince's Pine: Stems 1-1.5 dm H, few-branched; lvs not distinctly whorled, 1.5-3.5 cm L, serrulate, sometimes mottled, dark green above, paler beneath; peduncles mly 4-5 cm L; sepals about 5 mm L; petals sometimes pinkish in age, round, about 6 mm L. Shaded woods, 2500-8000', SN (all), Jun-Aug.

C. umbellata var. *occidentalis*, Western Prince's Pine: Stems 1.5-3 dm H; lvs in whorls of 3-8, 3-7 cm L, serrate, mly yellow-green beneath; peduncles mly 6-8 cm L; sepals ovate, about 3 mm L; petals 5-6 mm L. Dry, shrubby slopes in forest, below 10,000', SN (all), Jun-Aug.

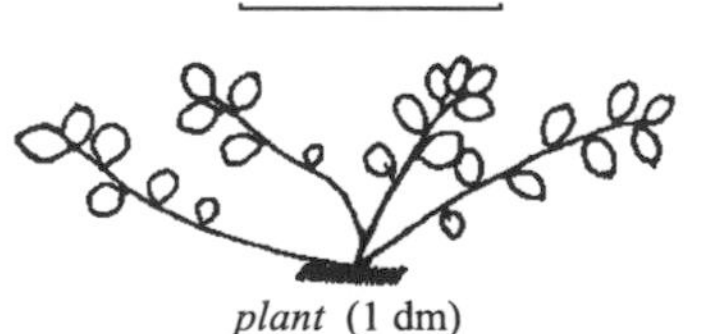

Gaultheria

Alpine Wintergreen

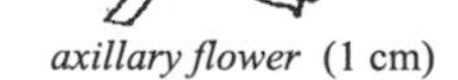

plant (1 dm) *axillary flower* (1 cm)

G. humifusa: Stems creeping, rooting at nodes, 1-2 dm L; lvs round to oval, subentire, mly 1-2 cm L; petioles 1-3 mm L; fls solitary, axillary; pedicels about 1 mm L; calyx glabrous, about 2.5 mm L; corolla slightly longer, white; fr 5-7 mm in diameter, red. Rare, in moist places, about 8000-10,500', Tuolumne Co. s., Jul. Both lvs and fr have wintergreen flavor and are edible raw or cooked.

Hemitomes

Gnome Plant

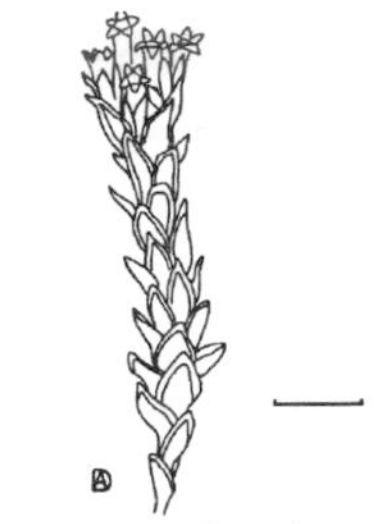
upper stem (2 cm)

H. congestum: Plants fleshy, white to pink, turning brown with age, 3-12 cm H; lvs scale-like, overlapping, ovate, blunt; fls in short terminal spike; sepals 2 or 4, bract-like; corolla 10-15 mm L, central fls with ovate lobes about 1/3 as long as tube; marginal fls more deeply lobed; stigma yellow, conspicuous. Wooded habitats, mly below 4000', SN (all).

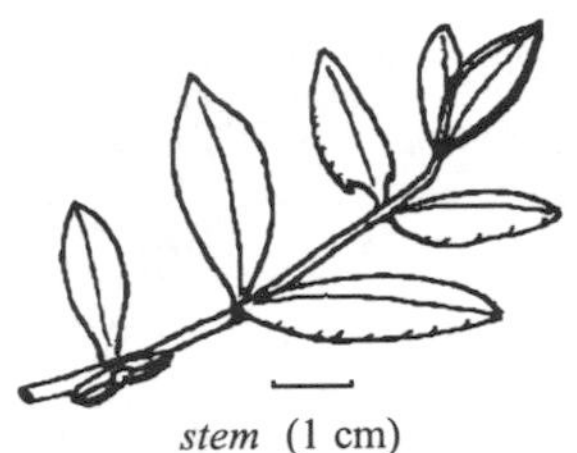
stem (1 cm)

Kalmia

Bog Kalmia

axillary flower (1 cm)

K. polifolia: Stems diffusely branched, 1-2 dm H; lvs dark green above, paler below; infl few-fld; pedicels slender, glabrous, 2-4 cm L; sepals 4, about 4 mm L, lanceolate; corolla rose-purple, 8-12 mm W, bowl-shaped. Boggy places and wet mdws, 7000-12,000', SN (all), Jun-Aug. All parts contain andromedotoxin which causes tear formation, nasal discharge, slow pulse, low blood pressure, convulsions and paralysis.

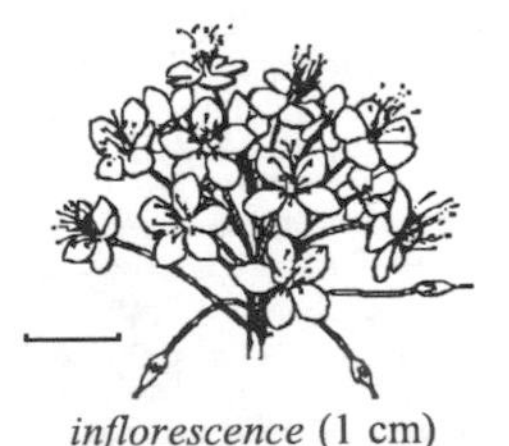
inflorescence (1 cm)

Ledum

Western Labrador Tea

leaves (1 cm)

L. glandulosum: Plants 5-15 dm H, with divaricate-ascending branches; twigs rather stiff, glandular-pubescent, yellow-green; lvs often clustered near tip of branch, 1.5-6 cm L, 5-10 mm W, fragrant, green above, mly yellow-green or whitish below, entire; petioles 5-10 mm L; pedicels puberulent, 1-2 cm L. Boggy and wet places, 4000-12,000', SN (all), Jun-Aug. Poison is andromedotoxin, a resinoid carbohydrate (see *Kalmia*), found in the lvs.

Leucothoe

Sierra-laurel

L. davisiae: Stems erect, 5-15 dm H, with glabrous, leafy twigs; lvs 2-6 cm L, 1-2 cm W, oblong, entire to somewhat serrulate; petioles 3-6 mm L; panicle terminal, 6-15 cm L; calyx of 5 almost distinct segments, whitish. Moist, springy places, 3200-8500, Fresno to Plumas Co., Jun-Aug. Herbage contains andromedotoxin (see *Kalmia*).

inflorescence (1 cm)

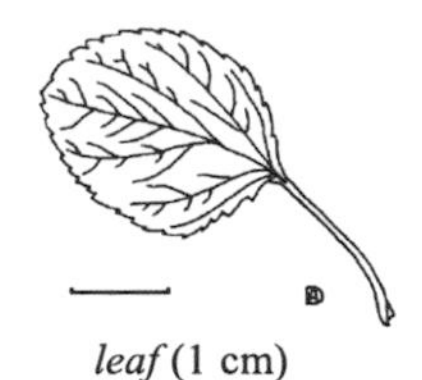

leaf (1 cm)

Moneses

Woodnymph

flower (1 cm)

M. uniflora: Lvs basal, thin, ovate, sharply serrulate, 1-3 cm L; petioles 1-2 cm L; scape 4-10 cm H, with 1-2 bracts,; fl solitary; sepals 3 mm L; petals white to pink, ovate, about 1 cm L; anthers 2-horned; stigma 5-lobed; capsule round, 6-8 mm in diameter. Uncommon in woods, Fresno Co. below 3500'.

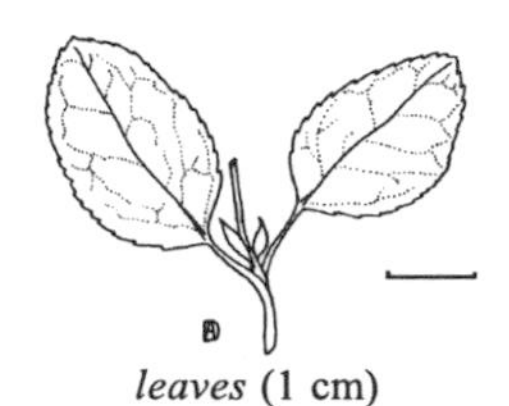

leaves (1 cm)

Orthilia

One-sided Wintergreen

inflorescence (1 cm)

O. secunda: Lvs shining, elliptic to ovate, ± serrate, the blades 2-5 cm L; petioles usu somewhat shorter; fl-stem 5-15 cm H; petals oblong, yellow-green, 4-5 mm L. Dry, shaded woods, mly 3000-10,500', SN (all), Jul-Sep.

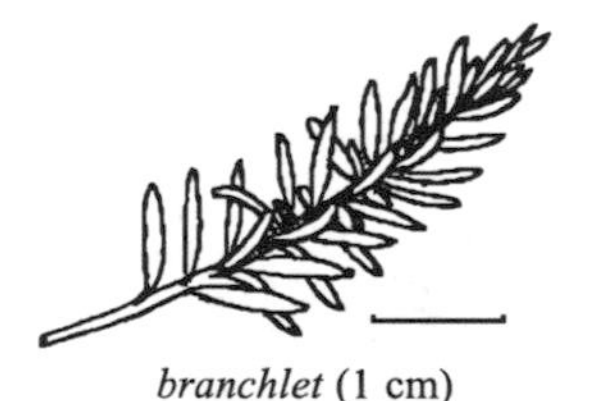

branchlet (1 cm)

Phyllodoce

Mountain-heather

inflorescence (1 cm)

P. breweri: Shrubs 1-3 dm H, evergreen, much-branched; lvs 6-15 mm L, somewhat glandular, crowded, ± revolute; infl umbel-like; pedicels 1-1.5 cm L, glandular-pubescent; calyx-lobes 3-5 mm L; corolla urn-shaped, 8-10 mm L, rarely white. Rocky, often moist places, 6000-12,000', SN (all), Jul-Aug. Herbage probably poisonous.

Pityopus

Pityopus

P. californicus: Plant waxy-white, subglabrous; stems unbranched, erect, 7-20 cm H, 1 to few in a cluster; lvs scale-like, crowded, 1-2 cm L; fls in a dense terminal spike-like raceme; sepals 2-5, about 12 mm L; petals 4-5, white, as long or longer than sepals. Rare, deep shade, below 5000', Fresno Co., May-Jul. Probably toxic.

upper stem (1 cm)

Pleuricospora

Fringed Pine-sap

P. fimbriolata: Stems white or brownish, rather stout, 1-2 dm H, glabrous, unbranched; lvs scale-like, 8-12 mm L; fls at first white, later yellow to brown, in terminal spike-like raceme with broad conspicuous bracts; sepals 4-5, 5-10 mm L; petals 4-5, about as long as sepals. Dry, deep humus, 4000-8500', SN (all), Jun-Aug. Best treated as potentially toxic.

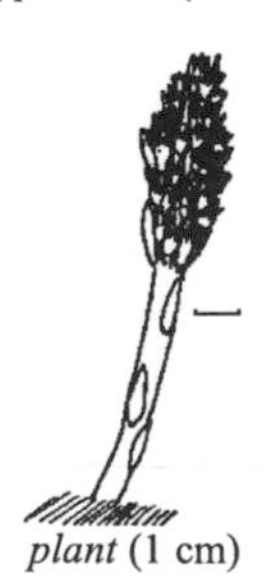

plant (1 cm)

Pterospora

Pinedrops

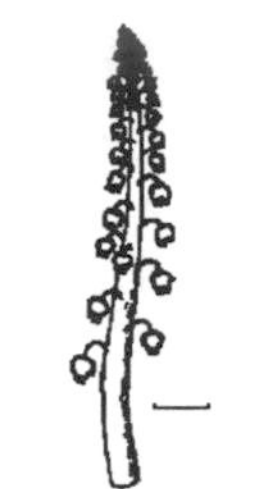
upper stem (1 cm)

P. andromedea: Stems unbranched, erect, 3-10 dm H, purple-brown to reddish, clammy, pubescent; lvs scale-like, crowded below, 1.5-3.5 cm L; fls in long terminal raceme, nodding on recurved pedicels, white to red; calyx 4-5 mm L; corolla urn-shaped, 7-8 mm L. Humus in forests, below 8500', SN (all), Jun-Aug. Best treated as potentially toxic.

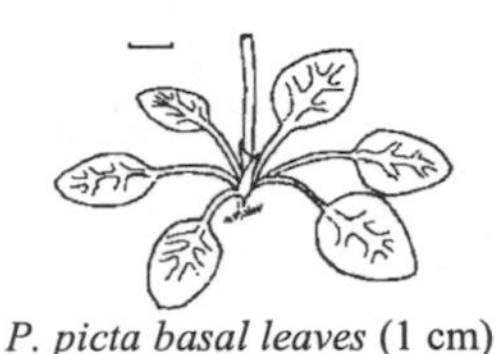
P. picta basal leaves (1 cm)

Pyrola

Wintergreen

P. picta flowers (1 cm)

Low perennials. Lvs evergreen, petioled. Fls in a simple raceme on end of leafless somewhat scaly stem. Fr a capsule. Widespread in SN. Best treated as potentially toxic.

A Lvs roundish, ± entire; plants of moist habitats
  B Lvs 3-8 cm L including petiole; petals red to purplish — *P. asarifolia*
  BB Lvs 1-3 cm L including petiole; petals white to pinkish — *P. minor*
AA Lvs usu elliptic and serrate or with white veins; petals mly greenish to white; plants of dry forest floor — *P. picta*

P. asarifolia, Bog Wintergreen: Plants creeping; lvs leathery, obovate to suborbicular, dull, entire to finely scalloped, 3-8 cm L; petioles about as long as blades; fl-stems slender, 2-4 dm H; fl-bracts 8-15 mm L, reddish; sepals 4-5 mm L, reddish; petals ovate to obovate, 6-8 mm L; style 6-8 mm L. Moist habits, 4000-9300', Jul-Sep.

P. minor, Common Wintergreen: Lvs rounded, slightly scalloped, the blades dull, 1-2(-3) cm L; petioles equally as long; fl-stem slender, 6-15 cm H; bracts at base of pedicels lance-oblong, 3-7 mm L; calyx-lobes triangular, scarcely 2 mm L; petals white to pink, 3-5 mm L. Occasional in boggy, shaded places, 7000-10,000', Jul-Aug.

P. picta and sspp, White-veined Wintergreen: Lvs various, usu mottled or veined with white, 2-7 cm L; petioles about as long; fl-stem 1-2 dm H; bracts lance-deltoid, 3-5 mm L; style sharply bent at base. Common on wooded slopes below 10,000', Jun-Aug.

inflorescence (1 cm)

Rhododendron

Western Azalea

leaves (1 cm)

R. occidentalis: Stems loosely branched, 1-3 m H, with shredding bark; lvs 3-10 cm L, 1-3 cm W, light green, oblanceolate, minutely denticulate; petioles 4-8 mm L; corolla funnelform, white, occasionally with yellow or orange on upper lobe. Stream banks and moist places, below 7500', SN (all), May-Aug. The lvs contain andromedotoxin (see *Kalmia*).

Sarcodes

Snow Plant

S. sanguinea: Stems unbranched, 2-4 cm thick, red; lvs scale-like, 2-8 cm L, ciliate; fls many in stout spicate racemes. Thick humus of forests, 4000-8000', SN (all), May-Jul. Stalks edible when cooked like asparagus; however this plant is rare and is protected by law.

plant (1 cm)

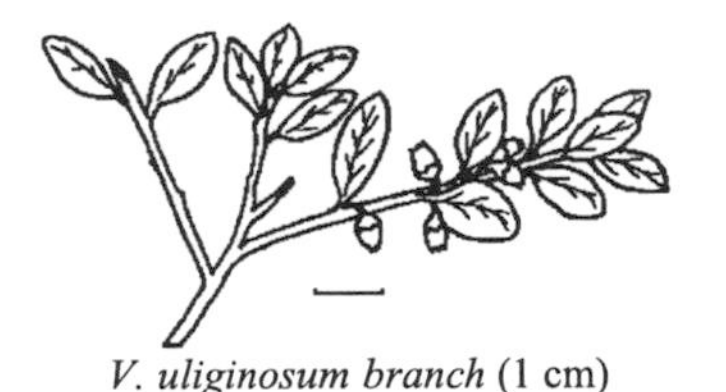

Vaccinium

Huckleberry

V. uliginosum branch (1 cm) *V. caespitosum leaves* (1 cm)

Lvs thin, deciduous, Calyx-tube adnate to ovary which in fr becomes a berry or drupe crowned with calyx-teeth. Berries edible raw, cooked, or after drying and storage.

A Lvs serrulate; stems depressed and usu less than 1 dm H *V. caespitosum*
AA Lvs mly entire; stem erect and usu over 3 dm H
B Branchlets ± angled; fls solitary *V. parvifolium*
BB Branchlets round; fls sometimes grouped *V. uliginosum*

V. caespitosum, Dwarf Bilberry: Branchlets glabrous, occasionally ± reddish; lvs obovate, glabrous, occasionally subglabrous to puberulent, serrulate, often in tufts at the tips of the stems, 1-3 cm L, 6-10 mm W; petioles 1-3 mm L; fls nodding, axillary; corolla ovoid, pink or white, 5-6 mm L; berry blue-black, with a bloom, globose, 5-7 mm in diameter. Wet mdws and near snow banks, 6700-12,000', SN (all), May-Jul.

V. parvifolium, Red Huckleberry: Stems erect, branched, 1-4 m H; branchlets green, glabrous or somewhat rough; lvs ovate to oblong, usu obtuse, subsessile, 1-3 cm L; corolla 4-6 mm L; berry 6-10 mm in diameter. Occasional, shady, moist habitats, mly below 7000', Fresno Co. n., May-Jun.

V. uliginosum ssp. *occidentale,* Western Blueberry: Plant glabrous, compact, 3-7 dm H, sometimes decumbent; lvs oblanceolate to obovate, 0.5-3 cm L, entire, gray-green, subsessile; corolla white or pinkish, about 4 mm L; berry blue-black, with a bloom, ellipsoid, about 6 mm L. Wet mdws, 5000-11,000', SN (all), Jun-Jul.

EUPHORBIACEAE - Spurge Family

Annuals with thick, white sap. Fls inconspicuous, without petals. The acrid, milky sap may burn the lips and mouth, and plants may cause intestinal upset and even death if eaten in quantity.

Stem short, lvs appearing tufted or whorled; herbage gray-pubescent *Eremocarpus*
Stems conspicuous; lvs clearly opp or alt; herbage glabrous
Stem prostrate, lvs opp, ± serrate *Chamaesyce*
Stem erect, lvs mly alt, entire *Euphorbia*

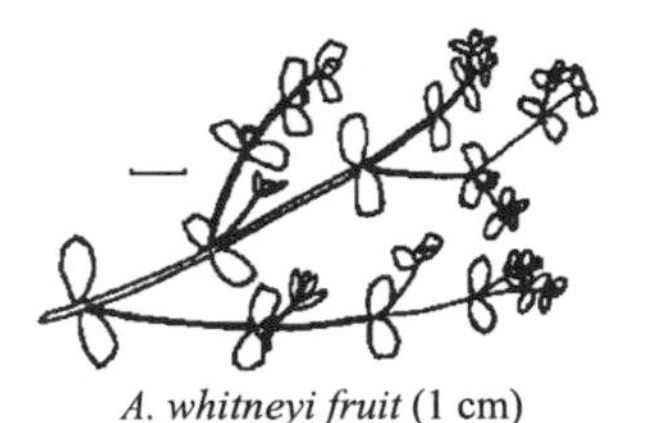

Chamaesyce

Thyme-leaved Spurge

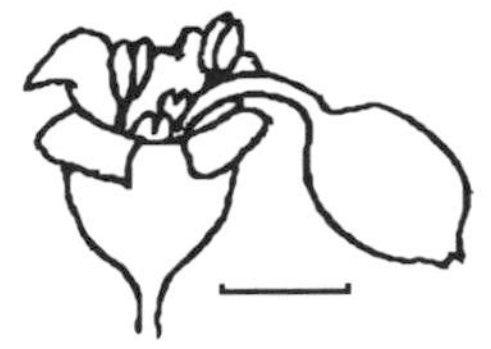

A. whitneyi fruit (1 cm) *cyanthium with flowers* (1 mm)

C. serpyllifolia: Stems 5-35 cm L; lvs ovate to obovate, 3-15 mm L; stipules distinct, linear, entire or few-parted; invol. a cyanthium, solitary, about 1 mm W. Common in dry disturbed areas below 7000', SN (all), Aug-Oct.

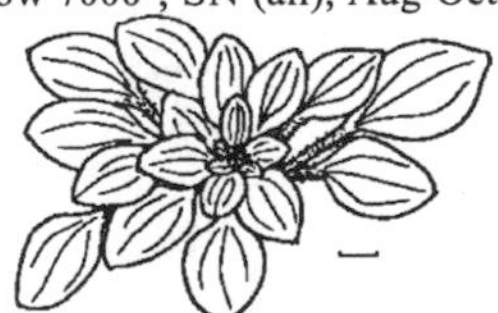

Eremocarpus

Turkey-mullein
Dove Weed

plant (1 cm) *staminate flower* (1 mm)

E. setigerus: Stem much-branched from base, forming dense rounded clumps, 3-20 cm H and 5-80 cm across; lvs ± round, 3-veined, 1-6 cm L; petioles about as long as lvs; staminate fls in terminal cymes; pistillate fls in lower axils. Common in dry, open habitats below 6000', SN (all), May-Oct. Herbage poisonous and was used by Indians to stupefy fish. Seeds eaten by birds.

Euphorbia

Chinese Caps
Spurge

plant (5 cm)

E. crenulata: Stems 1 to several, often branched, 2-6 dm H; stem lvs obovate to spatulate, obtuse, 1.5-3.5 cm L, subsessile; floral lvs opp or in 3's subcordate, often appearing perfoliate, 5-15 mm L; fls in a cyanthium (cup-like invol. resembling a calyx with united lobes) about 2 mm W. Common in dry places below 5000', SN (all), May-Aug.

FABACEAE - Pea Family (Leguminosae)

Lvs alt, usu with stipules, mly cmpd, the ultimate lfts usu entire. Fls mly perfect and irregular. Calyx 5-toothed or -cleft. Petals usu 5. Fr usu a pea-pod-like 'legume'.

A Lvs simple, palmately cmpd, or pinnately 3-foliolate
  B Lvs palmately 5- to many-foliolate
    C Lfts 6-12 mm L; stamens grouped as 9 and 1 — *Trifolium lemmonii* p. 117
    CC Lfts usu longer; stamens all in one group — *Lupinus* p. 113
  BB Lvs 3-4-foliolate or simple
    C Lvs or lfts 3-8 cm L
      D Lvs simple, roundish; plant a shrub — *Cercis* p. 110
      DD Lvs with 3 round lfts; plant herbaceous — *Hoita* p. 111
    CC Lfts mly 3, 1-2 cm L
      D Infl a raceme; fr indehiscent, appearing round; disturbed habitats at low elevations
        E Raceme long (5-10 cm) and slender with many white fls; fr not coiled, ovoid — *Melilotus* p. 116
        EE Raceme short (1-2 cm), roundish; fls 2-20, yellow; fr coiled and often bur-like — *Medicago* p. 116
      DD Infl a head, umbel, or fls solitary; plants widespread and common
        E Fls in heads, the head often subtended by leafy bracts; fr inconspicuous, rounded — *Trifolium* p. 117
        EE Fls solitary or in umbels; fr usu much longer than broad — *Lotus* p. 112
AA Lvs pinnate, the lfts more than 3
  B Rachis of lf prolonged into a tendril or at least a short bristle
    C Stems angled or winged; stipules entire to dentate — *Lathyrus* p. 111
    CC Stems usu not angled; stipules usu cut into narrow lobes — *Vicia* p. 119
  BB Rachis of lf not prolonged into a tendril
    C Fls solitary or umbellate, mly yellow or whitish — *Lotus* p. 112
    CC Fls racemose or spicate
      D Lfts more than 1 cm W; plants gland-dotted — *Glycyrrhiza* p. 111
      DD Lfts less than 2 cm W; plants not gland-dotted
        E Keel petals not produced into a beak; widespread — *Astragalus* p. 109
        EE Keel petals produced into a beak; e. SN of Inyo and Mono cos. — *Oxytropis* p. 116

A. whitneyi fruit (1 cm)

Astragalus

Milkvetch
Locoweed

A. bolanderi upper stem (1 cm)

Perennials. Lvs odd-pinnate, with stipules. Infl axillary. Calyx 5-toothed. Stamens 9 and 1. Style glabrous. Specimens in this genus are often difficult to key to sp, and there is occasional hybridization between spp. Related spp will accumulate selenium and become toxic, however none of the Sierran soils are high in selenium, nor are any of the following spp renowned as selenium accumulators. Unripe fr edible in related spp.

A Stems 15-45 cm H; herbage ± glabrous; lfts 17-25 — 2) *A. bolanderi*
AA Stems shorter; herbage usu gray-pubescent; lfts 7-19
  B Lfts long-acuminate, about 1 mm W, with spinule at tip — 4) *A. kentrophyta*
  BB Lfts broader, without spinule
    C Pods not inflated; stems mly less than cm L; petals over 8 mm L
      D Petals villous, peaks of n. SN — 1) *A. austinae*
      DD Petals glabrous; mly s. SN or lower habitats
        E Pods densely villous-hairy; widespread — 7) *A. purshii*
        EE Pods short-villous-tomentose; Tulare Co. — 9) *A. subvestitus*
    CC Pods strongly inflated; stems often over 10 cm L; petals usu less than 8 mm L
      D Lfts 7-13; petals whitish, occasionally purple-veined; plants
        E Lfts 1-3.5 mm L; above 10,000' e. Fresno Co. — 8) *A. ravenii*
        EE Lfts 2-12 mm L; below 6000'
          F Pubescence of fine, subappressed hairs; Plumas Co. n. — 6) *A. pulsiferae*
          FF Pubescence of long, shaggy hairs; Walker Pass, Kern Co. — 3) *A. ertterae*
      DD Lfts 9-21; petals usu purplish; mly e. SN
        E Lfts broad, usu over 5 mm W; pods acute — 5) *A. lentiginosus*
        EE Lfts narrow, 1-3 mm W; pods balloon-shaped — 10) *A. whitneyi*

1) *A. austinae*, Austin's Locoweed: Stems tufted or matted, 1-15 cm L; lvs 1-4 cm L, with 7-13 crowded elliptic lfts, 3-9 mm L; peduncles 1-4 cm L; racemes subcapitate, 4-10 fld; petals whitish or lilac-tinged, the banner 8-12 mm L; pod 6-7 mm L. Dry, exposed crests and ridges, 8800-10,500', Tinker Knob, etc., Jun-Aug.

2) *A. bolanderi*, Bolander's Locoweed; Stems mly subglabrous, 1.5-4.5 dm L; lvs 4-15 cm L; lfts ± oblong, 5-25 mm L; peduncles 3-8 cm L; racemes loosely 5-30-fld, 1-5 cm L in fr; petals whitish, rarely tinged with lavender, the banner 13-18 mm L; pod stipitate, the stipe 4-12 mm L, the body inflated, 13-30 mm L, 5-12 mm in diameter. Dry habitats, 5200-10,000', Tulare Co. to Lake Tahoe, Jun-Aug.

3) *A. ertterae,* Walker Pass Milkvetch: Stems 3-10 cm, prostrate; lvs 4-5, 3-7 cm L, crowded on upper stem; lfts 9-13, 5-12 mm L; fls 5-15; petals cream, banner 10 mm L; fr glabrous.

4) *A. kentrophyta* var. *danaus*, Alpine Spiny Rattleweed: Stems densely matted, 5-25 cm L; lvs less than 2 cm L, 3-5-foliolate; spinule 1 mm L; racemes 2-3-fld; petals whitish with purple keel-tip. Alpine summits on metamorphic bedrock, 11,000-12,000', Mono Co. , e. Fresno Co., Jul-Sep.

4) *A. lentiginosus* vars. Mottled Rattleweed: Stems erect or prostrate; fls loosely or densely racemose, small, the keel not more than 8 mm L. Up to 12,000', mly e. SN, May-Oct.

5) *A. pulsiferae*, Pulsifer's Locoweed: Stems softly gray-villous, 1-3 dm L; lvs 2-4 cm L, with 7-13 oblanceolate crowded lfts; peduncles 4-30 mm L; racemes short, loosely 3-12-fld, petals whitish, purple-veined, the banner 6-9 mm L; pod 0.8-2 cm L, 6-11 mm in diameter. Sandy and stony openings, 4300-5500', Jun-Aug.

6) *A. purshii* vars., Pursh's Woolly-pod: Stems low or none; plant usu matted, cobwebby-pubescent; lvs 2-13 cm L, with 3-13 ovate-elliptic lfts; peduncles leafless; racemes subcapitately 2-10-fld; fls various; calyx 6-15 mm L; corolla 8-19 mm L. At 6000-11,000', e. SN, Inyo to Alpine Co. and then n., mly below 7000', Apr-Aug.

7) *A. ravenii*, Raven's Locoweed: Stems very slender, silvery, long-hairy, 1.5-10 cm L, prostrate; lvs 0.5-2.5 cm L; lfts elliptic; peduncles 1-5 cm L; racemes loosely 2-6-fld; banner 6-9 mm L; pod contracted at apex into a laterally flattened deltoid beak. Open stony slopes, on metamorphic strata, about 11,250', n. of Sawmill Pass, Jul-Aug.

8) *A. subvestitus*, Bear Valley Woolly-pod: Much like *A. purshii*; pod ovoid, flattened in the lower half, 8-15 mm L, 4-6.5 mm in diameter. Sandy mdws and flats, about 8000', Jun-Jul.

9) *A. whitneyi*, Whitney's Locoweed: Stems decumbent or ascending, dwarfed at high altitudes, 0.3-2.5 dm L; lvs 8-8.5 cm L; lfts ± oblong, 4-20 mm L, well spaced; peduncles 2-10 cm L; racemes 4-15-fld, 1-4 cm L in fr; petals with white wing-tips, early-fading, sometimes whitish, the banner 8-14 mm L; pod pendulous, 1.5-3(-4) cm L. Dry, gravelly crests and slopes, 6800-12,000', Inyo to Alpine Co., May-Sep.

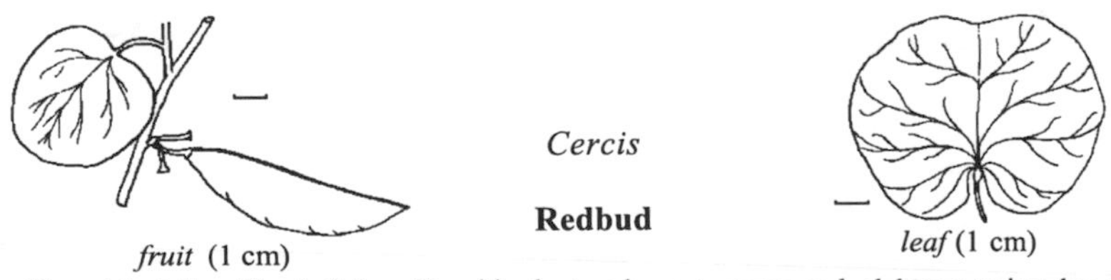

Cercis

Redbud

C. occidentalis: Shrub 2-5 m H, with clustered erect stems and glabrous twigs; lvs round to reniform, glabrous, glossy cordate at base, 7-9-veined, entire; petioles 1-3 cm L; fls purplish in small clusters; pods flat, 4-9 cm L, 2-2.5 cm W, often ± reddish purple. Dry slopes and canyons below 4000', Tulare Co. n., Mar-Apr. The buds, fls or young pods may be fried in butter or made into fritters; the buds and fls have an acid taste and are good in salads or the buds may be pickled. The bark may be used raw to treat diarrhea.

Glycyrrhiza

Wild Licorice

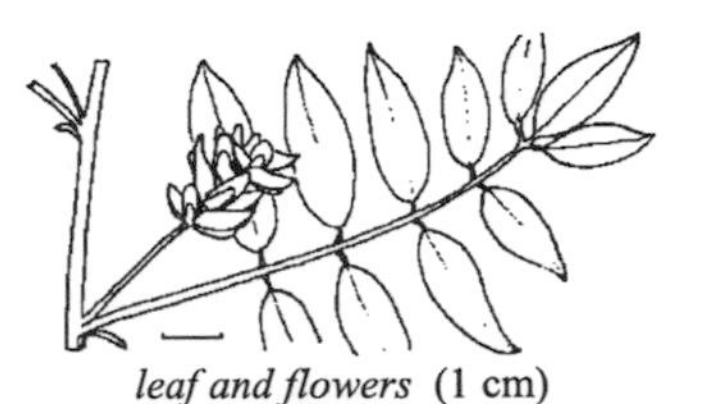

leaf and flowers (1 cm)

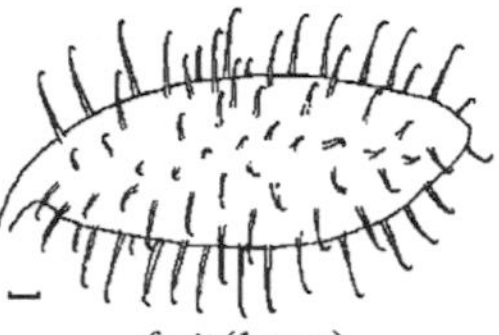

fruit (1 mm)

G. lepidota: Perennial herbs from thick rootstocks; stems erect, 3-10 dm H; lvs glandular; lfts 11-19, oblong to ovate lanceolate, 2-3 cm L, 1-2 cm W; fls in axillary spikes; corolla 8-12 mm L; pod oblong, 12-15 mm L, bur-like with hooked prickles. Occasional, as patches in low ground and moist waste places below 7500', May-Jul. The roots are good raw or added as flavoring to other food.

Hoita

Round-leaved Psoralea

inflorescence (1 cm)

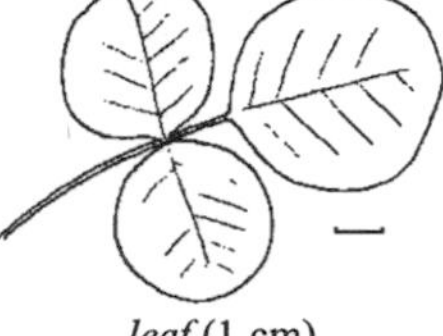

leaf (1 cm)

H. orbicularis: Stems prostrate; herbage heavy-scented, glandular; stipules large, petioles 5-25 cm L, erect; lfts round-obovate, 3-6 cm L, glabrous to short-pubescent; peduncles 10-30 cm H, erect; racemes dense, 5-20 cm L; calyx 15-20 mm L; corolla reddish purple, about 15 mm L, banner often with a white spot on each side; pod about 8 mm L, hairy. Moist places below 4000', Mariposa Co. n., May-Jul. The foliage of a related sp is used for tea, and the tubers of several similar spp are edible raw or cooked. Seeds of a related sp are reported to be poisonous.

Lathyrus

Sweet-pea

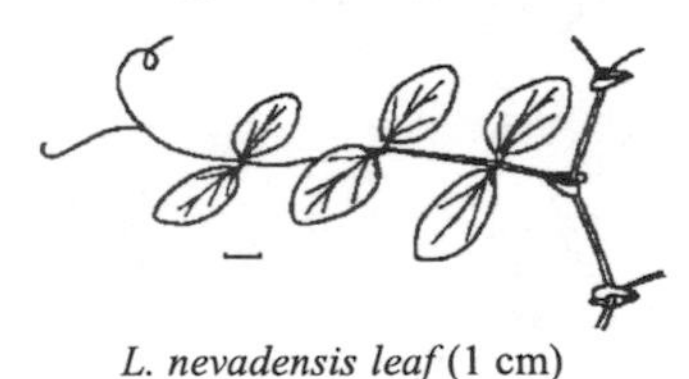

L. nevadensis leaf (1 cm)

L. jepsonii flowers (1 cm)

Erect or climbing perennials. Fls racemose, axillary; corolla showy. Stamens 9 and 1. Plants usu below 8000'. The frs of many members of the genus are edible in small amounts but cause paralysis and several secondary disorders if eaten in large quantities over a period of time. The paralysis is permanent though the other symptoms usu disappear with a change in diet. Caution should be used when eating the seeds or pods of any of the Sierran spp.

A Fls 10-20; stem usu 10 dm or more H
- B Stems winged; herbage puberulent; below 5000' — 2) *L. jepsonii*
- BB Stems angled, not winged; herbage glabrous — 5) *L. sulphureus*

AA Fls 2-8; stem usu 1-6 dm H
- B Lvs usu ovate, 8-10 mm W; calyx-lobes subequal — 4) *L. nevadensis*
- BB Lvs usu oblong to linear, mly 1-6 mm W; calyx-lobes unequal
  - C Banner reflexed at about 90 degrees; stipules often dentate — 1) *L. brownii*
  - CC Banner reflexed at more than 90 degrees; stipules ± entire — 3) *L. lanszwertii*

1) *L. brownii*, Brush Pea: Stems angled, erect, 2-6 dm H, sturdy; herbage glabrous except for calyx teeth; stipules 1/3-2/3 as long as lfts; lfts thickish, mly 8-10, linear to ovate, 1.5-3.5 cm L; tendrils often forked; fls purplish, aging bluish, 13-17 mm L; calyx 7-10 mm L; pods 3-5 cm L, 3-6 mm W, glabrous. Dry slopes, 4000-6000', SN (all), Apr-Jun.

2) *L. jepsonii* ssp *californicus*, Jepson's Pea: Stems mly 10-25 dm H, winged; stipules lanceolate, 1/4-1/2 the length of the lfts; lfts 10-14, ± scattered, elliptic to lanceolate, 4-7 cm L, tendrils well developed; fls mly crimson or rose-purple, about 2 cm L; calyx 13-16 mm L, the

teeth unequal; banner 18-22 mm L; pods 5-9 cm L, 6-9 mm W. Along watercourses and on sandy slopes, SN (all), Apr-Jun.

3) *L. lanszwertii*, Nevada Pea: Stems 2-8 dm H, angled, mly soft-pubescent; stipules mly narrow, 1/5-3/4 as long as lfts; lfts 4-10, usu paired, linear to oblong-elliptic, 1.5-3.5 cm L; tendrils well developed or short and scarcely prehensile; fls usu lavender to white, 13-16 mm L; calyx 5-8 mm L; pods glabrous, 4-6 cm L, 3-6 mm W. Dry slopes, 4000-6500', Eldorado Co. n., May-Jul.

4) *L. nevadensis*, Sierra Nevada Pea: Stems 1.5-4 dm H, angled, pubescent; stipules narrow, less than half as long as lfts; lfts 4-14, paired or scattered, linear to ovate, 2-10 cm L; tendrils well developed or lacking; fls mly pale to dark blue, reddish or purple, 15-25 mm L; calyx 6-10 mm L; banner obcordate, strongly reflexed; wings and keel usu paler than banner; pods 3-7 cm L, 4-9 mm W, glabrous. Dry slopes below 7000', Fresno Co. n., Apr-Jun.

5) *L. sulphureus*, Snub Pea: Stems 0.5-3 m H, angled, mly climbing, glabrous to pubescent; stipules lanceolate to ovate, often longer than lfts, ± dentate; lfts mly 8-10, narrowly lanceolate-elliptic to ovate, 2-5 cm L, the tendrils well developed; fls 10-15 mm L, tan to yellowish with some purple tinge or orange; calyx 9-12 mm L; pods 4-7 cm L, 4-6 mm W, glabrous. Dry slopes below 8000', SN (all), Apr-Jul.

L. micranthus upper stem (1 cm)

Lotus

Bird's Foot Trefoil

L. crassifolius inflorescence (1 cm)

Annual or perennials. Stamens 9 and 1. Calyx-teeth 5, subequal. Peduncles mly subtended by leafy bracts. Although not much is known about the Sierran spp, related spp contain alkaloids, saponins, and potentially toxic amounts of cyanogenic substances, and it is best to avoid eating any of the spp.

A Stipules expanded, not gland-like
  B Stipules green, resembling and almost as large lfts — 12) *L.. stipularis*
  BB Stipules membranous, not like the lfts
    C Herbage and stem densely villous; peduncles 1-3 cm L — 5) *L.. incanus*
    CC Herbage and especially stem ± glabrous; peduncles usu longer
      D Umbels 8-15-fld; stem 4-12 dm H; herbage often silver-gray — 2) *L . crassifolius*
      DD Umbels 1-10-fld; stem 1-5 dm H; herbage usu green
        E Fls whitish yellow to yellow or copper colored; claws usu included in calyx tube — 8) *L. oblongifolius*
        EE Banner and keel yellow, the wings white; claws exerted from calyx tube — 9) *L.. pinnatus*

AA Stipules reduced to dot-like often dark or reddish glands
  B Plants annual; pods usu over 1 cm L and easily observed
    C Lfts 6-10, rachis flattened; pod 2-3 cm L — 13) *L. strigosus*
    CC Lfts 3-5, rachis ± round; pods usu less than 2 cm L
      D Pods 0.5-1 cm L, densely villous; calyx and upper lvs usu villous — 4) *L. humistratus*
      DD Pods longer, glabrous; calyx and upper lvs glabrous to pubescent
        E Calyx-teeth shorter than tube; lfts mly 3-10 mm L — 6) *L. micranthus*
        EE Calyx-teeth longer than tube; lfts mly 10-15 mm L — 11) *L. purshianus*
  BB Plants perennial; pods often less than 1 cm L and inconspicuous
    C Stems erect, 2-6 dm H; lfts 7-9; pods 3-4 cm L, dehiscent — 3) *L. grandiflorus*
    CC Stems prostrate to erect, usu lower; lfts 3-7; pods less than 1 cm L, indehiscent
      D Infl 1-3-fld; lfts 3; plants gray with appressed hairs — 10) *L. procumbens*
      DD Infl 3-12-fld; lfts 3-7; plants silvery or green
        E Herbage densely silvery-tomentose; pod usu equal to calyx tube — 1) *L. argophyllus*
        EE Herbage pubescent but not silvery; pod clearly exceeding calyx — 7) *L. nevadensis*

1) *L. argophyllus*, Silver Lotus: Stems 2-10 dm L, much branched; lfts broadly oblanceolate to obovate, mly 4-12 mm L; umbels + sessile; bract unifoliolate; calyx densely woolly, about 4 mm L, the teeth half as long as tube; corolla yellow, 7-9 mm L (10-12 in var. fremontii), the banner brown or purple in age. Dry hills and slopes below 5000', Placer Co. s., Apr-Jul.

2) *L. crassifolius*, Broad-leaved Lotus: Stems stout, erect; stipules triangular-lanceolate; lvs 10-20 cm L; lfts 7-15, oval, 1-3 cm W; peduncles 3-8 cm L; bract usu present, remote from umbel, 1-5-foliolate; fls 8-15; pedicels 2-4 mm L; corolla 9-12 mm L, greenish yellow with some purplish red; pods 3.5-6.5 cm L. Dry bands and flats, 2000-8000', SN (all), May-Aug.

3) *L. grandiflorus*, Large-flowered Lotus: Stems with upwardly appressed hairs; lfts mly 7-9, obovate to elliptical, 7-20 mm L; peduncles exceeding lvs, 4-8 cm L; umbels 2-5-fld; bract 1-3-foliolate, 1-2 cm L; corolla yellow, aging red, 15-25 mm L. Mly dry slopes or disturbed sites (sometimes moist stream beads) below 6000', w. SN (all), May-Jul.

4) *L. humistratus*, Short-pod Lotus: Stems decumbent to ascending, 1-3 dm L; herbage densely villous; lfts obovate, entire, 5-15 mm L; fls subsessile, solitary in axils; corolla yellow, tinged red-purple in age, 5-7 mm L; pod 5-10 mm L. Common below 6000', w. SN (all), May-Jun.

5) *L. incanus*, Woolly Lotus: Stems several, ± erect, 1-3 dm H; herbage densely short-hairy, almost silky-tomentose; stipules ovate; lfts usu 7-9, elliptic to obovate, 7-15 mm L; fls few to many; bract usu of 5 lfts; petals 10-15 mm L, the banner reddish; wings white; pods reddish or dark yellowish brown, glabrous, 1.5-4 cm L. Below 5000', w. SN, Placer to Butte Co., May-Jun.

6) *L. micranthus*, Small-flowered Lotus: Stems slender, diffusely branched, 1-3 dm H; lvs 1-1.5 cm L; lfts oblong to oblanceolate or elliptical; peduncles 1-fld; bract of 1-3 lfts; corolla 4-5 mm L, pinkish or pale salmon, tinged or turning red; pods 15-20 mm L, constricted between seeds. Open flats and slopes below 5000', w. SN (all, Apr-May.

7) *L. nevadensis*, Sierra Nevada Lotus: Stems matted, 1-5 dm L; lfts 3-5, obovate, 5-10 mm L, silky hairy; peduncles 1-2.5 cm L on lower umbels, upper umbels often subsessile; corolla yellow, often tinged red; pod about 6 mm L, curved. Common, dry habitats, 3500-8500', May-Aug.

8) *L. oblongifolius*, Narrow-leaved Lotus: Stems erect or ascending, 2-5 dm H; stipules lanceolate; lfts mly 7-11, lance-linear to elliptical, 5-20 mm L; bract 1-3-foliolate, closely subtending the 1-5-fld umbel; corolla 10-15 mm L; banner yellow, often veined with purple; pods 2.5-4 cm L. Wet places below 8500', SN (all). May-Sep. A copper-flowered form (var. *cupreus*) is found at about 8000' in Tulare Co.

9) *L. pinnatus*, Pinnate-leaved Lotus: Stems 2-4 dm H; stipules narrowly ovate; lfts 5-9, oval to obovate, 1-2.5 cm L; peduncles 5-10 cm L; bract mly absent; fls 3-7; corolla 12-14 mm L; pods 4-6 cm L. Moist places, May-Jul.

10) *L. procumbens*, Procumbent Lotus: Stems usu much-branched, often woody at base, 4-8 dm H; lfts 5-12 mm L, oblanceolate to obovate; peduncle lacking or very short; calyx 2-6 mm L; corolla 6-12 mm L yellow or tinged reddish; fr exserted, pendent, 1-1.5 cm L. Dry flats and slopes in pine forests, below 7000', Tulare and Kern cos, May-Jul.

11) *L. purshianus* and var., Spanish-clover: Stems decumbent to erect, much branched; 1.5-8 dm H; lvs 1-2.5 cm L, mly 3-foliolate; lfts lance-oblong to elliptical; peduncles 10-15 mm L, 1-fld; bract 1-foliolate; corolla whitish, tinged with rose, 4-7 mm L; pods 1.5-2.5 cm L glabrous, deflexed. Common in dry fields and disturbed places below 7500', w. SN (all), May-Oct.

12) *L. stipularis*, Stipulate Lotus: Much like L. incanus, but 2-5 dm H; stipules lanceolate; lfts 9-19, 6-20 mm L, 2-5 mm W; umbels 4-10-fld; fls 10-12 mm L; banner and keel red-purple with white tips, wings white; pods 2-3 cm L. Dry, usu wooded slopes below 4000', SN (all), Apr-Jun.

13) *L. strigosus*, Strigose Lotus: Stems slender, decumbent to ascending; branches mly 0.5-3 dm L; herbage appressed stiff-hairy; lvs 1-2.5 cm L, the lfts linear-oblong to elliptic, acute, 5-12 mm L; peduncles mly exceeding lvs, the lower 1-fld, the upper 2-3-fld; bract none or 1-3-foliolate; corolla mly 6-10 mm L; pods 2-3 cm L. Common in dry, disturbed places below 5000', Tuolumne Co. s., May-Jun.

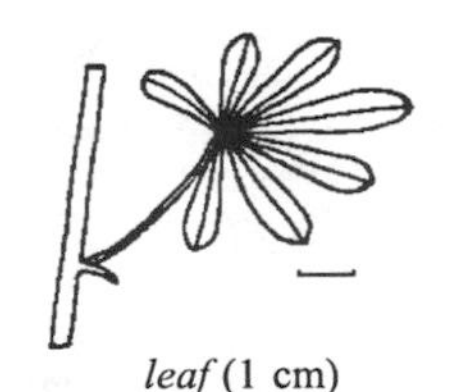
leaf (1 cm)

Lupinus

Lupine

inflorescence (1 cm)

Herbs or shrubs with distinctive palmately cmpd lvs. Fls perfect in terminal usu showy racemes. The seeds and herbage of older plants contain alkaloids producing a bitter taste and causing a weak pulse, convulsions, and paralysis if ingested in sufficient quantity. The lvs and unopened fls of young plants may be steamed and eaten with soup or stew. The genus is large and certain species are difficult to distinguish or of uncertain taxonomic status. Two species not included in the key deserve mention. *L. densiflorus* is a third annual species with blue fls. It is common in the foothill from Placer Co. s. and reaches 3500', in the Kern River Canyon. A rare visitor to the yellow pine forest in Merced Co. s. is *L. formosus* var. *robusta*. This taxon is usually associated with the drier oak savannah.

A Plants annual; below 5000'

B Fls blue (rarely white or pinkish) with white spot on banner; lfts often linear, 1-5 mm W ..... 7) *L. bicolor*

BB Fls yellow or orange; lfts oblanceolate to obovate, 3-15 mm W

C Fls orange; Fresno Co. 9) *L. citrinus*
CC Banner yellow, wings pink to purple; Butte Co. s. 22) *L. stiversii*
AA Plants perennial; banner usu blue, occasionally yellow or white
B Lfts glabrous on upper surface
C Plants 1-3 dm H; lvs ± basal; dry slopes, Plumas Co. n. 19) *L. onustus*
CC Plants usu 4-15 dm H; stems leafy; widespread in moist habitats
D Upper edges of keel ciliate on proximal 1/4-1/2 15) *L. latifolius*
DD Upper edges of keel glabrous 20) *L. polyphyllus*
BB Lfts ± hairy on upper surface
C Plants ± prostrate or in dense, low tufts; infl usu less than 8 cm L
D Infl dense, on ± naked stalk from base of plant; bracts persistent 16) *L. lepidus*
DD Infl ± open, on leafy stem; stem often prostrate or decumbent; bracts deciduous
E Lfts 7-9; stipules 2-5 mm L 8) *L. breweri*
EE Lfts 6-7; stipules 7-14 mm L 17) *L. obtusilobus*
CC Plants usu erect with leafy stems; infl mly over 8 cm L
D Plants shrubby; herbage silvery-silky; below 5000' 3) *L. albifrons*
DD Plants herbaceous throughout
E Stipules green, ± leafy, 4-8 mm W; Eldorado Co. s. 12) *L. fulcratus*
EE Stipules 1-2 mm W, often membranous; widespread
F Upper edge of keel glabrous
G Herbage usu green, minutely to generally pubescent
H Fls blue to violet; widespread 2) *L. albicaulis*
HH Fls yellow to orange; Plumas Co. n. 4) *L. angustiflorus*
GG Herbage white- to silvery-hairy, often densely so; fls often yellow
H Stems 5-8 dm, densely silvery- to white-woolly; Tulare Co. 18) *L. padre-crowley*
HH Stems mly shorter, with longer hairs
I Back of banner hairy; fls yellow; Plumas Co. 11) *L. dalesiae*
II Back of banner glabrous; fls yellow or blue; widespread 1) *L. adsurgens*
FF Upper edge of keel ciliate to densely hairy
G Stem prostrate to matted; upper edge of keel densely hairy 14) *L. grayi*
GG Stem ± erect; if not, upper edge of keel ciliate
H Calyx with 3 mm L spur at base; herbage green 5) *L. arbustus*
HH Calyx not spurred, sometimes gibbous at base
I Banner patch orange to red; keel densely hairy; bracts persistent 21) *L. pratensis*
II Banner patch yellow, white or absent; keel ciliate; bracts mly deciduous
J Stipules 12-30 mm L; bracts persistent 10) *L. covillei*
JJ Stipules 2-15 mm L; bracts ± deciduous
K Infl 3-7 cm L; peduncle 2-4 cm L; stem occasionally decumbent 17) *L. obtusilobus*
KK Infl 5-20 cm L, peduncle mly more than 4 cm L; stem erect
L Stipules 2-12 mm ; herbage usu silvery 6) *L. argenteus*
LL Stipules 10-15 mm; herbage usu green 13) *L. gracilentus*

1) *L. adsurgens*: Stems erect, 2-6 dm H; lvs cauline; stipules 5-15 mm L; petiole 2-6 cm; lfts 2-5 mm L; infl 2-20 cm L, peduncle 2-8 cm; pedicel 2-6 mm L; fls usu not whorled; bracts 2-8 mm L, deciduous; fls 10-14 mm L; petals pale yellow to violet; patch yellow to white; keel upcurved. Dry places below 12,000'.

2) *L. albicaulis*: Stems 4-10 dm H, often branched above; petioles 2-4 cm L; lfts 4-9, oblanceolate, 2-5 cm L; racemes 6-30 cm L; bracts 6-7 mm L; pedicels 4-6 mm L; fls 10-16 mm L (8-11 mm L in var. *shastensis*), whitish to purple, fading brown; pods 3-4 cm L, silky-villous. Dry slopes and openings, 2000-8500', Fresno Co. n., May-Sep.

3) *L. albifrons*, Silver Lupine: Rounded leafy shrub, appressed-silky, much-branched, 6-15 dm H; petioles 2-10 cm L; lfts 7-10, oblanceolate, 1-3 cm L; peduncles 5-13 cm L; racemes 8-30 cm L, the fls mly whorled, 10-14 mm L; pedicels 4-8 mm L; petals blue to red-purple or lavender; pods 3-5 cm L. Sandy to rocky places, SN (all). May-Sep.

4) *L. angustiflorus*, Christine's Lupine: Stems to 1 m H; lfts 5-9, 2-5 cm L; racemes 6-10 cm L; fls 10-12 mm L, yellow. Common in Lassen Volcanic National Park and vicinity.

5) *L. arbustus* sspp Crest Lupine: Stems several, 3-6 dm H, pubescent; the lower petioles 6-9 cm L, the upper shorter; lfts 7-9, oblanceolate, 2-4 cm L, 3-8 mm W; peduncles 2-4 cm L; racemes 6-12 cm L; upper calyx-lip 2-toothed, 3 mm L, lower about 4 mm L; fls 8-9 mm L, mly blue or violet; pods 2-2.5 cm L. Dry slopes and mdws, 5600-9600', SN (all), May-Jul.

6) *L. argenteus* var. *meionanthus*, Tahoe Lupine: Stems 2-5 dm H, white- to yellow-silky; lvs well distributed; lfts 6-9, 1.2-2.5 cm L, silky on both sides; racemes 5-14 cm L; pedicels 2-3 mm L; bracts deciduous; petals dull blue or lilac, the banner roundish; pods 1.5-2.5 cm L, hairy. Dry,

open places, 5000-9800', Madera and Mono cos. to Lassen Co., Jul-Aug.

var. *montigenus*, Alpine Lupine: Stems erect or decumbent, 3-7 dm L; lvs well distributed, the basal petioles 10-15 cm L, upper 2-5 cm L; lfts 7-9; peduncles 3-4 cm L; racemes 5-10 cm L; pedicels 4-5 mm L; calyx gibbous at base; petals 9-12 mm L, blue or violet, the banner longer than broad; pods 1.5-2 cm L. Inyo Co. Jul-Aug.

7) *L. bicolor*, Miniature Lupine: Stems 1-4 dm H, pubescent; petiole 1-7 cm, lfts 5-7, 1-4 cm L; infl 1-8 cm L, usu 5 whorls; peduncle 3-10 cm L; bracts 4-6 mm L, deciduous; pedicels 1-3 mm L; fl 4-10 mm L, usu blue with white banner spot; upper keel margins usu ciliate. Usu disturbed places below 5500', SN (all).

8) *L. breweri*, Brewer's Lupine: Stems matted; herbage silvery-silky; lvs crowded; petioles 1-5 cm L; racemes 3-5 cm L, usu dense; pedicels 1-3 mm L; fls violet with white or yellow center to banner; keel subglabrous; pods 10-15 mm L, silky. Common on dry slopes and flats, 4000-12,000', SN (all), Jun-Aug.

var. *grandiflorus*: like sp except peduncles 3-8 cm L; racemes 3-10 cm L; fls 9-11 mm L; keel usu densely ciliate. With sp, Tuolumne Co. s.

9) *L. citrinus*, Orange-flowered Lupine: Stems 1-2 dm H, white-villous, diffusely branched; lvs ± crowded, the lfts 6-8, oblanceolate, 1-2.5 cm L, on longer petioles; racemes 5-12 cm L; pedicels 2-5 mm L, deflexed after anthesis; banner suborbicular, slightly notched; pods 12-14 mm L. Rocky hills, 4000-5300', Apr-Jun.

10) *L. covillei*, Coville's Lupine: Stems 2-8 dm H; petioles 1-6 cm L; lfts 7-9, almost linear, acute, 3-10 cm L, loosely silky-hairy; peduncles 2-6 cm L; racemes 10-20 cm L; pedicels 2-4 mm L; bracts linear, hairy, persistent, 8-15 mm L; petals blue with darker veins, the banner about 8-9 mm L, wings and keel longer, the latter sparsely ciliate above near apex; pods densely hairy, 2.5-3 cm L. Rocky places, 8500-10,000', Tulare to Tuolumne Co., Jul-Sep.

11) *L. dalesiae*, Quincy Lupine: Stem erect, leafy; the lvs tomentose; stipules 5-15 mm L; petiole 1-3 cm; lfts 2-4 cm L; infl 5-15 cm L; peduncles 2-5 cm; fls ± whorled. Rare, pine forests, 3500-8000', n. SN.

12) *L. fulcratus*, California Green-stipuled Lupine: Much like and may hybridize with *L. albicaulis*; stipules leaf-like, lanceolate to ovate. Mly dry habitats, 6000-10,000', Tulare to Nevada Co., Jun-Aug.

13) *L. gracilentus*, Slender Lupine: Petioles very slender, 3-6(-12) cm L; lfts 5-8, ± linear, 3-8 cm L; peduncles 6-12 cm L; racemes 10-20 cm L; bracts 8-10 mm L, deciduous; pedicels 2-4 mm L; fls 10-13 mm L, usu whorled; petals blue; keel ciliate above on outer half; pods 2-3 cm L, villous. At 8000-10,500', Rock Creek Lake Basin to Yosemite National Park, Jul-Aug.

14) *L. grayii*, Gray Lupine: Stems usu several, 1-3 dm H, densely grayish pubescent; lfts 5-10; petioles 5-12 cm L; peduncles 5-15 cm L; racemes 5-15 cm L; fls 12-14 mm L; petals pale to deep purplish; pods 4-6-seeded, 2-4 cm L. Dry slopes below 7800', Plumas Co. s., May-Jul.

15) *L. latifolius*, Broad-leaved Lupine: Petioles, stems and even lvs often purplish; stems leafy; petioles 5-20 cm L; lfts mly 7-9, sometimes 5-12, broadly oblanceolate, 4-10 cm L; racemes 15-45 cm L, the fls whorled or scattered, 10-14 mm L; pedicels 6-12 mm L; upper calyx-lip notched, lower entire; petals whitish blue or purplish or with some pink, fading brown, the banner suborbicular, 9-10 mm W; pods brown, about 3 cm L; hairy. On moist slopes and along streams 3000-11,000', SN (all), May-Sep.

16) *L. lepidus*, Dwarf Lupine: Stems matted, less than 6 dm L; lvs basal; petiole 2-10 cm L; infl less than 30 cm L; bracts persistent; fls 5-10 mm L; banner back glabrous. Open slopes above 5000', SN (all). Includes several varieties previously recognized as spp:

var. *confertus* Sierra Lupine: Stems stout, several; herbage densely white- or golden-silky; lfts mly 7, elliptical-oblanceolate, 1.5-4 cm L; peduncles 1-7 cm L, surpassing foliage; racemes 5-30 cm L; bracts 8-9 mm L; fls 10-14 mm L, keel woolly-ciliate on upper edges; pods 10-18 mm L, silky-villous. Mdws and dampish places, 3000-8500', Plumas Co. s., Jun-Aug.

var. *culbertsonii*, Hockett Meadows Lupine: Stems to 3.5 dm H, hairy; petioles slender, 3-9 cm L; racemes 5-10 cm L with bracts 4-5 mm L; keel ciliate on upper surface near apex. Rare on rocky slopes, 8000-12,000', Tulare Co., Jul-Aug.

var. *lyallii*, Lyall's Lupine: Stems mly 5-12 cm L; petioles slender, 3-5 cm L; lfts oblanceolate, 4-12 mm L, appressed-silky on both sides; peduncles 5-10 cm L; racemes ± capitate; petals 8-12 mm L, mly ciliate; pods silky, 1-1.5 cm L. Dry ridges and summits, 8000-11,000', SN (all) Jul-Sep.

var. *sellulus* (including *L. lobbii*) Torrey's Lupine: Tufted herb, 1-2 dm H; petioles 2-8 cm L; lfts 5-8; peduncles usu 1-3 cm longer than lvs; infl dense; pedicels 1-2 mm L; fls 7-9 mm L; banner with whitish central spot. Moist to dry habitats, 4000-10,000', Mariposa Co. n., Jun-Aug.

17) *L. obtusilobus*, Obtuse-lobed Lupine: Stems ± ascending, 1.5-3 dm L; herbage silvery-silky; petioles 2-5 cm L; lfts 5-8, oblong-oblanceolate 4-6 mm W; peduncles 1-5 cm L; racemes 3-7 cm L; fls blue or lilac; pedicels 3-5 mm L; banner broader than long; keel ciliate; pods silky, 2.5-4 cm L. Gravelly summits, 5500-10,000', Jun-Sep.

18) *L. padre-crowley*, Father Crowley's Lupine: Stems erect from a woolly mat; stipules 5-10 mm L, petiole 2-3 cm L; lfts 6-9, 2-7 cm L; infl 5-20 cm L; peduncle 2-5 cm L; pedicels 2-4 mm L; fls 10-15 mm L, cream to pale yellow. Rare; rocky soil; above 8000'.

19) *L. onustus*, Plumas Lupine: Stems slender, decumbent at base, appressed-silky; lvs ± basal; petioles mly 8-13 cm L; lfts 5-9, oblanceolate, 2-4 cm L; peduncles 5-8 cm L; racemes 5-15 cm L; fls scattered, 8-12 mm L; pedicels 3-5 mm L; petals deep blue-violet; keel ciliate on upper edges; pods 3-4.5 cm L. Occasional, on dry slopes, 2900-5500', Plumas Co. n., Apr-Sep.

20) *L. polyphyllus*, Large-leaved Lupine: Stems stout, 5-15 dm H, petioles 1.5-3 dm L: lfts 5-9, 4-7 cm L, 0.5-3 cm W; racemes 15-60 cm L, lax, fls 10-14 mm L; pedicels 5-15 mm L; petals blue, purple or reddish; keel papillose but not ciliate; pods 5-9 seeded, 2.5-4 cm L. Wet places, 4000-8500', SN (all), May-Jul.

21) *L. pratensis*, Inyo Meadow Lupine: Stems clumped, 3-7 dm H; petioles slender, 2-8 cm L; lfts 5-9, linear to oblong, 3-6 cm L, 3-8 mm W, acute, peduncles 5-13 cm L; racemes 8-16 cm L; pedicels 2-3 mm L; bracts persistent 6-9 mm L; petals violet to blue or purple, the banner oval, about 10 mm L, wings slightly longer; pods about 2 cm L, loosely pubescent. Moist places, 4000-10,500', Tulare, Fresno, Mono, and Inyo cos., May-Aug.

22) *L. stiversii*, Harlequin Lupine: Stems 1-4.5 dm H, branched; lvs scattered; petioles 3-8 cm L; lfts 6-8, ± obovate, 1-4 cm L; peduncles 3-8 cm L; racemes 1-3 cm L; banner 10-15 mm L; pod about 2 cm L, glabrous. Sandy or gravelly habitats below 4600', May-Jul.

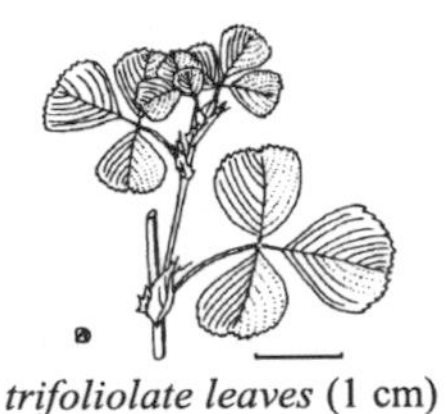

trifoliolate leaves (1 cm)

Medicago

Burclover

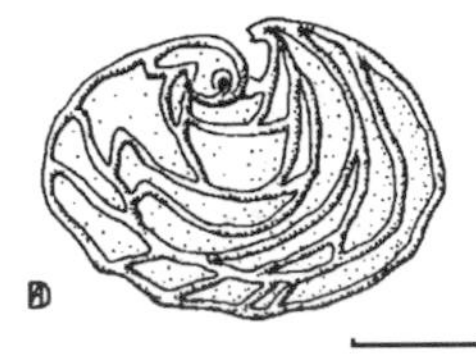

markings on pod (1 mm)

M. lupulina: Stems prostrate to ascending, 1-3 dm L, hairy; lfts orbicular, 1-2 cm W; infl spherical, 10-15 fld, dense; corolla 2-3 mm L, yellow; fr coiled, about 2 mm L, lacking prickles. Disturbed habitats to 8000', SN (all).

Melilotus

Sweet-clover

upper stem (1 cm)

M. albus: Stems erect, 1-2 m H; petioles 5-15 mm L; stipules 5-7 mm L; lfts mly lanceolate to oblanceolate, 1-2 cm L, serrate; peduncles usu 3-5 cm L; racemes 5-10 cm L, spike-like; fls white, 4-6 mm L; pods ovoid, glabrous. Occasional in disturbed habitats below 4000', SN (all), Jun-Sep. The young lvs may be eaten raw or boiled. Fr may be used as seasoning for soups. Mature lvs and stems contain coumarin (an anticoagulant) which is toxic in moderate amounts and should be avoided.

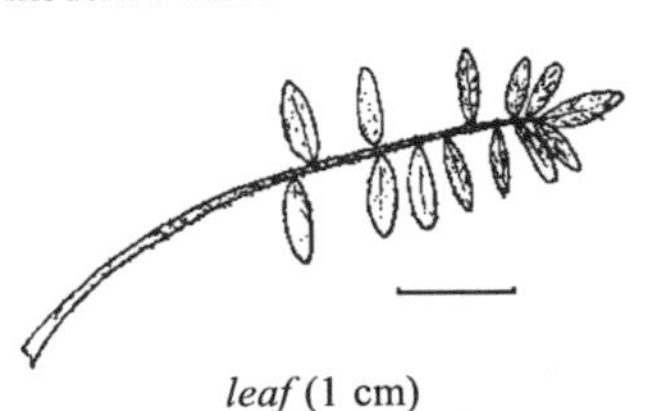

leaf (1 cm)

Oxytropis

Locoweed Oxytrope

fruits (1 cm)

Cushion-like perennials much like *Astragalus*. Lfts asymmetrical at base, about 1 cm L. Keel-petals abruptly narrowed into a cuspidate or mucronate apex. Neither sp is known to be poisonous (loco producing), but both should probably be treated as potentially toxic.

Plants green, the herbage glandular and resinous ... *O. borealis*
Plant silvery, the herbage silky-pilous, not glandular ... *O. parryi*

O. borealis, Sticky Oxytrope: Lvs 6-10 cm L, with 25-39 green, sparsely pubescent lfts less than 1 cm L; racemes about 10-fld, oblong, slightly exserted on erect or arched scapiform peduncles; petals whitish or red-purple, the banner about 12 mm L; pod about 15 mm L. Bare crests and talus slopes, 11,500-12,200', Inyo Co., Jul-Aug.

O. parryi, Parry's Oxytrope: Lvs 1-4 cm L, clustered in few rosettes, with 7-15 lfts 3-8 mm L;

racemes mly 2-fld, subcapitate, usu well exserted; pod erect, 13-20 mm L. Dry knolls and rocky ridges near timberline and above, 11,000-12,000', e. SN in Inyo and Mono cos., Jun-Jul.

Trifolium

Clover

T. tridentatum upper stem (1 cm) *T. longipes inflorescence* (5 mm)

Herbs with mly palmately trifoliate lvs and adnate stipules. Stamens 9 and 1. All species are edible raw but may cause bloat if eaten in large quantities. To avoid this effect, the lvs may be steamed or cooked or soaked for several hours in salt water. Prepared in these ways, clover makes an excellent protein supplement. The prepared lvs may be dried and stored for future use. A tea may be made from the dried fl-heads. Seeds are also edible.

A Heads without an invol. at base of fls
- B Lfts 3-9 (at least some lvs on plant with more than 3 lfts)
  - C Lfts 7-9; Infl 3-5 cm W — 14) *T. macrocephalum*
  - CC Lfts 3-7; infl 1.5-2.5 cm W
    - D Annual; fls erect at bloom; n. SN — 1) *T. andersonii*
    - DD Perennial; fls reflexed at bloom; rare, Sierra and Nevada cos. — 11) *T. lemmonii*
- BB Lfts 3; widespread
  - C Plants annual; herbage mly glabrous; fls pink to purple
    - D Stipules 8-10 mm L; calyx-lobes glabrous — 8) *T. gracilentum*
    - DD Stipules 15-30 mm L; calyx-lobes ciliate — 5) *T. ciliolatum*
  - CC Plants perennial; herbage often pubescent
    - D Peduncles axillary; pedicels mly 3-5 mm L
      - E Stems 3-6 dm H; lfts 2-3 cm L; stipules 1-2.5 cm L — 9) *T. hybridum*
      - EE Stems 1-3 dm H; lfts 0.5-2 cm L; stipules less than 1 cm L
        - F Peduncles 3-4 cm L; corolla rose to cream white, 5-6 mm L — 4) *T. breweri*
        - FF Peduncles 5-30 cm L; corolla white to pale pink, 7-10 mm L — 18) *T. repens*
    - DD Peduncles terminal or subterminal; pedicels short or none
      - E Pedicels recurved and fls reflexed in full bloom
        - F Rachis of infl prolonged beyond the head; peduncle often bent at tip — 10) *T. kingii*
        - FF Rachis of infl not apparent beyond head
          - G Herbage densely pubescent; peduncle often recurved at tip — 7) *T. eriocephalum*
          - GG Herbage glabrous; head erect, only fls reflexed — 13) *T. macilentum*
      - E Pedicels straight or none; fls erect in full bloom
        - F Calyx 3-4 mm L; corolla rose to purplish, 8-12 mm L — 3) *T. bolanderi*
        - FF Calyx 7-9 mm L
          - G Lfts usu lance-linear; corolla whitish or tinged purple, 8-12 mm L — 12) *T. longipes*
          - GG Lfts mly oblong; corolla red, 12-15 mm L; Nevada Co. n — 2) *T. beckwithii*

AA Heads with an invol. at base or fls solitary
- B Invol. bell-shaped to bowl-shaped
  - C Plant glabrous; lobes of invol. toothed; lvs rarely obcordate — 6) *T. cyathiferum*
  - CC Plant ± hairy; lobes of invol. entire; lvs usu obcordate — 15) *T. microcephalum*
- BB Invol. flat, rotate
  - C Plant perennial; fls often few in head
    - D Lvs usu less than 1 cm L; fls 1-8 in head — 16) *T. monanthum*
    - DD Lvs usu over 1.5 cm L; fls many in head — 20) *T. wormskioldii*
  - CC Plant annual; fls many in head; corolla often with a central dark spot
    - D Plant viscid-pubescent; lvs definitely spinose-dentate — 17) *T. obtusiflorum*
    - DD Plant glabrous; lvs shallowly dentate — 19) *T. wildenovii*

1) *T. andersonii*: Plant tufted or cushion-forming, lacking stem (peduncle ± erect); herbage silvery or gray; stipules entire, persistent; lfts entire, 5-20 mm L; infl 1-2.5 cm W, head-like. Many habitats, to 7,000'.

2) *T. beckwithii*, Beckwith's Clover: Glabrous with stout ascending stems, 0.5-3 dm L; petioles 2-10 cm L; stipules 1-2.5 cm L; lfts 0.5-4 cm L, serrate; peduncles 3-15 cm L; heads roundish, dense, many-fld, 1.5-3 cm across; pedicels reflexed in age. Moist mdws, 4000-7000', May-Aug.

3) *T. bolanderi*, Bolander's Clover: Stems glabrous, many decumbent or ascending, slender, 1-

2 dm L; petioles mly basal, 1-7 cm L; stipules entire, 5-10 mm L; lfts 0.6-1.5 cm L; peduncles 7-15 cm L; heads ovoid, 10-15 mm L; pedicels recurved; corolla lavender, narrow, about 8-9 mm L. Wet mdws, about 7000', Jun-Jul.

4) *T. breweri*, Brewer's Clover: Herbage glaucous, often pubescent; petioles slender, mly 1-3 cm L, stipules 6-8 mm L; lfts obovate, 0.4-1.5 cm L, denticulate, obtuse to notched at apex; peduncles slender, 2-4 cm L, recurved at top in age; heads 1-15-fld. Wooded slopes mly below 7000', Madera Co. n., May-Aug.

5) *T. ciliolatum*, Tree Clover: Stems 2-5 dm H; petioles to about 10 cm L; lfts oblong to ± obovate, 1-3 cm L, obtuse, entire to spinose-serrulate; peduncles 5-15 cm L; heads ovoid, 10-20 mm L; pedicels reflexed in age; corolla pinkish purple, 6-7 mm L, the banner inflated at the base. Common on open and grassy slopes below 5000', w. SN (all), Apr-Jun.

6) *T. cyathiferum*, Bowl Clover: Annual; stems erect or decumbent, 1-3 dm L; stipules narrowly cut or toothed; lfts obovate to elliptic-oblong, ± spinulose-denticulate, 1-2 cm L; peduncles slender, mly 2-9 cm L; invol. 8-10 mm W; calyx-teeth 1-3 times trichotomously forked, equaling the pinkish corolla which is 7-10 mm L. Moist places below 8000', SN (all), May-Aug.

7) *T. eriocephalum*: Stems ± erect, usu unbranched; lvs basal and cauline; lower stipules sheathing; lfts 1-4 cm L, elliptic to ovate; infl head-like, 1.5-3 cm across; peduncle 5-15 cm L; fls soon reflexed; corolla dull white to yellowish.

8) *T. gracilentum*, Pin-point Clover: Stems slender, erect to procumbent, 1-4 dm L; petioles usu 2-7 cm L; stipules entire; lfts obovate, 0.6-1.5 cm L, serrulate, notched at apex; peduncles slender, mly 2-6 cm L; heads 6-10 mm L; pedicels reflexed in age; petals pink to reddish purple, 5-6 mm L. Common in open, grassy places below 5000', w. SN (all), Apr-Jun.

9) *T. hybridum*, Alsike Clover: Petioles usu 2-8 cm L; lfts obovate to ovate, sharp-serrulate; peduncles to about 1 dm L; heads many-fld, globose, 1.5-2 cm across; pedicels 4-8 mm L; calyx 3-4 mm L; corolla pink, 7-9 mm L. Mly damp places below 6000', Mariposa Co. n., May-Oct.

10) *T. kingii*, Shasta Clover: Stems slender, ascending, glabrous, 1-4 dm H; petioles mly basal, to 12 cm L; stipules entire, 6-12 mm L; lfts 1-4 cm L; peduncles slender, 5-12 cm L; heads many-fld, short-ovoid, mly 1-2 cm L, with rachis produced above the fls; pedicels reflexed in age; corolla rose to purplish, 9-12 mm L. Moist ± wooded places, 3800-8000', Jun-Aug.

11) *T. lemmonii*, Lemmon's Clover: Stems 1.5-2 dm L; stipules coarsely few-toothed, to about 1 cm L; lfts coarsely toothed, 6-12 mm L; peduncles 5-12 cm L, far surpassing lvs; heads 1.5-2 cm across, many-fld; pedicels reflexed in age; corolla bright yellow, 1 cm L. Slopes and valleys, 5000-7000', Jun-Jul.

12) *T. longipes*, Long-stalked Clover: Stems decumbent to erect, 0.5-4 dm L, ± glabrous; stipules entire, 0.5-2 cm L; lfts occasionally obovate, 0.5-3(-6) cm L, serrulate; peduncles mly 3-10 cm L, heads mly ovoid, dense, 1-2 cm L. Moist places below 9000', SN (all), Jun-Sep.

13) *T. macilentum*, Dedecker's Clover: Stems tufted, glabrous, ascending; lvs mly basal; basal stipules clasping stem; lfts 5-40 mm, lanceolate, thick, usu serrate; infl head-like, 1.5-3 cm W; fls soon reflexed; corolla pink to pale violet. Rare, alpine habitats, 7000-11,500', s. SN.

14) *T. macrocephalum*:, Large-headed Clover: Stems ascending; herbage hairy; lvs basal and cauline; basal stipules membranous, the others green; lfts obovate, thick; infl head- or raceme-like, 3-6 cm L; corolla 2-3 cm L often bicolored. Rocky flats and ridges below 8000', n. SN.

15) *T. microcephalum*, Small-headed Clover: Mly slender-stemmed annual, the stems procumbent to ascending, 2-4 dm L; stipules 5-10 mm L; lfts serrate, 0.5-1.5 cm L; peduncles very slender, usu 3-7 cm L; invol. about 5-8 mm W, the lobes 7-10, lanceolate; corolla rose to white, about 6 mm L. Open, grassy places below 8500', w. SN (all), Apr-Aug.

16) *T. monanthum*, Carpet Clover: Stems slender, decumbent to suberect, 1-35 cm L; stipules mly subentire; lfts obcordate to oblanceolate, 0.4-2 cm L, ± toothed; infl 1-8-fld; invol. small, 2-4-lobed; peduncles often bent below the invol.; corolla 8-12 mm L, white toward tip, purple near base, keel purple-tipped. Mly wet places, 5000-11,500', Plumas Co. s., Jun-Aug.

17) *T. obtusiflorum*, Clammy Clover: Stems ± hollow, 3-5 dm L, decumbent to erect; stipules deeply cut or toothed; lfts lance-linear to narrow-obovate, 2-3 cm L; peduncles 3-8 cm L; invol. 12-16 mm W, deeply cut; corolla about 12-14 mm L. Moist places below 5000', w. slope of SN (all), Apr-Jul.

18) *T. repens,* White Clover: Stems glabrous, creeping, rooting at nodes; lfts 1-2 cm L, ± obcordate, serrulate; heads globose, 1.5-2.5 cm W; pedicels reflexed in age. Common in disturbed habitats, mly below 7000', SN (all), May-Aug.

19) *T. wildenovii,* Tomcat Clover: Stems erect to ± decumbent, 1-4 dm L; lower stipules entire, the upper deeply cut; lfts linear to lance-oblong, 1.5-3.5 cm L; peduncles 5-9 cm L; invol. unevenly deeply cut but not lobed, usu 10-15 mm W; corolla 12-15 mm L, red-purple, the banner pale toward tip, the wings dark. Common in grassy places below 5000', w. SN (all), Apr-Jun.

20) *T. wormskioldii*, Mountain Clover: Stems decumbent, 1-3 dm L, branched, glabrous; stipules deeply cut or toothed, 1-3 cm L; petioles usu 2-7 cm L; lfts oblanceolate to wider, 1-3 cm L, finely serrulate; peduncles mly 2-6 cm L; invol. mly 10-15 mm W, mly lobed and then toothed; corolla about 12 mm L, the banner white to light purple, darker at base; the wings and keel dark with light tips. Wet places below 10,000', SN (all), May-Oct.

Vicia

Vetch

leaves (1 cm) *flower* (1 cm)

V. americana: Perennial vines; stems 6-12 dm L, glabrous to pubescent; stipules incisely toothed; lfts 4-8 pairs, ovate- to oblong-elliptic, obtuse, 1-4 cm L; peduncles shorter than lvs, 4-9-fld; corolla 16-18 mm L, purplish aging blue, pods 3-4 cm L, 7-9 mm W. Open places, w. SN (all), Apr-Jun. Many related spp are toxic, containing alkaloids and cyanogenic substances.

FAGACEAE - Beech Family

Plants woody shrub or trees with simple, alt, petioled lvs. Lvs with small mly deciduous stipules. Plants monoecious.

Fr a spiny bur enclosing 1-3 nuts; lvs yellowish beneath *Castanopsis*
Fr an acorn with a cup-like invol.
  Lvs entire, ± prominently veined; invol. cup with recurving scales *Lithocarpus*
  Lvs usu serrate; invol. cup lacking recurving scales *Quercus*

Castanopsis

Chinquapin

branchlet and fruit (1 cm) *staminate catkins* (1 cm)

Evergreen shrubs. Staminate fls in 3's arranged in the axils of bracts along the elongate catkins. Calyx 5-6-parted. Pistillate fls 1-2 in an invol. Fr maturing in the second season; the spiny invol. enclosing 1-3 nuts. The nuts ripen in September and can be eaten raw or roasted.

Lvs acuminate, ± folded along midrib; Eldorado Co. *C. chrysophylla*
Lvs mly obtuse, flat; widespread *C. sempervirens*

C. chrysophylla var. *minor,* Golden Chinquapin: Shrub or small tree to about 5 m H; lvs lanceolate to oblong, 5-15 cm L. Infrequent, gravelly or rocky ridges and slopes below 6000', Jun-Sep.

C. sempervirens, Bush Chinquapin: Round-topped shrub, 0.5-2.5 m H, with smooth brown or gray bark; lvs oblong, mly obtuse, 3-7 cm L, 1-3 cm W, subentire, yellowish gray-green above, golden- or rusty-tomentose beneath; petioles 10-15 mm L; staminate fls ill-smelling; burs 2-3 cm thick, 4-valved; seeds 8-12 mm L, hard-shelled. Dry, rocky slopes and ridges, 2500-11,000', SN (all), Jul-Aug.

Lithocarpus

Tanbark-oak

acorn (1 cm) *leaves* (1 cm)

L. densiflora var. *echinoides*: Evergreen shrub to about 3 m H; lvs oblong, 1.5-6 cm L, pale beneath; staminate catkins many, erect, ill-smelling. Occasional on dry slopes below 8000', Mariposa Co. n., Jun-Aug. The nuts are edible when treated as in *Quercus*.

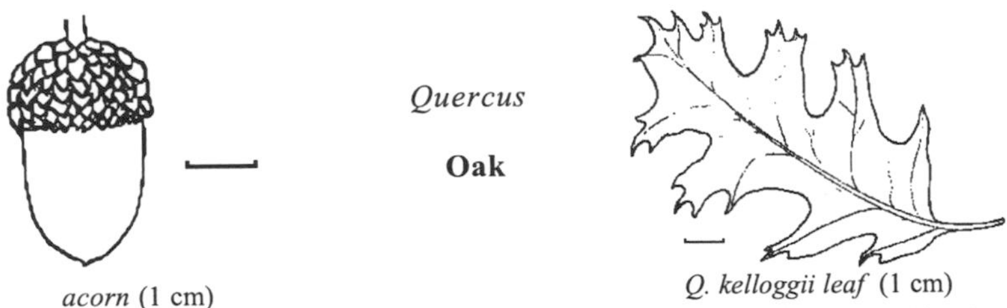

Quercus

Oak

acorn (1 cm) *Q. kelloggii leaf* (1 cm)

Deciduous or evergreen trees or shrubs with ± contorted branches. Staminate catkin slender, drooping or spreading, one or more from the lower axils of the current season's growth. Pistillate fls solitary in upper axils. The acorns of all spp are edible although some spp are better than others, the scrub oak being rarely used. In all Sierran spp the tannic acid must first be leached from the fr before the fr can be eaten. The nut is removed from the shell and dried. The dried fr is ground to a fine meal and soaked in a lye solution (hardwood ashes) or in water (hot works faster than cold) until the acid is removed (5-10 waters). Placing the meal in a cloth sack in flowing water for several days will also work. The meal is then dried and stored for use in soups (which need seasoning), mush or bread. Soaking the acorn overnight or drying the acorn will aid in the husking process. The meal has been leached of its tannic acid when it has lost its bitter taste. Mixing the acorn meal with corn meal or white flour makes a more palatable bread. The dried husks can be used as a coffee substitute. The herbage and bark also contain tannic acid which make these parts unpalatable.

A Lvs pinnately lobed, deciduous
  B Lobes with bristle-tipped teeth; plant usu a tree — 4) *Q. kelloggii*
  BB Lobes ± blunt; plant a shrub — 3) *Q. garryana*
AA Lvs entire to spinose, evergreen
  B Lf-blades shiny green above; below 5000'
    C Acorns ± globose; plant usu a shrub — 1) *Q. berberdifolia*
    CC Acorns elongate; plant usu a tree — 6) *Q. wislizenii*
  BB Lf-blades dull above
    C Acorns 1-1.5 cm L; cupules thin-walled; plant a shrub — 5) *Q. vaccinifolia*
    CC Acorns 2-3 cm L; cupules thick-walled; plant often a tree — 2) *Q. chrysolepsis*

1) *Q. berberdifolia*, Scrub Oak: Mly shrubby, sometimes arborescent, usu 1-3 m H; lf-blades oblong to elliptic, mly dentate, 1.5-2.5 cm L, paler and pubescent beneath; petioles mly 2-3 mm L; acorns ovoid, 1-3 cm L; cups 10-15 mm W, covering most of acorn, the walls thick, the scales usu rough. Common on dry slopes, Tehama Co. s., Apr-May.

2) *Q. chrysolepsis,* Canyon Oak, Maul Oak: Shrub or tree with pale gray, rather smooth, scaly bark and pubescent young twigs; lvs entire to spinose even on the same plant, usu oblong, flat, 2-6 cm L, grayish or yellowish tomentose beneath or later glabrate and glaucous; petioles mly 3-8 mm L; cup thick-walled, shallowly bowl-shaped, mly 2-2.5 cm W. Common in canyons and on moist slopes below 6500', SN (all), Apr-May. May hybridize with *Q. vaccinifolia.*

3) *Q. garryana* var. *breweri,* Brewer's Oak: Bark smooth; gray; lf-blades mly 5-9 cm L; petioles about 1 cm L. Dry slopes, SN (all), Apr-Jun.

4) *Q. kelloggii,* California Black Oak: Tree with broad, rounded crown, 10-15 m H, shrubby at higher elevations; bark dark, rough in age; lf-blades usu 3-lobed on each side, each lobe with 1-4 bristle-tipped teeth, bright green and mly glabrous above, paler beneath and often stellate-pubescent when young, 7-20 cm L; petioles 2.5-5 cm L; acorns oblong, 2.5-3 cm L, about 1.5 cm thick; cups 2-2.5 cm W; scales thin. Common, to 7000', SN (all), Apr-May.

5) *Q. vaccinifolia*, Huckleberry Oak: Stems spreading to about 12 dm H, often prostrate; branchlets flexible; lf-blades oblong-ovate, mly entire, 1.2-3 cm L; petioles mly 3-6 mm L; acorns round-ovoid, 1-1.5 cm L; cup 1-1.4 cm W, thin-walled, the scales white-tomentose. Dry ridges and rocky places, 3000-10,000', Fresno Co. n., May-Jul.

6) *Q. wislizenii*, Interior Live Oak: tree 10-20 m H, with round top and smooth bark 5-6 cm thick, broadly ridged below in age; lf-blades mly oblong, entire to spine-toothed, flat, 2-4 cm L; petioles 0.5-1 cm L; acorns slender, 2-4 cm L, 7-12 mm thick, often with longitudinal dark bands; cup 12-14 mm W. Dry habitats, SN (all), Apr-May.

GARRYACEAE - Silk Tassel Family

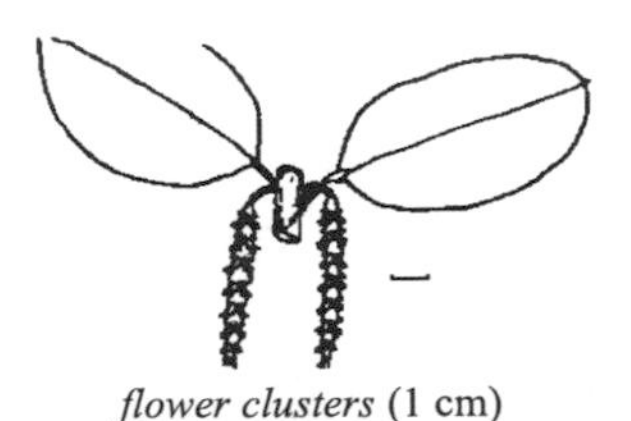

Garrya

Silk-tassel Bush

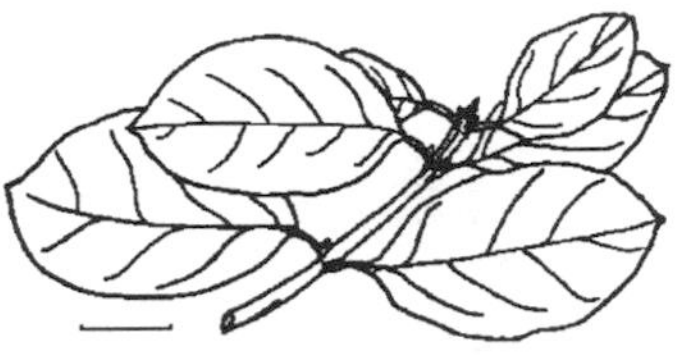

flower clusters (1 cm) *leaves* (1 cm)

Evergreen shrubs with opp, leathery, short-petioled lvs. Fls small in pendulous catkin-like clusters which are usu silky-hairy. Staminate fls pedicelled, in 3's in axil of each bract. Pistillate fls subsessile, solitary in axil of each bract. Fr a dark purple to black berry, the bitter pulp surrounding the 1-2 seeds. Lvs contain garryine, an alkaloid which is used medicinally, but is not good to eat.

Lvs grayish above, at least the younger pubescent below *G. flavescens*
Lvs glossy above, glabrous below *G. fremontii*

G. flavescens var. *pallida*, Pale Silk-tassel: Stems erect, 1.5-3.5 m H; lvs elliptic, 3-6 cm L, 15-35 mm W; petioles 3-10 mm L; staminate catkins 3-4 cm L; fruiting catkins densely silky, 3-5 cm L; fr 6-8 mm W, densely silky. Dry slopes, 3000-8000', Fresno Co. s., Mar-Apr.

G. fremontii, Fremont's Silk-tassel: Stems erect, 1.5-3 m H; the young twigs soon glabrate and red-brown; lvs oblong-elliptic, 2-5 cm L; petioles to about 1 cm L; staminate catkins 7-20 cm L, yellowish, pistillate 4-5 cm L; fr about 6 mm W. Dry, brushy slopes, mly below 7500', SN (all), Mar-Apr.

GENTIANACEAE - Gentian Family

Herbs with colorless bitter juice. Lvs usu opp, sessile and simple. Calyx persistent, 4-12-lobed or toothed. Corolla withering-persistent. Stamens inserted on corolla-tube or -throat, alt with lobes. Fr a capsule.

A Corolla nearly cleft to base, with fringed glands on upper surface *Swertia* p. 123
AA Corolla with definite basal tube, without such glands
B Corolla rose, sometimes white; fls in terminal clusters *Centaurium* p. 121
BB Corolla blue to white or yellow; fls usu solitary
C Perennial from stout root-crown; corolla with small 'appendages' between lobes *Gentiana* p. 122
CC Mly annuals from slender roots: corolla without appendages
D Corolla 1-2 cm L *Gentianella* p. 122
DD Corolla 2-5 cm L *Gentianopsis* p. 122

Centaurium

Canchalagua

C. venustum: Stems mly unbranched, 0.5-3 dm H; lvs broadly lanceolate to ovate, 1-2.5 cm L, acute; pedicels 2-25 mm L; calyx segments 6-9 mm L, corolla rose with red spots in the white throat, sometimes albino; corolla-lobes 5-10 mm L, about as long as tube; style long-exserted; stigmas fan-shaped, broader than long. Dry habitats, mly below 5000', SN (all), May-Aug. Herbage has a very bitter taste and may contain toxic quantities of alkaloids.

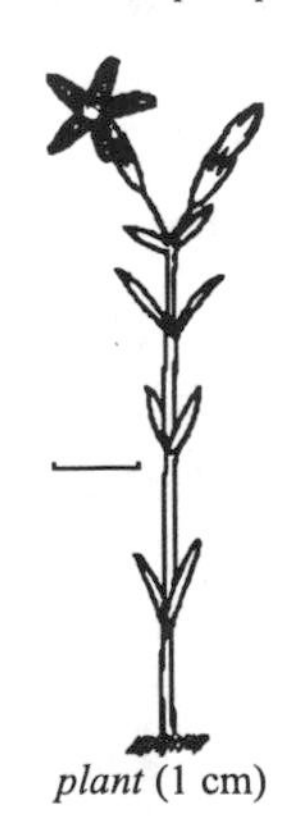

plant (1 cm)

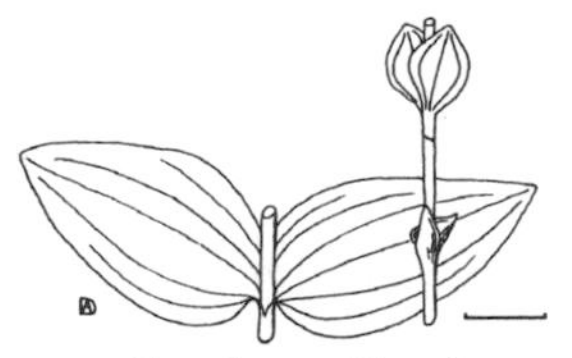

G. calycosa (1 cm)

Gentiana

Gentian

G. newberryi stem and flower (1 cm)

Glabrous perennials. Stems usu unbranched below infl. Style short or none. Fls 4-5-merous. Corolla tube bell-shaped, the lobes shorter than the tube and spreading. The sinus between adjacent lobes usu containing a small 'appendage' divided into 2 triangular parts. Herbage and roots of most spp bitter. The two spp hybridize in Tulare Co.

Stems 15-30 cm H; plant without basal rosette of lvs *G. calycosa*
Stems 4-12 cm H, arising laterally below a basal rosette of lvs *G. newberryi*

G. calycosa, Explorer's Gentian: Stems erect or ascending, from stout root-crown; lvs ovate to almost round, 2-4 cm L, ± cordate; fls mly 1-3, subtended by lance-ovate bracts; calyx-tube 6-9 mm L, the lobes ovate, unequal, 3-8 mm L; corolla deep blue, rarely violet, 1.5-3.5 cm L, the lobes 7-10 mm L. Moist places, 4000-10,500', SN (all), Jul-Sep.

G. newberryi, Alpine Gentian: Stems somewhat decumbent, 1 to few; lvs broadly spatulate, 2-6 cm L including the broad petiole; the upper narrower, shorter, subsessile; calyx-tube 10-25 mm L, the lobes 6-12 mm L, acute; corolla usu white within, with greenish spots, sometimes pale blue to purplish, the exterior usu with dark brownish purple bands on and below lobes, the tube 2-3 cm L, the lobes 5-8 mm L. Moist mdws and banks, mly 7000-12,000,' SN (all), Jul-Sep.

G. amarella (1 cm)

Gentianella

Little Gentian

cross-section of flower (5 mm)

Glaborous annuals with basal and cauline lvs. Calyx tube shorter than lobes; corolla with spreading lobes shorter than tube.

Fls often clustered in cymes, usu 5-merous; plants 5-50 cm H; common *G. amarella*
Fls solitary, usu 4-merous; plants 3-8 cm H; rare *G. tenella*

G. amarella, Felwort: Stems mly branched, slender, erect; lvs lanceolate to oblong or the lower spatulate, 1.5-3.5 cm L; fls axillary and terminal; pedicels slender, 5-50 mm L; calyx-tube 1-4 mm L, the lobes unequal, 2-10 mm L, lanceolate; corolla blue, the lobes lanceolate, acute. Moist places, 4500-11,000', SN (all), Jun-Sep.

G. tenella, Dane's Gentian: Stems usu much-branched at base, slender, 3-8 cm H; lvs oblong to spatulate, 8-15 mm L, the cauline somewhat reduced; fls terminal on slender elongate pedicels; calyx deeply cleft, 5-9 mm L; corolla whitish or greenish to bluish, 8-12 mm L, the lobes lanceolate, acute. Mdws, 8000-12,200', Rock Creek Lake Basin, Whitney Meadows, Jul-Aug.

G. holopetala (2 cm)

Gentianopsis

Fringed Gentian

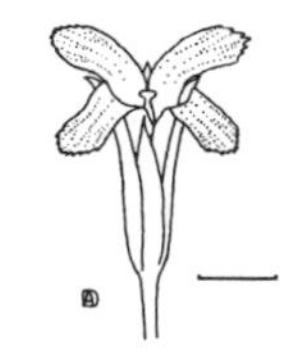

G. simplex (1 cm)

Glabrous annual or perennial. Fls solitary, 4-merous. Herbage bitter.

Corolla-lobes half as long as tube; stem with usu 1-3 pairs of lvs *G. holopetala*
Corolla-lobes as long as tube; stem with usu 3-6 pairs of lvs *G. simplex*

G. holopetala, Sierra Gentian: Stems 0.5-4 dm H; leafy below and terminating in long naked 1-fld peduncles; lower lvs spatulate-obovate, crowded, 1/4 cm L, those on lower stems ± linear; calyx mly 1-2.5 cm L, dark-ribbed; corolla blue, funnelform, the lobes 4. Wet mdws, 6000-11,000', Tulare to Tuolumne Co., Jul-Sep.

G. simplex, Hiker's Gentian: Stems erect, 0.5-2 dm H; lower lvs clasping, the upper sessile, linear-oblong to lance-oblong, 0.6-2.5 cm L; stem ending in a single fl; calyx 1.5-2 cm L; corolla blue, 2.5-4 cm L. Mdws, 4000-9500', SN (all), Jul-Sep.

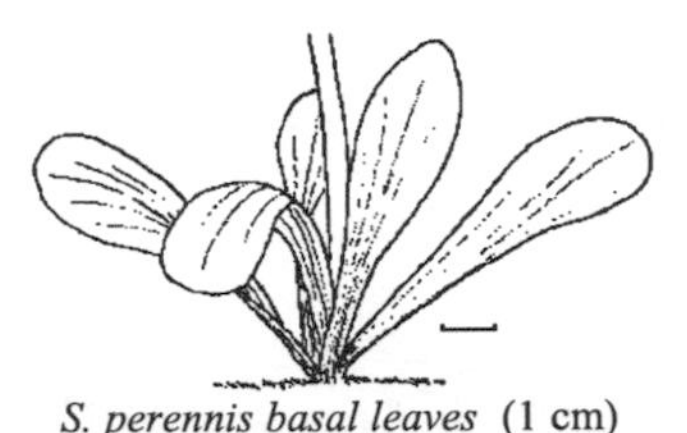

S. perennis basal leaves (1 cm)

Swertia

Green Gentian

S. radiata flower (1 cm)

Lvs longitudinally parallel-veined from base. Calyx 4-parted, the lobes deeply cleft. Corolla often speckled with dots. Stamens inserted at base of corolla. Herbage bitter and unpalatable. *S. radiata* grows from a large taproot which is edible raw, roasted or boiled, being an excellent source of carbohydrate in a stew.

A Fls 5-merous; lvs mly basal; widespread — 3) *S. perennis*
AA Fls 4-merous; stem lvs usu numerous (except in *S. tubulosa*)
  B Stem lvs opp, lvs with narrow white margins
    C Panicle narrow; Lassen Co. n., 4000-6000' — 1) *S. albicaulis*
    CC Panicle broad; e. SN of Mono and Inyo cos., 8000-11,000' — 2) *S. puberulenta*
  BB Stem lvs whorled
    C Stem 10-20 dm H; lvs not white-margined; Fresno Co. n. — 4) *S. radiata*
    CC Stem 2-7 dm H; lvs with narrow white margins; Tulare Co. s. — 5) *S. tubulosa*

1) *S. albicaulis*, White-stemmed Swertia: Stems 1 to several, ascending to erect, 2.5-4.5 dm H, ± glabrous; basal lvs spatulate-oblanceolate, 5-16 cm L, the base narrowed to a winged petiole; pannicle ± interrupted below, 3-16 cm L, of dense whorl-like cymes; pedicels 4-10 mm L; calyx-lobes 5-7 mm L; corolla-lobes 6-10 mm L, greenish white to bluish. Dry to moist places, May-Jul.

2) *S. puberulenta*, Inyo Swertia: Stems 1-2, stoutish, 1-3 dm H; lvs folded along midrib, mly 5-12 cm L, the basal oblanceolate, petioled, the stem lvs sessile; panicle open, comprising the upper half or more of the plant; pedicels 5-25 mm L; calyx-lobes 6-8 mm L; corolla-lobes about as long as calyx, greenish white with purple dots. Dry slopes, Jun-Aug.

3) *S. perennis*, Perennial Swertia Stems unbranched, glabrous, 1-3 dm H; lvs mly basal, obovate to elliptic, the lower 4-12 cm L with blades about equal to the broad petioles; stem lvs few, smaller, alt, or the upper opp and sessile; infl narrow, terminal, elongate, the lowest pedicels mly 1-4 cm L; calyx deeply lobed, these 4-5 mm L; corolla 8-10 mm L, greenish white or with a bluish purple tinge. Mdws and damp places, 7000-10,500', SN (all), Jul-Sep.

4) *S. radiata*, Giant Frasera: Stem stout, erect, 1-2 m H, about 2-4 cm thick; lvs lance-oblong or the lower oblanceolate to obovate, ± glabrous, acute, 3-7 at a node, 1-2.5 dm L; infl narrow, 3-6 dm L; pedicels stout, 3-8 cm L; calyx-lobes 1.5-2 cm L; corolla-lobes about as long as calyx, acute, greenish white, dotted with purple. Dry slopes, 6800-9800', Jul-Aug.

5) *S. tubulosa* , Kern Frasera: Stems erect, usu solitary, 2-7 dm H, mly glabrous; lvs mly basal, spatulate, 4-8 cm L, with petiole-like base and folded blades; stem lvs in whorls of 5-9, reduced; infl narrow, spicate, dense, interrupted below; pedicels 3-20 mm L; calyx-lobes 6-9 mm L, white-margined; corolla-lobes white with bluish veins, 8-10 mm L, the gland lacking. Dry, granitic and volcanic gravels and slopes, mly 6000-9000', Jun-Aug.

GERANIACEAE- Geranium Family

Annual or perennial herbs. Lvs various, with stipules. Fls 5-merous. Style-column usu elongate in fr.

Lvs pinnately cmpd — *Erodium*
Lvs palmately veined or divided — *Geranium*

Erodium

Storksbill
Filaree

inflorescence in fruit (1 cm) *leaves and flowers* (1 cm)

E. cicutarium: Stems slender, decumbent, 1-4 dm L; lvs opp, 3-10 cm L; lfts pinnately lobed or divided; peduncles 5-15 cm L; pedicels 8-18 mm L; petals rose-lavender, 5-7 mm L; style-column 2-4 cm L. Common in dry, open habitats below 6000', SN (all), Apr-May. Young lvs are edible raw or cooked.

Geranium

Cranesbill

flowers (1 cm) *G. richardsonii leaf* (1 cm)

Herbs with forking stems. Lvs with stipules. Fls bisexual. Sepals awn-tipped. Related spp used medicinally or as a tonic.

Annual from a slender base; lvs usu less than 3 cm W — *G. carolinianum*
Perennial from thick base; lvs usu over 3 cm W
  Free tips of styles 3-5 mm L — *G. californicum*
  Free tips of styles 6-9 mm L — *G. richardsonii*

G. californicum, California Geranium: Stems 1 to few, sparsely villous, 2-6 dm H; lvs 3-8 cm W, usu 5-parted; petioles 5-25 cm L; peduncles 2-8 cm L; pedicels 2-3, mly 3-12 cm L, glandular-villous; sepals 10-16 mm L; petals rose-pink to white, dark-veined, 12-20 mm L; style-column 2-2.5 cm L; carpel-bodies 4-5 mm L. Damp woods and mdws, 4000-8000', Tuolumne Co. s., Jun-Jul.

G. carolinianum, Carolina Geranium: Stems branched, erect or ascending, 2-4 dm H; lvs 2-6 cm W, 5-7-parted; lower petioles very long, upper short; peduncles mly 2-fld, 1-3 cm L, solitary or loosely aggregated; pedicels mly 2-7 mm L; sepals 5-7 mm L, 3-veined; petals pink or paler, about as long as sepals; style-column 10-14 mm L; carpel-bodies 2.5-3 mm L. Common in grassy and shaded places below 5000', SN (all), Apr-Jul.

G. richardsonii, Richardson's Geranium: Stems 1 to few, glabrous or sparsely pubescent, 3-9 dm H; lower petioles 0.5-3 dm L; blades 3-15 cm W, 5-7-parted; upper lvs reduced; peduncles 2-12 cm L; pedicels paired, slender, 1-2 cm L; sepals 6-12 mm L; petals 10-18 mm L, white or pinkish with purple veins; style-column 2-2.5 cm L; carpel-bodies 3-4 mm L. Moist places, 4000-9000', Nevada Co. s., Jul-Aug.

GROSSULARIACEAE - Gooseberry Family

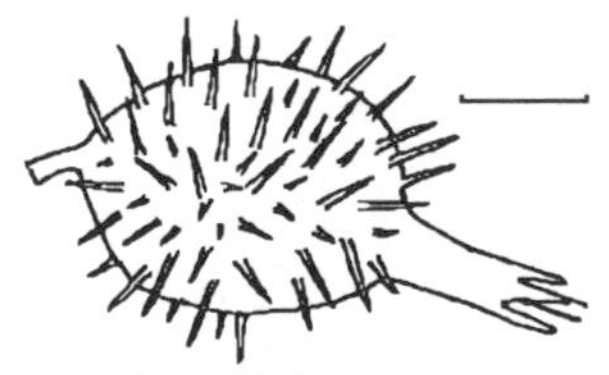

Ribes

Currant
Gooseberry

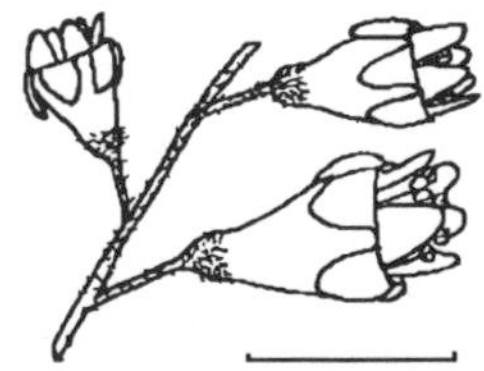

R roezlii fruit (1 cm) *R. viscosissimum flowers* (1 cm)

Shrubs with or without nodal spines. Lvs deciduous, alt, simple, usu palmately 3-5-lobed, the lobes blunt, toothed. Stipules lacking. Infl an axillary, usu pendant raceme. Fls 5-merous, perigynous. Fr a pulpy berry. The berries of all spp are edible raw, cooked, or dried.

A Nodal spines lacking; berry spineless; fls usu several to many in a raceme
  B Fr red; lvs 1-4 cm W, indistinctly lobed; fl-tube elongate — 2) *R. cereum*

BB Fr blue or black; lvs 3-8 cm W, definitely lobed; fl-tube about as long as broad
C Herbage glandular; fl-tube about 10 mm L; sepals spreading to reflexed, often greenish — 10) *R. viscosissimum*
CC Herbage glabrous to pubescent; lf-tube about 5 mm L; sepals erect, pink to red — 6) *R. nevadense*

AA Nodal spines present; berry often spiny; fls mly 1 to few
B Ovary and berry without spines or prickles
C Ovary and berry soft pubescent to glandular; e. SN — 9) *R. velutinum*
CC Ovary and berry glabrous
D Lvs ± glandular-pubescent on both sides; fls lemon-yellow — 4) *R. lasianthum*
DD Lvs glabrous above, subglabrous beneath; fls green except for white petals — 3) *R. inerme*

BB Ovary and berry with spines or bristles
C Plants usu straggling or trailing; twigs often prickly; sepals 3-5 mm L
CC Ovary and berry glabrous
D Petals purplish; berry 5 mm thick; 6400-12,500' — 5) *R. montigenum*
DD Petals whitish; berry 10 mm thick; 5000-6000', Tulare Co. — 8) *R. tularense*
CC Plants erect; twigs without internodal spines; sepals 7-10 mm L
D Berry with short gland-tipped bristles; lvs glandular beneath — 1) *R. amarum*
DD Berry pubescent and with long spines; lvs not glandular — 7) *R. roezlii*

1) *R. amarum*, Bitter Gooseberry: Stems 1-2 m H; herbage and young twigs pubescent and glandular; nodal spines to 1 cm L; lvs roundish, with cordate base, 2-3 cm W; petioles to about as long as blades; fls 1-3, purplish; free part of fl-tube 5-6 mm L; petals pinkish white, about 10 mm L; berry rounded, 1.5-2 cm in diameter. Wooded canyons, mly below 5000', Eldorado Co. s., Mar-Apr.

2) *R. cereum*, Squaw Wax Currant : Stems erect to 2 m H, much-branched, fragrant, with glandular-pubescent young growth; lvs puberulent and glandular or upper surface subglabrous and ± shining; petioles 5-15 mm L; bracts of infl dentate-truncate at summit; racemes mly 3-7 fld; fl-tube 6-8 mm l (including ovary), greenish white to reddish; petals minute; berry about 6 mm in diameter. Dry, rocky places, 5000-12,600', SN (all), Jun-Jul.

3) *R. inerme,* White-stemmed Gooseberry: Stems branched, spreading, 1-3 m H; nodal spines 1-3, short; lvs 2-5 cm W; petioles 1-3 cm L; racemes open, 1-6-fld; sepals greenish to purplish, about 5 mm L, spreading to reflexed; berry about 6 mm in diameter, purple. Moist, cool habitats, 3500-10,900', SN (all), May-Jul.

4) *R. lasianthum*, Alpine Gooseberry: Spreading shrub to 1 m H; spines 1-3 at each node; lvs 1-3 cm W, clustered on short spur-like branchlets; petioles pubescent, 3-6 mm L; fls 1-4 in a short cluster; fl-tube 5-6 mm L; sepals early reflexed, then erect, 2-4 mm L; petals shorter; berry 6-7 mm in diameter, dark red. Rocky places, 7000-10,000', Tulare to Nevada Co., Jun-Aug.

5) *R. montigenum*, Alpine Prickly Currant: Nodal spines 3-5; lvs 5-25 mm W, 3-5-cleft almost to base, very glandular-pubescent on both sides; racemes few-fld; fl-tube about 5 mm W; sepals 3-4 mm L, yellow; berries about 5 mm in diameter. Dry, rocky places, SN (all), Jun-Aug.

6) *R. nevadense*, Sierra Currant: Slender-stemmed open shrub 1-2 m H, glabrous to puberulent on young growth; lvs rather thin, 2-7 cm W, glabrous above, ± pubescent and paler beneath; petioles 1-4 cm L; racemes 8-10 fld, spreading or drooping; fls rose to deep red; sepals 4-5 mm L; petals white, shorter; berry about 8 mm in diameter with bloom. Common in moist places and along streams, 3000-10,000', SN (all), May-Jul.

7) *R. roezlii*, Sierra Gooseberry: Stems stout, 5-12 dm H, with spreading branches; herbage and young growth pubescent; nodal spines 1-3; lvs 1-2.5 cm W, dark green above, paler beneath; petioles 6-18 mm L; fls 1-2, the peduncles ± glandular; sepals dull purplish red; petals whitish, 3-5 mm L; berry purple or lighter, 14-16 mm in diameter. Common on dry, open slopes, mly 3500-8500', SN (all), May-Jul.

8) *R. tularense*, Sequoia Gooseberry: Branches 1 m L; nodal spines mly 3, to 2 cm L; lvs thin, 2-5 cm W, deeply lobed, pubescent and glandular; fls 1-3, greenish white; sepals 4-5 mm L, long-pubescent; petals 2-4 mm L; berry about 1 cm in diameter. In woods, May.

9) *R. velutinum*, Plateau Gooseberry: Stems stout, rigidly branched, 0.6-2 m H, branches and herbage soft-pubescent; nodal spines usu stout; lvs 1-2 cm W, deeply 5-cleft, the lobes often 3-cleft; fls 1-4, yellowish to whitish; sepals about 3 mm L; petals 2-2.5 mm L; berry dark purple, 4-6 mm in diameter. Dry slopes, 2500-8500', Inyo Co. n., May-Jun.

10) *R. viscosissimum*, Sticky Currant: Stems erect, 1-1.5 m H, with fragrant glandular foliage; lvs 3-8 cm W; petioles shorter than blades; racemes 3-10 cm L, 4-15-fld; pedicels to 1 cm L; sepals about 4 mm L; petals white, about half as long; berry black, 10-12 mm in diameter. Shaded woods and rocky places, 5000-9500', SN (all), Jun-Jul.

HALORAGACEAE - Water-milfoil Family

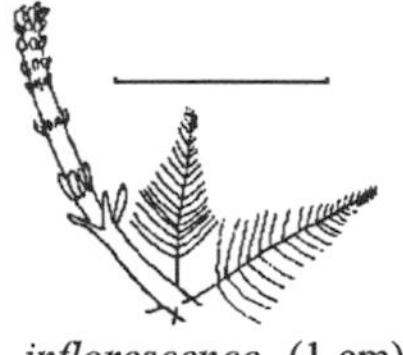

Myriophyllum

Water-milfoil

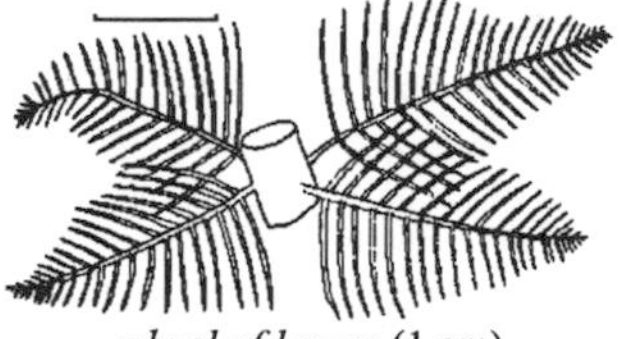

inflorescence (1 cm) *whorl of leaves* (1 cm)

M. spicatum ssp *exalbescens*: Aquatic perennial, stem often forked, to 1 m L, purple, whitish on drying; lvs whorled in 3's or 4's, 1-3 cm L, with 6-11 pairs of capillary segments; upper lvs reduced; submersed lvs pinnatifid, emersed lvs entire or toothed; infl an emersed almost naked interrupted spike, the lower fls pistillate, the upper staminate, fr indehiscent. In quiet water below 8000', SN (all), Jun-Sep. Edibility unknown.

HIPPOCASTANACEAE - Buckeye Family

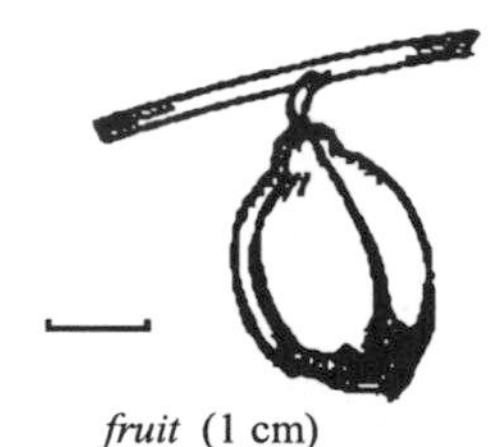

Aesculus

Buckeye
Horse-chestnut

fruit (1 cm) *leaf* (5 cm)

A. californica: Large bush or tree to 7 m H; lfts 5-7, 5-15 cm L; infl 1-2 dm L; fls white or pale rose; fr a leathery capsule containing usu 1 shiny seed, 2-3 cm in diameter. Common on dry slopes and in canyons below 4000', SN (all), May-Jun. The nuts are poisonous untreated, the poison causing nausea, vomiting and paralysis. The nuts can be rendered edible by steaming for several hours until the consistency of a boiled potato, and then either slicing or mashing them and leaching for several days depending on the thickness and bulk of the pieces. This treated starchy material may be eaten cold or made into cakes.

HIPPURIDACEAE - Mare's-tail Family

Hippuris

Mare's-tail

upper stem (1 cm) *whorl of leaves* (1 cm)

H. vulgaris: Aquatic perennials, stem unbranched, erect, 2-5 dm H, glabrous, usu the upper half or fourth emersed, the lower part rooting at the nodes; lvs whorled, 7-20 in a whorl, lanceolate to linear, sessile, 0.5-2.5 cm L; submersed lvs simple, entire, fr indehiscent. Uncommon, in quiet water below 8000', Jul-Sep. Whole plant edible as a potherb.

HYDROPHYLLACEAE - Waterleaf Family

Herbs or shrubs. Fls usu 5-merous, cymose or solitary. Calyx deeply lobed, the lobes often unequal. Fr a capsule.

A Lvs lanceolate, 5-15 cm L, entire to serrate; aromatic
  B Plant shrubby; stamens equal; SN (all) — *Eriodictyon* p. 127
  BB Plant herbaceous; stamens unequal; Fresno Co. s. — *Turricula* p. 132

AA Lvs usu smaller, if not then deeply lobed to pinnately cmpd
B Fls solitary in leaf-axils (see also *Nama densum*)
C Stem none; lvs in basal rosette — *Hesperochiron* p. 128
CC Stem present; at least some lvs opp
D Stems fleshy; petioles winged and clasping; below 4500' — *Pholistoma* p. 132
DD Stems thin; petioles not winged and clasping; widespread — *Nemophila* p. 129
BB Fls few to many in clusters
C Lvs mly pinnately cmpd, 10-20 cm L (including petiole), mly basal; infl spherical — *Hydrophyllum* p. 128
CC Lvs entire to pinnately lobed, or if cmpd then smaller
D Lvs all opp — *Draperia* p. 127
DD Lvs mly alt or basal
E Stamens unequal or attached to corolla at different levels — *Nama* p. 128
EE Stamens equally inserted subequal in length
F Corolla white to pink; lvs oblong, 1-3 cm W, often pinnately lobed — *Emmenanthe* p. 127
FF Corolla blue to violet; lvs various but not as above — *Phacelia* p. 130

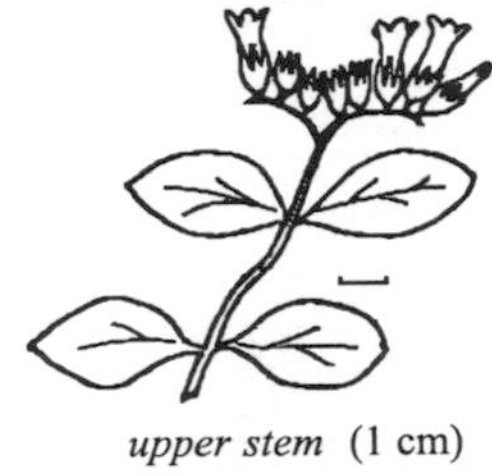

upper stem (1 cm)

Draperia

Draperia

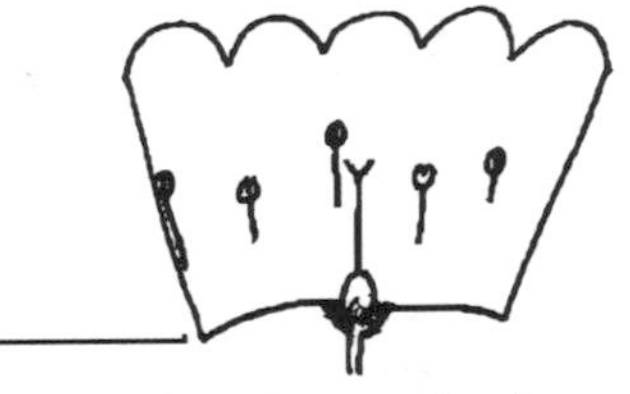

flower with corolla opened to show unequally inserted stamens (1 cm)

D. systyla: Low often matted perennial with slender stems, 1-4 dm L, hairy; lvs ovate, 2.5-5 cm L, entire, soft-hairy, sessile or petioled; fls in subcapitate terminal clusters of 10-20; calyx deeply divided, its lobes linear, about 10 mm L; corolla pale violet, 10-25 mm L, deciduous, the roundish lobes 2-3 mm L; stamens 2-3 mm L, unequally inserted; style included, 2-lobed. Dry slopes in woods, 2400-8000', SN (all), May-Aug. Edibility unknown.

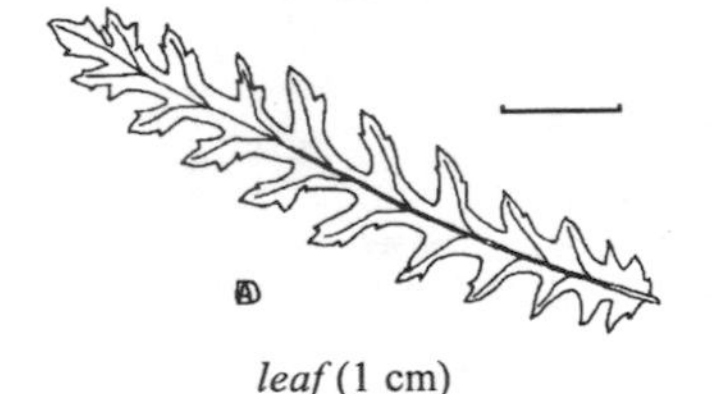

leaf (1 cm)

Emmenanthe

Whispering Bells

inflorescence (1 cm)

E. peduliflora: Aromatic, glandular annual; stems erect, 1-8 dm H; lvs basal and cauline, the upper sessile and usu clasping, 1-10 cm L, toothed to deeply pinnately lobed; infl terminal; fls drooping, on pedicels 5-15 mm L; calyx lobes 5-10 mm L; corolla 5-15 mm L, withering persistent. Dry slopes, common in disturbed areas below 7000', SN (all).

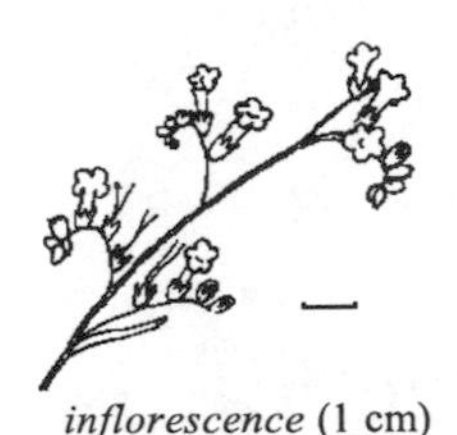

inflorescence (1 cm)

Eriodictyon

Yerba Santa

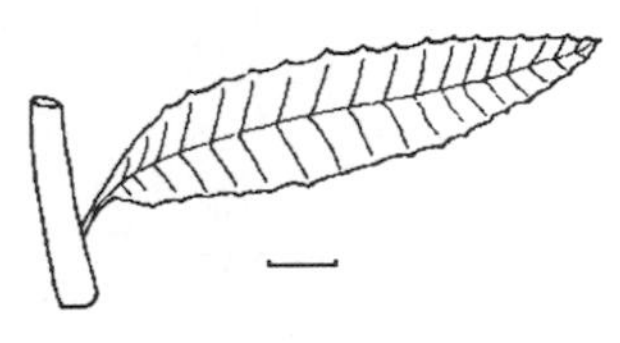

leaf (1 cm)

E. californicum: Evergreen shrub, 5-22 dm H, with sticky branches; lvs 1-5 cm W, glabrous and glutinous above, veiny and tomentulose below, tapering into short petioles; panicles 0.5-2 dm L; corolla lavender to white; tubular-funnelform, 9-15 mm L; style divided to base. Dry, rocky slopes and ridges below 5500', SN (all), May-Jul. Lvs can be brewed to make a bitter tea. A stronger version was used to treat coughs and sore throats.

Hesperochiron

Hesperochiron

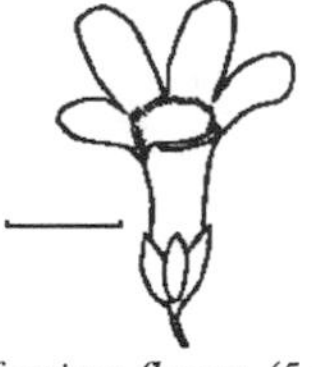

H. pumulis plant (1 cm) *H. californicus flower* (5 mm)

Glabrous to pubescent perennials from short thick vertical roots. Lvs entire, petioled. Calyx 5-parted, the lobes ± unequal. Corolla white to bluish. Stamens included, often unequal. Style included, shortly 2-cleft. Edibility unknown.

Fls usu several to many per plant, corolla glabrous to slightly hairy within *H. californicus*
Fls 1-few per plant; corolla densely long-hairy within *H. pumilus*

H. californicus, California Hesperochiron: Lvs many, narrow-oblong to oval, 1-5 cm L, on somewhat shorter petioles; pedicels usu many, 2-8 cm L, spreading; calyx lobes 4-7 mm L; corolla 1-2.5 cm L, 1-2 cm W, the lobes 3-6 mm L, capsule 5-20 mm L. Moist, often subsaline places, 4000-9000, mly w. SN (all), May-Jul.

H. pumilus, Dwarf Hesperochiron: Lvs mly 5-10, narrow-oblong to oblanceolate, 2.5-5 cm L; petioles 0.5-2.5 cm L; pedicels few, 2-5(-8) cm L; calyx lobes 4-9 mm L; corolla 6-15 mm L, 1-3 cm W, the roundish lobes 4-10 mm L; capsule 5-9 mm L. Moist, often subalkaline flats and mdws, below 9000', SN (all), Apr-Jul.

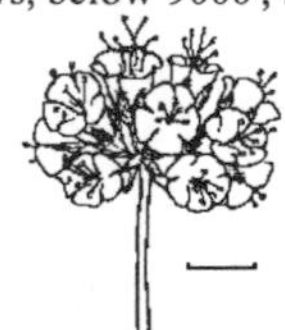

Hydrophyllum

Hydrophyllum

inflorescence (1 cm) *leaf* (1 cm)

Perennials with mly basal lvs. Lvs pinnatifid. Infl a many-fld globose cyme. Calyx divided nearly to base. Corolla deciduous, divided to at least middle, white to purple. Stamens and style exerted beyond petals, the style shallowly bifid. The young shoots and lvs may be eaten raw, or these and the roots may be cooked.

Stem very short; peduncles shorter than lvs; lvs usu not paler beneath *H. capitatum*
Stems 1-6 dm H; the infl above the basal lvs; lvs usu paler beneath *H. occidentale*

H. capitatum var *alpinum*, Woolen-breeches: Plants nearly lacking stem; lvs 4-7 cm L, longer than peduncle; pedicels 5-20 mm L; calyx densely soft-hairy; corolla 5-9 mm L. Moist habitats, 3000-7000', Placer Co. n. May-Jun.

H. occidentale: Stems 1-6 dm H, pubescent; lvs oblong in outline, 5-15 cm L, on shorter or equally long petioles, long-hairy, paler below, deeply lobed into 7-15 oblong lfts; pedicels 2-5 mm L; calyx-lobes 3-4 mm L; corolla 7-15 mm L, the lobes 4-6 mm L. Dryish or moist ± shaded slopes, 2500-9000', SN (all), May-Jul.

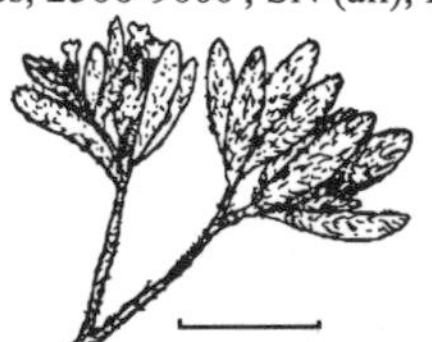

Nama

Purple Mat

N. densum upper stem (1 cm) *N. lobbii upper stem* (1 cm)

Annual or perennial. Lvs usu well distributed along stem. Fls in reduced terminal nonscorpioid cymes and axillary. Calyx deeply divided. Style included, usu bifid. Fr a ± ovoid capsule. Related spp are slightly toxic if eaten.

A Prostrate annuals, stems 3-8 cm L; fls solitary in axils; e. SN
 B Corolla 3-5 mm L, 1-3 mm W 2) *N. densum*
 BB Corolla 10-17 mm L, 7-12 mm W 1) *N. aretioides*
AA Plants ± woody perennials, 10-30 cm H; fls many in terminal cymes

B Lvs coarsely dentate; Eldorado Co. s. 4) *N. rothrockii*
BB Lvs entire; Eldorado Co. n. 3) *N. lobbii*

1) *N. aretioides*: Similar in habit to *N. densum*; calyx lobes 4-7 mm L; corolla purple to red. Below 6500' May-Jun.

2) *N. densum*, Matted Purple Mat: Stems hairy, branched from base, the branches densely leafy at tips, 2-8 cm L; herbage with a slight skunk-like odor; lvs narrow-oblong, 8-15 mm L, gray-hairy; calyx-lobes linear-lanceolate, 2.5-3.5 mm L; corolla lavender, tubular, 3-4 mm L; capsule 2-3 mm L. Dry, sandy or loose soil, 3000-11.700', May-Jul.

3) *N. lobbii*, Lobb's Purple Mat: Stems leafy, tomentose; lvs linear-oblong to oblanceolate, sessile, those of some shoots broader and plane, of others narrow and revolute; fls in densely leafy reduced cymes; calyx-lobes linear, 3-7 mm L; corolla purple, broadly funnelform, about 10 mm L, the round lobes 2-3 mm L; style about 3 mm L, deeply divided; capsule about 3 mm L. Dry, rocky or sandy slopes and ridges, 4000-7000', Jun-Aug.

4) *N. rothrockii*, Rothrock's Purple Mat: Stems hairy and glandular, often few-branched; lvs lanceolate to oblong, 2-5 dm L, short-petioled, coarsely sinuate-dentate, usu revolute; infl subcapitate; calyx-lobes linear, 10-25 mm L, hairy; corolla purplish lavender, funnelform, 10-15 mm L, 6-9 mm W, style 8-10 mm L, divided to base; capsule 4-5 mm L. Dry sandy flats and benches, 7000-10,000', Jul-Aug.

N spatulata upper stem (1 cm)

Nemophila

Nemophila

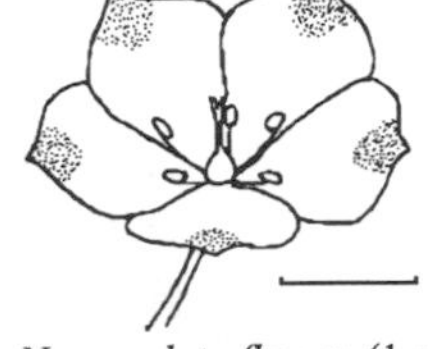

N. maculata flower (1 cm)

Annual; the stems usu branched and diffuse, sometimes prostrate, often weak. Lvs mly opp, on petioles about as long as blades. Fls pedicelled, in upper axils or opp the lvs. Calyx deeply divided, the sinuses usu with sepal-like spreading or reflexed auricles. Style usu deeply bifid. Fr a $\pm$ globose capsule. Edibility unknown.

A Corolla 1-4 cm W, each corolla-lobe with a purple blotch 2) *N. maculata*
AA Corolla mly 0.2-1 cm W, usu not blotched
  B Auricles 1-2 mm L, reflexed and conspicuous; lower lf-lobes not narrowed toward base
    C Lvs oblong to oval, deeply pinnately 5-9-lobed 4) *N. pedunculata*
    CC Lvs spatulate, shallowly palmately to pinnately 3-5-lobed 6) *N. spatulata*
  BB Auricles minute; lower lf-lobes not strongly narrowed toward base
    C Corolla 2-4 mm W; lf-lobes not strongly narrowed toward base 3) *N. parviflora*
    CC Corolla 5-10 mm W; lf-lobes strongly narrowed toward base
      D Corolla $\pm$ flat, blue to violet with white center or white; Madera Co. s. 5) *N. pulchella*
      DD Corolla bowl-shaped, white or bluish; Madera Co. n. 1) *N. heterophylla*

1) *N. heterophylla,* Variable-leaved Nemophila: Stems $\pm$ erect, 1-3 dm L; lower lvs 1.5-3.5 cm L, 5-7 pinnate into rounded rather remote divisions; upper lvs alt, entire to 3-5-lobed; pedicels slender, mly longer than lvs; calyx-lobes 2.5-3.5 mm L. Light shade, slopes and canyons below 5000', Apr-Jul.

2) *N. maculata*, Fivespot: Stems several, decumbent, 1-3 dm L; lvs oblong to oval in outline, 1-3 cm L, pinnately and deeply 5-9-lobed (or the upper entire to 3-lobed), hairy; pedicels stout, 2-5 cm L; calyx-lobes 5-8 mm L, the reflexed auricles 1-4 mm L; corolla bowl-shaped, 1.5-4.5 cm W, white and veined or dotted. Mly moist slopes and flats below 9000', w. SN, Plumas Co. s., Apr-Jul.

3) *N. parviflora* var. *austinae*, Small-flowered Nemophila: Stems mly 1-2 dm L; lower lvs 1-4 cm L, usu 5-pinnate, upper lvs subsessile. Between 3500-7000', Sierra Co. n., May-Jun.

var. *quercifolia*: At 2500-6500', Madera Co. s., May-Jun.

4) *N. pedunculata*, Meadow Nemophila: Stem 1-3 dm L; lvs 1-3 cm L; pedicels short; calyx-lobes 1-3 mm L; corolla bell-shaped, 3-6 mm W, each lobe with a purple botch. Moist habitats, mly below 5000', Calaveras Co. n., May-Aug.

5) *N. pulchella*, Pretty Nemophila: Stems 1-4 dm L; herbage usu with white hairs; lvs 2-4 cm L, pinnately divided into 5-7 remote divisions; uppermost lvs entire to shallowly lobed, sometimes alt; pedicels slender; calyx-lobes 2-4 mm L. Partial shade and moist places below 5500', Apr-Jun.

6) *N. spatulata*, Sierra Nemophila: Stems few, 0.5-2 dm L; lvs pubescent, 1-3 cm L; pedicels rather short; calyx-lobes 2.5-5 mm L; corolla white or bluish, shallowly bowl-shaped, often dotted, sometimes with purple blotches on lobes. In shaded, damp places, 4000-10,500', Plumas Co. s., May-Jul.

Phacelia

Phacelia

P. hastata basal leaves (1 cm) *P. hydrophylloides* (1 cm)

Usu pubescent herbs, often glandular. Calyx lobed almost to base. Stamens mly exserted. Style usu exserted, ± bifid. Exact sp often difficult to determine because of hybridization and polyploidy (*Magellanica* complex in the perennials), and except for the more common spp only brief descriptions will be given. At least one sp, *P. ramosissima*, may be eaten as cooked greens.

A Plants perennial or biennial from woody taproot or caudex
 B Lower stem lvs mly entire with prominent longitudinal parallel veins; basal lvs sometimes 2-4-lobed
  C Lvs all entire; herbage conspicuously silvery; montane habitats 7) *P. hastata*
  CC Lvs usu with basal lobes or lfts; herbage green or grayish, hardly silvery
   D Stems usu solitary; stem lvs mly lanceolate; corolla yellow to greenish 8) *P. heterophylla*
   DD Stems usu several; stem lvs often elliptic; corolla lavender 13) *P. mutabilis*
 BB Lower stem lvs dentate or more often pinnately lobed or divided
  C Stems erect, mly 5-10 dm H, unbranched; below 7000', Mono Co. n.
   D Lvs pubescent but green, mly over 2 cm W; corolla-margin revolute 15) *P. procera*
   DD Lvs grayish, ± densely hairy, mly less than 2 cm W; corolla margin plane
    E Stem lvs few, usu much reduced; corolla white, 7-9 mm L 5) *P. egena*
    EE Stem lvs many, well-developed; corolla yellow to greenish, 4-5 mm L 8) *P. heterophylla*
  CC Stems decumbent to erect, mly 1-5 dm H, usu branched
   D Lvs mly pinnate, the lfts lobed or divided; corolla margins plane 19) *P. ramosissima*
   DD Lvs toothed or lobed, the lobes entire; corolla-margins revolute 10) *P. hydrophylloides*

AA Plants annual
 B Lvs pinnately cmpd; lfts ± pinnately lobed
  C Corolla 4-7 mm L; stems erect 1.5-5 dm H 3) *P. cryptantha*
  CC Corolla 7-18 mm L; stems often decumbent
   D All lvs pinnately to bipinnately cmpd; e. SN 2) *P. bicolor*
   DD Only lower lvs cmpd, the upper entire to lobed; c. & s. SN 4) *P davidsonii*
 BB Lvs simple, entire to pinnately lobed
  C Infl usu lax, scarcely scorpioid; lower lvs opp
   D Pedicels short, 0-3 mm L
    E At least larger lvs dentate to pinnately lobed; rare 20) *P. stebbinsii*
    EE Lvs usu entire; widespread
     F Corolla bluish, 2-4 mm L; calyx-lobes slightly pubescent 18) *P. racemosa*
     FF Corolla violet, 4-6 mm L; calyx-lobes white-hairy 9) *P. humilis*
   DD Pedicels 5-10 mm L
    E Corolla lavender or paler, 2-4 mm L; Tulare to Eldorado Co. 6) *P. eisenii*
    EE Corolla violet, 4-6 mm L; near Mineral King 14) *P. orogenes*
  CC Infl usu dense, scorpioid; corolla mly over 4 mm L; only lowest lvs opp
   D Herbage with ± skunk-like odor; corolla tardily deciduous, 4-10 mm L
    E Lvs elliptic to oval; Nevada to Placer Co. 12) *P. marcescens*
    EE Lvs linear to oblong; Fresno to Eldorado Co. 17) *P. quickii*
   DD Herbage without skunk-like odor; corolla promptly deciduous, 4-10 mm L
    E Lvs linear to lanceolate or oblanceolate, ± sessile
     F Lvs 2-8 cm L; 3000-6000', Plumas Co. n. 11) *P. linearis*
     FF Lvs 1-3 cm L; 4000-9000', Tulare Co. s. 1) *P. austromontana*
    EE Lvs oblong to elliptic, petioled
     F Corolla 4-5 mm L; stamens and style included 21) *P. vallicola*
     FF Corolla 6-7 mm L; stamens and style exserted 16) *P. purpusii*

1) *P. austromontana*, Southern Sierra Phacelia: Stems mly under 10 cm H; corolla lavender, 3-5 mm L. Dry slopes, 6000-9000', s. SN, Jun-Jul.

2) *P. bicolor*: Stem prostrate to erect, 5-40 cm L, usu branched, short-hairy; lvs 2-6 cm L; pedicels 1-4 mm L; corolla 5-15 mm L, the tube yellow, the lobes violet; style 3-7 mm L. Below 11,000', e. SN (all).

3) *P. cryptantha*, Limestone Phacelia: Stem 1-5 dm H; lvs usu with 7-11 lfts. Reported from Tehipite Valley, normally a desert sp

4) *P. davidsonii*: Stem prostrate to erect, 5-20 cm L, few branched, sparingly hairy; lvs 1-7 cm L; pedicels 3-8 mm L; corolla 5-15 mm L, the tube white, the lobes violet; style 4-8 mm L. Open rocky slopes below 8000', mly e. slope SN, Mono Co. s.

5) *P. egena*: Stems several, ascending to erect, rather slender, 2-6 dm H; lvs usu 2-7-pinnatifid into acute divisions. Dry places, mly below 5000', SN (all), May-Jun.

6) *P. eisenii*, Eisen's Phacelia: Stems 3-15 cm H, ascending, usu branched; lvs mly entire, 1-2 cm L, on petioles about as long. Mly gravelly places, 4000-11,000', Jun-Aug.

7) *P. hastata*, Silverleaf Phacelia: Stems 1 to several, ascending, grayish with spreading white hairs 1-2 mm L; lf-blades lanceolate to narrow-ovate, 3-6 cm L; infl of 10-20 cymes remote below to densely clustered at tip. Dry, rocky places, below 10,500', Plumas Co. n., May-Jul.

var. *compacta* : Stems decumbent to erect; lower lvs clustered at base; petioles 2-5 cm L; stem lvs few, reduced; fls many, in dense short cymes; calyx lobes linear, 3-4 mm L, often purplish, 6-8 mm L in fr; corolla lavender to white, 4-6 mm L; stamens 6-8 mm L; style about 8 mm L. Common, gravelly to rocky places, mly 7000-13,000', SN (all), Jul-Sep.

8) *P. heterophylla*, Virgate Phacelia: Stems unbranched, coarse, erect, 3-12 dm H, hairy; lower lf-blades 5-10 cm L, petioled; upper lvs reduced, passing into leafy bracts in the infl; infl 1-5 dm L; calyx-lobes linear to linear-oblong, 4-7 mm L. Dry, rocky places below 9000', Mono Co. n., May-Jul.

9) *P. humilis*, Low Phacelia: Stems 5-20 cm H; lvs green, linear-oblong to ovate, 1-4(-10) cm L, entire, short-petioled or the upper subsessile; racemes 1-3, the fls many; corolla 4-6 mm L. Flats and borders of mdws, 5000-9400', Mly e. SN from Inyo Co. n., May-Jul.

10) *P. hydrophylloides*, Waterleaf Phacelia: Stems few to several, spreading to decumbent, 1-3 dm L; lvs well distributed, ovate in outline, the blades 1.5-6 cm L, 1-3 cm W, the basal sometimes pinnatifid, petioled; fls in subcapitate cymes 1-3 cm L, on peduncles 1-7 cm L; corolla violet-blue to whitish, 5-6 mm L. Occasional, dry, open woods, 5000-9800', SN (all), Jun-Aug.

11) *P. linearis*, Linear-leaf Phacelia: Stems stiff, erect, mly unbranched, 1-6 dm H, leafy; lvs linear to lanceolate, mly erect, entire or sometimes pinnately lobed, the lobes linear. Dry, gravelly slopes and flats, mly 3000-6000', May-Jun.

12) *P. marcescens*, Persistent-flowered Phacelia: Stems 5-20 cm H, glandular-pubescent; lvs 1-3.5 cm L; petioles of lower lvs as long as blades; racemes dense. Dry, gravelly habitats, 4000-7000', May-Jul.

13) *P. mutabilis*, Changeable Phacelia: Stems 1-4.5 dm H; lower lvs ± tufted, lanceolate to ovate, rather thin, 2-8 cm L, petiole usu at least as long as blade; stem lvs often broad, petioled; fls many in mly few cymes; calyx-lobes linear, 3-4 mm L, 6-10 mm L in fr; corolla 4-6 mm L, tardily deciduous; stamens 6-8 mm L; style 6-8 mm L. Intergrades with *P. hastata*. Rocky places, 4000-10,000', SN (all), Jun-Aug.

14) *P. orogenes*, Mountain Phacelia: Stems 2-10 cm H, lower lvs lance-linear, 1-3 cm L, entire, petioled; fls few, in open cymes. Mdws, 8500-10,300', Jul-Aug.

15) *P. procera*, Tall Phacelia: Stems stout, 5-20 dm H, leafy, lvs ovate, the lower blades 5-12 cm L, incised to pinnatifid into few lanceolate mly entire lobes; stem lvs gradually reduced upward; infl terminal, 5-12(-16) cm L; corolla purplish to greenish white, 5-6 mm L. Rocky habitats, 4000-7000', Placer Co. n., Jun-Aug.

16) *P. purpusii*, Purpus' Phacelia: Stems 1-4 dm H, finely pubescent; lower lvs 2-5 cm L, entire or with 1-3 teeth or lobes on each side, the petioles 1-4 cm L; fls many, subsessile; calyx-lobes very unequal; corolla lavender to violet. Occasional, gravelly or sandy places, 3000-7000', w. SN (all), May-Jul.

17) *P. quickii*, Quick's Phacelia: Much like *P. marcescens*: corolla blue to lavender, drying whitish. Open, sandy places, 4000-7500', w. SN, May-Jun.

18) *P. racemosa*, Racemose Phacelia: Stems slender, erect, usu few-branched above, often reddish, glandular-pubescent above, 5-20 cm H; lower lvs lance-oblong, 1-3 cm L, short-petioled, entire; stem lvs few. Dry, rocky habitats, 5000-9000', Fresno to Lassen Co., Jun-Aug.

19) *P. ramosissima*, Branching Phacelia: Stems 5-10 dm L, branching, glandular-pubescent; lvs oblong to broadly ovate in outline, 3-10 cm L, pinnate, the lobes oval to oblong, toothed to pinnatifid; petioles mly short; infl of few rather dense short cymes; calyx-lobes mly spatulate, 5-6 mm L; corolla dirty white to bluish, 6-8 mm L, deciduous; style deeply cleft. Rocky places below 10,000', SN (all), May-Jul.

20) *P. stebbinsii*, Stebbins' Phacelia: Stems erect, few branched; lower lvs opp, 10-50 mm L, dentate to lobed; upper lvs reduced, often entire; fls ± sessile; corolla 4-5 mm L, white to pale blue. Rare in woods and mdws, about 4000', Eldorado Co.

21) *P. vallicola*, Mariposa Phacelia: Stems 1-3 dm H, densely pubescent; lvs ± ovate, entire, 1-3 cm L, on somewhat shorter petioles; cymes many-fld; corolla purple. Rare, rocky habitats 5000-7000', central SN, May-Jun.

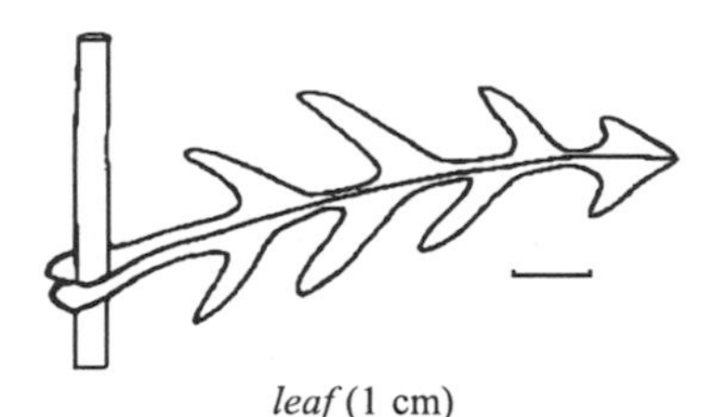

Pholistoma

Common Fiesta-flower

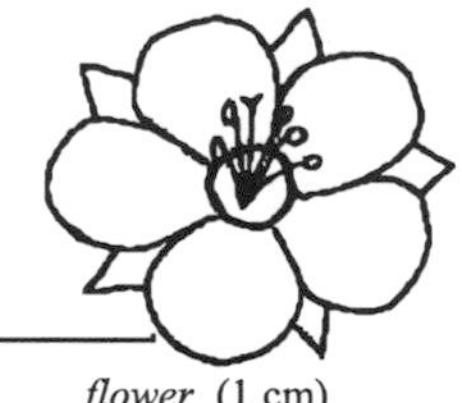

leaf (1 cm) *flower* (1 cm)

P. auritum: Annual, straggling, with coarse loosely branched stems 3-10 dm L; lower lvs ± oblong in outline, 6-15 cm L, with 7-13 divisions; fls cymose and terminal or solitary and axillary; pedicels 2-3 cm L; calyx 5-10 mm L; corolla lavender to purple with darker markings, 1.5-3 cm W; style about 5 mm L, cleft less than half its length. Shaded slopes and deep canyons, Calaveras Co. s., Mar-May. Edibility unknown.

Turricula

Poodle-dog Bush

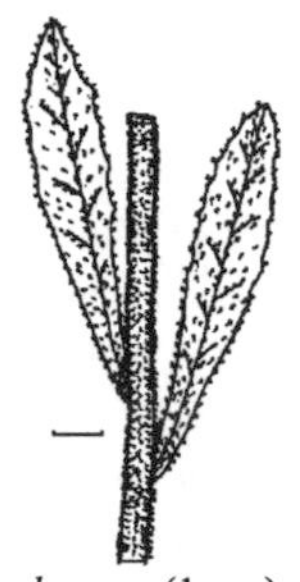

T. parry: Coarse, stout perennial, 1-2.5 m H, usu branched from base; lvs ± erect; fls many in terminal panicle; calyx deeply divided, the lobes 3-4 mm L; corolla purple, deciduous, funnelform, shallowly lobed, pubescent, 13-18 mm L; stamens unequal, included; style deeply divided. Occasional, in dry places, especially after a fire, 1000-8000', Fresno and Kern cos., Jun-Aug. Edibility unknown.

leaves (1 cm)

HYPERICACEAE - St. John's Wort Family

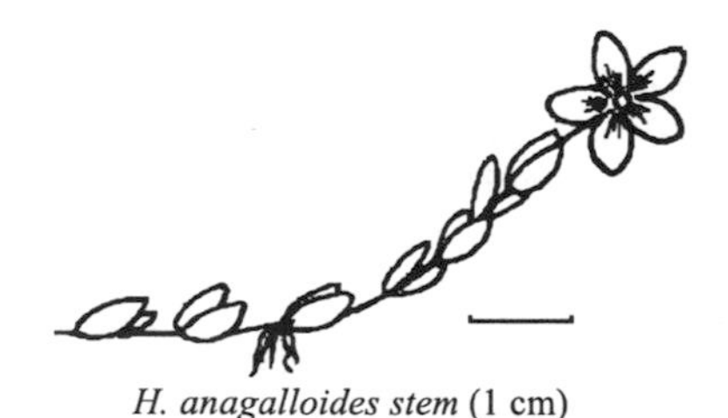

Hypericum

St. John's Wort

H. anagalloides stem (1 cm) *H. formosum flower* (1 cm)

Plants usu perennial, glabrous. Lvs entire, several-nerved from base, sessile, without stipules, glandular. Fls 4-5-merous, often yellow, in terminal clusters. Stamens usu many, distinct or ± united into 3-5 clusters. Ovary superior, with 3-styles. Lvs may be eaten fresh or may be dried and ground to a flour that can be used like acorn meal, however only small amounts of the herbage should be consumed at any one time, for the lvs also contain a photosensitive agent and alkaloids which may be toxic if ingested in quantity.

Plants procumbent, often in mats; infl few-fld; stamens distinct — *H. anagalloides*
Plants erect; infls many-fld; stamens in groups
  Lvs black-dotted along lower margin; sepals ovate, obtuse, 3 mm L — *H. formosum*
  Lvs revolute; sepals linear-lanceolate, acute, 4-5 mm L — *H. perforatum*

H. anagalloides, Tinker's Penny: Stems rooting at nodes, 5-15(-20) cm L; lvs elliptic to round, 5-7-nerved, 4-12 mm L; fls golden to salmon-color; sepals 2.5-3 mm L; petals slightly shorter to longer; stamens 15-25. Wet places, 4000-10,000', SN (all), Jun-Aug.

H. formosum var. *scouleri*, Scouler's St. John's Wort: Stems slender, 2-7 dm H, often branched above; lvs oblong-ovate, gray-green, 1-2.5 cm L; sepals black-dotted on margin; petals yellow, 7-10 mm L, obovate; stamens conspicuous; capsule 3-lobed, 6-7 mm L. Common, wet mdws and banks, 4000-7500', occasionally lower, SN (all), Jun-Aug.

H. perforatum, Klamath Weed: Stems tough, 3-10 dm H, much branched; lvs ± oblong, 1.5-2.5 cm L, subtending short, leafy branchlets; petals orange-yellow, often black-dotted, 8-12 mm L, twisting after anthesis; capsule 7-89 mm L. Weedy, below 4500', Tuolumne Co. n., Jun-Sep.

LAMIACEAE - Mint Family

Herbs mly with 4-angled stems and simple lvs. Fls bilabiate, variously clustered. Calyx persistent, 2-lipped or regular, mly 5-toothed or -lobed. Stamens on the corolla-tube, mly 4 and in two pairs of different lengths.

A Fls few in axils or in open axillary cymes
  B Corolla showy, usu 15-20 mm L; calyx 2-lipped; stamens included — *Scutellaria* p. 135
  BB Corolla 2-6 mm L; calyx 5-lobed
    C Stamens barely exerted beyond corolla tube; plants of wet habitats — *Lycopus* p. 133
    CC Stamens much exerted and arched, conspicuous; dry habitats — *Trichostema* p. 136
AA Fls in dense clusters or whorls
  B Lvs coarsely serrate, petioled; fls in bractless whorls
    C Corolla 2-6 mm L, nearly regular — *Mentha* p. 134
    CC Corolla over 10 mm L, strongly asymmetric
      D Stamens conspicuously exserted; fl-whorls usu clustered — *Agastache* p. 133
      DD Stamens ± included in hood-like upper corolla-lip; at least lower fl-whorls usu remote — *Stachys* p. 136
  BB Lvs finely serrate to entire, often subsessile; fls often in bracted heads
    C Bracts of infl conspicuous and often colored; infl dense
      D Corolla 2-lipped; moist habitats — *Prunella* p. 135
      DD Corolla ± regular; dry habitats — *Monardella* p. 134
    CC Bracts small or none; infl often interrupted
      D Lvs long-petioled — *Salvia* p. 135
      DD Lvs sessile — *Pycnanthemum* p. 135

Agastache

Agastache

Tall perennials. Lvs ovate, truncate or subcordate at base. Fls in dense sessile whorls which are mly in continuous or interrupted cylindrical or tapering spikes. Stamens 4, the lower pair shorter. Seeds may be eaten raw or cooked; lvs may be used for flavoring stews, etc.

inflorescence (1 cm)

Stems 4-8 dm H; lvs mly 1-1.5 cm W; Lassen Co. n. — *A. parvifolia*
Stems mly 1-2 m H; lvs mly 3-4 cm W; widespread — *A. urticifolia*

A. parvifolia, Small-leaved Agastache: Stems slender, ± branched, thinly pubescent; lf-blades deltoid to deltoid-ovate, the median 2-3.5 cm L; petioles 1-2 cm L; infl 5-9 cm L; calyx rose, the tube 5-6 mm L, the teeth 4-7 mm L; corolla rose, tubular, 10 mm L. Lava rock, 3000-5000', Jun-Aug.

A. urticifolia, Horse-mint: Stems several, branched above, subglabrous; lf-blades ovate or deltoid-ovate, the median 3.5-8 cm L; petioles 1-2.5 cm L; infl 4-15 cm L; calyx green or rose, the tube 4-7 mm L, the teeth 2.5-5 mm L; corolla rose or violet, 10-15 mm L. Moist places below 10,000', SN (all), Jun-Aug.

Lycopus

Northern Bugleweed

L. uniflorus: Stems erect, 1-5 dm H, pubescent; lvs 2-6 cm L, usu short petioled and serrate; elliptic to lanceolate; calyx lobes ovate; corolla 2-4 mm L, white. Uncommon, moist to wet places, below 6500', c. and n. SN.

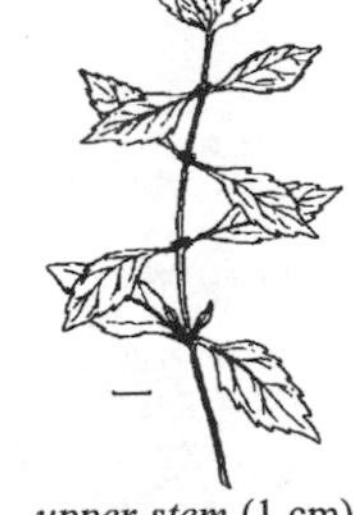

upper stem (1 cm)

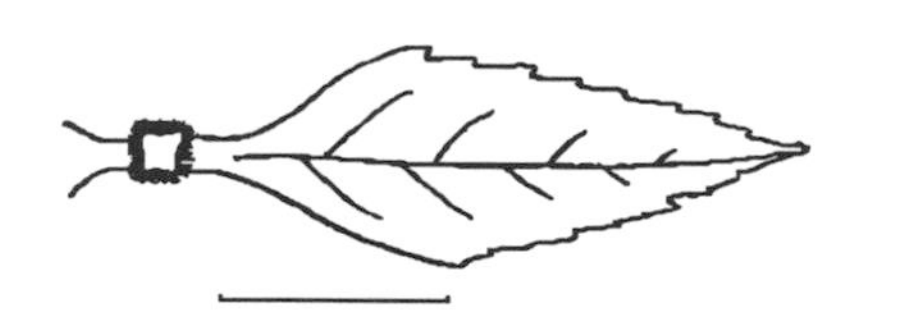

Mentha

Mint

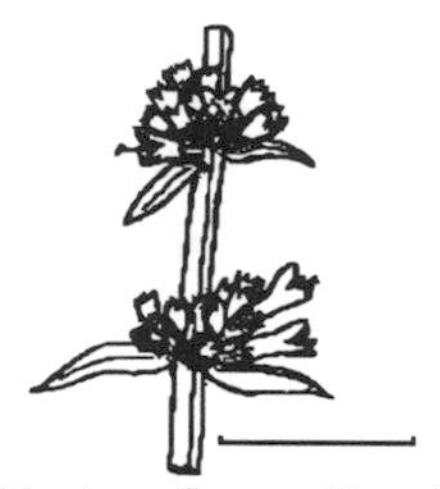

M. arvense leaf on stem (1 cm) *M spicata flowers* (5 mm)

Aromatic perennials. Calyx 10-nerved, 5-toothed. Corolla-tube shorter than the calyx, upper lip entire or slightly notched, lower 3-lobed. Stamens 4, equal, included or exserted. The fresh or dried lvs may be soaked in hot water to make a tea.

Stems pubescent; fl-whorls axillary, remote — *M. arvensis*
Stems glabrous; fl-whorls in terminal ± interrupted spike — *M. spicata*

M. arvensis, Field Mint: Stems usu branched, 1-8 dm H; lvs lanceolate to oblong or ovate, 2-5 cm L, petioled; calyx about 3 mm L, pubescent; corolla lilac-pink to purplish, 5-6 mm L. Moist places below 7500', SN (all), Jul-Oct.

M. spicata, Spearmint: Stems 3-12 dm H, often purplish; lvs oblong- or ovate-lanceolate, ± rounded at base, pubescent to villous, 2-6 cm L, subsessile; spikes slender, to 4 or 6 cm L in fr, the bracts 4-8 mm L; calyx about 1.5 mm L; corolla pale lavender, 2.5-3 mm L. Moist fields and marshes below 5000', Jul-Oct.

Monardella

Monardella

upper stem (1 cm) *flower* (5 mm)

Fls borne in terminal heads subtended by broad involucral bracts which are often colored. Calyx narrow, about equally 5-toothed, 10-15-nerved. Upper lip of corolla 2-lobed, the lower 3-lobed, horizontal or declined. Style unequally 2-cleft at apex. *M. odoratissima* is a complex of species which are difficult to identify in the field. Medicinal tea can be made from infl.

A Plant annual with mly slender base; stems mly naked — 2) *M. lanceolata*
AA Plant perennial, ± woody at base, stems leafy throughout
  B Lf-margins wavy; stems matted; rare, s. SN — 1) *M. beneolens*
  BB Lf-margins not wavy; stems various; widespread
    C Outer bracts erect or absent, not leaf-like; lvs entire — 3) *M. odoratissima*
    CC Outer bracts reflexed and leaf-like; lvs often serrate — 4) *M. sheltonii*

1) *M. beneolens*, Sweet-smelling Monardella: Stems tufted or matted, 10-30 cm L with long soft hairs; lvs less than 10 mm L, often ± serrulate; head 10-20 mm W; fls violet to rose. Subalpine forests, 8000-11,000', Fresno Co. s.

2) *M. lanceolata*, Mustang-mint: Stems erect, 2-5 dm H, puberulent above; lf-blades lanceolate to lance-oblong, 1.5-5 cm L; petioles 5-15 mm L; heads 1.5-3 cm W; bracts lance-ovate, acute, often purplish toward tips, 5-15 mm L; calyx 6-8 mm L; corolla rose-purple or paler, 12-15 mm L; stamens well exserted. Locally common in dry places, especially disturbed areas, below 8000', SN (all), May-Aug.

3) *M. odoratissima* complex, Mountain Monardella: Stems branched; lvs lanceolate to ovate, 1-4 cm L, pubescent (if not then ssp *glauca*), sometimes purplish; petioles 1-8 mm L; bracts often purplish, sometimes inconspicuous; calyx 5-8 mm L, pubescent; corolla reddish purple to whitish, 1-2 cm L; heads 1-3 cm W. Dry slopes, 3000-11,400', SN (all), Jun-Sep.

4) *M. sheltonii*: Lvs mly 2-2.5 cm L, occasionally serrate; petioles 2-5 cm L; corolla 1-2 cm L. Dry habitats below 6000', SN (all) Jun-Aug.

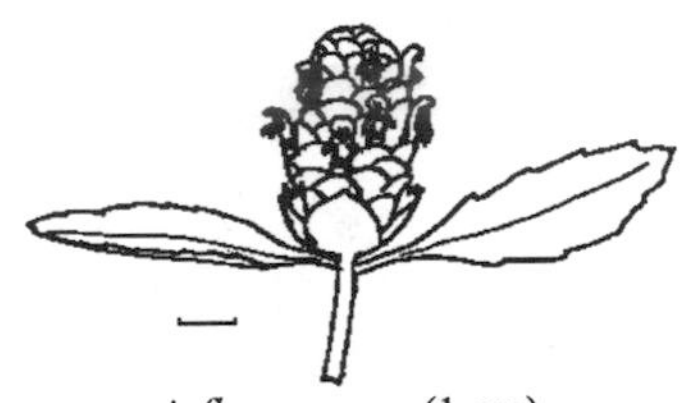

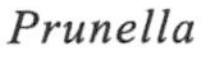

Prunella

Selfheal

inflorescence (1 cm) *flower* (1 cm)

P. vulgaris ssp *lanceolata*: Stems and lvs commonly pubescent; lvs lance-ovate to -oblong, 2.5-5 cm L, 1-3 cm W; bracts deltoid, 5-15 mm L, often tinged purple; fls small, in 3's in bract-axils; calyx 2-lipped; the corolla 10-20 mm L, mly violet. Moist habitats below 7500', SN (all), May-Sep. Chopped or powdered lvs soaked in cold water make a refreshing drink.

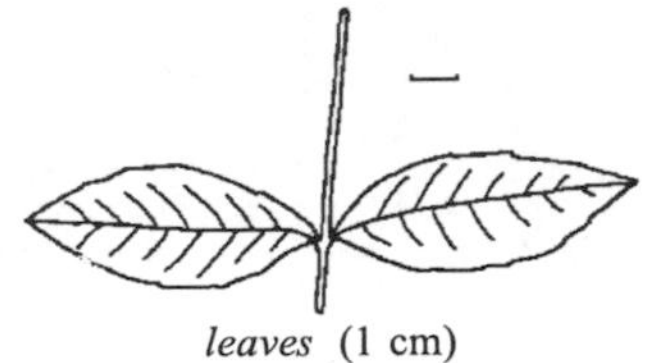

Pycnanthemum

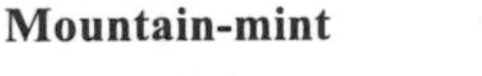

Mountain-mint

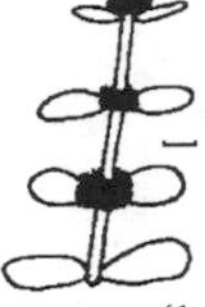

leaves (1 cm) *upper stem* (1 cm)

P. californicum: Stems erect, ± white-woolly, often few-branched; lvs ovate to lance-ovate, finely pubescent, the upper serrulate, 3-8 cm L; heads 1-4, compact; calyx 4-5 mm L, pubescent; corolla 6-7 mm L, white. Moist slopes and canyons, below 5500', Mariposa Co. n., Jun-Sep. Herbage can be used to flavor stews or soups.

Salvia

Sage

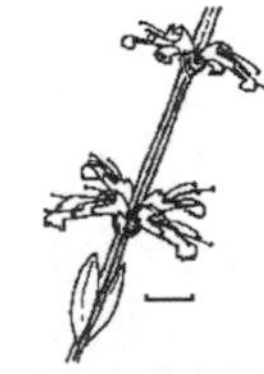

basal leaves (1 cm) *whorls of flowers* (1 cm)

S. pachyphylla: Prostrate shrub, 3-8 dm H; lf-blades oblanceolate to obovate, 2-4 cm L; petioles 5-15 mm L; infl of crowded whorls forming a dense spike; bracts purplish, 1-2.5 cm L; calyx about 1 cm L; corolla blue-violet, 15-20 mm L. Dry slopes 5000-10,000', Kern Co. Seeds may be eaten raw or parched and ground into flour; they are quite nutritious.

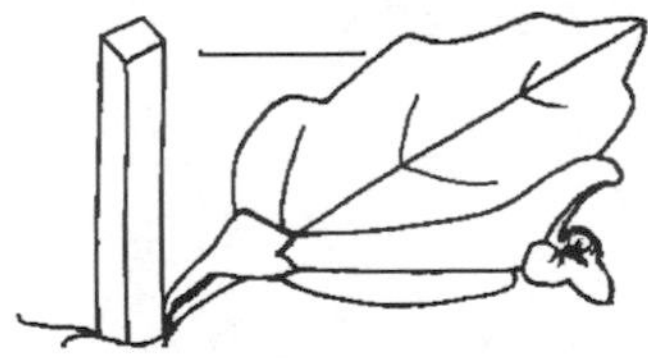

Scutellaria

Skullcap

S. tuberosa flower (1 cm) *inflorescence* (1 cm)

Rhizomes often forming lateral tubers. Fls 1-3 in the axils or in bracted racemes or spikes. Calyx gibbous, 2-lipped, both lips entire, the upper with a crest-like projection on the back. Corolla blue, violet or white, well exserted, the upper lip entire or slightly notched, the lower with small lateral lobes and a large middle one. Style unequally 2-cleft at apex. Herbage of related sp toxic in large quantities.

A Lvs ± sessile, the blades broad and truncate-cordate at base
  B Corolla blue; lvs acute — 3) *S. galericulata*
  BB Corolla whitish; lvs obtuse — 1) *S. bolanderi*
AA Lvs mly petioled or narrow, occasionally the lowermost truncate-cordate
  B Lvs mly coarsely dentate, ovate — 6) *S. tuberosa*
  BB Lvs ± entire, mly linear to oblong
    C Hair of stem appressed; stems 4-8 cm H — 4) *S. nana*
    CC Hair of stem not appressed; stems 10-40 cm H

D Corolla white with ± yellow tinge; Tuolumne Co. n. — 2) *S. californica*
DD Corolla deep violet-blue — 5) *S. siphocamyploides*

1) *S. bolanderi*, Bolander's Skullcap: Stems 2-4 dm H; lvs thin, spreading, deltoid-ovate, 1-3 cm L; fls few, solitary; calyx 4-5 mm L; corolla 16-18 mm L, the lower lip blotched with violet. Moist, ± gravelly places below 4000', Plumas Co. s., Jun-Jul.

2) *S. californica*, California Skullcap: Stems 1-3 dm H, puberulent; lvs linear-oblong to oblong-ovate, 15-25 mm L, on petioles 5-20 mm L, obtuse; pedicels about 3 mm L; calyx about 4 mm L; corolla 15-20 mm L. Drying, gravelly places below 7000', Jun-Jul.

3) *S. galericulata*, Marsh Skullcap: Stems mly 3-6 dm H; lvs ovate-oblong, 3-5 cm L, scalloped-serrate; pedicels to 2.5 mm L; calyx about 4 mm L; corolla 14-20 mm L. Swamps and wet places, 4000-7000', Eldorado Co. n., Jun-Sep.

4) *S. nana*, Dwarf Skullcap: Stems branching at base; lvs oblong-elliptical, entire, 1-1.5 cm L, obtuse, short-petioled; calyx 3-5 mm L; corolla 16-10 mm L. Dry slopes, 4000-6500', Plumas Co. n., Jun-Aug.

5) *S. siphocamyploides*, Gray-leaf Skullcap: Stems 1-4 dm H, glabrous to glandular- pubescent; lvs oblong-lanceolate to -linear, 15-25 mm L, the tube slender, curved upward. Dry, gravelly or rocky places below 8000', SN (all), May-Jul.

6) *S. tuberosa*, Dannie's Skullcap: Stems 0.5-2 dm H, usu branched at base, ± viscid with rather long spreading hairs; lvs 1-2 cm L, the lower on petioles 5-15 mm L; fls solitary; pedicels 2-3 mm L; calyx 3-5 mm L; corolla 15-20 mm L. Borders of brush and open woods below 5000', Fresno Co. n., May-Jul.

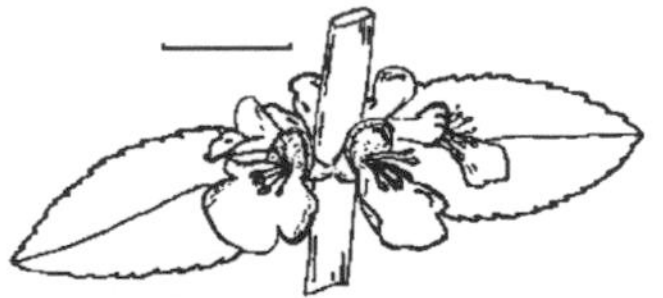
flower whorl (1 cm)

Stachys

Hedge-nettle

leaves (1 cm)

Fls few to many, borne in whorls in usu interrupted terminal spikes or also in upper axils. Calyx 5-toothed, the teeth subequal. Tubers of all spp are probably edible cooked, but this remains unconfirmed.

Herbage mly greenish with straight hairs — *S. ajugoides*
Herbage mly white-woolly with cobweb-like hairs — *S. albens*

S. ajugoides: Stem erect to decumbent, usu less than 6 dm H; lvs oblong, 2-7 cm L, on petioles to 3 cm L; fls in whorls of 6; calyx 5-8 mm L, woolly; corolla white to pale rose, veined with purple. Moist habitats.

S. albens, White Hedge-nettle: Stem erect, usu branched, 5-25 dm H; lvs 3-12 cm L, lance-oblong to broadly ovate, acute, scalloped-serrate, the lower on petioles to 5 cm L; spikes 1-2 dm L, ± interrupted in age; calyx 5-8 mm L, woolly; corolla white to rose purple, veined with purple, 12-15 mm L. Wet places below 9000', SN (all).

flower (1 mm)

Trichostema

Bluecurls

inflorescence (1 cm)

Strong-scented annuals with entire lvs. Calyx equally 5-lobed to unequally so. Corolla 5-lobed, usu blue. Fresh lvs were mashed by Indians and thrown into streams to stupefy fish.

Stems with straight, erect hairs; to 10,000', SN (all) — *T. oblongum*
Stems with downwardly curled, appressed hairs; Plumas Co. n. — *T. simulatum*

T. oblongum, Mountain Bluecurls: Stems 1-3(-5) dm H; lvs oblong to lance-oblong, 2-3 cm L, mly obtuse; petioles to 5 mm L; calyx 3-5 mm L in fr, the teeth lanceolate, subequal; corolla 5-6 mm L, lower lip 2-4 mm L. Dry margins of mdws, etc., SN (all), Jun-Sep.

T. simulatum, Siskiyou Bluecurls: Stems erect, 1-3 dm H, mly branched below, ± reddish, glandular-pubescent; lvs lanceolate to ovate-lanceolate, 2-5 cm L; calyx 4 mm L, the lobes deltoid; corolla about 4 mm L, the lower lip 2-3 mm L. Dry, open places, 2000-6000', Jun-Aug.

LAURACEAE - Laurel Family

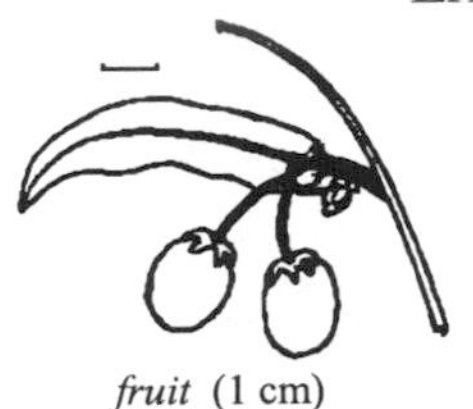

Umbellularia

Pepperwood
Bay Tree
California-laurel

fruit (1 cm) *leaves* (5 cm)

U. californica: Evergreen, pungently aromatic tree (or shrub in drier places); bark grayish to reddish brown, smooth; lvs oblong to oblong-lanceolate, 3-8 cm L, 1.5-3 cm W, short-petioled, yellowish green; fls 6-10 in an umbel, yellow-green; drupe usu solitary, round, 2-2.5 cm L, greenish becoming dark purple when ripe. Common, canyons and valleys below 5000', SN (all), Mar-May. The thin shelled nuts may be parched or roasted in the ashes of a fire and then cracked and eaten or ground and molded into small cakes. The fresh nuts store well, but must be parched or roasted before using to remove the bitter taste. The leaves may be used as a spice, the bay leaf, but a strong or prolonged whiff of the scent may cause headache and nausea.

LENTIBULARIACEAE - Bladderwort Family

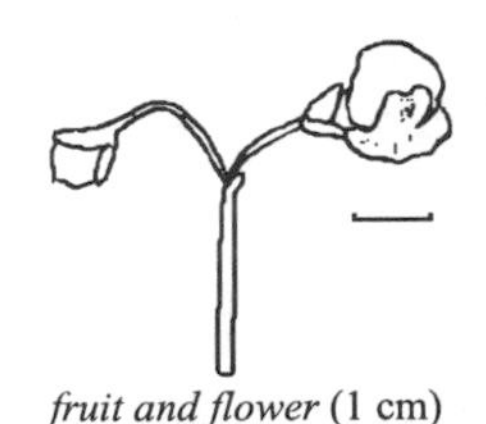

Utricularia

Bladderwort

fruit and flower (1 cm) *U. vulgaris stems habit* (5 mm)

Stems mly submersed. Lvs much dissected and bearing small urn-shaped bladders which possess a valve-like opening to trap insects and minute crustaceae. Fl-stems emergent, 1- to several-fld. Fls perfect, racemose, yellow; the calyx with 2 entire lips; the corolla 2-lipped, the lower lip with a narrow basal spur, the upper lip erect. Stamens 2, twisted, flattened. Fr a capsule. Edibility unknown.

Lf-segments flattened *U. intermedia*
Lf-segments round or capillary
  Lvs mly tripinnately divided, 1.5-4.5 cm L; corolla 10-18 mm W *U. vulgaris*
  Lvs mly with 2 filiform segments, less than 1 cm L; corolla 4-6 mm W *U. gibba*

U. gibba, Slender Bladderwort: Branches delicate, filiform, creeping or floating, to 2 dm L; lvs with scattered traps; fl-stems slender to 1 dm H; fruiting pedicels ascending, less than 1 cm L. Shallow water below 5200', Tuolumne Co., Jul-Sep.

U. intermedia, Flat-leaved Bladderwort: Stems creeping under water, 2-5 dm L; lvs 4-10 mm L, mly 3 times forked; bladders borne on branches separate from the lvs; fl-stems to 2.5 dm H, 1-4-fld; pedicels ascending; corolla about 1.5 cm L, the spur almost as long. Shallow water, 4000-7500', Fresno and Plumas cos., Jul-Aug.

U. vulgaris, Common Bladderwort: Immersed stems 3-10 dm L, coarse, few-branched; lvs with numerous traps about 2 mm L; fl-stems stout, 1-3 dm L; lower lip of corolla 12-15 mm L, yellow with brown stripes; fruiting pedicels mly arched-recurving, 1-1.8 cm L. Ponds and quiet water, scattered localities, Jul-Sep.

LIMNANTHACEAE - False Mermaid Family

Low, tender glabrous herbs of moist places. Lvs alt, without stipules. Sepals persistent; petals contorted, white.

Petals 3, shorter than sepals; 5000-9500' *Floerkea*
Petals usu 5, longer than sepals; below 5500' *Limnanthes*

Floerkea

False Mermaid

F. proserpinacoides: Stems slender, 3-10 cm H; lvs once-pinnate; lfts 3-7, 4-15 mm L; pedicels 1-3 cm L; petals about 1-2 mm L; sepals 2-4 mm L. Habitats moist in early season, SN (all), May-Jul. The stems and lvs have a spicy flavor and may be eaten raw in salads.

upper stem (5 mm)

Limnanthes

Meadow-foam

L alba leaves and flowers (1 cm)

L montana flower (1 cm)

Stems from fibrous roots. Lvs usu pinnate then pinnatifid. Petals sometimes 4 or 6. Fr 1 to several nutlets. Edibility unknown.

Corolla bowl-shaped; Tuolumne Co. n.	*L. alba*
Corolla bell-shaped; Mariposa Co. s.	*L. montana*

L. alba, White Meadow-foam: Stems 1-3 dm H, mly erect; lvs 3-10 cm L (including petioles), usu glabrous; pedicels spreading-ascending, slender, 2-8 cm L; calyx 5-8 mm L or longer in fr; petals white, sometimes rose-pink at apex in age, truncate or very slightly notched, 10-14 mm L; nutlets obovoid, dark brown, 3 mm H. Moist habitats, Tuolumne Co. n., May-Jun.

L. montana, Mountain Meadow-foam: Stems slender, usu several, ascending, 1-2 dm H; lvs 4-9 cm L, ultimate lobes oblanceolate, sometimes hairy; pedicels 3-7 cm L, slender; sepals 4-5 mm L; corolla white, round-truncate, 7-10 mm L; nutlets scarcely 2 mm H. Seeps and moist habitats, Mar-May.

LINACEAE - Flax Family

Stems ± glabrous, with tough fibrous cortex. Lvs sessile, usu entire, without stipules. Fls in terminal or axillary racemes, 5-merous. Fr a capsule. Seeds mucilaginous.

Plants perennial; fls showy, blue or rarely white; styles 5	*Linum*
Plants annual; fls small	
Fls yellow; lvs oblong to elliptic, mly opp	*Sclerolinon*
Fls pale pink to white; lvs linear , alt	*Hesperolinon*

Hesperolinon

Small-flowered Flax

fruit and flower (1 cm)

upper stem (1 cm)

H. micranthum: Freely branched at or usu above the base, 1-4 dm H, the branches very slender; lvs 1-2.5 cm L; pedicels 5-15 mm L, filiform; petals mly 2-3 mm L; styles 3, about 1 mm L. Often on serpentine, open slopes and ridges, mly below 5500', SN (all), May-Jul. Seeds probably edible after cooking.

Linum

Western Blue Flax

L. lewisii: Stems several, leafy, 1.5-7.5 dm H; lvs linear to lance-linear, acute, ascending-erect, 1-2 cm L, 4-6 mm W; fls in leafy 1-sided racemes; pedicels mly 1-3 cm L; petals 1-1.5 cm L; styles about 8 mm L. Dry slopes and ridges, mly 4000-11,000', SN (all), May-Sep. Seeds are edible after cooking (removal of cyanide) and have a high oil content.

upper stem (1 cm)

Sclerolinon

Northwest Yellow Flax

S. digynum: Stems mly 6-20 cm H, slender, branched above; lvs erect or ascending, 5-15 mm L, the lower entire: pedicels 1-2 mm L, subtended by leafy bracts; styles 2, partly united, about 1 mm L. Moist, grassy mdws, 3500-4700', Fresno Co. n., Jun-Jul. Seeds probably edible after cooking.

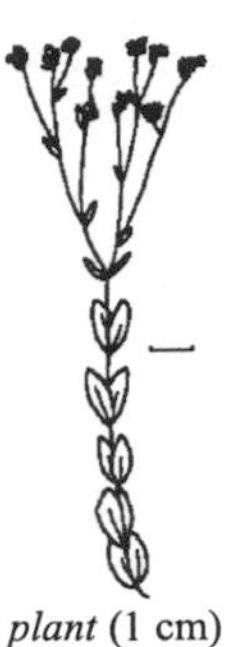

plant (1 cm)

LOASACEAE - Loasa Family

M dispersa inflorescence (1 cm)

Mentzelia

Blazing star

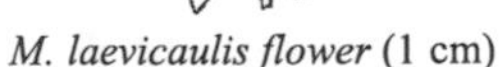

M. laevicaulis flower (1 cm)

Mly annual herbs, freely branched, covered with various types of rigid or barbed hairs. Lvs alt, brittle at least in age, sessile. Sepals 5, petals 5-10, mly free, yellow. Seeds are edible when parched and ground.

Lower lvs dentate to entire, usu not lobed or only 2 lobes at base of blade; petals 2-6 mm L
- Bracts of infl green, not white at base — *M. dispersa*
- Bracts of infl whitish base — *M. montana*

Lower lvs pinnately lobed or divided
- Petals 4-8 mm L, usu orange, blunt; n. SN — *M. veatchiana*
- Petals over 10 mm L, bright yellow, acute; widespread — *M. laevicaulis*

M. dispersa, Nevada Stickleaf: Stems slender, whitish, 1-3 dm H; lower lvs lanceolate, 3-10 cm L, the upper ovate to lanceolate, entire to sinuate-pinnatifid, 1-5, cm L; fls axillary, mly near ends of branches; sepals 1-3 mm L; petals 3-4 mm L with basal orange spots; capsule 15-25 mm L. Dry, disturbed sandy places, below 8500', SN (all), May-Aug.

M. laevicaulis, Blazing Star: Stems erect, stout, 4-16 dm H, branched, shining white, subglabrous below, stiff-hairy above; rosette-lvs oblanceolate, usu pinnately lobed, 1-3 dm L, the cauline sinuate-dentate, ovate-lanceolate, 1-10 cm L; fls in terminal clusters of 1-3; floral tube 1.5-3 cm L; sepals lanceolate, acuminate, 2-4 cm L, reflexed in fr; petals 5-8 cm L, lanceolate; capsule 3-4 cm L. Dry, gravelly or stony places below 8500', Tulare to Plumas Co., Jun-Oct.

M. montana: Stems erect to 40 cm H; lvs to 12 cm L; bracts ovate, toothed below middle; sepals 1-4 mm L; base of petals often with orange spot; fr 5-20 mm L; 4000-8500', SN (all).

M. veatchiana: Stems to 40 cm H; lvs to 15 cm L, the basal lobed, the cauline toothed or lobed; bracts usu toothed, ovate; sepals 1-5 mm L; fr 10-30 mm L. Open slopes and flats below 6,000'.

MALVACEAE - Mallow Family

Perennials usu with stellate or branched pubescence. Lvs alt, palmately lobed, with small deciduous stipules. Fls usu 5-merous, bisexual. Calyx 5-lobed. Stamens united into tube around style for about half their length. Fr a capsule.

Petals yellowish; fls axillary, solitary or in small clusters *Malvella*
Petals pink to purple; fls in terminal racemes or spikes
  Lvs densely pubescent, faintly lobed Sphaeralcea
  Lvs glabrous or nearly so, often deeply lobed *Sidalcea*

Malvella

Alkali-mallow

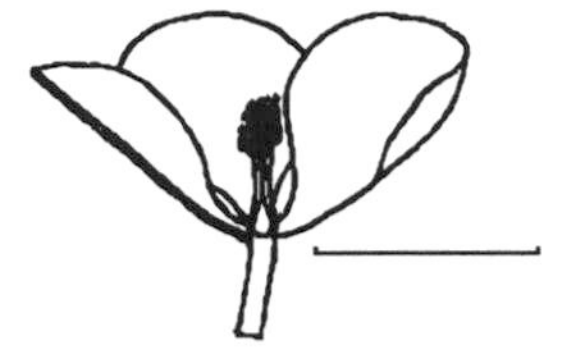

leaf with axillary flower (1 cm) *cross-section of flower* (1 cm)

M. leprosa: Stems decumbent or prostrate, whitish stellate, 1-4 dm L; lvs round-reniform to broadly deltoid, dentate, 1.5-4.5 cm W, on petioles 1-3 cm L; calyx 6-7 mm L; petals 10-12 mm L. Moist places below 6000', SN (all), May-Oct. Herbage toxic to sheep.

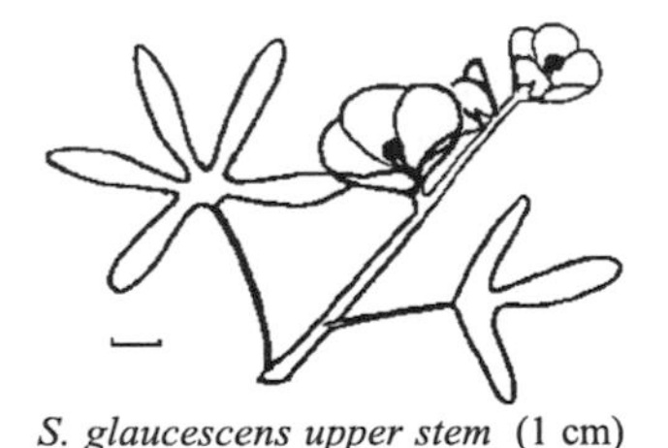

Sidalcea

Checker

S. glaucescens upper stem (1 cm) *S. oregana upper stem* (1 cm)

Lvs of lower stem usu shallowly and broadly lobed, the lobes often toothed. Upper lvs mly deeply divided into oblong to linear segments. Related sp edible when cooked as greens.

A Infl long and narrow, ± spike-like; fls usu densely clustered; wet habitats 4) *S. oregana*
AA Infl not spike-like, more open or spherical; usu dry habitats
  B Plants glaucous, nearly glabrous; lvs deeply lobed to cmpd
    C Infl 1-sided 1) *S. glaucescens*
    CC Infl not 1-sided 3) *S. multifida*
  BB Plants not clearly glaucous; lvs toothed or lobed but not deeply so
    C Lower lvs not lobed, on long petioles 5) S. reptans
    CC Lower lvs usu lobed 2) S. malvaeflora

1) *S. glaucescens*, Glaucous Sidalcea: Stems slender, procumbent to erect, 3-7 dm L; lvs 5-7-lobed or -parted, the divisions entire or 3-5-lobed; infl a slender, open raceme, ± stellate-pubescent; calyx 5-7 mm L, the lobes veiny in fr; petals pink to purple, 8-20 mm L. Dry, grassy places or open woods, 3000-11,000', SN (all), May-Jul.

2) *S. malvaeflora* ssp *asprella*, Checker Mallow: Gray-glaucous perennial; stems decumbent to 6 dm L; upper and lower lvs ± lobed; infl dense to open; calyx 5-10 mm L in fl,; petals 10-20 mm L, bright pink. Open woodlands below 8000', Mariposa Co. n.

3) *S. multifida*: Gray-glaucous perennial 1-6 dm H, pubescent; lvs deeply 7-lobed, each lobe ternately divided into linear or oblong segments; infl open; 3-9-fld; calyx 5-10 mm L in fl; petals 10-25 mm L, rose. Dry habitats 6500-8000', Mariposa and Mono cos. s.

4) *S. oregana* ssp *spicata* , Spicate Checker: Stems 3-9 dm H; lvs 3-10 cm W; petioles 6-15 cm L, hairy; fls ± connate; petals 10-20 mm L, rose, usu notched at apex. Moist habitats below 8500', Jun-Aug.

5) *S. reptans*, Creeping Checker: Stems 1 to few, decumbent at base, 2-5 dm H; lvs mly from lower stem, sparsely hairy, coarsely dentate to lobed or incised, 2-7 cm W; calyx 8-10 mm L, long-hairy; petals purple to deep pink, 12-18 mm L. Moist mdws, 4000-7600', Tulare to Amador Co., Jul-Aug.

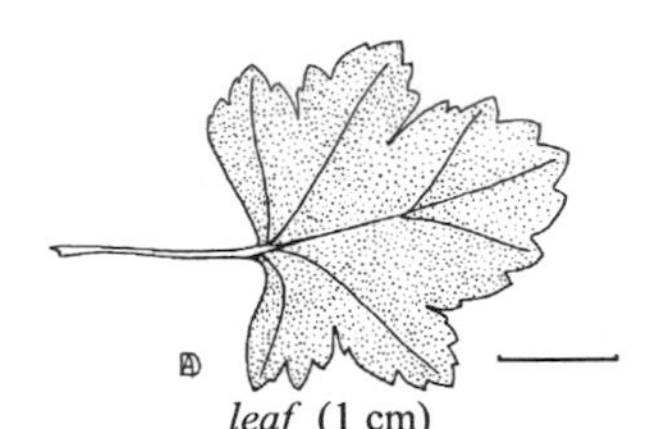

leaf (1 cm)

Sphaeralcea

Globemallow

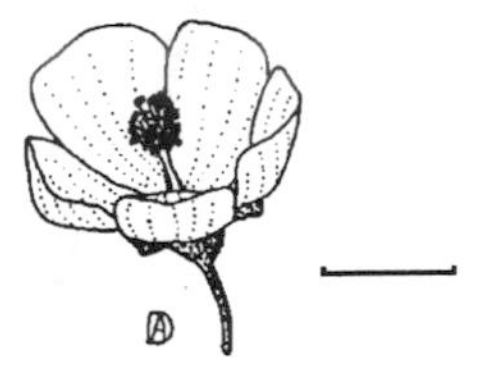

flower (1 cm)

S. munroana: Stem erect, to 8 dm H, densely gray tomentose; lvs triangular to faintly 5-lobed, to 4 cm W, coarsely dentate; fls in raceme-like clusters; petals 10-15 mm L, red-orange; anthers yellow. Uncommon in dry, open habitats below 6500', Placer Co. n.

MENYANTHACEAE - Buckbean Family

Aquatic perennials from thick rhizomes. Lvs us alt, the petiole sheathing at base. Fls bisexual, regular, 5-merous.

Lvs with 3 lfts; fls white to pink — *Menyanthes*
Lvs simple; fls yellow; rare — *Nymphoides*

Menyanthes

Buckbean

M. trifoliata: Glabrous perennial from a creeping rhizome; lfts ± oblong, 2-8 cm L, mly entire, sessile; petioles 5-20 cm L; peduncles 10-30 dm H; raceme 10-15 cm H; pedicels 5-25 mm L; fls 5-merous; corolla white or pink, the lobes about 5 mm L; capsule about 8 mm L. Occasional; 3000-10,500', SN (all), May-Aug. Herbage and rhizome bitter; the rhizome can be made palatable when collected in early season and boiled in several waters.

plant (1 cm)

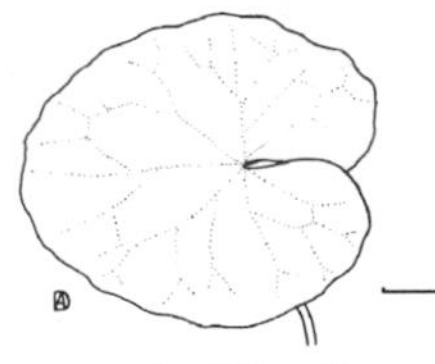

leaf (2 cm)

Nymphoides

Water Fringe

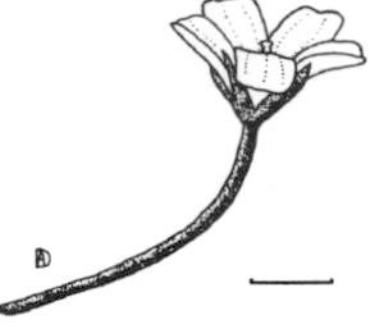

flower (1 cm)

N. peltata: Stem creeping; fl stems and lvs floating, the lvs deeply cordate, opp on fl stem, 3-10 cm W, entire or with wavy margin; infl umbellate, 2-5-fld, axillary; pedicels 3-10 cm, corolla 3-4 cm diameter. Trout Lake, Eldorado Co.

MYRICACEAE - Wax-myrtle Family

branchlet (5 cm)

Myrica

Sierra Sweet-bay

leaves and fruit (1 cm)

M. hartwegii: Shrub usu 1-2 m H; young branches pubescent and with resin globules, becoming glabrate; lvs aromatic, broadly oblanceolate, 4-8 cm L, appearing after fls, pubescent; petioles 4-12 mm L; staminate aments 1-2 cm L; pistillate aments about 1 cm L in fr. Stream banks, 1000-5000', Fresno to Yuba Co., Jun-Jul. Fr of related spp edible.

NYCTAGINACEAE - Four-O'clock Family

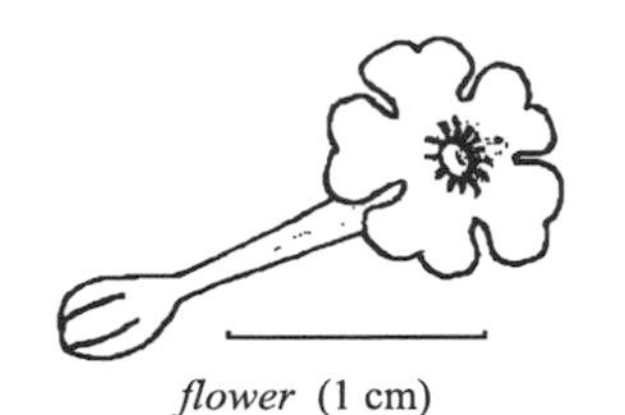

Abronia

Sand-verbena

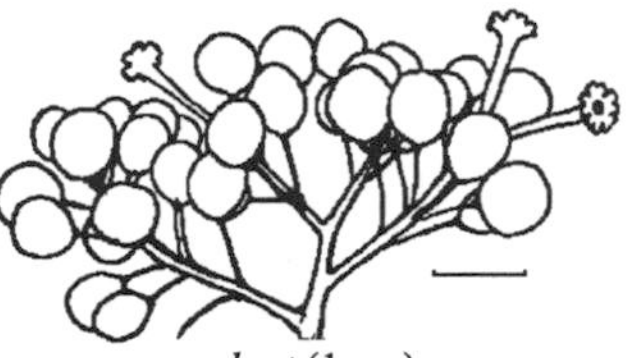

flower (1 cm) *plant* (1 cm)

A. alpina: Matted perennial, 1-2 dm across, sticky-hairy; lvs opp; lf-blades 5-10 mm L, ± round, on petioles 1-2 cm L; peduncles shorter than petioles, slender; fls 1-5 in a capitate cluster, the perianth lavender-pink, 10-15 mm L, the limb 6-8 mm W. Rare, sandy mdws, 8000-9000', Tulare Co., Jul-Aug. Plant too rare to consider edibility.

NYMPHAEACEAE - Water-lily Family

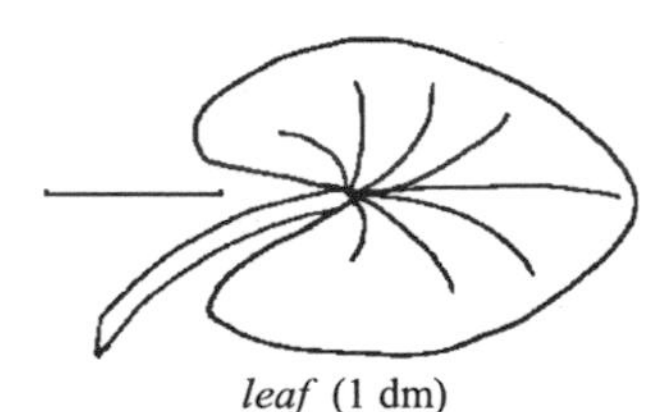

Nuphar

Cow-lily
Yellow Pond-lily

leaf (1 dm) *flower* (1 cm)

N. luteum ssp: Lvs simple, floating or emergent, broadly oval, 1-4 dm L; submersed lvs thin, delicate; fls yellowish to purple, axillary, solitary, peduncled, usu standing above the water; sepals 7-9, yellow, often tinged red, rounded, thick, the inner to 5 cm L; petals narrow, mly hidden by the stamens, both numerous. Ponds and slow streams below 7500', Mariposa Co. n., Apr-Sep. The globular seed vessels mature in midsummer and may be collected with their seeds. When dry, the seeds are easily removed and will keep indefinitely. The seeds may be treated like popcorn and after popping may be eaten or ground into meal. The starchy rootstocks should be boiled and then peeled. The core can then be eaten, placed in soup or stew, or dried, ground into meal and used as flour.

OLEACEAE - Olive Family

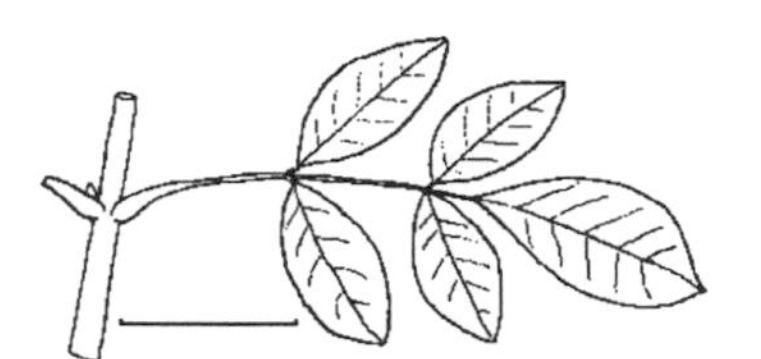

Fraxinus

Oregon Ash

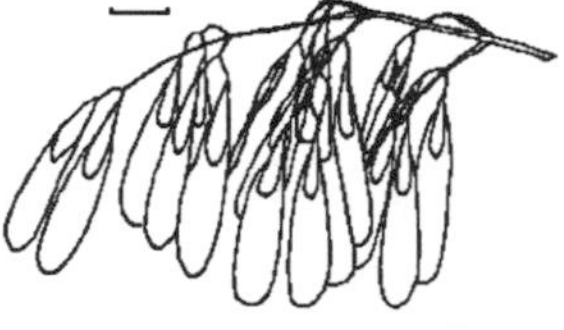

leaf (5 cm) *fruits (samaras)* (1 cm)

F. latifolia: Tree 10-25 m H; lfts oblong to oval, the terminal 6-10 cm L, the lateral smaller, ± sessile; samaras mly 3-5 cm L, 5-9 mm W. In canyons and near streams, SN (all), Apr-May. A smaller sp (*F. dipetala*) is found below 5000' on dry slopes. Related spp have edible inner bark and sap.

ONAGRACEAE - Evening-primrose Family

Lvs simple, alt or opp. Stipules none or ± glandular. Sepals usu deciduous after anthesis.

A Lvs mly opp, usu broad; plants perennial
  B Lvs long-petioled; petals 2, minute — *Circaea* p. 143
  BB Lvs ± sessile; petals 4, conspicuous — *Epilobium* p. 144
AA Lvs mly alt; plants often annual

B Sepals mly reflexed (rarely erect in *Gayophytum*); petals rarely lobed
  C Stems capillary; fls small, white, axillary — *Gayophytum* p. 146
  CC Stems stouter or none: fls mly showy, if small then yellow
    D Petals mly purplish, sometimes white or yellow with purple spots — *Clarkia* p. 143
    DD Petals yellow, aging reddish
      E Stigma capitate, not lobed — *Camissonia* p. 143
      EE Stigma 4-lobed — *Oenothera* p. 147
BB Sepals not reflexed; petals often lobed
  C Lvs ± sessile; petals often lobed; fr elongate, dehiscent — *Epilobium* p. 144
  CC Lvs usu ovate, petioled; petals not lobed; fruit round, indehiscent — *Clarkia heterandra* p. 143

Camissonia

Sun Cup

C. boothii inflorescence (1 cm) *C. sierrae upper stem* (1 cm)

Annual with basal or alt lvs. Infl bracted, nodding. Sepals 4, reflexed. Petals 4, usu fading reddish. Fr cylindric. Edibility unknown.

Petals white, opening at dusk; s. SN — *C. boothii*
Petals yellow, opening at dawn; c. SN
  Basal lvs usu many; stems + fleshy — *C. ignota*
  Basal lvs usu none; stems slender — *C. sierrae*

C. boothii, Booth's Sun Cup: Basal rosette of lvs well developed; stem 10-30 cm H; lvs 10-40 mm L, lanceolate to ovate; fr usu curved. Mly a desert sp but occasional in s. SN below 7000'.

C. ignota: Herbage often reddish; stems ascending to erect, 10-50 cm H; lvs 10-50 mm L, lanceolate to ovate, minutely serrate; sepals 2-6 mm L; petals 5-8 mm L; fr 20-30 mm L. Heavy soils below 3500', Madera Co. Apr-Jun.

C. sierrae, Sierra Sun Cup: Stem decumbent to ascending, 5-15 cm H; lvs 5-15 mm L, lanceolate or wider; sepals 2-4 mm L, adhering in pairs; petals 2-5 mm L; fr 20-30 mm L. Rocky outcrops below 8000', Madera and Mariposa cos., May-Jul.

Circaea

Enchanter's Nightshade

upper stem (1 cm)

C. alpina var. *pacifica*: Stem unbranched, 1-4 dm H, slender, sparsely hairy; lvs remotely denticulate, thin, sometimes truncate or subcordate at base, the blades 2-6 cm L, 1-4 cm W; petioles 1-4 cm W; sepals and petals about 1 mm L; capsule obovate, about 1 mm L, fruiting racemes a small version of pepper-grass. Deep woods below 8000', SN (all), Jun-Aug. Edibility unknown..

Clarkia

Clarkia

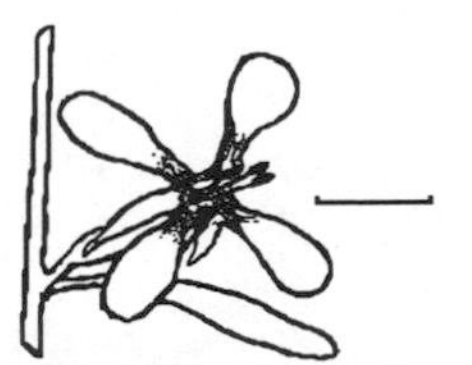

C. rhomboidea flower (1 cm)

C. williamsonii flower (1 cm)

Stems erect, slender except sometimes in *C. unguiculata*. Lvs alt. Infl a spike or raceme often recurved in bud and straightening as the fl opens. Sepals 4, green, reflexed (often in pairs or as a single group attached by their tips). Petals 4, pink to purple. Stamens 8. Stigma often 4-lobed. Fr a many-seeded capsule. Seeds small but edible raw or after grinding.

A Petals about 5 mm L and 2-3 mm W; fr ovoid, indehiscent — 3) *C. heterandra*
AA Petals 10-30 mm L, 5-20 mm W; fr elongate, splitting lengthwise
  B Petals shield-shaped, about as wide as long
    C Fl-buds and infl erect; petals usu with dark spot on distal half
      D Petals 1-2 cm L, usu purple with darker central spot — 5) *C. purpurea*
      DD Petals usu 2-3 cm L, usu with white central area and distal spot — 8) *C. williamsonii*
    CC Fl-buds and apex of infl drooping; petals lacking darker spot on distal portion
      D Petals lavender with whitish streaks; Tuolumne Co. s. — 2) *C. dudleyana*
      DD Petals reddish-purple without streaks; Sierra Co. n. — 4) *C. lassenensis*
  BB Petals narrowed toward base, 2-3 times as long as wide
    C Petals with 2 lobes — 1) *C. biloba*
    CC Petals not lobed
      D Petioles 10-30 mm L; fr banana-shaped; usu above 3000' — 6) *C. rhomboidea*
      DD Petioles 5-10 mm L; fr cigar-shaped; below 5000' — 7) *C. unguiculata*

1) *C. biloba*, Two-lobed Clarkia: Stems to 10 dm H; lvs linear to lanceolate, 2-6 cm L, entire to denticulate, with petioles 5-15 mm L; infl drooping in bud; fl-tube 1-4 mm L; sepals 5-15 mm L, united and deflexed to one side; petals 10-25 mm L, pale pink to magenta; capsule 1-2.5 cm L, straight. Open, drying slopes below 4000', Eldorado to Mariposa Co., May-Jul.

2) *C. dudleyana*, Dudley's Clarkia: Stems to 7 dm H; lvs lanceolate, denticulate, 2-7 cm L with petioles 3-10 mm L; fl-tube 1-3 mm L; sepals 10-20 mm L, united and deflexed to one side; petals 10-25 mm L; capsule 1-3 cm L, straight. Open woodland below 4500', May-Jul.

3) *C heterandra*: California Gaura: Stem erect, 1-4 dm H; lvs oblong-ovate to lanceolate, entire to denticulate, the blades 2-5 cm L, on petioles 5-10 mm L; pedicels about 1 mm L; sepals 2-3 mm L; petals pink. Dry, shaded places, 2000-5000' Placer Co. s., May-Jun.

4) *C. lassenensis*, Lassen Clarkia: Stems usu unbranched, to 5 dm H; lvs linear-lanceolate 2-5 cm L with petioles to 10 mm L; fl-tube 3-5 mm L; sepals 5-15 mm L, united and deflexed to one side; petals 5-10 mm L, often red-purple at base; capsule 1-3 cm L, straight, beaked. Occasional on open slopes below 7000', May-Jun.

5) *C. purpurea* sspp, Purple Clarkia: Stems unbranched, usu 1-3 dm H; lvs linear to lanceolate, 1-5 cm L, sessile or nearly so; fl-tube 2-7 mm L; sepals 5-15 mm L, reflexed individually or in pairs; petals 5-15 mm L, occasional deep wine-red and lacking darker spot; capsule 1-2 cm L, often densely pubescent. Common on dry slopes below 6000', w. SN (all), May-Jul.

6) *C. rhomboidea*, Rhomboid-leaf Clarkia: Stems to 10 dm H; lvs usu ovate and entire, 1-6 cm L, on petioles 10-25 mm L; fl-tube 1-3 mm L; sepals 5-10 mm L, reflexed individually; petals lavender, 5-12 mm L, with small projections on each side near base; capsule 1-2.5 cm L, curved, glabrous, smooth. Dry slopes below 8000', w. SN (all), May-Jul.

7) *C. unguiculata*, Elegant Clarkia: Stems often stout, glaucous, to 10 dm H; lvs lanceolate to ovate, entire to denticulate, 1-6 cm L; fl-tube 1-5 mm L; sepals 10-15 mm L, united and deflexed to one side; petals 10-20 mm L, the claw as long as the expanded portion and lacking projections; capsule 1-3 cm L, straight, pubescent, grooved. Moist to dry habitats, Eldorado Co. s., May-Jun.

8) *C. williamsonii*, Williamson's Clarkia: Stems to 10 dm H; lvs linear or slightly broader, 2-5 cm L, subentire, sessile to short-petioled; fl-tube 5-13 mm L; sepals 5-18 mm L, reflexed individually or in pairs; petals 1-3 cm L, lavender near tip and base; capsule 1-3 cm L, straight. Dry places below 5000', Fresno to Nevada Co., May-Jul. Plants from Yosemite often have wine-red fls.

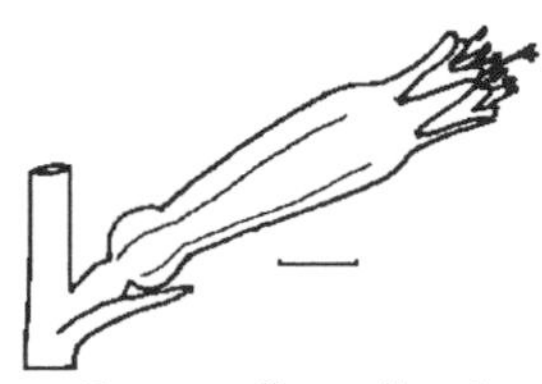

E. canum flower (1 cm)

Epilobium

Willow-herb

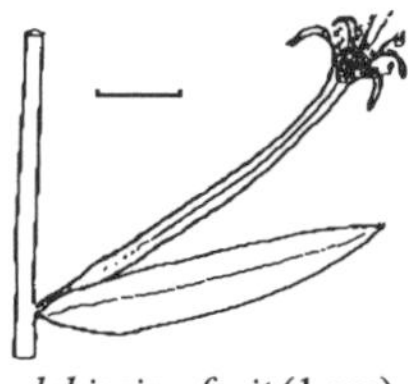

dehiscing fruit (1 cm)

Annuals or perennials over-wintering by turions (succulent artichoke-like buds which persist about the base of the next year's stem) or rosettes (at first fleshy) or slender stolons. Lvs nearly or quite sessile. Petals usu lobed. Stamens 8. Fr a capsule. The young stems and lvs of *E. angustifolium* may be cooked like asparagus or eaten raw in salads. The older lvs fresh or dried make a good tea. The pith of the stems may be added to soups.

A Lvs alt; plant annual
  B Plant 3-20 dm H; infl open; bark shreddy — 3) *E. brachycarpum*
  BB Plant less than 5 dm H, or if more the infl dense
    C Infl dense

D Lvs usu folded along midrib — 8) *E. foliosum*
DD Lvs usu flat — 7) *E. densiflorum*
CC Fls in axils
D Petals rose-purple — 18) *E. torreyi*
DD Petals reddish — 17) *E. pallidum*
AA Lvs opp; plant perennial
B Corolla-tube 2-3 cm L; sepals usu red — 4) *E. canum*
BB Corolla-tube much shorter; sepals usu green
C Petals usu 12-20 mm L; stigma lobed
D Plant 6-25 dm H — 2) *E. angustifolium*
DD Plant less than 5 dm H; above 7000'
E Fls solitary in upper axil; plant often matted — 15) *E. obcordatum*
EE Fls in short terminal racemes; rare — 14) *E. latifolium*
CC Petals 2-12 mm L; stigma usu oblong
D Rootstocks bearing globose or ovoid turions — 10) *E. halleanum*
DD Rootstocks not turioniferous, instead having stolons, etc.
E Petals 2-4 mm L, mly white to pinkish; stem mly green above
F Stems ± erect — 5) *E. ciliatum*
FF Stems clumped or tufted
G Stems densely glandular-hairy — 12) *E. howellii*
GG Stems hairy, not glandular — 13) *E. lactiflorum*
EE Petals 4-8 mm L
F Plants subglabrous
G Pedicels 5-25 mm L; infl crowded; stems tufted, 2-9 dm H — 9) *E. glaberrimum*
GG Pedicels 20-60 mm L; infl open; stems matted, 1-3 dm H — 16) *E. oregonense*
FF Plants variously hairy
G Plant loosely clumped, to 5 dm H; fr 40-65 mm L — 11) *E. hornemannii*
GG Plant densely tufted, less than 2 dm H; fr 20-40 mm L
H Infl ± nodding; pedicel in fr 10-55 mm L — 1) *E. anagallidifolium*
HH Infl erect; pedicel in fr 2-15 mm L — 6) *E. clavatum*

1) *E. anagallidifolium*, Alpine Willow-herb: Stems stoloniferous, unbranched, erect or decumbent at base, many, 0.3-1(-1.5) dm L; lvs well distributed, mly oblong, obscurely petioled; fls few, nodding in bud; petals lilac to purple, 4-5 mm L; pedicels 5-15 mm L in fr; capsule linear, purplish, 2-4 cm L. Moist rock slides and stony places, 8000-11,500', SN (all), Jul-Sep.

2) *E. angustifolium*, Fireweed: Stems few, usu unbranched, rather densely leafy; lvs alt, lanceolate, subentire, sessile or nearly so, 7-20 cm L; fls many in long terminal racemes; pedicels 5-12 mm L; sepals 8-12 mm L; petals lilac-purple, rose, rarely white; capsule 5-8 cm L. Mly below 9000', SN (all), Jul-Sep. Exceedingly variable with an alpine form, 1-4 dm H with middle stem lvs 3-6 cm L, infl to 1.3 dm L, petals 1-1.5 cm L; mly 9000-11,000'.

3) *E. brachycarpum*: Stems erect, highly branched, glabrous except for the tips of the infl; lvs ± linear, 2-5 cm L, short petioled; fls in open racemes; pedicels 5-15 mm L; sepals 2-3 mm L; petals pink to almost white, 3-6 mm L, rotate, deeply 2-cleft; capsule 2-2.5 cm L, 4-angled, linear. Open, usu dry, disturbed ground below 7500', SN(all), Jun-Sep.

4) *E. canum* , California Fuchsia: Stems 1-5 dm L; herbage green to gray green, pubescent, sometimes silky; lvs ± ovate, 1-5 cm L, 5-20 mm W; fl-tube scarlet, globose at base; sepals erect, lanceolate, 8-10 mm L; petals 2-cleft, 8-15 mm L; capsule 4-angled. Dry slopes and ridges below 10,000', SN(all), Aug-Sep.

5) *E. ciliatum*, Northern Willow-herb: Stems erect, from fleshy rosettes, glabrous below with some hair on decurrent lines below nodes; lvs ovate- to elliptic-lanceolate, 3-6 cm L, serrulate, upper lvs gradually reduced, ± pubescent; sepals 2 mm L; fruiting pedicels 3-8 mm L; capsule ± reddish, 4-6 cm L. Moist places below 11,000', SN (all), Jul-Sep.

6) *E. clavatum*, Plant tufted, less than 20 cm H, hairy; stolons wiry; lvs 10-30 mm L, elliptic to ovate; petiole to 3 mm L; infl erect, sepals 2-5 mm L; petals 3-6 mm L; pink to purple; capsule 20-40 mm L. Rocky places, 4000-12,000', c. SN.

7) *E. densiflorum*, Dense-flowered Epilobium: Stems erect, mly 3-10 dm H, usu hairy, leafy; lower lvs lance-linear to lanceolate, entire to denticulate, acute, 2-5 cm L; bracts mly entire, 0.5-2 cm L; infl dense, long spicate in fr; sepals 2-4 mm L; petals rose-purple, occasionally whitish, 6-12 mm L; capsule stout, 8-10 mm L. Moist places below 8500', SN (all), May-Oct.

8) *E. foliosum*: Lvs 5-30 mm L, linear to lanceolate; petiole 3-12 mm L; infl dense, long-hairy; sepals 1-3 mm L; petals 1-3 mm L, white. Usu dry, disturbed habitats below 5000', SN (all).

9) *E. glaberrimum*, Glaucous Willow-herb: Stems several usu unbranched; lvs glabrous, ascending, oblong-lanceolate, entire or minutely denticulate; sepals 1-2 mm L; petals 4-7 mm L, purplish to almost white; fruiting pedicels 1-2 cm L; capsule slender, suberect, 4-7 cm L. Stream banks and wet places, 3000-11,500', SN (all), Jul-Aug.

10) *E. halleanum*, Hall's Willow-herb: Stems slender, erect, 1-4 dm H, glandular-pubescent in

upper parts; turions small; lvs mly opp, lance-linear, erect, 1.5-4 cm L, some with clasping base, entire to serrulate; fls small; petals white to purplish; fruiting pedicels 3-5 mm L; capsule 2-5 cm L. Wet places, 5000-9000', SN (all), Jul-Aug.

11) *E. hornemannii*, Hornemann's Willow-herb: Stems slender, erect except at very base, unbranched, with some subterranean scaly branches, glabrous except for the pubescence on the decurrent lines below the lvs; lvs ± ovate, 1.5-4 cm L, subentire; fls few, erect; sepals 3-4 mm L; pedicels 1-2 cm L in fr; capsule erect, linear, 4-5 cm L. Moist, rocky and mossy places, 6000-11,500', SN (all), Jul-Aug.

12) *E. howellii*, Stems loosely clumped, less than 20 cm H, densely glandular; stolons thread-like, short; lvs sessile, 5-20 mm L, round to lanceolate; fl-buds usu nodding; sepal and petals less than 3 mm L, the petals white. Wet habitats, 6500-7500', c. and n. SN (Yuba Pass).

13) *E. lactiflorum*, White-flowered Willow-herb: Much like *E. hornemannii* but glabrous; the basal stolons above ground; lvs thin, 2-5 cm L. Moist habitats, 6000-11,000', SN (all), Jun-Aug.

14) *E. latifolium*, Broad-leaved Willow-herb: Stems several, usu ascending, 1-5 dm L, glabrous below, puberulent above; lvs elliptic to lanceolate, subopposite, fleshy, glaucous, entire, sessile or nearly so, 2-6 cm L; pedicels 5-10 mm L; sepals purplish, 13-18 mm L; petals purple, rose or white, purple-veined, 15-25 mm L; capsule 5-8 cm L. Wet, stony places 7600-11,400', Inyo, Fresno, Tuolumne, and Mono cos., Jul-Aug.

15) *E. obcordatum*, Rock-fringe: Stems several from a ± woody base, mly 0.5-1.5 dm L, the branches mly glabrous below, often puberulent at summit, leafy; lvs crowded, glabrous and glaucous, ovate, obscurely and remotely denticulate, mly 6-12 mm L, sessile or nearly so; fls 1 to few; pedicels 2-20 mm L; sepals purplish, 9-12 mm L; petals rose-purple, broadly obcordate, 1-2 cm L; capsule 2.5-3.5 cm L. Dry ridges and flats, 7000-13,000', SN (all), Jul-Sep.

16) *E. oregonense*, Oregon Willow-herb: Stems stoloniferous, glabrous except in infl; lvs oblong-linear to -ovate, entire to ± denticulate, somewhat crowded on lower stem, remote above, 1-2.5 cm L; fls 1 to few; sepals often purplish, 1-3 mm L; petals cream to purplish, 4-7 mm L, deeply notched; pedicels 1-3.5 cm L in fr; capsule erect, 2-5 cm L, often purplish. Stream banks and wet places, 3000-11,500', SN (all), Jul-Aug.

17) *E. pallidum*, Pale Boisduvalia: Stems 1-4 dm H, mly branched from base, soft-hairy; lvs not crowded, reduced above, 1.5-5 cm L, 3-6 mm W, lanceolate, subentire; sepals 2-4 mm L, capsule 1.5-3 cm L, 4-angled. Damp places 4000-6000', Fresno Co. n. Jun-Jul.

18) *E. torreyi*, Narrow-leaved Boisduvalia: Stems 1-5 dm H, erect, often branched from near base, with slender, straight branches, soft-hairy throughout; lvs linear to lance-linear, 1-4 cm L, entire to sharply denticulate; sepals 1 mm L; capsule 8-10 mm L. Moist places below 8500', SN (all), May-Jul.

Gayophytum

Gayophytum

Stems erect, very slender. Lvs alt, entire, linear or lanceolate and subsessile, or the lower may be opp and short-petioled. Fls blooming in summer months. Sepals 4, usu reflexed; petals 4, often drying pink or red. Most spp. found throughout SN. Edibility unknown. Difficult genus to determine to sp in the field; only abbreviated descriptions of the spp will be given following the key.

plant (1 cm)

A Petals 1 mm or less L; plants usu with some fls in lower axils
  B Plants branched only below; lvs 1-3 cm L, upper lvs well developed
    C Plants glabrous; capsule flattened — 5) *G. humile*
    CC Plants subglabrous to strigulose; capsule subterete, linear — 6) *G. racemosum*
  BB Plants diffusely branched; lower lvs 2-5 cm L, upper lvs bract-like
    C Petals about 0.5 mm L; capsule 2-5 mm L — 7) *G. ramosissimum*
    CC Petals about 1 mm L; capsule 4-12 mm L — 1) *G. decipiens*
AA Petals mly 2-7 mm L; plants rarely flowering near base
  B Petals 4-7 mm L; sepals 3-4 mm L; Eldorado Co. s. — 3) *G. eriospermum*
  BB Petals 1-4 mm L; sepals 1-3 mm L
    C Lower lvs 6-7 cm L; capsule irregularly lumpy — 4) *G. heterozygum*
    CC Lower lvs 2-4 cm L; capsule often constricted between seeds — 2) *G. diffusum*

1) *G. decipiens*, Gravel Gayophytum: Stems 1-3 dm H, the branches mly separated by 2-8 nodes; lower lvs 2-3 cm L, the uppermost 0.5 -1.2 cm L. Occasional in dry, sandy and gravelly places below 7000'.

2) *G. diffusum* and ssp, Diffuse Gayophytum Stems usu branched, only at successive nodes, 1-5 dm H; pedicels 1.5-8 mm L, ascending to erect; capsule 5-12 mm L. Dry to moist habitats, 3000-9000'.

3) *G. eriospermum*, Coville's Gayophytum: Much like *G. diffusum* except as in key.

4) *G. heterozygum*, Varied-seeded Gayophytum: Much like *G. diffusum*; stems much branched. Infrequent. below 9000'.

5) *G. humile,* Low Gayophytum: Stems 0.5-1.5 dm H, glabrous, reddish; lvs short-petioled; pedicels scarcely evident; capsule erect, 10-15 mm L. Common on drying flats, mly 3000-9000'.

6) *G. racemosum*, Black-foot Gayophytum: Stems 1-2 dm H; capsule erect, not markedly constricted between seeds, 6-14 mm L. Drying slopes and flats, 5000-11,000'.

7) *G. ramosissimum*, Much-branched Gayophytum: Stems 1-3 dm H; lvs mly 1-2 cm L but not obvious, short-petioled; pedicels 3-5 mm L, mly spreading-deflexed; capsule plump, 1-5 mm L. Dry slopes and ridges, 4500-8000', along e. SN.

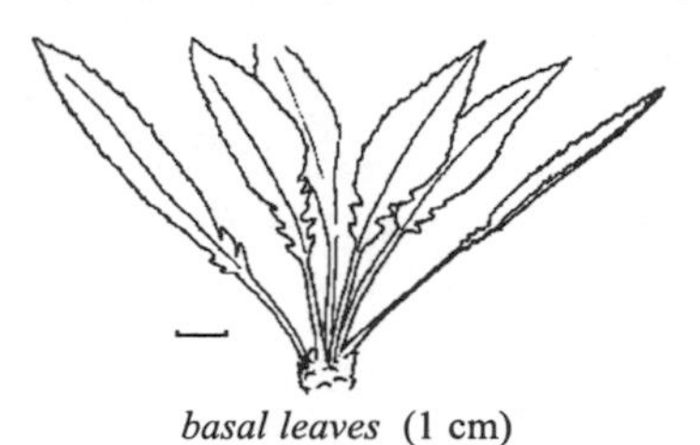

basal leaves (1 cm)

Oenothera

Evening-primrose

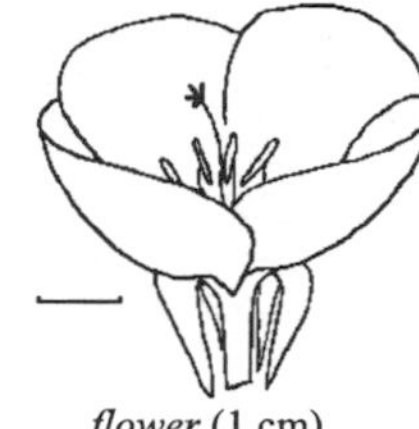

flower (1 cm)

Herbs mly without stems. Fls 4-merous, axillary. The roots of at least *some spp are* edible.

Fr winged; herbage green; n. SN *O. flava*
Fr. sharply angled but not winged; herbage gray-green; c. and s. SN *O. xylocarpa*

O. flava: Lvs oblong, irregularly pinnately lobed, 5-30 cm L; sepals 10-30 mm L; petals 10-30 mm L, yellow fading orange; fr 10-40 mm L; wings 2-6 mm W. Mly sagebrush and juniper woodland, below 5000'.

O. xylocarpa, Woody-fruited Evening-primrose: Lf-blade ± pinnately parted, often spotted red, 2-7 cm L, densely soft pubescent, the terminal lobe much the largest; petioles about as long as blades; fl-tube slender, 2.5-4.5 cm L; sepals 2-3 cm L; petals 2.5-3 cm L with broad sinus 4-5 mm deep; stigma-lobes 4-5 mm L; capsule somewhat woody, 3.5-6 cm L. Dry benches, 7000-9800', Jul-Aug.

OROBANCHACEAE - Broom-rape Family

Rather fleshy herbs without chlorophyll, having alt scales in place of lvs. Fls with persistent calyx; corolla tubular ± 2-lipped. Stamens 4, in 2 pairs. Fr a capsule.

Fls in an elongate spike; pedicels none *Boschniakia*
Fls pedicelled; infl not an elongate spike *Orobanche*

Boschniakia

Boschniakia

B. stobilacea: Stem unbranched, brown, thick, 1-2.5 dm H, from a corm-like basal thickening; spike 3-6 cm thick through flowering parts, dark reddish brown; bracts 1.5-2 cm L, covering most of stem; fl 1.5 cm or more long; capsule 1-1.5 cm in diameter. Below 10,000', SN (all), May-Jul. Edibility unknown

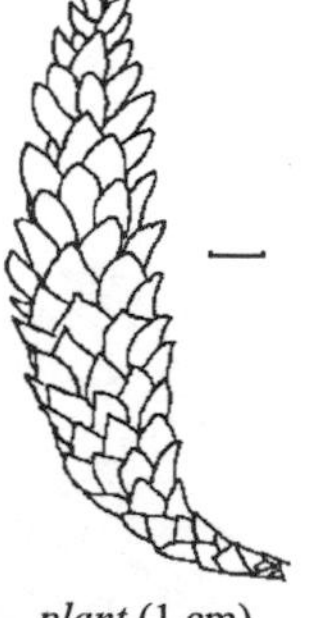

plant (1 cm)

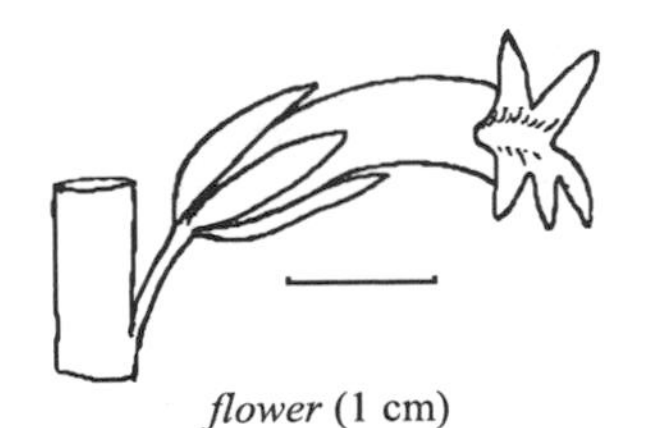

Orobanche

Broom-rape

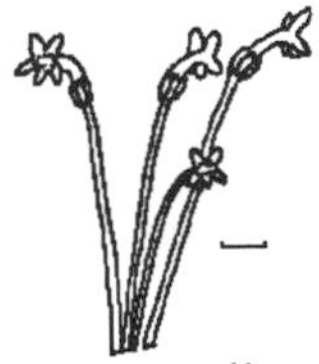

flower (1 cm) *upper stems* (1 cm)

Fls purplish to yellowish. Calyx 5-cleft. Stamens mly included. Capsule 2-valved. The entire plant is edible raw though bitter. The bitterness is somewhat removed by cooking.

Fls on pedicels 5-25 mm L, subtended by bracts — *O. californica*
Fls on elongate scape-like pedicels, the pedicels without bracts
  Pedicels few, 1-3 times as long as the stem — *O. uniflora*
  Pedicels many, not exceeding length of stem — *O. fasciculata*

O. fasciculata, Clustered Broom-rape: Stems 3-10 mm thick, 3-12 cm L; calyx 6-8 mm L; the lobes not longer than tube; corolla 15-22 mm L, constricted at base, 2-5 mm W at throat. Occasional, mly at 4000-10,000', on *Artemisia, Eriogonum, Eriodictyon* and others, SN (all), May-Jul.

O. californica, California Broom-rape: Stems 5-10 cm H, glandular-puberulent throughout; bracts 5-12 mm L; calyx 12-16 mm L; corolla 25-30 mm L, strongly 2-lipped, pale with darker veins, the lips 10-14 mm L, the upper reflexed, the lower spreading. At 6000-9100', Yosemite National Park n., Jun-Sep.

O. uniflora vars., Naked Broom-rape: Stems slender, 2-4 mm thick, 0.5-5 cm L, mly subterranean; pedicels slender, 2-10 cm L; calyx-lobes usu longer than tube; corolla deep purple to straw-colored, 15-30 mm L, constricted at base of tube and 3-8 mm W at throat. Mly 3000-8500', SN (all), May-Aug.

PAEONIACEAE - Peony Family

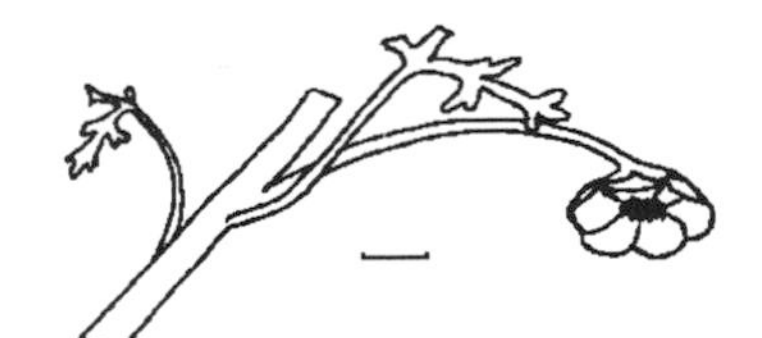

Paeonia

Wild Peony

axillary flower (1 cm) *leaf* (1 cm)

P. brownii: Somewhat fleshy perennial; stems 1 to several, 2-4 dm H; lvs 5-8 per stem, basal and alt, glaucous, mly 3-6 cm L, 2-5 cm W, the primary segments abruptly contracted at base; petioles mly 2-7 cm L; fls on recurved leafy peduncles; sepals 5 or 6, persistent, green; petals mly 5 or 6, rounded, 8-13 mm L, usu slightly wider, maroon to bronze in center, yellowish or green at margin; pistils 2-5, free, becoming fleshy follicles 2-4 cm L. Dry slopes, 3000-8600', Tuolumne Co. n., May-Jun. Related spp contain alkaloids, and it is probably best to avoid eating any part of this plant.

PAPAVERACEAE - Poppy Family

Herbs or shrubs. Lvs alt, without stipules, usu lobed or dissected. Fls bisexual. Sepals 2-3, sometimes. dehiscent. Fr a capsule.

A Fls radially symmetric (each petal having the same shape), solitary, white or yellowish
  B Plant prickly; sap yellow; petals white; sepals 3 — *Argemone*
  BB Plant ± glabrous; sap clear; petals yellow to orange; sepals 2 — *Eschscholzia*
AA Fls not radially symmetric, the shapes of the petals differing
  B One outer petal spurred — *Corydalis*
  BB Outer petals both saccate or spurred at base — *Dicentra*

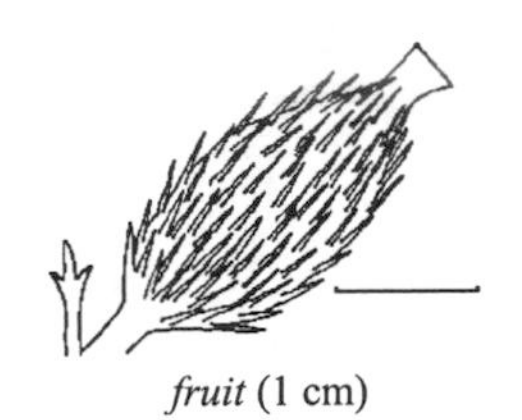

fruit (1 cm)

Argemone

Prickly Poppy

flower (1 cm)

A. corymbosa: Perennial; stems 4-10 dm H, often tinged with purple; lf-surfaces prickly above and below; lvs lobed about halfway to midrib below, more shallowly above; fls 5-13 cm in diameter; petals white. Dry habitats, e. SN (all), 4000-8500', Jun-Sep. Plant with unpleasant taste. Herbage and seeds probably contain alkaloids which are toxic if eaten in quantity.

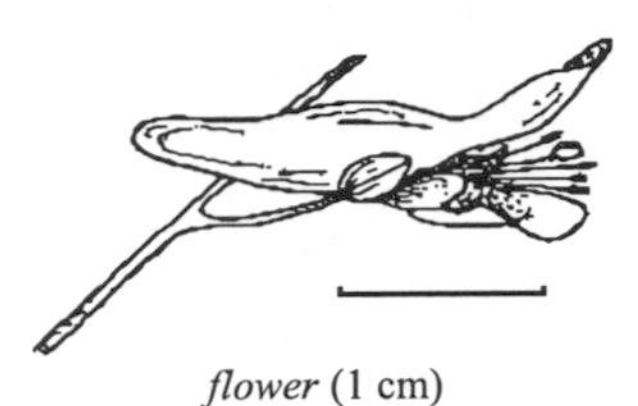

flower (1 cm)

Corydalis

Sierra Corydalis

section of leaf (5 cm)

C. caseana: Stems 1 to several, stout, glaucous, 5-10 dm H; lvs about 5, bipinnate, 15-35 cm L, the ultimate segments 1-2.5 cm L; infl dense, narrow, 5-12 cm L; fls white or pinkish with purple petal-tips; sepals 2-4 mm L; both outer petals hooded at apex but only one spurred; capsule 10-15 mm L. Moist, cool habitats, 4000-9000', Placer Co. n., Jun-Aug. Another sp, *C. aurea*, with yellow fls and a shorter infl is occasionally found in Mono Co. below 7500'. Both spp are toxic, containing alkaloids which may be lethal if eaten in sufficient quantities.

flowers (1 cm)

Dicentra

Bleeding Heart

leaf (1 cm)

Perennials with basal and sometimes stem lvs. Plants contain poisonous alkaloids similar to those in *Corydalis*.

A Stem 5-15 dm H, leafy on lower half; fls bright yellow, erect — 1) *D. chrysantha*
AA Stem less than 5 dm H; lvs all basal; fls white to rose-purple, nodding
  B Lvs about 1 dm L on petioles 2-5 dm L; fls several in compact infl — 2) D. formosa
  BB Lvs 4-6 cm L; petioles about 1 dm L; fls usu 1-2
    C Recurving tips of outer petals longer than body of the petals — 4) *D. uniflora*
    CC Recurving tips of outer petals shorter than the body — 3) *D. pauciflora*

1) *D. chrysantha*, Golden Ear-drops: Stems several, erect, glaucous, 5-15 dm H; lvs bipinnate, rather stiff, 15-30 cm L; infl narrow, 2-5 dm L, terminal, many-fld; fls erect; sepals oval, 3-4 mm L; corolla bright yellow, 12-15 mm L, the outer petals saccate below the tip, spreading in upper half; capsule mly 15-25 mm L. Common on burns and in disturbed places below 5000', Calaveras Co. s., May-Sep.

2) *D. formosa*, Bleeding Heart: Stems slender, 2-4.5 dm H; lvs biternately cmpd, glaucous beneath or also above, the ultimate lfts oblong, 2-5 cm L, pinnately cleft into divisions 1-4 mm W; sepals 3-5 mm L; corolla rose-purple to whitish, 14-18 mm L, cordate at base, the outer petals with spreading tips; capsule 14-20 mm L. Damp, ± shaded places below 7000', SN (all), May-Jul. A closely related sp (*D. nevadensis*) with base of central stamen of each set bulging out from lateral 2 is found above 7000' in the s. SN.

3) *D. pauciflora*, Few-flowered Bleeding Heart: Lvs 4-6 cm L, ternately dissected, the divisions narrowly linear, acute, glaucous beneath; fl-stems slender, slightly exceeding lvs, 1-3 fld; corolla about 20 mm L, white to pink, deeply cordate at base. Gravelly habitats, 6000-10,000', Tulare and Plumas cos., Jun-Jul.

4) *D. uniflora*, Steer's Head: Fl-stems 3-7 cm H; lvs 4-6 cm L, 2-3-ternate into oblong lobes, glaucous beneath; sepals 5-6 mm L; fls white or pink to lilac, about 15 mm L, cordate at base; capsule about 12 mm L. Gravelly or rocky places, 5400-12,000', Fresno Co. n., May-Jul.

Eschscholzia

California Poppy

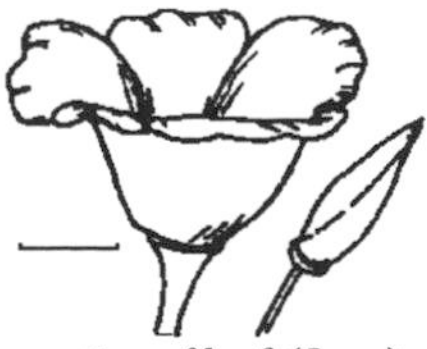

leaf (1 cm) *section of leaf* (5 cm)

E. californica: Stems becoming 2-6 dm L and falling over in age; lvs finely dissected; capsule 3-8 cm L. Grassy and open places up to 6500', SN (all), Apr-Sep. Herbage contains toxic alkaloids.

PHILADELPHACEAE - Mock Orange Family

Shrubs, often with peeling bark. Lvs simple, opp, deciduous, without stipules. Fls 4-5-merous.

Lvs coarsely serrate; plants 3-10 dm H; petals pink *Jamesia*
Lvs entire to minutely serrulate; plants 10-30 dm H; petals white *Philadelphus*

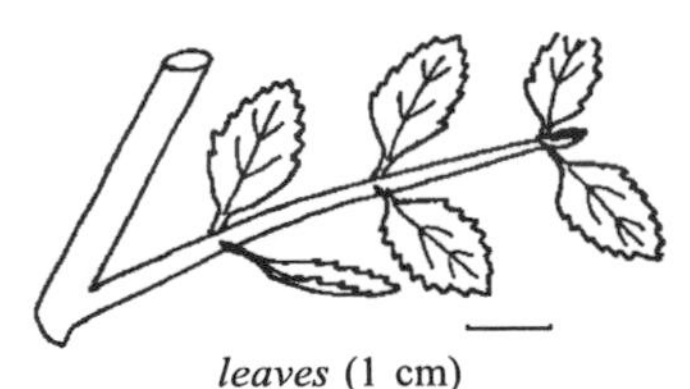

Jamesia

Cliff Bush

leaves (1 cm) *leaves and inflorescence* (1 cm)

J. americana var. *rosea*: Stems with shreddy bark; twigs pubescent; lvs oblong to roundish, 1-2.5 cm L, green and pubescent above, densely gray-hairy beneath, prominently veined; petioles mly 2-6 mm L; fls few, in terminal clusters. About rocks, 7800-12,000', Mono and Fresno Co. s., Jul-Aug. Edibility unknown.

Philadelphus

Mock-orange
Syringa

leaves and inflorescence (1 cm) *maturing fruits* (1 cm)

P. lewisii: Loosely branched shrub or tree; lf-blades ovate, 2.5-8 cm L; petioles about 8 mm L; infl a 5-40-fld terminal clusters; sepals 4-5 mm L; petals ovate, 11-16 mm L. Rocky slopes and canyons below 6000', SN (all), May-Jul. Poisonous, has been known to cause death if eaten.

PLANTAGINACEAE - Plantain Family

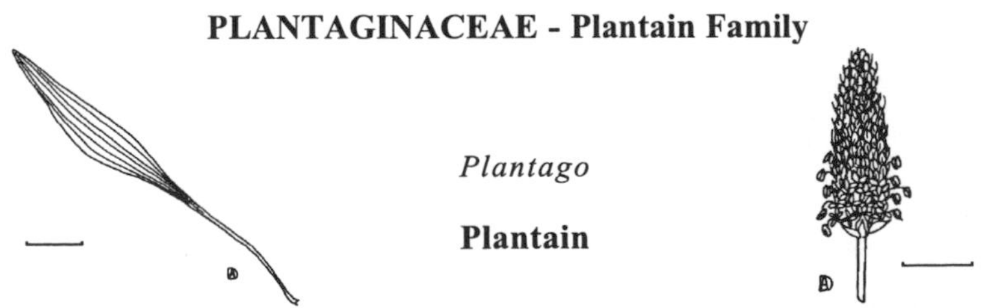

Plantago

Plantain

leaf (1 cm) *inflorescence* (1 cm)

P. major, Common Plantain: Perennial from thick caudex; lvs all basal, petioled; the blade widely elliptic to cordate with several veins from base, 5-18 cm L; thick, usu rough to touch; infl a dense linear-cylindric spike 5-20 cm H, on a scape equally as long; fls inconspicuous, 4-merous. Common in moist, disturbed habitats below 4000', SN(all).

POLEMONIACEAE - Phlox Family

Fls mly 5-merous as to perianth and stamens. Style usu 3-cleft. Ovary usu 3-loculed. Capsule usu 3-valved. Little is known regarding the edibility of these genera.

A Infl a spiny-bracted head
  B Bracts with linear lobes; heads greenish — *Navarretia* p. 154
  BB Bracts not lobed; heads white-woolly — *Eriastrum* p. 152
AA Infl various, not a spiny-bracted head
  B Lvs absent; head-like infl subtended by a whorl of leafy bracts — *Gymnosteris* p. 153
  BB Lvs present, often lobed
    C Lvs, at least lower, opp
      D Lvs simple, linear to lanceolate — *Phlox* p. 155
      DD Lvs palmately lobed, usu appearing like remote whorls of linear lvs — *Linanthus* p. 154
    CC Lvs all alt or basal
      D Plants ± woody; lvs palmately cleft, the lobes sharp-pointed — *Leptodactylon* p. 153
      DD Plants herbaceous; if lvs palmately lobed the lobes not sharp-pointed
        E Perennials; lvs pinnately cmpd or lobed, or if palmate then infl ± capitate
          F Lvs cmpd into ± broad lfts which may be further divided — *Polemonium* p. 156
          FF Lvs pinnately to bipinnately or palmately lobed — *Ipomopsis* p. 153
        EE Annuals, rarely perennials in *Collomia*; lvs simple to pinnately or palmately lobed
          F Upper lvs mly 3-lobed with a long middle lobe — *Allophyllum* p. 151
          FF Upper lvs simple or much reduced
            G Calyx growing with capsule and not ruptured by it — *Collomia* p. 151
            GG Calyx ruptured by maturing capsule — *Gilia* p. 152

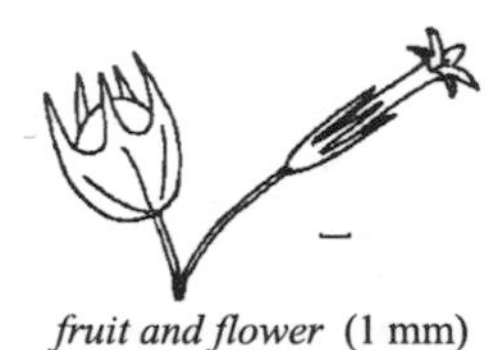

fruit and flower (1 mm)

Allophyllum

Allophyllum

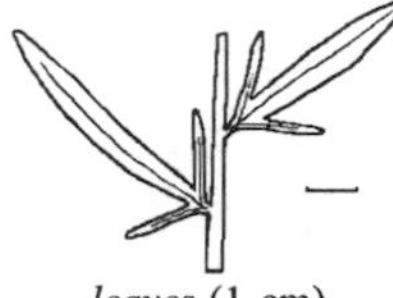

leaves (1 cm)

Stems erect, usu well branched, leafy. Fls in loose to ± congested 2-8-fld clusters, each cluster subtended by a lf; pedicels usu much elongate in fr. Calyx glandular-pubescent, the lobes joined by a sinus-membrane to about middle, this splitting in fr. Corolla funnelform.

Segments of stem lvs usu less than 4 mm W; corolla dark blue-violet — *A. gilioides*
Segments of stem lvs 3-15 mm W; corolla pink-purple to white
  Corolla-tube 6-16 mm L; fls colored — *A. divaricatum*
  Corolla-tube 5-8 mm L; fls white — *A. integrifolium*

A. divaricatum, Divaricate Allophyllum: Stems usu 1-10 dm H, pubescent to viscid-villous and with a strong skunk-like odor; lower lvs entire to pinnately lobed with 1-6 pairs of lobes; infl congested in early stages, but with pedicels elongating in fr; calyx about 6 mm L; corolla regular, 10-20 mm L, the tube red-violet and lobes pink or tube purple and lobes pink-violet, the tube 2-5 times as long as lobes. In open, often disturbed places below 7000', SN (all), May-Jun.

A. gilioides, Straggling Allophyllum: Stems usu branched above, 1-4 dm H; herbage puberulent throughout, ± glandular in the infl; lower lvs with 2 pairs of linear lobes; each fl-branch terminated by 2 or 4 fls; corolla 5-10 mm L. Dry slopes and flats below 8000', s. SN, Apr-Jun.

A. integrifolium, Entire-leaved Allophyllum: Stems 1-3 dm H, villous; lower lvs oblong to lanceolate, entire or irregularly toothed or lobed; pedicels slender, unequal, 1-4 mm L in fl and 5-10 mm L in fr; corolla 6-11 mm L. Open, sandy or rocky habitats, 4500-9000', SN (all), Jun-Aug.

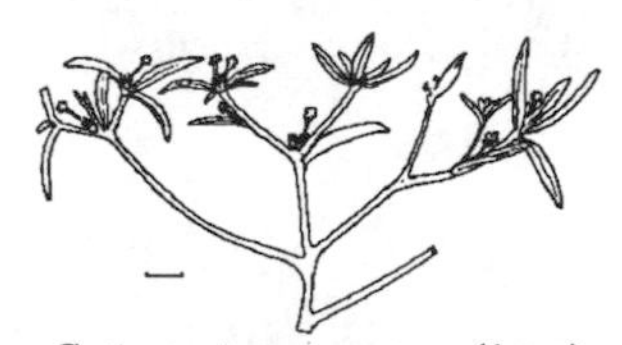

C. tinctoria upper stem (1 cm)

Collomia

Collomia

C. grandiflora upper stem (1 cm)

Calyx herbaceous or rarely with a membrane-like area below the sinuses. The sinuses in age distended at base into a projecting fold. Corolla narrow-funnelform to trumpet-shaped.

A Plants perennial: lvs forked or dentate
  B Stems 2-7 cm H; lvs 2-3 times forked; Lassen Peak — 2) *C. larsenii*
  BB Stems 10-60 cm H; lvs coarsely dentate; Madera Co. — 4) *C. rawsoniana*
AA Plants annual; lvs entire
  B Stems usu branched above; at least some fls 1-3 in axils — 5) *C. tinctoria*
  BB Stems usu unbranched; fls in terminal or sometimes axillary many-fld heads
    C Corolla pink to purplish, 8-15 mm L — 3) *C. linearis*
    CC Corolla yellow-salmon to white, 15-30 mm L — 1) *C. grandiflora*

1) *C. grandiflora*, Large-flowered Collomia: Stems 1-10 dm H, sometimes branched from base or even above, leafy throughout; lvs lanceolate to linear, entire, sessile, 3-5 cm L, passing upward into ovate leafy bracts; fls sessile; calyx 7-10 mm L, densely glandular-pubescent; style included. Dry, open and wooded slopes below 8000', SN (all), May-Jul.

2) *C. larsenii*, Talus Collomia: Stems tufted, branched from base; herbage pubescent; lf-blades 5-15 mm L, round-ovate in outline; style exserted. Loose volcanic soil, 10,400', Jul-Sep.

3) *C. linearis*, Narrow-leaved Collomia: Stem 0.5-6 dm H, sometimes branched from base or even above, leafy throughout; lvs lanceolate to linear, entire, sessile, 3-5 cm L, passing upward into ovate leafy bracts; fls sessile; calyx slopes below 8000', SN (all), May-Jul.

4) *C. rawsoniana*, Flaming Trumpet: Stems slender, 1-4 dm H, glandular-pubescent, strong-scented, leafy; lf-blades lance-ovate to elliptic, mly 4-8 cm L, petioles 3-20 mm L; fls crowded at ends of branches, short-pedicelled, subtended by a few leafy bracts; calyx 8-12 mm L; corolla-lobes pink to violet, the tube very narrow; style exserted. Local along streams, 3700-7000', Jul-Aug.

5) *C. tinctoria,* Yellow-staining Collomia: Stems erect to spreading, mly cymosely forked, glandular-viscid; lvs linear to lanceolate, sessile or the lower short-petioled, the upper surpassing the fls; fls 1-3 in axils or in forks, those at ends of branches in subcapitate clusters; calyx 5-8 mm L, the lobes bristle-tipped; corolla 8-12 mm L, pink with red-violet tube; style barely exserted. Dryish, open, often disturbed places below 9000', Mariposa Co. n., Jun-Jul.

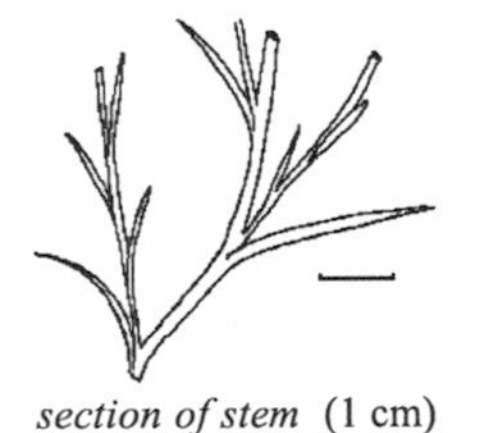

section of stem (1 cm)

Eriastrum

Few-flowered Eriastrum

flower cluster (1 cm)

E. sparsiflorum: Stems 1-4 dm H, mly woolly, often purplish on lower half, usu much branched; lvs 0.5-3 cm L, linear or with a pair of short lobes at base, sessile; fl-clusters 2-5-fld at the ends of slender branches 1-10 cm L; calyx 5 mm L; corolla 7-8 mm L, pale blue to whitish or pinkish. Dry slopes below 8000', Fresno Co. s., and along e. slope to Oregon, Jun-Jul.

G. leptantha lower stem (1 cm)

Gilia

Gilia

G. leptalea upper stem (1 cm)

Erect herbs. Genus usu divided into two parts: cobwebby gilias with mly basal, pinnately lobed lvs, and glandular. gilias with leafy stems, the lvs often linear. Seeds usu gelatinous when wet.

A Plants almost always with a basal tuft of pinnately-lobed lvs
  B Infl a dense spherical head of 25-100 fls — 3) *G. capitata*
  BB Infl a loose cyme; plants cobwebby-pubescent below
    C Corolla-tube purple, throat yellow; e. SN of Inyo and Mono cos. — 1) *G. cana*
    CC Corolla-tube and throat yellow; localized areas in s. and n. SN — 5) *G. leptantha*
AA Plants without basal tuft, stem lvs well developed, linear
  B Fls 10-20 mm L; calyx less than 1/4 corolla length, glabrous; style exserted — 4) *G. leptalea*
  BB Fls 3-10 mm L; calyx about half corolla length, glandular; style included — 2) *G. capillaris*

1) *G. cana*, Canescent Gilia: Stems 1 to many, rather stout, erect, 1-3 dm H; lvs bipinnately lobed or toothed; uppermost lvs reduced to leafy bracts; pedicels glandular-pubescent, 1-10 mm L; calyx glandular-pubescent, 3-6 mm L; corolla 15-25 mm L. Dry, gravelly places, 6000-10,000', Jun-Aug.

2) *G. capillaris*, Smooth-leaved Gilia: Stems erect, glandular throughout; lvs simple or rarely with a pair of basal lateral lobes; the lower and middle lvs 1-4 cm L; calyx 3-4 mm L, glandular; corolla funnelform, pale violet to pink of white, often streaked purple, the tube sometimes yellow, the throat sometimes with purple spots. Sandy slopes and flats, 2500-10,500', SN (all), Jun-Aug.

3) *G. capitata* sspp, Blue Field Gilia: Stems 1.5-9 dm H, mly glandular; lower lvs mly pinnately to bipinnately dissected; heads 1.5-3 cm in diameter; calyx lightly to densely white-woolly; corolla 5-12 mm L, pale to light blue-violet. Dry, sandy or rocky places below 7000', Plumas Co. s., Apr-Jul.

4) *G. leptalea*, Bridge's Gilia: Stems erect, 0.5-3.5 dm H, ± glandular throughout; lvs narrow, entire, 1-3 cm L; the fls mly in pairs on pedicels 5-40 mm L; calyx slender, 2-4 mm L; corolla funnelform, throat 4-7 mm L, white, yellow, pink or violet; lobes oval, pink or violet, 3-5 mm L. Openings in woods, 4500-9700', SN (all), Jun-Sep.

5) *G. leptantha* sspp, Broad-flowered Gilia: Stems erect, to 6 dm H, usu glandular above; basal lvs 2-7 cm L; infl usu diffuse with 2-3 fls above a bract; pedicels very unequal, 1-25 mm L; calyx about 4 mm L at anthesis; corolla 10-30 mm L (ssp *purpusii*) or 6-9 mm L (ssp *salticola*). Open, dry places, 2500-8800', s. SN, May-Jul.

Gymnosteris

Gymnosteris

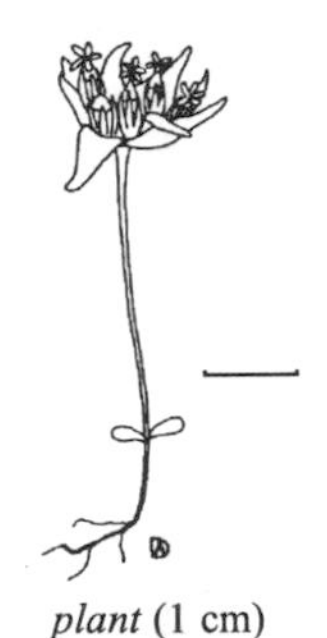
plant (1 cm)

G. parvula: Glabrous annual with erect, solitary stem less than 7 cm H; lvs actually absent, but fls subtended by a whorl of leaf-like ovate bracts; bracts sometimes tinged purple, 5-12 mm L; fls 1-5; calyx urn-shaped, membranous, awned; corolla 3-7 mm L, white to purplish with yellow throat; stamens without filaments. Moist, gravelly or sandy soils, 8000-12,000', Mono Co. n.

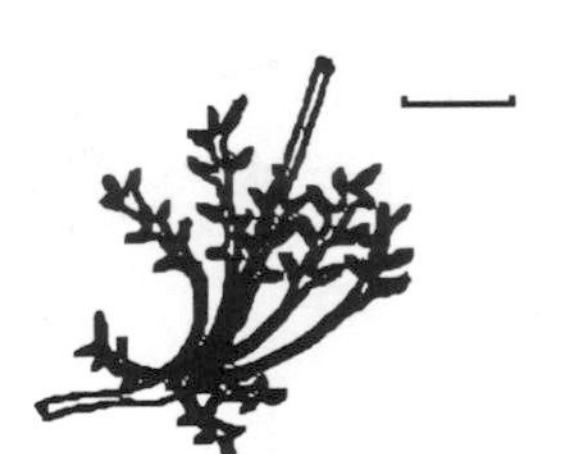
I. congesta basal leaves (1 cm)

Ipomopsis

Scarlet-gilia

I. aggregata flowers (1 cm)

Herbage pubescent, sometimes glandular; lf-segment tips with horny mucros. Infl cymose, each fl usu subtended by a bract. Calyx or equal herbaceous mucronate lobes joined by broad membranous sinuses.

Infl an elongate panicle; corolla 20-35 mm L — *I. aggregata*
Infl subcapitate; corolla 4-6 mm L — *I. congesta*

I. aggregata, Scarlet Gilia: Biennial; stems erect, mly 3-8 dm H, pubescent; lvs mly 3-5 cm L, the lobes 1-2 cm L; fls short-pedicelled; calyx 6-9 mm L; corolla tubular-funnelform, bright red often with yellow mottling, rarely pink or yellow; the lobes lanceolate, about 1/3 as long as tube. Open places like sandy flats, 3500-10,300', SN (all), Jun-Sep.

I. congesta, Many-flowered Gilia: Stems erect, several, 1-2.5 dm H, with woolly pubescence and a persistent tuft of lvs; lvs pinnately or bipinnately lobed from a broad petiole, 1-4 cm L, gradually reduced upwards and reduced to subpalmate leafy bracts in infl; calyx woolly, about 5 mm L; corolla white, the tube yellow. Dry, cinder slopes, etc., 4000-7000', Mono Co. n., Jun-Jul.

ssp *montana*: Densely matted plants forming cushion-like rosettes; basal lvs palmately divided into short broad segments. Mly at 7000-12,000', SN (all), Jul-Sep.

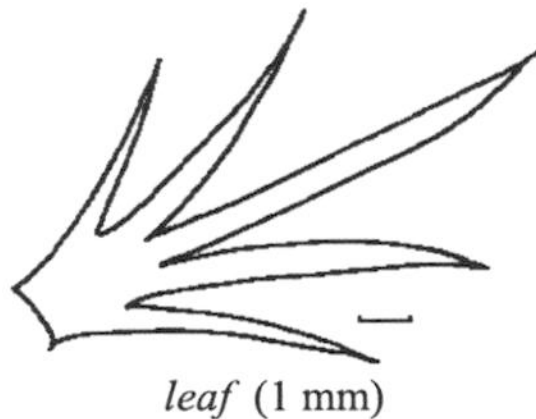

Leptodactylon

Granite-gilia

leaf (1 mm) *flowers* (1 cm)

L. pungens: Low branching shrub 1-2 dm H, glandular-villous throughout; lvs many, usu slightly longer than internodes, palmately cleft into 3-7 lobes, the middle lobe 8-15 mm L; fls sessile or nearly so, tending to open in evenings, in upper axils and in terminal or subterminal glomerules; calyx 8-10 mm L; corolla narrow-funnelform, 15-25 mm L, white to purplish. Dry, rocky and sandy habitats, mly 5000-12,000', SN (all), May-Aug.

Linanthus

Linanthus

upper stem (1 cm) *leaf* (1 cm)

Erect usu pubescent and mly annual herbs. Nodes usu remote; lvs often appearing as dense remote tufts of linear, simple lvs. Fls usu sessile in bracteate heads. Calyx deeply cleft with or without hyaline membrane.

A Infl an open panicle; pedicels 5-20 mm L; corolla about 2 mm L — 2) *L. harknessii*
AA Fls sessile in compact, leafy heads; corolla larger
  B Corolla 25-30 mm L; mly below 5000' — 3) *L. montanus*
  BB Corolla mly 10-20 mm L
    C Calyx about 5 mm L; Fresno Co. s.
      D Lf-lobes mly linear-oblanceolate, 1-2 mm W; above 9000’ — 5) *L. oblanceolatus*
      DD Lf-lobes linear, mly about 1 mm W; below 8000’ — 4) *L. nudatus*
    CC Calyx 7-10 mm L; widespread
      D Perennial; corolla lobes about 1/2 as long as tube, white — 6) *L. pachyphyllus*
      DD Annual; corolla lobes about 1/5 as long as tube; pinkish — 1) *L. ciliatus*

1) *L. ciliatus*, Whisker-brush: Stems erect, rather stiff, 1-3 dm H, stiff-pubescent; lvs 5-11-cleft into hairy lobes 5-20 mm L; bracts subtending heads hairy; corolla rose to white, 12-25 mm L, the tube slender, long-exserted, the throat yellow, short, the lobes 2-4 mm L. Common in dry, open places, mly below 10,000', SN (all), May-Jul.

2) *L. harknessii*, Harkness' Linanthus: Stems slender, erect, 5-15 cm H, branching above base, subglabrous to puberulent; lvs 3-5-parted into linear lobes 5-15 mm L, subglabrous to pubescent; pedicels filiform, subtended by simple or parted bracts; corolla white to pale blue. Open, sandy and gravelly places, 3000-10,400', Fresno Co. n., Jun-Aug.

3) *L. montanus*, Mustang-clover: Stems erect, 1-6 dm H, pubescent; lvs 5-11-cleft into linear ± hairy lobes 1-3 cm L; bracts subtending heads coarsely bristly-ciliate; calyx about 1 cm L; corolla lilac-pink or white, the tube long exserted, pubescent, the throat short, yellow, the lobes with a purple spot at base. Dry, gravelly places below 6700', Nevada Co. s., May-Aug.

4) *L. nudatus*, Tehachapi Linanthus: Stems erect, 1-2.5 dm H, slender; lvs 5-11-cleft, lobes 3-12 mm L, the upper lvs and bracts densely hairy; corolla-tube pubescent. Open slopes, 2000-7000', Tulare Co. s., May-Jul.

5) *L. oblanceolatus*, Tulare Linanthus: Stems 2-12 cm H, slender, puberulent; lvs 3-5-cleft, slightly hairy, the lobes green, 5-15 mm L; fls sometimes 1 or 2 in upper axils; calyx about 5 mm L; corolla white, 8-12 mm L, tube very slender, throat yellow, the lobes about 2 mm L. Mountain flats near edges of mdws, 7500-11,000', Jun-Aug.

6) *L. pachyphyllus*, Bush-gilia: Stems erect, mly 1-2 dm H, from a woody base; lvs 5-9-cleft into linear-oblanceolate lobes 1-1.5 cm L and with fascicled smaller lvs in axils; fls sessile or subsessile in upper axils, forming a subcapitate infl; calyx 8-9 mm L; corolla 12-15 mm L, the usu yellowish tube about 8 mm L, the lobes white or nearly so, 5-10 mm L. Dry, woody places, 4000-12,000', May-Aug.

Navarretia

Navarretia

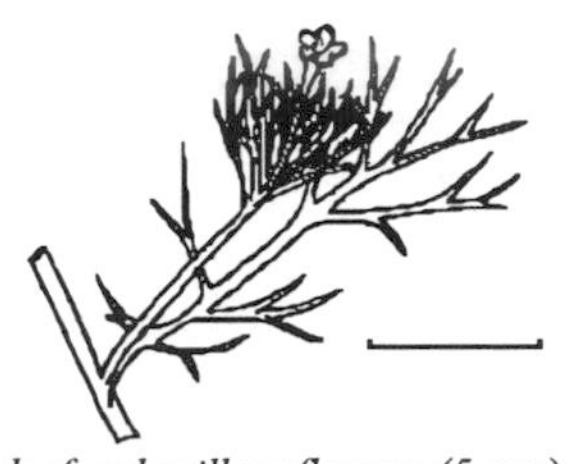

N. minima flower (1 mm) *leaf and axillary flowers* (5 mm)

Annuals with rigid stems. Fls sessile or nearly so in spiny densely bracted heads. Calyx cleft to base, the lobes united by a membrane extending part way up clefts. Stigma 2- or 3-lobed.

A Main stems ± retrorse-pubescent with white hairs; stigma 2-lobed; plants of moist areas
  B Corolla-tube short, barely exserted — 4) *N. minima*
  BB Corolla-tube long-exserted — 3) *N. intertexta*
A Main stems glabrous to puberulent; stigma 3-lobed; plants usu of dry areas
  B Infl-bracts pinnately parted; corolla yellow; mly e. SN — 1) *N. breweri*
  BB Infl-bracts ± palmately parted; w. SN
    C Corolla 10 mm L, yellow; below 4500' — 5) *N. prolifera*
    CC Corolla 3-4 mm L, purple to white; 3000-8500', — 2) *N. divaricata*

1) *N. breweri*, Brewer's Navarretia: Stems erect, 2-12 cm H, slender, puberulent throughout; lvs filiform, 1-3 cm L, entire or with 1-4 short slender lobes; bracts 4-10 mm L, pinnately 3-5-cleft; calyx 6-10 mm L; corolla 6-7 mm L, the lobes about 1 mm L. Damp to dryish flats and valley, 4000-11,000', SN (all), Jun-Aug.

2) *N. divaricata*, Mountain Navarretia: Stems erect or spreading, 3-15 cm H, often branched from below; lvs 1-2 cm L, simple or with few pairs of linear lateral lobes; bracts 3-5-lobed; calyx 4-7 mm L; corolla 3-4 mm L, the lobes pink to purple or white, scarcely 1 mm L. Dry, open places and at edge of mdws, SN (all), Jun-Aug.

3) *N. intertexta*, Needle-leaved Navarretia: Stems 5-20 cm H; lvs 1-5 cm L, 1-2-pinnate, the rachis linear; bracts leafy, pinnate, 1-2 cm L; corolla pale blue to white, 5-9 mm L, the lobes 2-6 mm L. Moist places, below 7000', w. SN (all), May-Jul.

4) *N. minima*, Least Navarretia: Stems 3-10 cm L, prostrate to almost erect, slender, mly whitish; lvs usu pinnately lobed, the lobes linear, spine-tipped; bracts ± ovate, to 1.5 cm L, pinnate; corolla white, 4-6 mm L, the lobes 1-2 mm L. Moist to dry places, below 7000', Placer Co. n. Jun-Aug.

5) *N. prolifera*, Bur Navarretia: Stems erect, 5-20 cm H, divaricately branched, brown, glabrous to puberulent, almost leafless; lvs 2-4 cm L, entire to pinnate with few pairs of short remote lobes; bracts to about 1 cm L. Dry slopes, w. base of SN in Tulare, Eldorado, Amador and Lassen cos. May-Jun.

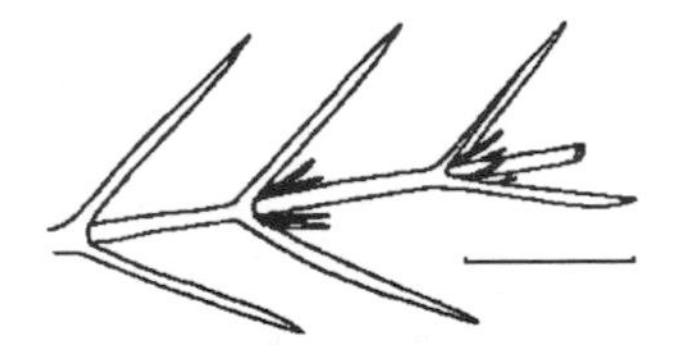

Phlox

Phlox

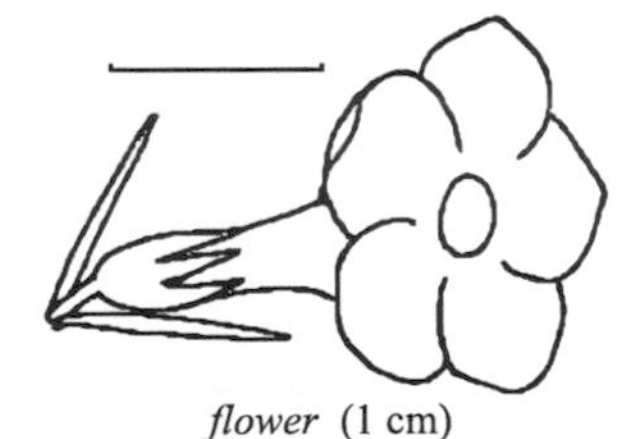

P. diffusa leaves (1 cm) *flower* (1 cm)

Plants often densely tufted, usu woody at base. Fls solitary or in a few-fld cyme, mly 5-merous.

A Plants openly branched, mly 1-4 dm H
  B Upper lvs alt; corolla 8-12 mm L; stems slender, herbaceous — 4) *P. gracilis*
  BB Lvs all opp; corolla over 15 mm L; stems from woody base
    C Lvs linear, sharp-pointed; corolla-lobes blunt at apex — 2) *P. diffusa*
    CC Lvs lanceolate, 4-10 mm W; corolla-lobes notched at apex — 6) *P. speciosa*
AA Plants densely tufted or matted, less than 1 dm H; high alpine
  B Lvs mly 10-15 mm L, linear, not glandular; SN (all) — 2) *P. diffusa*
  BB Lvs mly less than 10 mm L, lanceolate, usu densely glandular-hairy; s. SN
    C Plants forming spreading mats from slender underground rootstocks — 3) *P. dispersa*
    CC Plants densely cespitose; lf-margins revolute or thickened

D Lvs 3-5 mm L, concave above; styles 2-3 mm L — 1) *P. condensata*
DD Lvs 4-8 mm L, flat above; styles 3-6 mm L — 5) *P. pulvinata*

1) *P. condensata*, Cushion Phlox: Plants cushion-like; branchlets 1-3 cm L, subglabrous; lvs appressed or ascending, thick, the margins thick, rib-like, ciliate, the surfaces glandular-puberulent, the lower surface grooved; fls mly solitary, sessile, terminal; calyx 5-6 mm L; corolla white or pale pink, the tube 8-10 mm L, the lobes round-obovate, 4-6 mm L. Dry slopes and benches, 6000-10,000'(-12,000'), e. SN, mly in Mono and Inyo cos., Jun-Aug.

2) *P. diffusa*, Spreading Phlox: Stems 1-3 dm L, prostrate or decumbent, loosely branched; fls mly solitary at ends of short, leafy branches, on very short pedicels; calyx 8-10 mm L; corolla pink to lilac or white, the tube 9-13 mm L, the lobes broadly to narrowly obovate, 6-7 mm L. Dry slopes and flats, 3300-13,300', SN (all), May-Aug.

3) *P. dispersa*, Matted Phlox: Stems slender, tufted, mly 2-5 cm H; lvs sharp-pointed, 5-10 mm L, sessile; fls few, subsessile, fragrant; calyx about 7 mm L; corolla white, the tube about 1 cm L, the lobes obovate, irregular-margined, 5-6 mm L. Dryish flats of loose disintegrated granite, mly 11,000-12,500', Tulare and Inyo cos., Jul-Aug.

4) *P. gracilis*, Slender Phlox: Stems usu erect, 1-2 dm H, often branched and glandular-pubescent above; lvs 1-3 cm L, 1-5 mm W; calyx 5-8 mm L; corolla 8-12 mm L; tube yellowish, the lobes rose to white or violet. Common in open habitats below 10,000', SN (all), May-Aug.

5) *P. pulvinata*, Clustered Phlox: Much like *P. condensata*, forming clumps 3-7 cm H with short divergent branches; calyx 7-8 mm L. Dry, stony places, 10,000-13,000', mly Inyo and Mono cos., less common in Fresno and Tuolumne cos., Jul-Aug.

6) *P. speciosa* ssp *occidentalis*, Western Showy Phlox: Stems 2-4 dm H, leafy, glandular-puberulent above; leafy-bracted; pedicels 3-20 mm L; calyx mly 7-10 mm L; corolla bright pink, the tube 1-1.5 cm L, the lobes 7-12 mm L. Rocky hillsides and wooded slopes below 7000', Fresno Co. n., Apr-Jun.

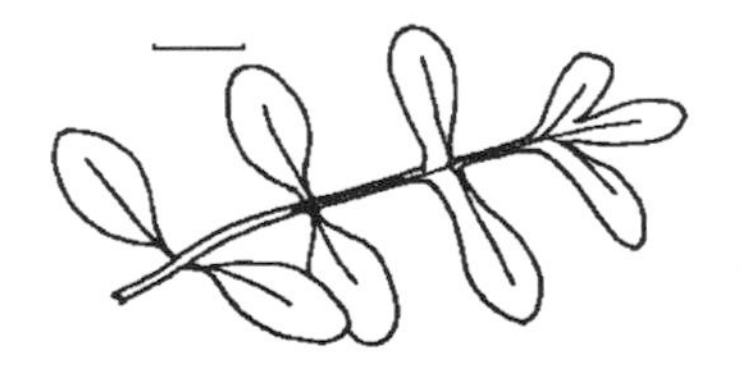

P. californicum leaf (1 cm)

Polemonium

Jacob's Ladder

P. eximium inflorescence (1 cm)

Perennials from woody caudex or rootstocks. Fls in terminal or axillary cymes. Calyx green without membranes. The old custom of picking a sprig of *P. eximium* when one first climbs a peak on which it grows has been abandoned; it is illegal in most localities, and with the large number of climbers in the Sierra such a practice could have a disastrous impact on the sp.

A Lfts deeply lobed or divided, 1-5 mm L; infl subcapitate — 2) *P. eximium*
AA Lfts entire, mly 5-20 mm L; infl cymose
  B Stems 2-8 dm H; lfts usu more than 20, 10-35 mm L; fl-tube blue — 3) *P. occidentale*
  BB Stems 1-3 dm H; lfts usu less than 20; fl-tube white or yellow
    C Lfts mly over 10 mm L, the terminal 3 often ± fused at base — 1) *P. californicum*
    CC Lfts 4-8 mm L, the terminal separate — 4) *P. pulcherrimum*

1) *P. californicum,* Low Polemonium: Stems solitary to subcespitose, glandular-pubescent; lvs largest at base of plant, the lfts 11-23, lanceolate to ovate, 5-20 mm L; pedicels about as long as calyx; calyx 5-8 mm L; corolla 8-15 mm W, the blue lobes about twice as long as the white tube; style exserted. Moist, shaded places, mly 6000-10,000', Fresno Co. n., Jun-Aug.

2) *P. eximium*, Sky Pilot: Stems 1-3 dm H, tufted; herbage glandular with a strong, musky odor; petioles expanded at base; lfts many, 3-5-parted, 1-5 mm L; pedicels 2-5 mm L; calyx 5-10 mm L; corolla 12-15 mm L, tube white, the lobes about 5 mm L; style included. Dry, rocky ridges and slopes, 10,000-14,200', Tuolumne to Tulare and Inyo cos., Jul-Aug.

3) *P. occidentale*, Western Polemonium: Stems solitary, erect, glandular-pubescent above; lfts mly 19-27, lanceolate to ovate, 5-10 mm L; pedicels about as long as calyx; calyx 5-8 mm L; corolla 8-15 mm W, the blue lobes about twice as long as the white tube; style exserted. Moist, shaded places, mly 6000-10,000', Fresno Co. n., Jun-Aug.

4) *P. pulcherrimum*, Showy Polemonium: Stems suberect; herbage glabrous to minutely glandular-puberulent; lvs mly basal, the lfts 11-23, ovate to round; pedicels slender, often longer than calyx; calyx 4-6 mm L; corolla 5-8 mm L, the tube yellowish, ± equal to the blue or purplish ovate lobes; style about as long as corolla. Dry, rocky, often volcanic slopes, mly 8000-11,000', Mono and Mariposa cos. n., Jun-Aug.

POLYGALACEAE - Milkwort Family

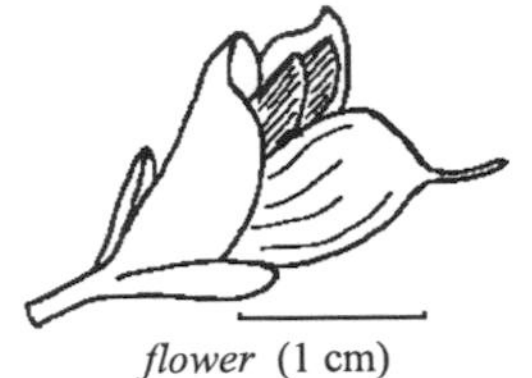
flower (1 cm)

Polygala

Milkwort

upper stem (1 cm)

P. cornuta: Stems many, slender, 6-9 dm H, subglabrous, woody below; lvs alt, ovate to almost linear, 2-4 cm L, short-petioled; fls 4-10, racemose, yellowish to greenish white, sometimes with dull plum color; pedicels 2-4 mm L; sepals 5, distinct, irregular, the outer ovate, about 4 mm L, densely puberulent; petals 3, the upper 2 ± united with the lowermost (the keel); the boat-shaped keel 11-12 mm L, including the straight, about 2 mm L beak. Rocky or gravelly slopes, mly 1000-5000', Fresno Co. n., Jun-Aug. Herbage of related spp very bitter and perhaps toxic.

POLYGONACEAE - Buckwheat Family

Herbs or shrubs. Lvs usu alt. Stipules often forming a membranous sheath (ocrea) around stem above base of petiole. Fls mly perfect, small. Calyx usu persistent. Stamens usu 6-9, inserted near base of calyx. Fr an akene.

A Lvs without ocreae, usu basal; the fls enclosed in an invol.
  B Invol. with 3-8 teeth or lobes, these not awned; fls many per invol. — *Eriogonum* p. 157
  BB Invol. with 3-6 awns, ± spiny; fls 1 per invol. — *Chorizanthe* p. 157
AA Lvs with ocreae; fls not enclosed in an invol.
  B Calyx 5-parted, usu brightly colored — *Polygonum* p.159
  BB Calyx 4- or 6-parted, brownish red
    C Lvs reniform; sepals 4; styles 2, short — *Oxyria* p. 158
    CC Lvs not reniform; sepals 6; styles 3 — *Rumex* p. 160

Chorizanthe

Spineflower

Annual with ± decumbent and scapose stems with grayish to reddish pubescence. Lvs usu oblanceolate; stipules lacking. Invol. bracts 6. Invol. tube urn-shaped. Perianth white, 2-3 mm L. Stamens 3. Edibility unknown

plant (3 cm)

Longest invol. awn hooked, about 2 times longer than others — *C. clevelandii*
Longest invol. awn straight, about 5 times longer than others — *C. uniaristata*

C. clevelandii, Cleveland's Spineflower: Stems 2-10 cm H; lvs 5-20 mm L; awns hooked. Common in chaparral below 6500', s. SN.

C. uniaristata, One-awned Spineflower: Stems 2-25 cm H; lvs 5-20 mm L; the shorter invol. spines hooked. Sandy or gravelly habitats below 7000', s. SN.

inflorescence (1 cm)

Eriogonum

Wild Buckwheat

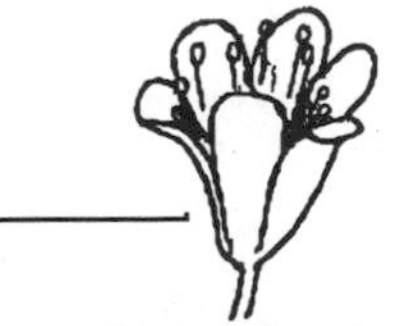
flower, calyx not elongate at base (5 mm)

Annual or perennial herbs or subshrubs. Stems usu with whorled scale-like or leafy bracts. Lvs entire and without stipules. Invol. bell-shaped to cylindric, sessile or peduncled. Pedicels ±

exserted. Calyx brightly colored. Stems of most spp edible before flowering. Seeds edible when ground and used as flour. An interesting, but often difficult genus to determine to sp. Only a key to the Sierran spp will be presented.

A Calyx narrowed and elongate to base (see also *E. saxatile*); bracts leafy, 2 to several
B Calyx pubescent externally
C Infl capitate; bracts 5; 6500-11,000', e. SN of Mono and Inyo cos. — *E. latens*
CC Infl subcapitate to umbellate; bracts 2 to several; Lassen Co. n.
D Calyx bright sulfur-yellow; dry, rocky habitats, below 7000' — *E. sphaerocephalum*
DD Calyx whitish to rose; above 8500', Lassen Peak and vicinity — *E. pyrolifolium*
BB Calyx glabrous externally
C Calyx white to rose, never yellowish
D Fl-stems erect; infl less than 2 cm W; mly Tulare Co. — *E. polypodum*
DD Fl-stems usu prostrate; infl 3-4 cm W; Inyo and Mariposa cos. n. — *E. lobbii*
CC Calyx yellow or yellowish
D Bracts in a whorl near middle of fl-stem which bears a single invol.; 3000-5000', Fresno to Nevada Co. — *E. prattenianum*
DD Bracts subtending a cluster of invols.; mly above 5000'
E Invols. deeply lobed, the lobes reflexed; plants usu not matted; common, below 10,000' — *E. umbellatum*
EE Invol. shallowly lobed; plants forming mats
F Lvs densely white-tomentose on both surfaces; 7000-12,000', Alpine Co. s. — *E. incanum*
FF Lvs ± glabrate and green above, densely tomentose below; above 3500'
G Petioles as long as lvs; invol. sparsely pubescent; widespread — *E. marifolium*
GG Petioles short; invol. densely woolly; Placer Co. n. — *E. ursinum*
AA Calyx not narrowed or elongate at base; bracts scale-like, regularly 3 in number
B Plants perennial; stem lvs often present
C Calyx 5-7 mm L, elongate and narrowed at base; s. SN — *E. saxatile*
CC Calyx 2-5 mm L, not elongate or narrowed at base
D Lvs glabrous above, white-woolly beneath, margins plane — *E. nudum*
DD Lvs pubescent above or with revolute margins
E Fl-stems unbranched, terminated by a dense, spherical infl; 7000-12,000'
F Lvs round, calyx white to rose — *E. ovalifolium*
FF Lvs oblanceolate, calyx yellowish — *E. rosense*
EE Fl-stem branched at least in infl; the infl ± open, not spherical
F Infl dense above with few to several lower branches — *E. wrightii*
FF Infl open, diffusely branched — *E. microthecum*
BB Plants annual; lvs mly basal
C Lvs not all basal
D Basal lvs linear, revolute; above 4000' — *E. spergulinum*
DD Basal lvs wider, not revolute; below 6000'
E Stems 5-25 cm H; calyx reddish — *E. hirtiflorum*
EE Stems 10-60 cm H; calyx yellow to pink or white — *E. roseum*
CC Lvs all basal
D Invol. 4-lobed; w. SN, Tuolumne Co. s. — *E. inerme*
DD Invol. 5-lobed; mly e. SN
E Lvs 2-6 cm L; 4000-9000', Tulare and Inyo cos. — *E. parishii*
EE Lvs 1-2 cm L; 7000-10,000', e. SN — *E. cernuum*

Oxyria

Mountain Sorrel

O. digyna: Stems 6-25 cm H, erect; lvs mly basal, round-reniform, 1.5-4 cm W; petioles 5-12 cm L; pedicels slender, recurved; infl compact, ± oblong in outline; sepals about 2 mm L, the inner becoming 4-6 mm L in fr, the outer reflexed. Rocky places, 7000-13,000', SN (all), Jul-Sep. The lvs and stem are high in vitamin C and can be eaten raw or cooked. They may also be stored for future use.

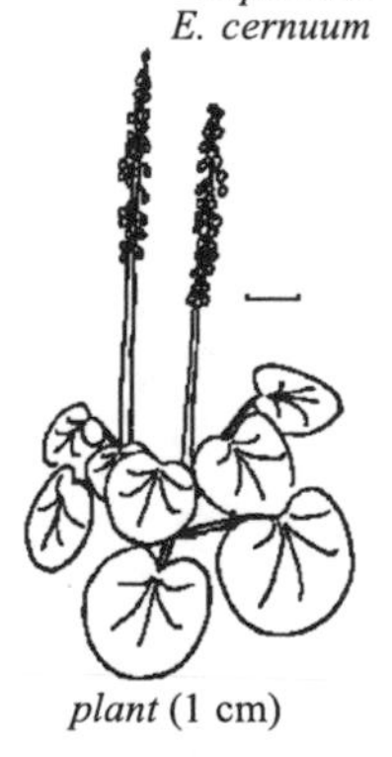

plant (1 cm)

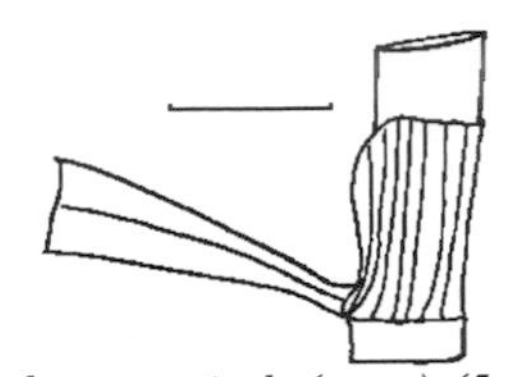

membranous stipule (ocrea) (5 mm)

Polygonum

Knotweed
Smartweed

flower (1 mm)

Annuals to woody perennials. Often reddish when young or dry. Joints often swollen. Various parts of several spp are known to be edible and all are probably edible. The sap is often very acid, however, and if eaten in large quantities could cause stomach and intestinal irritation. Edible parts include seeds (ground as flour), herbage raw or cooked, roots raw, boiled or roasted.

A Plants perennial, woody, or from rhizome or taproot
  B Fls 2-3 in lf-axils; stems and root woody — 10) *P. shastense*
  BB Fls in terminal spikes or panicles (occasionally some also axillary)
    C Lvs basal and long-petioled or cauline and ± sessile — 2) *P. bistortoides*
    CC Lvs all cauline and similar
      D Ocreae funnelform, oblique and somewhat open on side facing leaf
        E Plant 10-20 dm H, erect; fls in a loose leafy infl — 8) *P. phytolacceafolium*
        EE Plant 1-4 dm L, decumbent to ascending; fls mly axillary — 4) *P. davisiae*
      DD Ocreae cylindric, truncate; infl usu dense, not leafy — 1) *P. amphibium*
AA Plants annual
  B Fls solitary in axils; lvs mly 3-nerved; ocreae divided into bristle-like segments
    C Stems 6-20 cm L; below 4500', Mariposa Co. n. — 3) *P. californicum*
    CC Stems usu 2-5 cm L; below 6000', uncommon — 7) *P. parryi*
  BB Fls mly 2 or more in axils; lvs 1-nerved; stems angled
    C Plants usu less than 2 dm H; infl densely leafy, almost hidden by these lvs
      D Lvs ovate to obovate; pedicels about 2 mm L — 6) *P. minimum*
      DD Lvs linear or lance-linear; pedicels about 1 mm L — 9) *P. polygaloides*
    CC Plants mly over 1 dm H; infl conspicuous, with reduced leafy bracts — 5) *P. douglasii*

1) *P. amphibium*, Water Smartweed: Stems elongate, leafy, glabrous, unbranched; lvs lanceolate to lance-oblong, or even broader in the floating form, 5-10 cm L, petioled, usu obtuse; fls rose; calyx 4-5 mm L. Ponds and lakes below 10,000', SN (all), Jul-Sep.

2) *P. bistortoides*, Snakeweed: Stems several, erect, glabrous, slender, 2-7 dm H; lvs mly near base, the lower ± lanceolate, entire to undulate, 1-2.5 dm L, 2-5 cm W, with broad midrib; upper lvs reduced; ocreae narrowly cylindric, 3-7 cm L; spikes thick-cylindric, 1-6 cm L; calyx pink or white, 4-5 mm L. Wet mdws and along streams, mly 5000-10,500', SN (all), Jun-Aug.

3) *P. californicum*, California Knotweed: Stems glabrous, wiry, slender, diffusely branched; lvs linear to filiform, 1-3 cm L; calyx about 2 mm L, white with pink midveins. Dry, sandy and gravelly flats and bars, May-Oct.

4) *P. davisiae*, Davis' Knotweed: Stems often branched, glaucous, glabrous, or nearly so, usu several to many from a stout much-branched taproot; lvs ovate to oblong-ovate, sessile or nearly so, rough-pubescent, 1-4 cm L; fls in short 3-4-fld terminal and axillary clusters; calyx whitish to purplish green, about 3 mm L. Talus and rocky places, 5000-9000', Alpine Co. n., Jun-Sep.

5) *P. douglasii* and vars., Douglas' Knotweed: Stems erect, slender, loosely few-branched, 1-4(-6) dm H, subglabrous; lvs linear to ovate, 0.5-4 cm L, subsessile; fls 1-3 per axil, drooping, white to reddish or greenish; calyx 2-4 mm L Fairly dry areas as at edge of mdws, 4000-9000', SN (all), Jun-Sep.

6) *P. minimum*, Leafy Dwarf Knotweed: Stems 5-15 cm H, often branched; lvs subsessile, 5-15 mm L; fls greenish white, 2-3 in axils of most lvs; calyx about 2 mm L. Damp mdws and banks, 5000-11,200', SN (all), Jul-Sep.

7) *P. parryi*, Parry's Knotweed: Stems tufted; lvs narrowly linear, 0.5-2 cm L, rigid-tipped; ocreae ± concealing fls; calyx 1-5 mm L. Dry, sandy places, SN (all), May-Oct.

8) *P. phytolacceafolium*, Alpine Knotweed: Stems subglabrous; lvs lanceolate to lance-ovate, 3-15 cm L, 1-6 cm W, short-petioled; ocreae 1-3 cm L, brown, deciduous; infl sometimes with few lvs; pedicels slender, 2-3 mm L; calyx white or greenish white, about 3 mm L. Moist, often rocky places, 5000-9000', Yosemite n., Jun-Sep.

9) *P. polygaloides*: Stems glabrous, often tufted, 3-8 cm H, with very short internodes; lvs linear or lance-linear, 0.5-1 cm L, spreading, sessile; infl spicate, 0.5-3 cm L; calyx about 2 mm L, green with white margins. Damp, silty or gravelly places, 4500-11,500', SN (all), Jun-Sep.

10) *P. shastense*, Shasta Knotweed: Stems 1-3 dm L, prostrate or ascending, wiry, rough, branched; lvs oblong to subovate, revolute in age, 5-15 mm L; calyx white to rose with dark midveins, mly 5-8 mm L. Rocky or gravelly slopes, 7000-11,000', SN (all), Jul-Sep.

Rumex

Dock
Sorrel

plant (1 cm)

Perennials with grooved stems. Fls numerous, small, usu crowded in panicles. Calyx of 6 sepals, the 3 outer herbaceous, usu appressed to the margins of the inner; the 3 inner larger, somewhat colored, veiny. Stamens 6. The lvs and stems of all spp are edible after removal of tart acids which are poisonous if eaten in quantity. Removal of the acids may be accomplished by boiling in as many waters as is necessary to remove the bitter taste.

A Stems usu about 1 cm thick at base; fr 2.5-5 mm L — *R. salicifolius*
AA Stems slender at base; fr small, 1-3 mm L
  B Lvs hastate at base; mly below 7000' — *R. acetosella*
  BB Lvs not hastate at base; 5000-12,000' — *R. paucifolius*

R. acetosella, Sheep Sorrel: Stems tufted, erect, 1-4 dm H, glabrous; lvs lanceolate or linear, the blades 2-6 cm L, lower petioles often longer; fls nodding, yellow, reddish in age. Common in open habitats below 7000', SN (all), Apr-Oct.

R. paucifolius, Alpine Sheep Sorrel: Stems few, erect, 1.5-7 dm H, glabrous; lvs glabrous, broadly lanceolate, entire, 4-10 cm L; petiole about as long; sepals in fr 3-4 mm L. Damp places, 4000-9000', typical only in Fresno and Lassen cos., Jul-Sep.

ssp *gracilescens*: Stems numerous, 0.5-2 dm H; basal lvs linear-lanceolate to linear. At 10,000-12,250', Tulare to Alpine Co., Aug-Sep.

R. salicifolius, Willow Dock: Lvs lanceolate, 6-12 cm L, 2-2.5 cm W; sepals ovate to ovate-lanceolate in fr. Infrequent, moist habitats below 6500', Nevada to Plumas Co., May-Sep.

PORTULACACEAE - Purslane Family

Annual to perennial herbs, ± fleshy. Lvs entire. Styles 2-8, but usu 3. Fr a capsule.

A Stamens often more than 5 and different from petal number; petals sometimes more than 5
  B Annuals; petals red; sepals 2 — *Calandrinia*
  BB Perennials from a ± thick base; petals rose to white; sepals often more than 2 — *Lewisia*
AA Petals and stamens 5, sometimes 4; sepals 2
  B Sepals with broad, membranous margins; fl-stem often prostrate, rarely erect — *Calyptridium*
  BB Sepals not membranous-margins; fl-stem mly erect
    C Perennials; stem lvs often in several pairs — *Claytonia*
    CC Mly annuals from fibrous roots; if perennial then stem lvs in one pair — *Montia*

Calandrinia

Red Maids

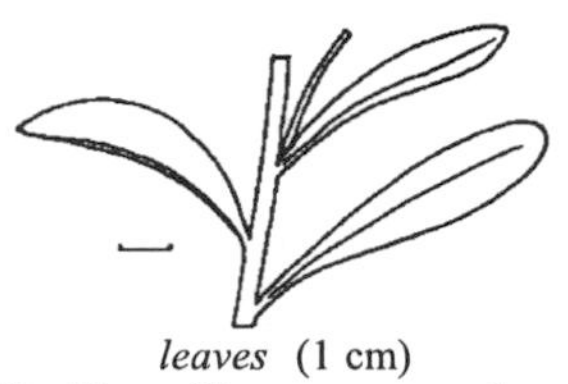
leaves (1 cm)

inflorescence (1 cm)

C. ciliata: Stems usu several, ascending, 1-4 dm H, ± glabrous; lvs alt, narrowly oblanceolate to linear, 2-8 cm L, well distributed along stem; pedicels 0.4-2 cm L, ± erect; petals 5, rarely white, 4-10 mm L; stamens usu more. Common in open habitats below 6000', SN (all), Apr-May. Herbage edible raw or cooked. Seeds also edible raw or can be ground to a flour.

Calyptridium

Calyptridium

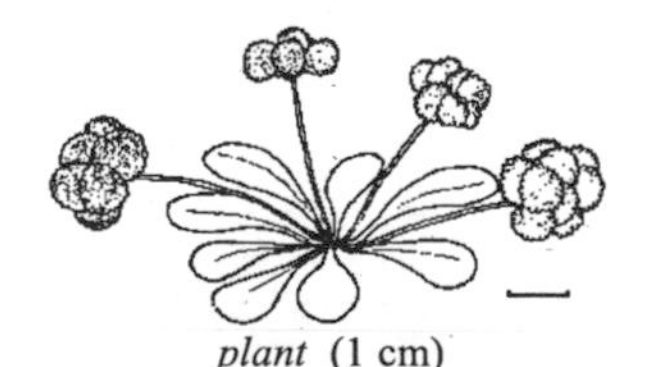
plant (1 cm)

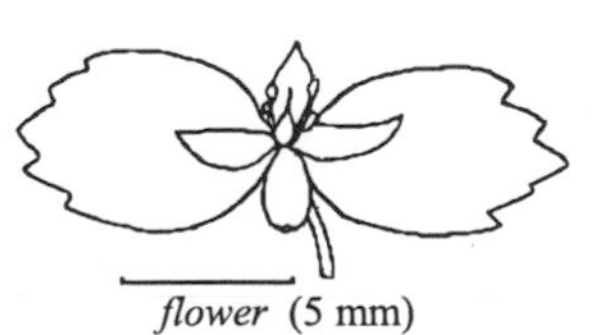
flower (5 mm)

Lvs alt or basal, spatulate. Fls small, perfect. Sepals membranous or membranous-margined. Petals 2 or 4. Style simple. Edibility unknown.

A Style long, thread-like; widespread
  B Infl terminal, 1 per plant; bracts of infl about equal to sepals 5) *C. umbellatum*
  BB Infl axillary, 2 or more per plant; bracts of infl smaller than sepals 1) *C. monospermum*
AA Style short; stems 1-10 cm L; e. SN
  B Lvs mly basal, less than 1 cm L 3) *C. pygmaeum*
  BB Lvs basal and cauline, usu 1-4 cm L
    C Fls on pedicels 1-3 mm L; petals 2 4) *C. roseum*
    CC Fls sessile; petals 3 2) *C. parryi*

1) *C. monospermum*: Annual from slender taproot; stems spreading, 2-15 cm L, leafy; lvs 1-5 cm L; infl usu open, 1-4 cm L; bracts + ovate; sepals 1-2 mm L; petals 3, 1-3 mm L, pink to reddish. Sandy, open flats below 10,000', SN(all). Forms hybrids with *C. umbellatum*.

2) *C. parryi*, Parry's Calyptridium: Annual from slender taproot; stems spreading, 2-10 cm L, leafy; infl 1-3 cm L, axillary; sepals 2-5 mm L; petals 1-3 mm L, usu white. Open areas below 11,000', Fresno and Tulare cos.

3)*C. pygmaeum,* Dwarf Calyptridium; Stems 1-2.5 cm H; fls few, in 1-sided racemes; sepals ovate, about 2 mm L, scarcely scarious-margined; petals 4, about 3 mm L. Rare, sandy or gravelly places, 7500-11,500', Inyo Co., Jun-Jul.

4) *C. roseum*, Rosy Calyptridium: Stems several, 2-10 cm L; lvs few; fls in panicles of short scorpioid clusters; sepals round, 2.5-4 mm W; petals 2, about 1 mm L. Moist, often alkaline places, 5000-10,500', Lassen, Sierra, Mono and Inyo cos., Jun-Aug.

5) *C. umbellatum*, Pussy Paws: Annual to perennial; stems several; lvs mly basal in a dense rosette, 2-7 cm L; stem lvs reduced; infl a ± umbellate cluster of scorpioid spikes; sepals red-violet to white, orbicular-reniform, mly 5-8 mm L; petals 4, oblong to ovate, 3-6 mm L. In loose sandy or gravelly places, 2500-12,000', SN (all), May-Aug.

var. *caudiciferum*: Root-crown much-branched, woody, clothed with dead lvs; lvs 5-10 mm L; stems 5-12 mm L; infl capitate, scarcely 1 cm thick. Dry, gravely places, 11,000-14,200', Jul-Aug.

Claytonia

Spring Beauty

C. lanceolata upper stem (1 cm) *C. perfoliata plant* (1 cm)

Glabrous perennials. Basal lvs 1 to several; stem lvs mly opp. Infl racemose, the lowest pedicel often subtended by a small bract. Sepals herbaceous, persistent. Petals usu 5, rose to white. Style branches 3. The bulbs of *C. lanceolata* may be eaten raw or cooked. The stems and lvs *of C. perfoliata* are edible raw or cooked.

A Annual; stem leaf usu perfoliate 6) *C. perfoliata*
AA Perennials from rhizomes, caudex or tuber
  B Lvs mly cauline; stem from a spherical tuber 2) *C. lanceolata*
  B Lvs mly basal, from a thick caudex or rhizome
    C Root caudex 5-30 mm across; rocky, alpine habitats, c. and n. SN 3) *C. megarhiza*
    CC Root caudex 1-5 mm across; wet habitats
      D Petals pink; stems less than 10 cm H; widespread 4) *C. nevadensis*
      DD Petals usu white; stems usu over 10 cm H; c. and n. SN
        E Fl bracts absent; petiole on basal lvs linear, distinct 1) *C. cordifolia*
        EE Each fl subtended by a bract; basal lvs tapering gradually to petiole 5) *C. palustris*

1) *C. cordifolia*, Truncate-leaved Claytonia: Stem erect, to 40 cm H; basal lvs 2-25 cm L, the blade widely ovate with a truncate or cordate base; cauline lvs 1-4 cm L, usu sessile; sepals 3-5 mm L, petals 8-12 mm L. Seeps and wet mdws, 4000-8000', Jun-Aug.

2) *C. lanceolata*, Western Spring Beauty: Stems 1 to several, 6-15 cm L; basal lvs mly 1-2, petioled, oblanceolate, 5-8 cm L stem lvs 3-7 cm L, 2-10 mm W; pedicels 1-2.5 cm L, recurved in fr; sepals ovate, 3-5 mm L; petals usu pink, 6-12 mm L, notched at apex. Moist woods and along streams, 4500-8800', Eldorado Co. n., May-Jul.

3) *C. megarhiza*, Fell-fields Claytonia: Basal lvs spatulate to broadly oblanceolate; stem lvs near infl, subopposite, narrow, bract-like; fls 2-6; pedicels 1-1.5 cm L; sepals ovate, 5-7 mm L; petals white, 6-10 mm L. Rare, on talus and loose rock or gravel, 9000-11,000', Yosemite National Park, Jul-Aug.

4) *C. nevadensis*, Sierra Claytonia: Stems 4-10 cm L, from slender, tangled rootstocks; basal lvs fleshy; stem lvs ovate, sessile, 8-15 mm L; fls 2-6; pedicels mly 1-2 cm L; sepals ovate, acute, 5-6 mm L; petals white with some pink, 5-8 mm L. Infrequent, gravelly, wet places, 5000-12,000', Lassen Peak and Alpine Co. s., Jul-Aug.

5) *C. palustris*, Marsh Claytonia: Stems erect, to 60 cm H; basal lvs 10-30 cm L, the blade 2-8 cm L; cauline 2-9 cm L, oblanceolate to elliptic, sessile to with winged petiole; sepals 3-4 mm L; petals 5-10 mm L. Rare below 8000'.

6) *C. perfoliata*, Miner's Lettuce: Stems 1-3 dm H, occasionally reddish; basal lvs variable; stem lvs usu fused at base to form a connate disk 1-8 cm W; racemes terminal; sepals 2-3 mm L; petals white. Moist habitats below 7500', SN (all), Apr-Jun.

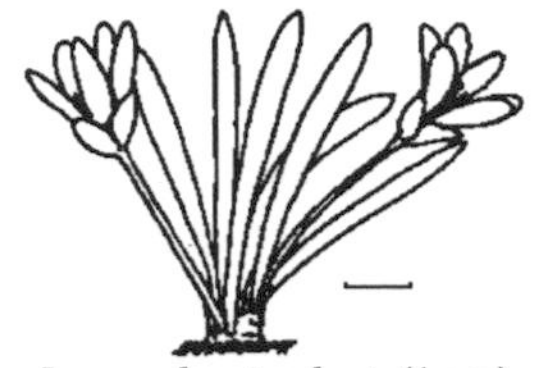

L. nevadensis plant (1 cm)

Lewisia

Lewisia

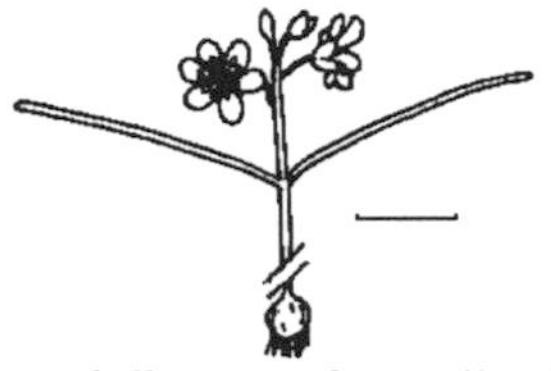

L. triphylla stem and corm (1 cm)

Mly fleshy perennials. Lvs mly in a basal rosette, entire; petioles mly membranous-margined at base. Petals 4-18, often unequal. Stamens 5 to many. Styles 3-8, united at base. Probably every sp has an edible root or corm, although the common bitterroot (*L. rediviva*) is rarely found in the Sierra. The bitter taste of the roots is much diminished by removing the brownish outer coating and cooking the starchy inner root.

A Plant without basal lvs, but with 2-5 linear stem lvs; from a globose corm — 8) *L. triphylla*
AA Plant with several to many basal lvs; from fleshy root and short caudex
  B Stems 1-4 dm H, much surpassing basal rosette; rare
    C Lvs ± round in cross-section, 2-6 cm L; 9000-10,000', Fresno Co. — 4) *L. leana*
    CC Lvs flattened, 5-10 cm L; below 6000', Mariposa and Fresno cos. — 1) *L. congdonii*
  BB Stems less than 1 dm H; scarcely if at all surpassing the basal lvs; widespread
    C Sepals petaloid, becoming membranous in age; infl readily disjointing in fr
      D Sepals 2, 7-8 mm L; bracts 2-3, ovate — 2) *L. disepala*
      DD Sepals several, 10-25 mm L; bracts 5-8, narrower — 7) *L. rediviva*
    CC Sepals herbaceous, not petaloid; infl not disjointing in fr
      D Sepals 5-10 mm L; petals mly over 10 mm L, white
        E Lvs spatulate; bracts green and closely subtending fl, often appearing like 2 additional sepals — 3) *L. kelloggii*
        EE Lvs linear; bracts membranous and below middle of stem — 5) *L. nevadensis*
      DD Sepals 3-5 mm L; petals usu less than 10 mm L, often pink or with pink veins; above 9000' — 6) *L. pygmaea*

1) *L. congdonii*, Columbia Lewisia: Lvs many, mly basal, oblanceolate; petioles 5-10 cm L; infl widely branched, rather few-fld; sepals about 2 mm L; petals rose. Rocky places, May-Jun.

2) *L. disepala*, Yosemite Bitterroot: Stems 2-3 cm L, 1-fld; lvs 8-20 mm L; pedicels 1-2 mm L; petals pinkish, 13-18 mm L. Rocky places, 6500-8500', Tulare and Mariposa cos., May-Jun.

3) *L. kelloggii*, Kellogg's Lewisia: Stems 1-5 cm H, 1-fld; basal lvs many, 3-6 cm L; fls sessile between a pair of sepal-like bracts; sepals lance-ovate, dentate; petals creamy-white, 8-15 mm L. Sandy places on ridges, 4500-7700', Mariposa to Plumas Co., Jun-Jul.

4) *L. leana*, Lee's Lewisia: Stems 1-4, 10-15 cm L; basal lvs many, linear, glaucous; stem lvs much reduced; infl many-fld; sepals rounded, about 2 mm L, dentate with dark glands on teeth; petals 6-8, red or white with red veins, 5-7 mm L. Cliffs and rocks, 9000-10,000', Fresno Co.

5) *L. nevadensis*, Nevada Lewisia: Stems several, usu recurved in fr; lvs mly 3-8 cm L; bracts 6-18 mm L, lanceolate; fls usu solitary; sepals ovate, obscurely toothed; petals 6-10 variable in length on the same fl. Moist to wet habitats, 4500-12,000', SN (all), May-Jul.

6) *L. pygmaea*, Alpine Lewisia: Stems several to many, sometimes tufted, 2-6 cm L, partly underground; bracts membranous, lanceolate, 6-10 mm L, opp, near middle of stem; fls 1-3 per stem; petals rarely 12-18 mm L (ssp *longipetala*). Damp gravel, 9000-12,750', SN (all), Jul-Aug.

7) *L. rediviva*, Bitterroot: Stems many, 1-3 cm L, 1-fld; lvs basal, linear, 2-5 cm L; bracts whorled; pedicels mly 10-15 mm L; sepals entire, rose to white; petals rose or white, many, 20-25 mm L. Loose, gravelly slopes and rocky places, mly 2500-6000', Mono Co. n., Apr-Jun.

8) *L. triphylla*, Three-leaved Lewisia: Corm 3-6 mm in diameter; stems slender, 1 to few, 3-10 cm H but half underground; stem lvs 2-3, whorled, 2-5 cm L; infl subumbellate, mly 3-15-fld; pedicels slender, 5-10 mm L; sepals entire, oval, 3-4 mm L; petals 5-8 white or pink, 4-5 mm L. Damp, gravelly places, 5000-11,200', SN (all), Jun-Aug.

M. parvifolia base of stem (1 cm)

Montia

Montia

M. chamissoi plant (1 cm)

Annual and perennial glabrous herbs, often fleshy, occasionally aquatic. Lvs entire. Infl a 1-sided raceme. Herbage of many spp edible.

A Lvs opp
  B Perennial with stolons; petals 5-10 mm L — 1) *M. chamissoi*
  BB Annual, often matted or aquatic; petals 1-2 mm L — 2) *M. fontana*
AA Lvs alt
  B Stems ± scapose; petals 7-10 mm L — 4) *M. parvifolia*
  BB Stems leafy, often much branched; petals 4-6 mm L — 3) *M. linearis*

1) *M. chamissoi*, Toad Lily: Stems creeping or floating with bulblet-bearing runners, and ascending to erect branches or tips usu 5-15 cm L; lvs spatulate, 1.5-4 cm L; racemes 3-8-fld; pedicels slender, recurved in fr, 1-2.5 cm L; sepals about 2 mm L; petals pink or white, 5-8 mm L. Wet mdws 4000-11,000', SN (all), Jun-Aug.

2) *M. fontana*, Water chickweed, Blinks: Stems prostrate to erect, sometimes floating, 1-30 cm L; lvs linear to oblanceolate, 3-20 mm L, the base tapered. Common in streams and pools below 10,000', SN (all).

3) M. linearis, Linear-leaved Montia: Stems ± erect; lvs somewhat fleshy, alt or in 2 to several pairs, 2-4 cm L; racemes mly 2-7-fld; pedicels 6-15 mm L, ± recurved; sepals 2, persistent, about 4 mm L; petals 5, unequal, white, about 5 mm L; stamens many, adhering to base of petals. Moist habitats, below 7500', SN (all), Apr-Jun.

4) *M. parvifolia*, Small-leaved Montia: Stems short, ascending or erect, 1.5-3 dm L, stolons often present; basal lvs to 3 cm L, petioled; fls 1-10; pedicels 5-15 mm L; sepals about 2 mm L; petals white with pink or lavender veins or often pink, slightly notched. Moist, rocky places below 8500', Tuolumne Co. n., May-Jul.

PRIMULACEAE - Primrose Family

Lvs simple, without stipules. Fls perfect, 4-7-merous. Calyx deeply lobed. Stamens same number as corolla-lobes and inserted opp them on the corolla tube. Style 1; stigma capitate. Fr a capsule.

A Fl-stem without lvs, the lvs all basal; fls in bracteate umbels
  B Corolla-lobes reflexed, acute to obtuse — *Dodecatheon*
  BB Corolla-lobes spreading or erect, notched or obcordate
    C Loves of corolla 8-10 mm L; ± woody perennial — *Primula*
    CC Lobes of corolla 1-2 mm L; annual — *Androsace*
AA Plants with leafy stems; fls not in umbels
  B Stems 2-8 dm H; lvs lanceolate — *Lysimachia*
  BB Stems 0.5-2 dm H; lvs ovate to elliptic — *Trientalis*

Androsace

Androsace

Mly annuals. Umbel subtended by a whorl of bracts. Edibility unknown.

Plants 3-6 cm H; pedicels slender, 5-15 mm L — *A. occidentalis*
Plant usu 1-3 cm H; pedicels stout, 10-20 mm L — *A. septentrionalis*

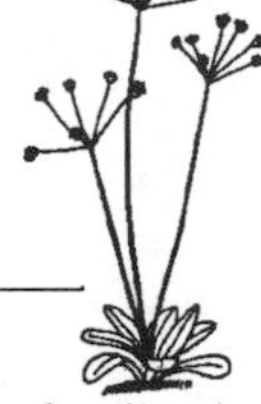
plant (1 cm)

A. occidentalis var. *simplex*, Western Androsace: Fl-stem 1 to several, branched; lvs elliptic-lanceolate, 5-10 mm L; bracts 2-5 mm L; fls 1-4, about 3 mm L, white. At about 5500', Emigrant Gap, Placer Co., Aug-Sep.

A. septentrionalis ssp *subumbellata*, Northern Androsace: Lvs linear-lanceolate, 5-20 mm L, entire or weakly denticulate; fl-stems erect or spreading; bracts linear-lanceolate, 2-3 mm L; fls about 4 mm L. Dry, rocky places, 10,000-13,200', scattered in SN, Jul-Aug.

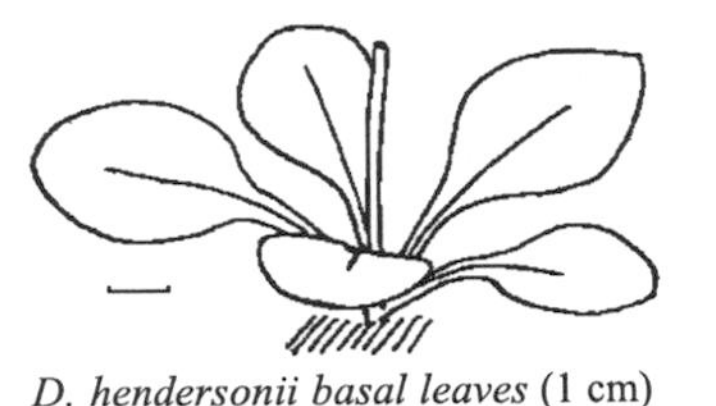

Dodecatheon

Shooting Star

D. hendersonii basal leaves (1 cm) *inflorescence* (1 cm)

Perennials with entire lvs. Fls nodding on slender pedicels. Calyx deeply 5-cleft, persistent, the lobes reflexed at anthesis, later erect. Stamens and style exserted. The roots and lvs of *D. hendersonii* may be eaten roasted or boiled. Those of the other spp are also said to be edible but should be tried with caution.

A Anthers 4; petals usu 4
  B Herbage usu glabrous; lvs ± linear; roots with small bulblets — 1) *D. alpinum*
  BB Herbage glandular-pubescent at least in infl; lvs oblanceolate; roots without bulblets — 3) *D. jeffreyi*

AA Anthers 5; petals 5
  B Lvs 10-50 cm L; plant heavily glandular-pubescent — 4) *D. redolens*
  BB Lvs mly shorter; plants glabrous
    C Herbage and roots reddish; above 7000' — 5) *D. subalpinum*
    CC Herbage usu greenish; below 7000' — 2) *D. hendersonii*

1) *D. alpinum*, Alpine Shooting Star: Plant 0.4-1.5 dm H; umbels 1-4-fld; pedicels 1-2 cm L at anthesis; calyx-tube 1.5-2.5 mm L, the lobes 4-5 mm L; corolla-tube maroon, yellow above, the lobes 8-11 mm L, magenta to lavender: anthers 5-6 mm L, linear. Boggy mdws and wet banks, 6400-12,000', SN (all), Jul-Aug.

ssp *majus*: Plant 1.5-3 dm H, ± glandular above; umbel 4-10-fld; pedicels 1-3 cm L; corolla-lobes 9-16 mm L; anthers 8 mm L. More common, 4000-11,000', SN (all), May-Aug.

2) *D. hendersonii*, Henderson's Shooting Star: Roots with bulblets at flowering time; lvs 5-15 cm L, spatulate to elliptic, obtuse to truncate at tip; fl-stem 10-50 cm H, glandular to glabrous; the umbel 3-12-fld; pedicels 2-7 cm L at anthesis; calyx-tube 2 mm L, the lobes 3-5 mm L; corolla-tube maroon, yellow above, the lobes 5-20 mm L, deep lavender or white (without intermediates); anthers 4-5 mm L, the pollen-sacs dark red to black. Mly in shaded places, SN (all), Mar-May.

3) *D. jeffreyi*, Jeffrey's Shooting Star: Plant 1.5-6 dm H; lvs 10-50 cm L; the umbel 3-18-fld; pedicels 3-7 cm L at anthesis; calyx-tube 2-5 mm L, the lobes 5-10 mm L corolla-tube closely reflexed, with a maroon ring below and yellow above, the lobes 10-25 mm L; anthers 7-10 mm L, maroon to yellow on pollen sacs. Wet places, 2300-10,000', SN (all), Jun-Aug.

4) *D. redolens*, Mountaineer Shooting Star: Lvs including petiole 20-40 cm L, oblanceolate; fl-stem 2.5-6 dm H; umbel 5-10-fld; pedicels 3-5 cm L; calyx tube 3-5 mm L, the lobes 5-12 mm L; corolla-tube yellow, the lobes 15-25 mm L, magenta to lavender; anthers 7-11 mm L, dark maroon to black. Moist places, 8000-11,500', Fresno Co. s., Jul-Aug.

5) *D. subalpinum*, Sierra Shooting Star: Lvs including petioles 3-7 cm L, oblanceolate to spatulate; fl-stem 0.7-1.5 dm H; umbel 1-5-fld; pedicels 10-15 mm L; calyx-tube 2-3 mm L; corolla-lobes 5-9 mm L. Moist, shaded places, Tulare to Tuolumne Co., May-Jul.

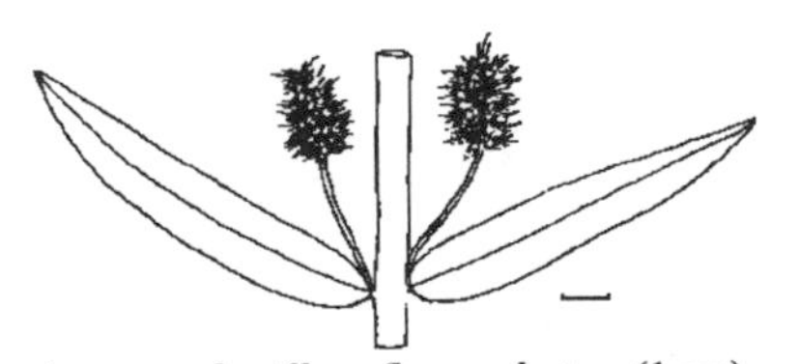

Lysimachia

Tufted Loosestrife

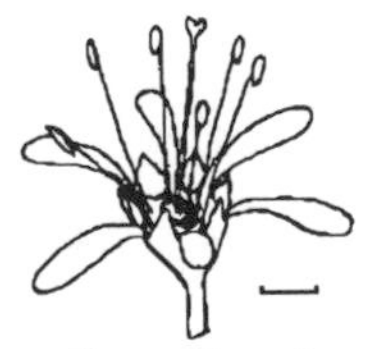

leaves and axillary flower clusters (1 cm) *flower* (1 mm)

L. thyrsiflora: Erect perennial; stems glabrous; lvs entire, opp, 5-12 cm L, sessile, gland-dotted; fls in short axillary spike-like racemes; corolla yellow, deeply 5-7-parted into linear segments 3-5 mm L. Occasional in wet places, 3000-4500', Plumas and Shasta cos., Jun-Aug. Edibility unknown.

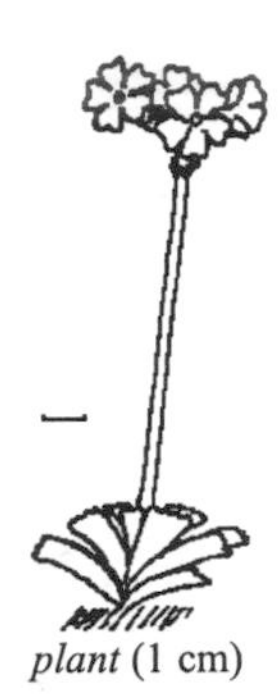
plant (1 cm)

Primula

Sierra Primrose

P. suffrutescens: Stems branched, creeping; lvs glabrous, crowded, spatulate, 1.5-3.5 cm L, the apex rounded and dentate, the base gradually narrowed to broad petiole; fl-stem 4-10 cm H, glandular-puberulent above; pedicels to 1 cm L; corolla magenta with yellow throat. Under overhanging rocks and about cliffs, mly 8000-13,500', SN (all), Jul-Aug. Lvs of related spp edible raw or as potherbs

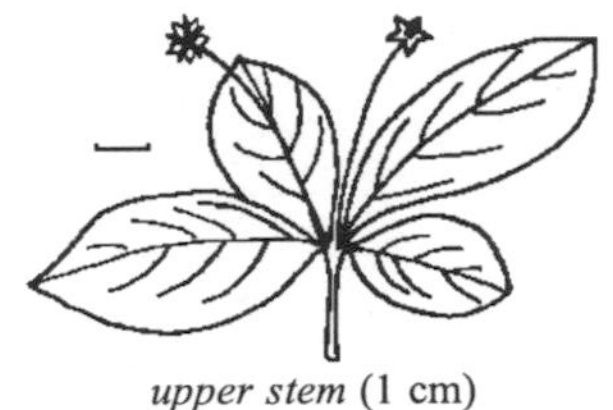
upper stem (1 cm)

Trientalis

Star-flower

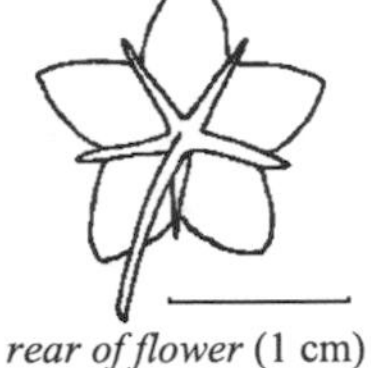
rear of flower (1 cm)

T. latifolia: Stems erect, slender, glabrous, from tuberous rootstocks; lower lvs few, scale-like; upper lvs in a single whorl of 4-6, 3-8 cm L, 2.5-5 cm W; petioles 1-4 mm L; pedicels about half as long as lvs; corolla white to pinkish, 8-15 mm W, parted almost to base into usu 6 lobes. Shady habitats mly below 4500', Mariposa Co. n., Apr-Jul. Edibility unknown.

RANUNCULACEAE - Buttercup Family

Herbs or woody vines. Lvs usu petioled, without stipules. Fl-parts usu distinct. Fr. various. Most members of this family are at least slightly toxic and should be avoided as a food source.

A Petalloid parts present and showy (fr also showy in the case of *Actaea*)
  B Fls irregular or with long spurs on petalloid parts
    C Petals not spurred; fls irregular, purplish blue; plants 5-20 dm H — *Aconitum* p. 165
    CC Fls with at least one spur; plants usu smaller
      D Fls with 5 spurs, reddish to pale yellow — *Aquilegia* p. 166
      DD Fls with 1 spur, usu violet — *Delphinium* p. 167
  BB Fls regular and radially symmetric
    C Stem lvs opp or whorled
      D The stem lvs usu forming a single whorl; herbs — *Anemone* p. 165
      DD The stem lvs many, opp; woody vines — *Clematis* p. 166
    CC Stem lvs alt or lvs all basal
      D Fls yellow; fr an akene — *Ranunculus* p. 169
      DD Fls white to greenish; fr various
        E Fr a reddish berry, these arranged in a showy raceme — *Actaea* p. 165
        EE Fr dry, not showy
          F Lvs simple, basal, long-petioled; blade round to reniform, often shallowly lobed
            G Lvs usu palmately 3-lobed; petalloid sepals 5-10 mm L; fr an akene — *Kumlienia* p. 168
            GG Lvs entire to dentate; petalloid sepals 10-20 mm L; fr a follicle — *Caltha* p. 166
          FF Lvs cmpd, cauline, sometimes finely dissected
            G. Plants ± aquatic, at least submersed lvs finely dissected — *Ranunculus aquaticus* p. 169
            GG Plants of dry habitats; lvs 2-3-ternate — *Isopyrum* p. 168
AA Neither fls nor fr showy or colorful
  B Lvs cmpd, lfts many — *Thalictrum* p. 170
  BB Lvs simple
    C Lvs deeply palmately lobed, the lower 10-20 cm W — *Trautvetteria* p. 170
    CC Lvs linear, grass-like — *Myosurus* p. 168

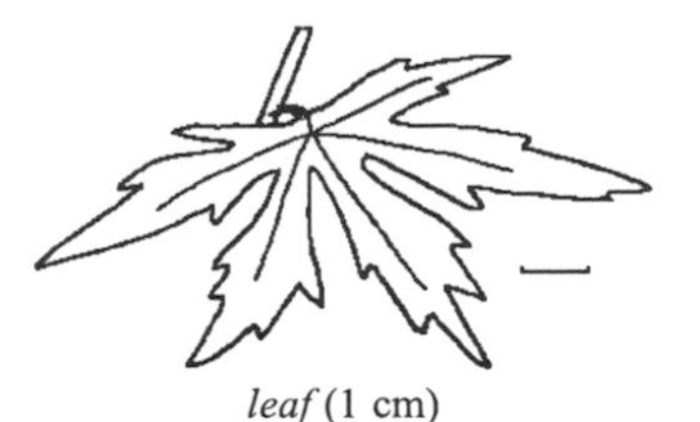

Aconitum

Aconite
Monkshood

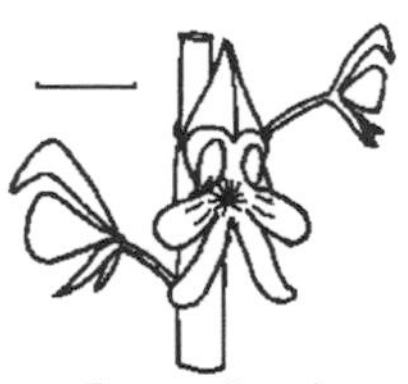

leaf (1 cm) *flowers* (1 cm)

A. columbianum: Stems mly erect and stout, 5-20 dm H, glabrous below, pubescent above; lvs thin, 5-12 cm W, palmately deeply 3-5-cleft; lower petioles longer than blades; fls purplish blue, sometimes pale; sepals 5, the upper (helmet) arched, 15-25 mm H; petals 2-5, the upper 2 concealed in the helmet. Moist places, especially in willow thickets, 4000-8700', SN (all), Jul-Aug. All parts of the plant contain an alkaloid called aconitine, which may cause restlessness, nausea, vertigo and impairment of speech and vision.

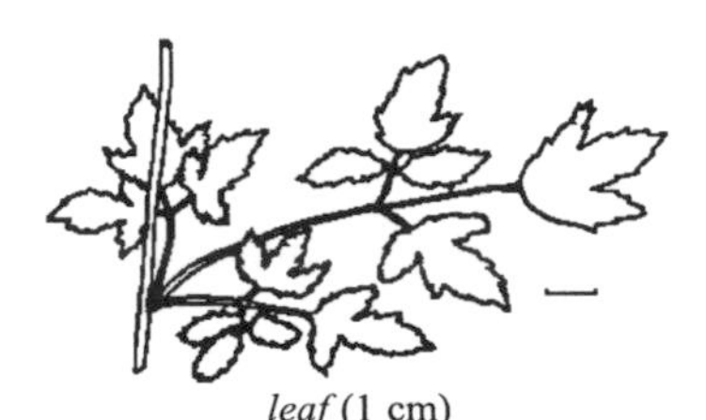

Actaea

Baneberry

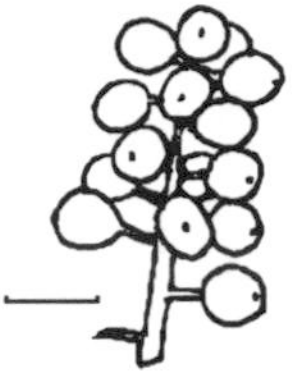

leaf (1 cm) *cluster of fruits* (1 cm)

A. rubra: Stem 1 to several, erect, branching above, sparsely pubescent; lvs 1-3, all cauline, 2-3-ternate, the lowest 1-2 dm W, with thin, ovate, serrate lfts 2.5-6 cm L; petioles to about 1 dm L; racemes dense, 6-10 cm L in fr; pedicels 4-8 mm L; sepals and petals 2-3 mm L; berry red or white, 6-8 mm L. Moist, fertile soil in woods, below 10,000', SN (all), May-Jun. Oil in the berry produces severe intestinal disturbances, vertigo and even death.

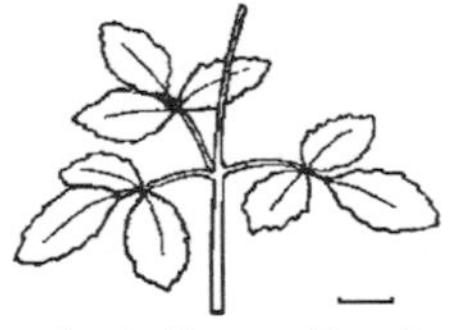

Anemone

Anemone
Windflower

whorl of leaves (1 cm) *A. occidentalis flowers* (1 cm)

Perennials. Lvs mly basal; stem lvs 2 or more together, forming an invol. subtending the fl. Fls showy. Stamens and pistils many. Akene-clusters densely hairy. Herbage slightly toxic, containing ranunculin which is converted to protoanemonin, a slightly toxic alkaloid, when eaten.

Lvs trifoliate; below 5000' — *A. oregana*
Lvs dissected; above 5000'
  Basal lvs 4-8 cm W; sepals 2-3 cm L; style plumose in fr — *A. occidentalis*
  Basal lvs 2-5 cm W; sepals 0.8-1.6 cm L; akenes woolly at maturity — *A. drummondii*

A. drummondii, Drummond's Anemone: Stems 1 to several, 1-3 dm H, hairy; basal lvs several, the ultimate divisions linear; petioles 2-10 cm L; sepals densely woolly; styles slender, 1.5-3 mm L. Talus and gravelly or rocky slopes, 5000-10,600', Inyo Co. n., May-Aug.

A. occidentalis, Pasque Flower: Stems 1 to several 1-6 dm H, long-hairy when young, often glabrate in age; basal lvs ternate, the divisions twice pinnately dissected into linear or lance-linear segments; petioles 3-10 cm L; invol.-lvs similar, sessile or nearly so; fls solitary; sepals 5-8, white or purplish, hairy without; heads large in fr because of the plumose 2-3.5 cm L. styles. Dry, rocky slopes, 5500-10,000', SN (all), Jul-Aug.

A. oregana, Windflower: Stem slender, 1-3 dm H, glabrous; stem lvs whorled; the lfts 1-2.5 cm L, broadly ovate, rounded at apex; peduncle 1, appressed-hairy; sepals mly 5, white to rose, 4-10 mm L. Shaded places, forest floor below 5000', Plumas Co. n., Apr-May.

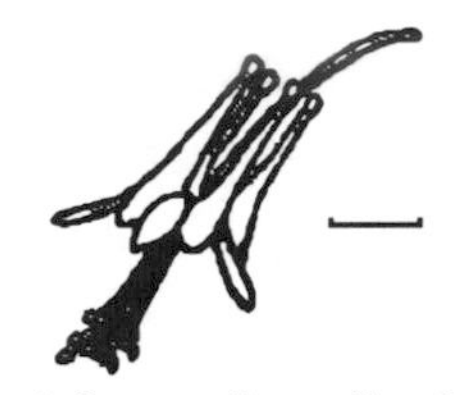

Aquilegia

Columbine

A. formosa flower (1 cm) *stem leaf* (1 cm)

Perennials. stems usu several, erect, branched. Lvs usu basal, 2-3-ternate, the stem lvs gradually reduced. Fls 5-merous, each petal spurred. Fr a follicle. The frs contain an alkaloid which causes a weak pulse and a burning sensation of the mouth, and if sufficient quantities are eaten, death. Lvs are edible, but grow bitter with age. Hybrids between the two species listed occasionally occur.

Fls nodding, red and yellow; mly below 9000' *A. formosa*
Fls ± erect, cream to yellow or pink; above 9000' *A. pubescens*

A. formosa, Crimson Columbine: Stems mly 5-10 dm H, openly branched and glandular-pubescent above; petioles of basal lvs 1-2 dm L; lfts mly 2-4 cm L; sepals red, wide-spreading to reflexed, 15-25 mm L; spurs red; petals yellow. Moist woods, 3000-9700', SN (all), May-Aug.

A. pubescens, Coville's Columbine: Stems tufted, 2-4.5 dm H, glandular-pubescent above or throughout; petioles 5-20 cm L; lfts crowded, deeply cleft, 1-2(-3) cm W; stem lvs few; sepals spreading, 1.5-2 cm L; spurs 2.5-4 cm L, yellow. Rocky places and talus, 9000-12,000', Tulare to Tuolumne Co., Jul-Aug.

Caltha

Marsh-marigold

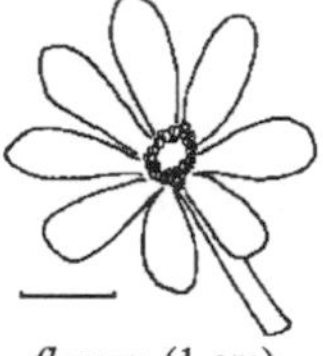

basal leaves (1 cm) *flower* (1 cm)

C. leptosepala: Fleshy perennial; lvs round-reniform, all basal, the blades 3-10 cm W, glabrous; petioles 5-30 cm L; sepals white, 6-10, 12-16 mm L; fr a follicle. Marshy and boggy places, 4500-10,000', SN (all), May-Jul. The lvs contain a poisonous substance called helleborin. Cooking destroys the poison. Lvs are best when collected in the spring and served like spinach.

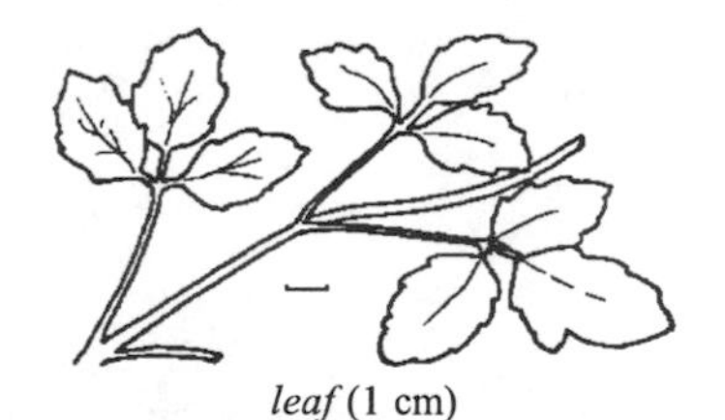

Clematis

Clematis
Virgin's-bower

leaf (1 cm) *flower* (1 cm)

Half-woody vines which climb over bushes and in trees by clasping or twining petioles. Lvs pinnate. Sepals silk-woolly. Styles plumose and 20-40 mm L in fr. Plants may contain the slightly toxic alkaloid, protoanemonin.

Fls mly solitary; lfts mly 3 *C. lasiantha*
Fls many in panicles; lfts 5-7 *C. ligusticifolia*

C. lasiantha, Chaparral Clematis: Stems to 4 or 5 m H; lfts mly broad-ovate, 2-5 cm L, coarsely toothed to 3-lobed; petioles usu 2-5 cm L; petiolules 3-20 mm L; peduncles 4-12 cm L; sepals 15-25 mm L. In canyons and near streams below 6000', SN (all), Apr-Jun.

C. ligusticifolia, Western Virgin's-bower: Stems climbing to a height of 6 m; lfts lanceolate to lance-ovate, entire to 3-lobed or coarsely toothed from about the middle, 2-8 cm L; petioles usu 3-7 cm L; petiolules 10-30 mm L; peduncles 3-10 cm L; sepals about 10 mm L. Mly along streams and in moist places below 7000', SN (all), Apr-Aug.

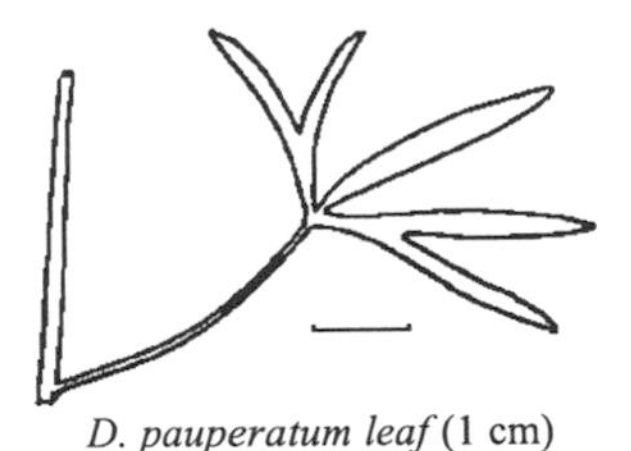

Delphinium

Larkspur

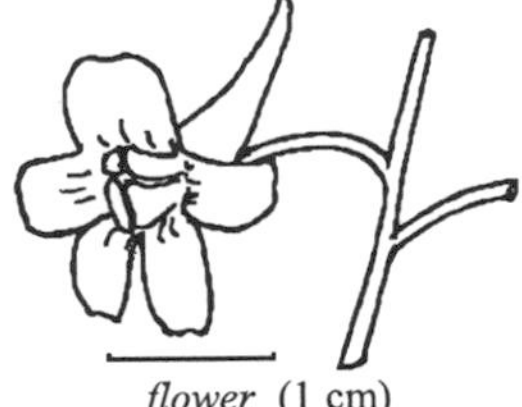

D. pauperatum leaf (1 cm) *flower* (1 cm)

Erect, branching perennials. Lvs palmately lobed or divided. Fls in showy spikes or racemes. Sepals usu blue to purple, posterior sepal spurred; petals 4; stamens many; pistils 1-5. At least several spp are known to be poisonous and the others should be regarded as so until shown otherwise. The poisonous alkaloid, delphinine, is found in all parts of the plant and causes vomiting, weak pulse, coma and death.

A Stems very slender at base and easily separated from the tuberous roots
  B Petioles 5-20 cm L — 4) *D. gracilentum*
  BB Petioles 3-10 cm L
    C Follicles with few hairs; sepals 12-16 mm L; dry places — 7) *D. nuttallianum*
    CC Follicles glabrous; sepals 8-10 mm L; damp places — 2) *D. depauperatum*
AA Stems not attenuate at base and rather firmly attached to woody or fibrous roots
  B Stems appressed-pubescent; lvs hairy — 5) *D. hansenii*
  BB Stems glabrous; lvs glabrous to pubescent
    C Fls orange or dull red — 6) *D. nudicaule*
    CC Fls bluish
      D Stems 10-25 dm H, glaucous — 3) *D. glaucum*
      DD Stems usu 2-8 dm H, not glaucous
        E Lvs glabrous; 7500-11,000', Eldorado Co. s. — 8) *D. polycladon*
        EE Lvs usu pubescent; 5000-7500', mly e. SN, Mono Co. n. — 1) *D. andersonii*

1) *D. andersonii*, Anderson's Larkspur: Stems hollow, glabrous, unbranched; lvs basal, rounded, 1-3 cm W; petioles 4-10 cm L; racemes mly 5-12-fld, with ascending, glabrous pedicels 1-6 cm L; sepals 10-14 mm L, somewhat hairy; spur thick, ± straight, 10-12 mm L; upper petals whitish, slightly notched; lower petals rounded; follicles 15-25 mm L, ± pubescent. Sandy and volcanic soil, among shrubs and pines, 5000-7500', Mono Co. n., Apr-Jun.

2) *D. depauperatum*, Dwarf Larkspur: Stems very slender, 0.6-3 dm L, glabrous; lvs few, near the base, subglabrous, 2-5 cm W, palmatifid into 3 or 5 divisions; pedicels slender, divaricate; sepals mly glabrous, 5-8 mm L; spur slender, straight, 10-15 mm L; upper petals narrow, white-edged; lower petals round-ovate, with central cluster of hairs; follicles ovoid, erect, glabrous, 8-12 mm L. Damp thickets and edge of woods, 4000-9500', May-Jul.

3) *D. glaucum*, Mountain Larkspur: Stems coarse, glaucous, leafy; lvs pubescent beneath; petioles 3-15 cm L; racemes 1-4 dm L, many-fld; pedicels divaricate, 1-4 cm L; sepals 8-12 mm L, puberulent on back; spur rather stout, 8-10 mm L. Wet mdws and near streams, 5000-10,000', Madera Co. n., Jul-Sep.

4) *D. gracilentum*, Slender Larkspur: Stem often procumbent, 2-4 dm L, unbranched, glabrous except for glandular-pubescent infl, ± glaucous; lvs few, mly basal, thin, equally 5-cleft, 4-5(-10) cm W, pubescent especially beneath; petioles 1-2 dm L; racemes open, narrow, 8-20-fld; pedicels filiform, lower petals blue, round-ovate; follicles glabrous, somewhat spreading, 8-12 mm L. Shady and damp places, below 8000', Butte Co. s., May-Jul. Plants of s. SN usu with pink fls.

5) *D. hansenii*, Hansen's Larkspur: Stems 4-9 dm H, greenish, ± soft-hairy at base; lvs withering early, mly basal, the lower hairy, 4-9 cm W, shallowly palmatifid, the upper smaller, divided into narrow segments; petioles ascending, about twice as long as blade; raceme dense, elongate; pedicels suberect, 1-2 cm L; sepals 6-8 mm L, pubescent; spur slender, curving, 7-10 mm L; follicles erect, 10-14 mm L, hairy. Open, grassy places below 4000', Butte Co. s., Apr-Jul.

6) *D. nudicaule*, Red Larkspur: Stems erect or ascending, 2-6(-8) dm H; lvs 3-10 cm W, 3-5-cleft, the ultimate lobes shallow, rounded, glabrous to sparsely pubescent; petioles 5-15 cm L; racemes 3-20 fld, open, with long ascending pedicels; fls cornucopia-shaped; sepals 10-12 mm L, glabrous or nearly so; spur 15-20 mm L; upper petals yellow with red tips; follicles divergent, glabrous, 1.5-2 cm L. Dry slopes among shrubs and in woods, below 6500', Mariposa Co., Butte and Plumas cos. n., Apr-Jun.

7) *D. nuttallianum*, Nuttall's Larkspur: Stems subglabrous below, pubescent above, often glandular, 1-3(-4) dm H; lvs few, mly cauline, lowest withering early, rounded in outline, 3-5 cm W, with 3-5 primary divisions, ± glabrous; petioles 5-7 cm L; upper lvs 3-parted, bract-like; racemes few-fld, crowded or open; fls sometimes pink; pedicels ascending; sepals pubescent; spur

slender, 10-15 mm L; upper petals whitish, lower petals translucent; follicles erect or flaring, 1-2 cm L. Grassy and brushy places and open woods, 5000-10,000', May-Jul.

8) *D. polycladon*, Dissected-leaf Larkspur: Stems several, glabrous, unbranched, mly 5-8 dm H; lvs mly basal, round-reniform in outline, 3-8 cm W, palmatifid into wedge-shaped divisions, these 3-5-lobed; petioles 5-18 cm L; racemes open, 1-2 dm L, few-fld; pedicels 1-5 cm L, ascending; sepals glabrous to hairy, 9-14 mm L; spur slender, 12-17 mm L; upper petals pale; lower petals hairy; follicles pubescent, 12-18 mm L. Moist to wet habitats, 7500-11,150', Jul-Sep.

leaf (1 cm)

Isopyrum

Isopyrum

upper stem (1 cm)

I. occidentale: Glabrous perennial with 1-3 erect stems 10-25 cm H, often branched above; lvs 2-3-ternate on petioles 5-10 cm L; lfts 1-2 cm L; cauline lvs reduced, sometimes appearing opp; sepals white to purplish, 7-10 mm L; fr 10-12 mm L. Uncommon in dry habitats below 5000', Butte Co. s., Apr.

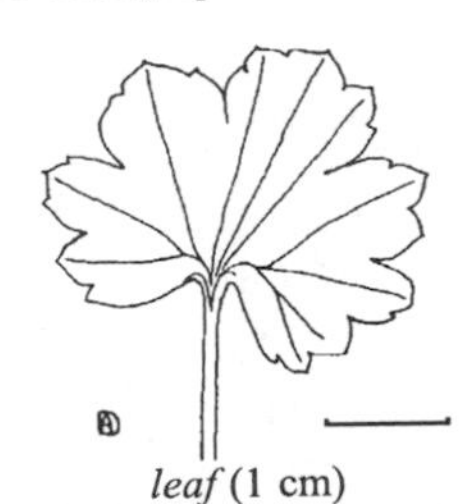

leaf (1 cm)

Kumlienia

Waterfall-buttercup

inflorescence (1 cm)

K. hystricula: Glabrous perennial with basal lvs and a 1-3-fld scape 15-40 cm L; lvs on petioles 2-10 cm L; blade simple, reniform to round , 2-4 cm L, 3-5 cm W, shallowly 3-5-lobed, these lobes again lobed; scape usu reclining; sepals 5-10 mm L, white; petals 2-5 mm L, greenish. Wet places among rocks below 6000', Butte Co. s. Apr-Jun.

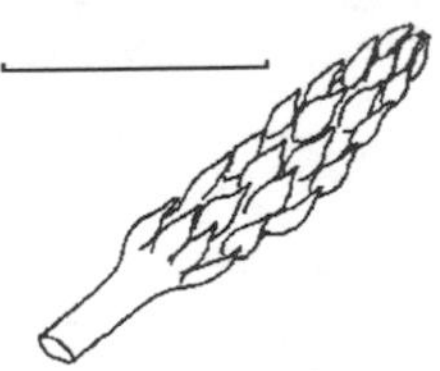

spike (1 cm)

Myosurus

Sedge Mouse-tail

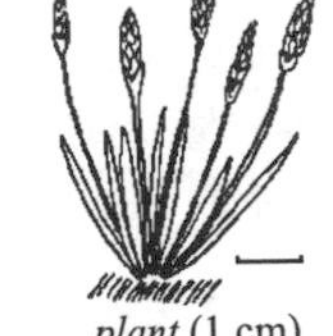

plant (1 cm)

M. apetalus: Small tufted grass-like annuals with fibrous roots; lvs ± linear, entire, 1-8 cm L; fl-stems 2-5 cm H, slender; fls minute, greenish yellow to whitish, solitary on naked stalks; sepals 5, sometimes 6-7, about 2 mm L, spurred, the spurs almost as long; pistils many, on a cylindrical axis which is long and spike-like in fr. Occasional, moist places, 4500-10,200', central SN, May-Jul. Plant probably toxic.

R. eschscholtzii flower (1 cm)

Ranunculus

Buttercup

R. flammula typical plant (1 cm)

Usu perennial herbs. Lvs alt. Fls usu 5-merous with glossy, mly yellow petals. Akenes mly in a rounded head. The seeds may be parched and ground into meal; roots may be boiled and eaten.

A few buttercups contain such toxins as protoanemonin and aenonal; buttercups with an acrid, burning juice should be avoided.

A Aquatic plants with submersed lvs dissected into linear lobes
  B Petals white, 4-8 mm L; widespread — 2) *R. aquatilis*
  BB Petals yellow, 7-15 mm L; Plumas Co. — 5) *R. flabellaris*
AA Terrestrial plants with all lfts ± broad
  B Lower lvs 3-7-parted or -lobed
    C Petals usu 2-3 mm L; stem 2-4.5 dm H — 11) *R. uncinatus*
    CC Petals 6-10 mm L; stem often shorter
      D Stems mly hairy; stem lvs often cmpd; below 7000'
        E Akene-beaks short or none; petals mly 5-8 mm L, 2-3 mm W — 8) *R. occidentalis*
        EE Akene beak about 3 mm L; petals 8-20 mm L, 4-7 mm W — 9) *R. orthorhynchus*
      DD Stem glabrous; stem lvs usu simple, sometimes lacking — 4) *R. eschscholtzii*
  BB Lower lvs entire to shallowly lobed or dentate
    C Lvs lanceolate to linear
      D Petals 6-10 mm L; stems not rooting at nodes — 1) *R. alismaefolius*
      DD Petals 3-6 mm L; stems rooting at nodes — 6) *R. flammula*
    CC Lvs oblanceolate to oval
      D Lvs irregularly dentate to lobed; stems rooting at nodes; Tuolumne Co. s. — 3) *R. cymbalaria*
      DD Lvs entire to minutely toothed; stems not rooting at nodes; below 6000', c. SN n.
        E Akenes in large globose heads 1-2 cm W; sepals tinged lavender — 7) *R. glaberrimus*
        EE Akenes in heads about 0.5 cm W; sepals green — 10) *R. populago*

1) *R. alismaefolius* and vars., Water Plantain Buttercup: Stems erect to reclining, 2-8 dm L, glabrous, branched above, several- to many-fld; basal lvs 2-10 cm L, entire or serrulate, tapering into a broad petiole; pedicels to 10 cm L; sepals yellowish green, spreading; petals 5, 6-10 mm L; akenes 10-50. Common in wet mdws to 12,000', SN (all), May-Jul.

2) *R. aquatilis* vars., Water Buttercup: Stems submersed, 2-6 dm L, often in large clusters; submersed lvs 2-4 cm L, floating lvs simple, 1-2 cm W, 3-lobed or -parted, the lobes again forked or parted; sepals 2-3 mm L; petals yellow at base, obovate, about 10 mm L. Ponds and slow streams below 10,000', SN (all), May-Aug.

3) *R. cymbalaria* var. *saximontanus*, Desert Buttercup: Stems ± glabrous, several dm L; basal lvs 1-4 cm L, 1-3 cm W; petioles 2-5 cm L; fl-stems erect, 5-30 cm H, 1-several-fld; sepals 4-8 mm L; petals 5-12, obovate, 4-8 mm L; akenes many, in an elongate head. Muddy places below 10,500', Mono and Tuolumne cos. s., Jun-Aug.

4) *R. eschscholtzii*, Eschscholtz' Buttercup: Stems erect or decumbent, 3-15 cm L; lvs rounded in outline, subcordate or truncate at base, 1.5-3 cm W, deeply 3-parted, the middle lobe 3-lobed or entire; petioles 2-6 cm L; stipular lf-bases sometimes thickened and persistent (var. *oxynotus*); pedicels 1-10 cm L; sepals tinged lavender, 4-8 mm L; petals 5, 7-11 mm L. Mdws and about rocks, 8000-13,300', SN (all), Jul-Aug.

5) *R. flabellaris*, Yellow Water Buttercup: Stems floating or reclining, rooting at lower nodes, 3-7 dm L, branched; lvs all cauline, triternately dissected, 2-12 cm W; petioles very short; sepals 5-8 mm L; petals 5-8. Mud or shallow water below 6000', Jun-Aug.

6) R. flammula vars., Creeping Buttercup: Stems reclining, subglabrous, 1-4 dm L; lvs entire, 1.5-5 cm L; petioles 1-6 cm L, occasionally none and lvs sheathing; pedicels 1-8 cm L; sepals 3-5 mm L; petals 5 or 10, obovate, 3-6 mm L. Marshy places below 7500', Fresno Co. n., Jul-Aug.

7) *R. glaberrimus* vars., Sagebrush Buttercup: Stems prostrate or ascending, 1-6-fld; basal lvs rounded to oblanceolate, usu entire; petioles 3-9 cm L; pedicels 1-10 cm L; sepals 5-8 mm L; petals usu 5, yellow or white in age, obovate, 5-15 mm L. Sandy places or mdws at about 5000', Mono and Nevada cos. n., Apr-Jun.

8) *R. occidentalis* vars., Western Buttercup: Stems erect to reclining, 1-7 dm L; lvs 2-6 cm W, 3-parted or with 3-5 lfts, often silky-villous beneath; petioles 3-12 cm L, hairy; pedicels 2-10 cm L; sepals 5; petals usu 5, elliptic. Mdws and moist places, below 6500', SN (all), May-Jul.

9) *R. orthorhynchus*, Straight-beaked Buttercup: Stems 1.5-5 dm H, glaucous, branched; basal lvs 3-8 cm W, pinnate, with 3-7 lfts, these again lobed into broad or narrow segments; petioles 3-15 cm L; sepals 6-8 mm L petals 5, ± truncate. Mdws below 7000', SN (all), Jun-Jul.

10) *R. populago*, Cusick's Buttercup: Stems declined or erect, glabrous, 1-4 dm L, 6-10-fld; basal lf-blades usu denticulate; petioles 4-12 cm L; pedicels 1-5 cm L; sepals 1.5-2.5 mm L; petals 5, light yellow, obovate, 3-6 mm L. Mdws and boggy places 5000-6000', Butte Co., Jun-Jul.

11) *R. uncinatus*, Bongard's Buttercup: Stems glabrous to sparsely hairy with white hairs (with stiff reddish brown hairs in var. *parviflorus*); basal lvs cordate-reniform in outline, 2-9 cm W, 3-parted, the divisions lobed; petioles 5-10 cm L; stem lvs often larger than basal; pedicels to 2 cm L in fl, twice that in fr; sepals reflexed, 3 mm L; petals 5, 2-3 (-6) mm L. Occasional, moist, shaded places below 8000', SN (all), May-Jul.

Thalictrum

Meadow-rue

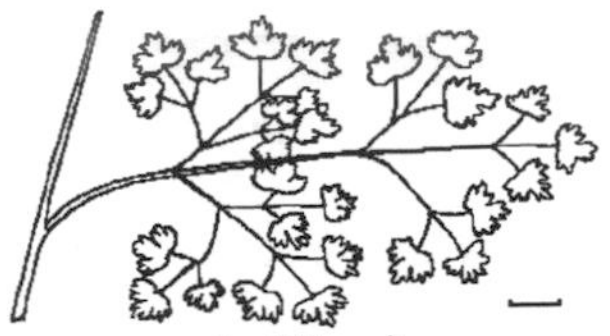

akenes and staminate flowers (1 cm) *leaf* (1 cm)

Erect perennials. Petioles dilated and clasping at base. Lvs 2-4-ternate. Fls often unisexual, numerous. Sepals 4(-7), dropping early, greenish or petaloid. Stamens numerous, exceeding the sepals at anthesis, the filaments often colored, erect or pendant. Lvs very bitter, containing alkaloids which may be toxic.

Lvs all basal; stems mly less than 2 dm H; above 10,500' *T. alpinum*
Stem leafy, mly over 3 dm H; below 10,500'
  Fls with both stamens and pistils *T. sparsiflorum*
  Fls with only stamens or pistils; plants with only one type of fl *T. fendleri*

T. alpinum, Dwarf Meadow-rue: Stems ± tufted, glabrous throughout; lvs 1-6, 2-5 cm L, the lfts thick, strongly veined, glaucous, dull on both surfaces, the margin revolute; raceme nodding; sepals 1.5-2 mm L. Moist mdws and bogs, Rock Creek Lake Basin and Convict Creek Lake Basin, Jun-Aug.

T. fendleri, Fendler's Meadow-rue: Stems branched above, 5-15 dm H; lfts mly 3-lobed, the lobes often toothed; sepals 2-3 mm L; staminate-fls with drooping clusters of stamens 10-15 mm L; pistillate fls with about 10 densely clustered frs 5-6 mm L. Moist habitats or occasionally on dry slopes, 4000-10,000', SN (all), May-Aug.

T. sparsiflorum, Few-flowered Meadow-rue: Much like *T. fendleri*; stems 3-12 dm H; upper lvs subsessile, the lfts 1-2 cm L; panicle leafy. Moist stream banks and bogs, often in willow thickets, 5000-11,000', SN (all), Jul-Aug.

Trautvetteria

False Bugbane

leaf (1 cm) *inflorescence* (1 cm)

T. caroliniensis: Stems slender, erect, glabrous or nearly so, 5-10 dm H, branching above; lower lvs long-petioled, 1-2 dm W, deeply 5-11 palmately lobed, the lobes irregularly and sharply toothed; fls terminal, clustered; sepals 3-5, early deciduous, 3-6 mm L; petals none. Swamps and along streams, 4000-5000', Placer Co. n., Jul-Aug. Edibility unknown.

RHAMNACEAE - Buckthorn Family

Shrubs or small trees. Lvs simple with small deciduous stipules or these sometimes thick, with corky persistent bases. Fls small, regular, usu in small umbels, these often in larger clusters. Calyx ± tubular at base, 4-5-lobed, lined with a disk on edge of which are inserted the petals and stamens

Fr dry, capsular; lvs alt or opp, often 3-veined from base *Ceanothus*
Fr fleshy, drupe-like; lvs alt, pinnately veined *Rhamnus*

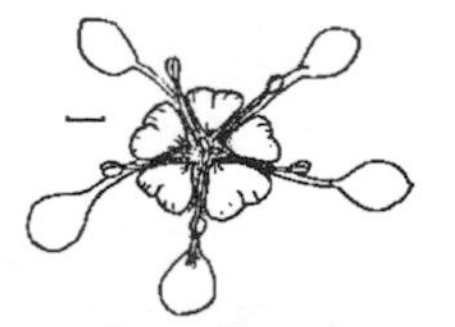

Ceanothus

Ceanothus

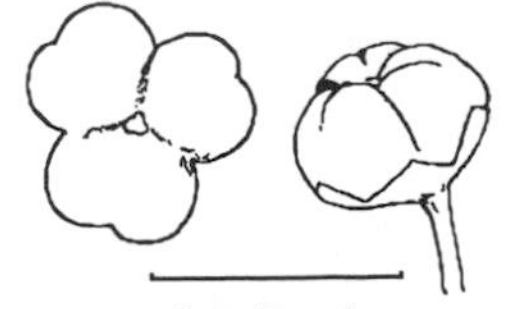

flower (1 mm) *fruit* (1 cm)

Lvs 3-nerved from base or pinnately veined, ± petioled. Fls 5-merous. Ovary 3-loculed, 3-lobed, with a short 3-cleft style. Fr a 3-lobed capsule separating at maturity. A tea can be made from the lvs or fls; some spp make a better tea than others.

A Lvs alt; stipules thin and deciduous; fls in panicles
  B Ultimate branches rigidly divaricate and spinose; lvs persistent, plant
    C Lvs mly serrulate; mly below 4000' — 6) *C. leucodermis*
    CC Lvs ± entire; mly above 4000' — 1) *C. cordulatus*
  BB Ultimate branchlets flexible, not rigidly divaricate and spinose
    C Lvs entire, ± deciduous
      D Lvs mly over 2 cm L; peduncles ± leafy; fls usu white — 5) *C. integerrimus*
      DD Lvs mly less than 2 cm L; peduncles naked; fls blue — 7) *C. parvifolius*
    CC Lvs serrate, serrulate, or glandular-denticulate
      D Plant forming mats 1-3 dm H — 3) *C. diversifolius*
      DD Plant erect, usu over 10 dm H
        E Lvs 1-2.5 cm L; fls blue to white — 10) *C. tomentosus*
        EE Lvs 2.5-8 cm L; fls white — 11) *C. velutinus*
AA Lvs opp, often tufted; stipules with thick corky persistent bases; fls in lateral umbels
  B Lvs entire to minutely dentate, dull above, mly 0.5-1.5 cm L
    C Stems erect; fls white — 2) *C. cuneatus*
    CC Stems forming mats with some ± erect branchlets; fls mly blue — 4) *C. fresnensis*
  BB Lvs conspicuously spinose-dentate, shining above, mly over 1 cm L
    C Branchlets red-brown; Calaveras and Alpine cos. n. — 9) *C. prostratus*
    CC Branchlets brown or gray; Tulare and Inyo cos. s. — 8) *C. pinetorum*

1) *C. cordulatus*, Snow Bush: Stems intricately branched, spreading, 1-2 m H, with smooth whitish bark; the whole plant grayish glaucous; lvs ovate to elliptic, 3-veined; petioles 3-6 mm L; fls white, the clusters dense, 1.5-5 cm L. Dry, open flats and slopes, 3000-9500', SN (all), May-Jul.

2) *C. cuneatus*, Buck Brush: Stems and branches rigid, 1-3.5 m H; lvs on spur-like branchlets, obovate to spatulate, 1-3 cm L, minutely but closely white-tomentose between veins beneath, gray-green and glabrous above; capsule with short, erect horns near the top. Common on dry slopes and fans below 6000', SN (all), Apr-May.

3) *C. diversifolius*, Pine Mat: Trailing shrub with long flexible hairy branches; lvs ovate to obovate, mly 1-3 cm L, pinnately veined with 1 central vein from base, densely soft-pubescent on both sides; petioles 3-12 mm L; fl-clusters rather few-fld, about 1 cm L, on longer peduncles; fls blue to almost white. Occasional on dry flats in pine forests, 3000-6000', SN (all), May-Jun.

4) *C. fresnensis*, Fresno Mat: Plants forming mats to 6 m W; branchlets to 3 dm H; lvs much as in *C. cuneatus* but usu less than 1 cm L and occasionally minutely toothed; infl a few-fld, sessile or short-peduncled umbel. Dry ridges, 3000-6700', Fresno to Eldorado Co., May-Jun.

5) *C. integerrimus*, Deer Brush: Stems loosely branched, 1-4 m H; lvs ovate to ± oblong, 2.5-7 cm L, thin, variously veined; petioles 6-12 mm L; fl-clusters mly branched, 4-15 cm L; peduncles about as long; fls white to dark blue or pink. Various habitats below 7000', SN (all), May-Jul.

6) *C. leucodermis*, Chaparral Whitethorn: Evergreen shrub with pale green bark and short spreading subglabrous spinescent branchlets; lvs elliptical-oblong to ovate, 3-veined from base, usu glabrous and glaucous on both surfaces, 1-2.5 cm L; petioles 2-3 mm L; fl-clusters mly unbranched, 3-8 cm L, the fls white to blue. Dry, rocky slopes, Eldorado Co. s., Apr-Jun.

7) *C. parvifolius*, Littleleaf Ceanothus: Spreading shrub 6-12 dm H with slender greenish to reddish twigs, ± glabrous throughout; lvs ± elliptic, 6-20 mm L, 3-veined from near base, green above, paler beneath, entire except sometimes near apex; petioles 2-5 mm L; fl-clusters rarely branched, 3-7 cm L, on peduncles about as long; fls pale to deep blue. Wooded slopes, 4500-7000', Tulare to Plumas Co., May-Jul.

8) *C. pinetorum*, Kern Ceanothus: Stems erect or semiprostrate, 1.5-10 dm H, with divergent branches that may root at nodes and become several m L; lvs often clustered at nodes, broadly obovate to roundish, 1-3 cm L, glabrous and light green above, paler beneath, slightly revolute, regularly pinnately veined, coarsely 6-9-toothed on each side; fls whitish to blue, in few-fld umbels. Dry slopes, 5400-9000', May-Jul.

9) *C. prostratus*, Squaw Carpet, Mahala Mats: Stems prostrate, the branches rooting and forming mats 1-2 m W; lvs ± obovate, 0.8-2.5 cm L, leathery, glossy light green above, paler beneath; petioles 1-3 mm L; fls deep or light blue, in umbellate clusters on short peduncles. Open flats in forests, 3000-7800', Apr-Jun.

10) *C. tomentosus*, Woollyleaf Ceanothus: Evergreen shrub, 1-3 m H with grayish brown or reddish bark, the branches long and slender; lvs ovate to grayish brown or reddish bark, the branches long and slender; lvs ovate to elliptic, glandular-serrulate, dark green and finely pubescent above, whitish or brownish tomentose beneath, veination various; petioles 1-5 mm L; fl-clusters branched, 2.5 cm L, on shorter peduncles. Scattered on dry slopes below 5000', Mariposa to Placer Co., Apr-May.

11) *C. velutinus*, Tobacco Brush: Mly a spreading round-topped shrub 1-2 m H, stout, much-branched, evergreen, puberulent on twigs and lower lf-surfaces; lvs ovate-elliptic, rounded or subcordate at base, closely glandular-serrulate, dark green and shiny above, paler beneath;

petioles 5-20 mm L; fl-clusters branched, 5-10 cm L, on ± angled peduncles 2-5 cm L. Open, wooded slopes, 3500-10,000', SN (all), Apr-Jul.

Rhamnus

Buckthorn
Cascara

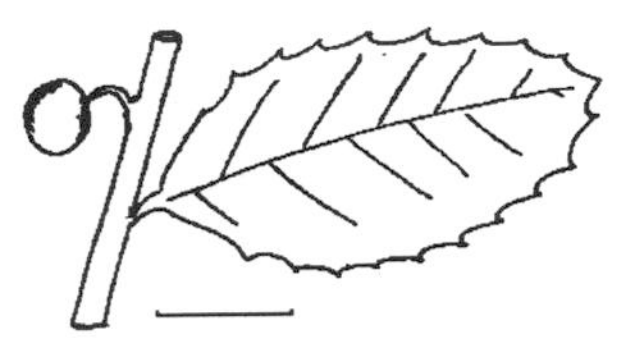

R. rubra leaves and fruits (1 cm) *R. ilicifolia leaf and fruit* (1 cm)

Lvs alt. Infl mly umbellate. Fls greenish, 4-5-merous, in axillary clusters. Pistil 1; ovary 2-4 loculed. The berries of *R. crocea* are edible cooked; they may temporarily turn the skin red if eaten in quantity. The bark and raw berries of *R. californica* and *R. purshiana* may be used as laxatives. *R. californica* seeds can be dried and ground for coffee.

A Plants of wet habitats; lvs thin, mly over 5 cm L
 B Large shrub or tree over 3 m H; petals present; below 5000' — 3) *R. purshiana*
 BB Shrub less than 1.5 m H; petals none; near Lake Tahoe — 1) *R. alnifolia*
AA Plants of dry habitats; lvs mly less than 5 cm L
 B Fls not in peduncled umbels; lvs with spin-tipped teeth — 2) *R. ilicifolia*
 BB Fls mly in peduncled umbels; lvs not spinose
  C Lvs persistent, ± thick, mly pubescent on both surfaces — 5) *R. tomentella*
  CC Lvs deciduous, thin, ± glabrous — 4) *R. rubra*

1) *R. alnifolia*, Alderleaf Coffeeberry: Lvs elliptical to ovate, long-pointed, serrulate; petioles 5-15 mm L; umbels 1-3-fld, appearing with the lvs; fls mly 5-merous, unisexual,; pedicels 2-6 mm L; calyx about 3 mm L; berry black, 6-8 mm in diameter, seeds 5-6 mm L. Swampy or boggy places, 4500-7000', Placer to Plumas Co., May-June.

2) *R. ilicifolia*, Hollyleaf Coffeeberry: Shrub or small tree, 1.5-4 m H; branchlets rigid, often spinescent; lvs often fascicled; lf-blades 2-4 cm L, oval to roundish; petioles 2-8 mm L; berry red, 5-6 mm L; nutlets 4 mm L. Dry slopes mly below 5000', SN (all), Apr-Jun.

3) *R. purshiana*, Cascara: Stems 5-12 m H; bark gray; lvs ± tufted at ends of branches; petioles 6-20 mm L; lf-blades 5-15 cm L, broadly elliptical or obovate, obtuse, mly irregularly serrate; fls 5-merous, in umbels of less than 25 fls; peduncles to 25 mm L; fls 4-5 mm L; berry black, about 10 mm L. Moist places below 5000', Placer Co. n., May-Jul.

4) *R. rubra* sspp, Sierra Coffeeberry: Low shrubs 1-1.5 dm H, spreading or rounded; lvs variable, 1.5-7 cm L; petioles 3-8 mm L; fls 5-merous; pedicels 2-8 mm L; calyx about 3 mm L; petals 1 mm L; berry black, less than 10 mm L; nutlets 6-8 mm L. Dry slopes, 2000-8000', Fresno Co. n., May-Aug.

5) *R. tomentella*, Chaparral Coffeeberry: Stems erect, to 6 m H, young growth densely pubescent and with long hairs; lvs oval to elliptical, serrate, 2-6 cm L, bright or dull green above, white-woolly beneath; umbels on peduncles 4-18 mm L, 6-50-fld; fls 2-3 mm L; petals present; berries black or red when ripe, 10-12 mm L; nutlets 7-9 mm L. Dry slopes and canyons, 2000-7500', Madera Co. s., Jun-Jul.

ROSACEAE - Rose Family

Trees, shrubs or perennial herbs. Lvs usu alt and with stipules. Sepals and petals at edge of fl-tube (perigynous) which is lined or rimmed with a glandular disk. Fls usu 5-merous.

A Lvs simple, sometimes lobed or cleft; plants mly woody
 B Lvs palmately lobed or cleft
  C Lvs wedge-shaped, 0.5-3 cm L, lobed at apex — *Purshia* p. 181
  CC Lvs roundish in outline, 3-15 cm L, broadly lobed
   D Foliage stellate-pubescent; petals 3 mm L — *Physocarpus* p. 179
   DD Foliage not stellate-pubescent; petals 15-20 mm L — *Rubus parviflorus* p. 182
 BB Lvs serrate or entire, not lobed
  C Fr fleshy; plants over 1 m H, often trees
   D Twigs with hollow, chambered center; lvs entire — *Oemleria* p. 179
   DD Twigs solid; lvs usu serrate
    E Ovary superior; fr 1-seeded — *Prunus* p. 181
    EE Ovary inferior; fr usu several seeded
     F Infl a 1-3-fld umbel; lvs entire or evenly serrulate — *Peraphyllum* p. 179

FF Infl a 4-8-fld raceme; lvs usu serrate only towards apex — *Amelanchier* p. 175
CC Fr dry; plants often under 1 m H
D Plants erect shrubs or trees; lvs tufted on spur-like branchlets
E Lvs linear, 4-8 mm L, about 1 mm W — *Adenostoma* p. 174
E Lvs broader, 1-3 cm L, 3-20 mm W — *Cercocarpus* p. 175
DD Plants shrubby or matted, lvs ± evenly spaced on branches
E Plants matted; lvs ± 3-nerved, entire — *Petrophytum* p. 179
EE Plants erect; lvs 1-veined from base, serrate
F Fls rose; lvs glabrous at least above — *Spiraea* p. 184
FF Fls whitish; lvs pubescent above — *Holodiscus* p. 177
AA Lvs pinnately cmpd (occasionally palmately so in *Potentilla*)
B Plants shrubby
C Foliage aromatic; low shrubs; lvs 2-3-pinnate, fern-like
D Lvs mly 2-pinnate; pistils 5; fr a follicle; e. SN — *Chamaebatiaria* p. 176
DD Lvs mly 3-pinnate; pistil 1; fr an akene — *Chamaebatia* p. 175
CC Foliage not aromatic; lvs 1-pinnate
D Stems with prickles
E Fr a cluster of drupelets; fls white — *Rubus* p. 182
EE Fr an akene; fls pinkish — *Rosa* p. 181
DD Stems lacking prickles
E Lfts 3-7, entire; infl 1- to few-fld — *Potentilla fruticosa* p. 179
EE Lfts 7-13, serrate; infl many-fld — *Sorbus* p. 183
B Plants herbaceous
C Lfts 3, not lobed (see also *Potentilla*)
D Lfts with 3 teeth; petals yellow — *Sibbaldia* p. 183
DD Lfts with many teeth; petals white — *Fragaria* p. 176
CC Lfts 5 to many, usu lobed
D Ultimate lfts 2-3 mm L, often with linear segments
E Stems leafy; lfts not especially crowded; leafy stipules evident; sepals 4 — *Sanguisorba* p. 183
EE Lvs mly basal, ± cylindric; lfts usu less than 0.5 cm L, crowded; sepals usu 5
F Petals white; herbage ± green; infl open — *Horkeliella* p. 177
FF Petals usu yellow; if not, herbage silvery-pubescent or infl capitate *Ivesia* p. 178
DD Ultimate lfts larger; lvs usu flat;
E Infl 20-40 cm L, narrow, spike-like, the pedicels recurved in fr — *Agrimonia* p. 174
EE Infl much shorter, not spike-like
F Styles many, persistent; petals 4-12 mm L; infl 1- to few-fld — *Geum* p. 176
FF Styles deciduous; infl usu several to many-fld
G Petals mly less than 5 mm L, mly white, inconspicuous — *Horkelia* p. 177
GG Petals mly over 5 mm L, usu yellow and conspicuous — *Potentilla* p. 179

Adenostoma

Chamise
Greasewood

A. fasciculatum: Much-branched shrub 0.5-3.5 m H, with well developed basal burl; bark reddish, becoming shreddy with age; lvs glabrous, rigid, resinous; panicles 4-12 cm L, terminal. Common dominant on dry slopes and ridges below 5000', foothills of SN, May -Jun. Seeds edible but tedious to collect.

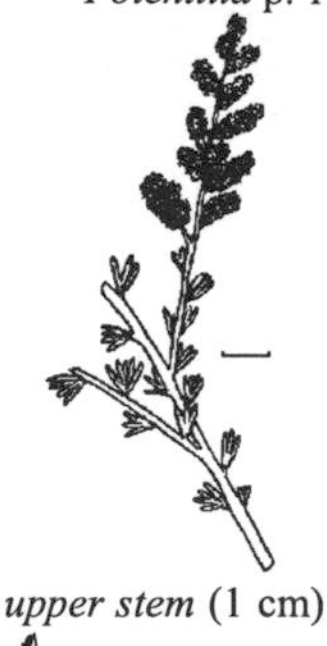

upper stem (1 cm)

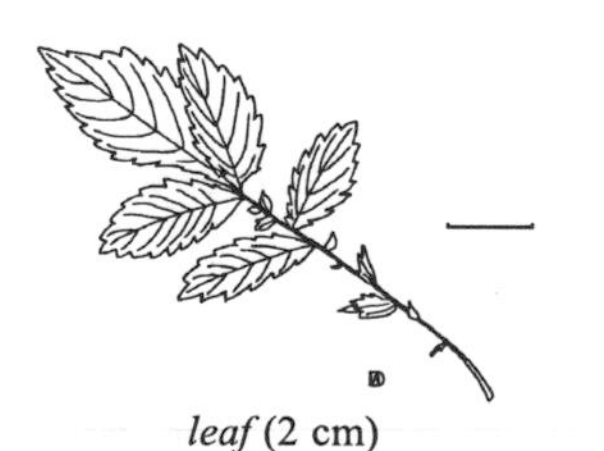

leaf (2 cm)

Agrimonia

Common
Agrimony

inflorescence (1 cm)

A. gryposepala: Glandular perennial; stems 3-12 dm H, hairy and glandular; lvs well-spaced, 10-25 cm L, pinnately cmpd into 5-9 lfts; lfts evenly serrate; infl a raceme 2-4 dm L, pedicels 2-10 mm L; sepals 1-4 mm L; petals 2-4 m L, yellow. Plumas Co. Jul-Aug. Previously used to treat eye disease.

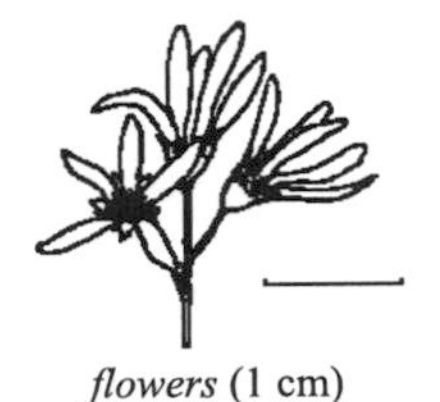

flowers (1 cm)

Amelanchier

Service-berry

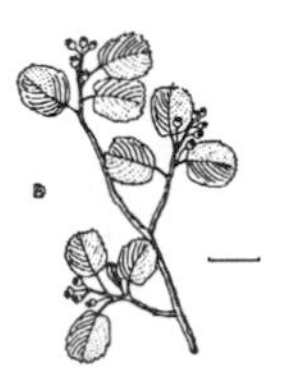

branchlet (3 cm)

Deciduous shrubs with slender branchlets. Stamens 10-20, short. Ovary inferior. Both spp have edible berries which may be eaten raw, cooked or dried. The berries ripen in late spring and throughout the summer.

Petioles and lvs glabrous *A. alnifolia*
Petioles and usu lvs somewhat pubescent *A. utahensis*

A. alnifolia, Smooth Service-berry: Stems 1-5 m H, glabrous throughout; lvs 1.5-3 cm L, oval to roundish, dark green above, light green below; petioles 6-10 mm L; racemes 4-8-fld, 2-4 cm L; sepals lanceolate, 3 mm L; petals 8-12 mm L; mature frs dark purple, 8-9 mm in diameter. Damp woods, 4500-8000', Eldorado to Nevada Co., May-Jun.

A. utahensis, Utah Service-berry: Stems 1-4 m H, glabrous to white-woolly; lvs 1-4 cm L, ovate; racemes 1-4 cm L, 3-6-fld; sepals lanceolate, 2-3 mm L; petals ovate, 5-10 mm L; fr purplish black, 5-6 mm across. Rocky slopes and forests to 11,000', SN (all), Apr-Jun.

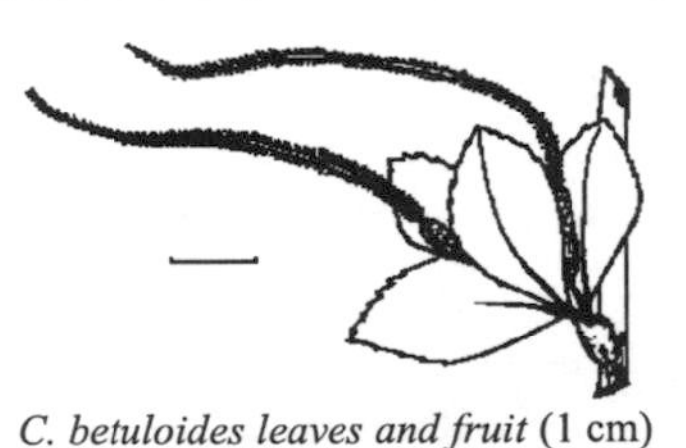

C. betuloides leaves and fruit (1 cm)

Cercocarpus

Mountain-mahogany

C. ledifolius branchlet (1 cm)

Evergreen shrubs or low trees with straight-veined lvs. Akene with a terminal elongate silk-plumose style. The wood is very hard and makes excellent coals. Herbage may contain cyanogenic substances, though neither of these sp has been reported as toxic.

Lvs toothed on distal half, not strongly revolute or resinous *C. betuloides*
Lvs entire, revolute, resinous *C. ledifolius*

C. betuloides, California Mountain-mahogany: Stems 2-7 m H; bark smooth gray; lf-blades obovate to oval, mly 1-2.5 cm L, wedge-shaped and entire below middle, dark green and glabrous on upper surface, paler and somewhat pubescent beneath; petioles 3-6 mm L; fls mly in clusters of 2-3; styles 4-9 cm L in fr. Dry slopes and flats below 6000', SN (all), Apr-May.

C. ledifolius, Curl-leaved Mountain-mahogany: Stems 2-9 m H, with red-brown furrowed bark; lvs lance-elliptic, 1-3 cm L, acute, short-petioled; fls 1-3, 4-5 mm W; tails of fr 4-7 cm L. Dry, rocky slopes 4000-10,500', SN (all), Apr-May.

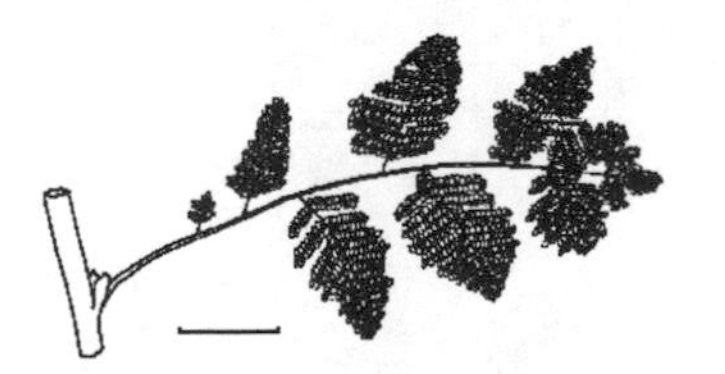

leaf (1 cm)

Chamaebatia

Mountain Misery

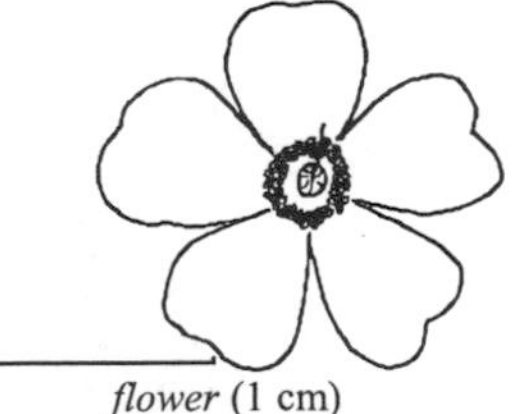

flower (1 cm)

C. foliolosa: Evergreen shrub, 2-6 dm H, with many leafy branches; lvs 2-10 cm L, sticky, with ultimate divisions minute; sepals about 4 mm L; petals white, 6-8 mm L. Open forests, 2000-7000', SN (all), May-Jul. Name comes from the plant's sticky black gum that gets on all clothing; resinous lvs make it a fire hazard.

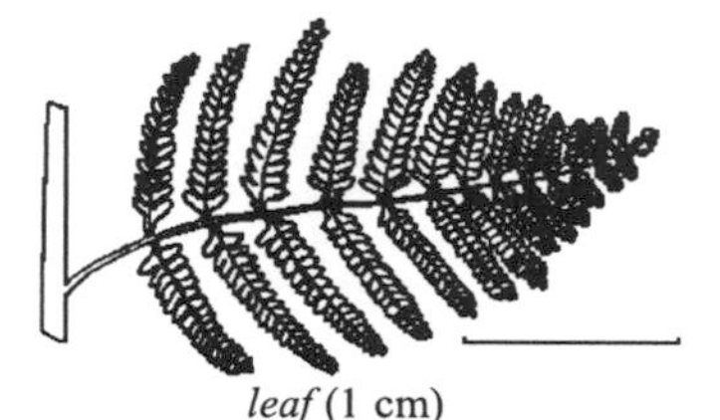
leaf (1 cm)

Chamaebatiaria

Fern Bush
Desert Sweet

follicles in flower tube (1 cm)

C. millifolium: Stems stout, densely branched, 0.6-2 m H; young growth and herbage ± glandular; lvs 2-4 cm L, ultimate segments about 1 mm L; panicles 3-10 cm L, heavily glandular; sepals 3-5 mm L; petals white, 5 mm L; follicles 5 mm L. Dry, rocky slopes, 3400-10,200', Jun-Aug. Plant probably inedible.

tufts of leaves connected by runners (1 cm)

Fragaria

Strawberry

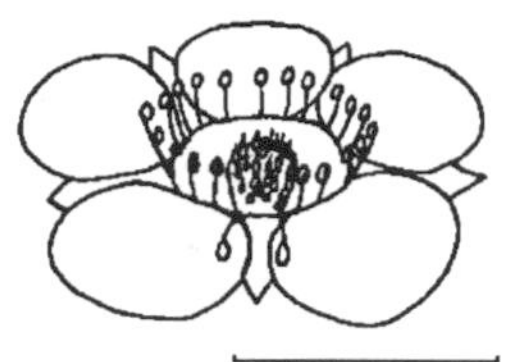
flower (1 cm)

Plants with conspicuous reddish runners which root at nodes. Lvs and fls in basal tufts. Stipules membranous. Fls white to pinkish, borne in clusters on a naked stalk, 5-merous. The enlarged receptacle or "strawberry" is edible in both spp. The green lvs also may be used to make a tea.

Lvs densely silky below; lfts subsessile — *F. vesca*
Lvs slightly silky to glabrate; lfts slightly petioluled — *F. virginiana*

F. vesca, California Strawberry: Lvs rather few; petioles slender, 3-12 cm L, sparingly villous; terminal lfts 2-5 cm L, obtuse, coarsely serrate; lateral similar, shorter; peduncles usu several, slender, few-fld, hairy; petals 5-8 mm L, not much longer than sepals; fr to 1 or 1.5 cm thick. Shaded fairly damp places below 7000', SN (all), Apr-Jun.

F. virginiana, Broad-petaled Strawberry: Lfts rather firm, 2-8 cm L, coarsely serrate above the middle; petioles 2-20 cm L, silky-hairy; peduncles 5-10 cm H, several-fld, often leafy-bracteate; petals 6-10 mm L, longer than sepals; fr 1-1.5 cm in diameter. Damp banks and woods, 4000-10,000', SN (all), May-Jul.

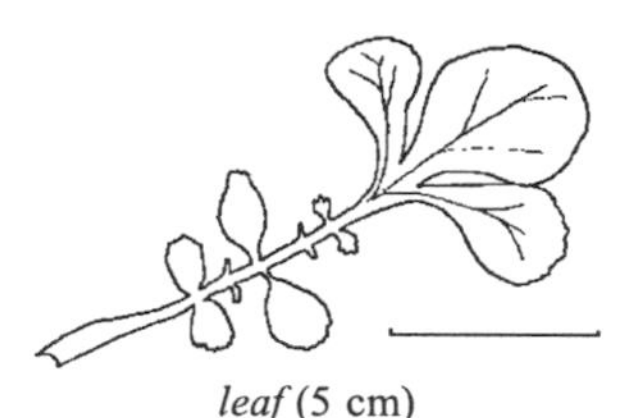
leaf (5 cm)

Geum

Avens

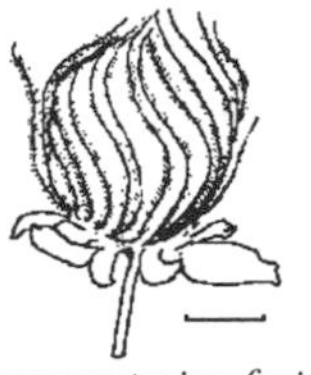
G. canescens maturing fruits (1 cm)

Herbage pubescent. Stipules adnate to petioles which are dilated and sheathing at base. Fl-tube persistent, usu with 5 bractlets. Stamens many. Pistils many. Styles filiform, elongate and conspicuous in fr. The roots of *G. ciliatum* and a related sp may be boiled to make a tea.

Terminal lft of basal lvs much larger than others and only shallowly lobed or incised; petals 4-8 mm L — *G. macrophyllum*
Terminal lft of basal lvs not much larger, deeply lobed; petals 8-12 mm L — *G. triflorum*

G. macrophyllum, Large-leaved Avens: Stems erect, 3-10 dm H; stipules broad, leafy; basal lvs 1-4 dm L including petioles; middle stem lvs sessile or broad, leafy; basal lvs 1-4 dm L including petiole; middle stem lvs sessile or short-petioled, mly with 3 lfts; infl open; sepals 3-5 mm L; petals yellow, 4-8 mm L; akenes puberulent below, bristly above, the persistent part of the style hooked, not plumose, 6-8 mm L. Moist places, 3500-10,500', SN (all) May-Aug.

G. triflorum, Old Man's Whiskers: Stems erect, 2-4 dm H; basal lvs 8-15 cm L including the petiole; lfts ± wedge-shaped, 3-5-lobed; stem lvs few, reduced; infl mly 3-fld; bractlets linear or lanceolate, 8-20 mm L; sepals about 1 cm L, purple-tinged; petals slightly longer than sepals. Styles plumose, 2-3 cm L in fr. Dryish to moist slopes and flats, 8500-11,000', Alpine Co. **n.**

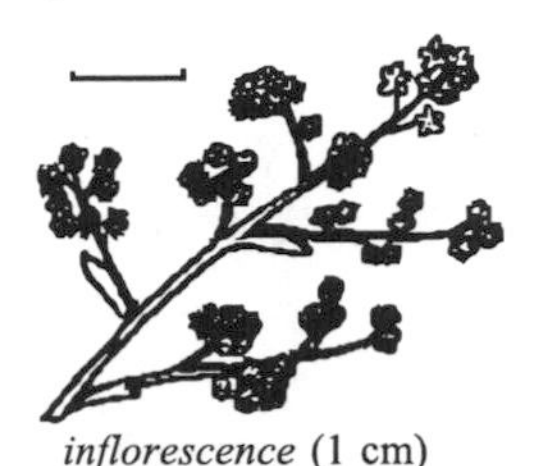

inflorescence (1 cm)

Holodiscus

Cream Bush

H. discolor leaves (1 cm)

Lvs without stipules. Infl terminal, villous. Fls whitish to pinkish, small, 5-merous. Frs of a related sp edible raw or cooked.

Lvs toothed along sides to below middle; petioles 2-3 mm L *H. discolor*
Lvs toothed at top, rarely to middle; petioles 1-2 mm L *H. microphyllus*

H. discolor, Ocean Spray: Stems to 1 m H, compact; young twigs often angled, pubescent; lf-blades broadly obovate to roundish, the teeth broad, rounded blades green or gray-green above, grayish beneath, often glandular, 1-3 cm L; infl 2.5-8 cm L; sepals 1-2 mm L; petals 2 mm L. Dry, rocky slopes, 4000-9600', SN (all), Jun-Aug.

H. microphyllus, Small-leaved Cream Bush: Stems spreading, bushy, 0.2-2 m H; lvs obovate to spatulate; infl 2.5-3.5 cm L, 1-3 cm W; sepals narrow, about 1 mm L; petals about 2 mm L. Dry, rocky places, 5500-11,000', Plumas Co. s., Jun-Aug.

Horkelia

Horkelia

Lfts toothed to divided, uppermost ± united. Fls white, rarely cream or pink, 5-merous. Styles glandular-thickened at base. Edibility unknown.

Lfts 5-10 pairs, toothed to lobed or incised *H. fusca*
Lfts 2-5 pairs, short-toothed at apex only or entire *H. tridentata*

upper stem (1 cm)

H. fusca sspp, Dusky Horkelia: Stems erect or ascending, 1-5 dm H; herbage pubescent; lvs various; lfts 5-10 pairs on basal lvs, fewer on cauline; infl usu dense, sometimes capitate. Moist to dry habitats, below 10,500', SN (all), May-Aug.

H. tridentata, Three-toothed Horkelia: Stems decumbent or ascending, sparingly leafy, usu purplish, 1.5-3 dm H; herbage silky or cobwebby-hairy, upper surface of lvs often glabrous; basal lvs 3-8 cm L; lfts not crowded 8-10 mm L, spatulate to broadly oval; infl compact; fl-tube 2-5 mm W; sepals 2-3 mm L; petals sometimes pinkish, narrow, about as long as sepals. Woods, 2000-6500', SN (all), May-Jul.

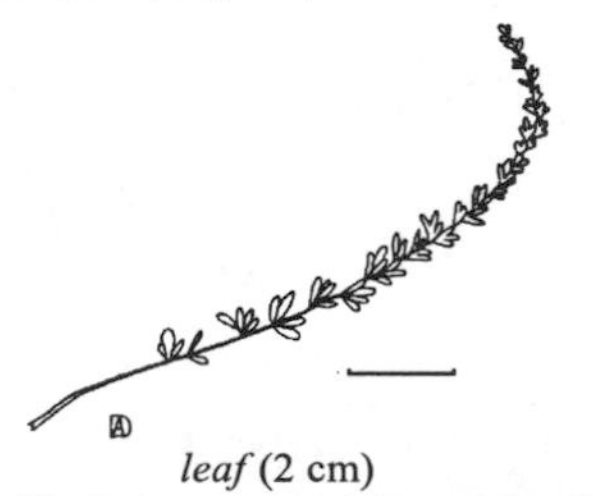

leaf (2 cm)

Horkeliella

Horkeliella

inflorescence (5 cm)

H. purpurascens: Perennial with erect stem, 2-6 dm H; herbage pubescent, ± green, usu glandular; lvs 7-20 cm L; lfts 15-30 pairs, 4-10 mm L, crowded; infl few-fld, narrow, often subcapitate; pedicels stout; fl-tube often purplish, 3-5 mm W; sepals lanceolate, about 4 mm L; petals oblong, about 5 mm L; stamens 10; pistils 25-50. Mdw borders and stream banks, 6000-9000', Tulare to Mono and Inyo cos., Jul-Aug.

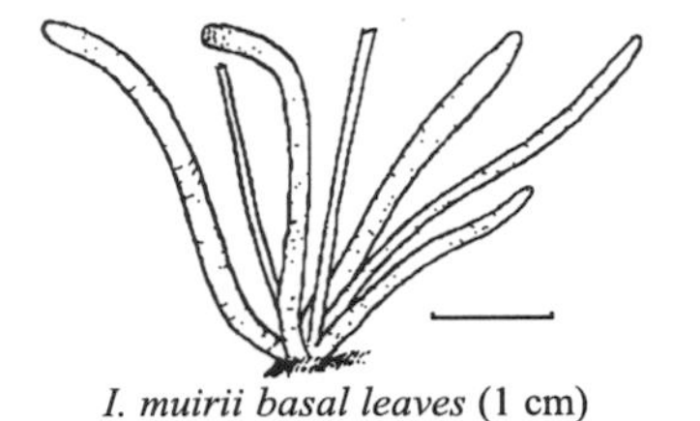

I. muirii basal leaves (1 cm)

Ivesia

Ivesia

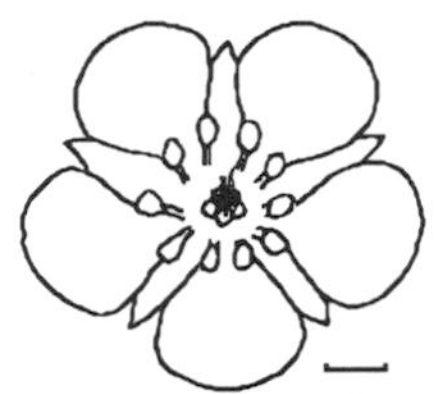

I. campestris flower (1 mm)

Lvs mly basal, narrow, and often ± cylindric. Lfts often divided to base into 3 or more segments. Fls yellow, white or purple, mly crowded, usu 5-merous. Edibility unknown.

A Stamens 5
  B Lfts 7-10 pairs; infl open, few-fld; petals pale yellow — 7) *I. shockleyi*
  BB Lfts 10-40 pairs; infl usu dense, many-fld; petals yellow
    C Herbage densely silvery-silky; lvs worm-like — 4) *I. muirii*
    CC Herbage pubescent to glabrate; lvs flat to cylindric
      D Pistils 1-6; petals ± linear; inconspicuous — 2) *I gordonii*
      DD Pistils 8-18; petals obovate to rounded conspicuous — 3) *I. lycopodioides*
AA Stamens 10 or more
  B Petals yellow; Fresno, Inyo and Tulare cos.
    C Fls usu 4-merous; stamens 15-20; stems leafy, 15-35 cm L — 1) *I. campestris*
    CC Fls 5-merous; stamens 10; stems leafless, 4-12 cm L — 5) *I. pygmaea*
  BB Petals white; plants widespread; lfts sometimes to 1 cm L
    C Infl ± capitate, dense; stems leafy — 8) *I. unguiculata*
    CC Infl usu open; lvs mly basal — 6) *I. santolinoides*

1) *I. campestris*, Field Ivesia: Stems decumbent to ascending, about 1 dm L; herbage silky-villous, greenish; basal lvs 5-12 cm L; lfts 15-20 pairs, crowded, mly 3-5 mm L; infl dense; fl-tube about 3 mm W; sepals 2-3.5 mm L; petals oblanceolate, somewhat longer. Mdws, 6500-11,000', Tulare Co., Jun-Aug.

2) *I. gordonii*, Gordon's Ivesia: Stems ± leafless, 0.5-2 dm L; lvs 3-18 cm L; lfts 10-25 pairs, 2-8 mm L; infl ± capitate, many-fld; fl-tube about 3 mm L, yellowish; sepals deltoid-lanceolate, yellow, erect, 2-4 mm L; petals yellow, usu as long as sepals. Dry, rocky places, 7500-13,000', Tuolumne and Mono cos. n., Jul-Aug.

3) *I. lycopodioides*, Club-moss Ivesia: Stems to 2.5 dm L; herbage glabrous to sparsely hairy, often glandular; lvs 2-6 cm L; lfts thick, 1-2 mm L; infl ± capitate; fl-tube about 2 mm W; sepals deltoid-lanceolate, about as long as petals; petals yellow, 2- 3 mm L. Moist places, 7500-12,500', Eldorado Co. s., Jul-Aug.

4) *I. muirii*, Muir's Ivesia: Stems slender, 1-2-bractate, purplish, 0.5-1.5 dm L; lvs basal, many, 2-6 cm L; lfts 25-40 pairs, densely imbricated, about 1 mm L; fls many, in congested clusters; fl-tube 2 mm W; sepals deltoid, about 2 mm L; petals linear, shorter; pistils 1-4. Gravelly slopes, 9500-12,000', Fresno to Tuolumne and Mono cos., Jul-Aug.

5) *I. pygmaea*, Dwarf Ivesia: Stems 0.4-1.2 dm L; herbage ± puberulent, densely glandular; lvs 2-8 cm L; lfts 10-20 pairs, crowded, 1-5 mm L; fl-tube about 3 mm W; sepals deltoid-lanceolate, shorter than petals; petals broadly spatulate to broadly obovate, 2-4 mm L; pistils 12-30. Rocky slopes, 9500-13,000', Jul-Aug.

6) *I. santolinoides*, Mouse-tail Ivesia: Stems erect, slender, 1-4 dm H, diffusely branched above, subglabrous; lvs worm-like, 3-10 cm L, stem lvs 1-3; lfts very numerous, minute; infl many-fld, diffuse; sepals less than 2 mm L; petals obovate to roundish, slightly longer than sepals. Dry, gravelly flats and ridges, 5000-12,000', Eldorado Co. s., Jun-Aug.

7) *I. shockleyi*, Shockley's Ivesia: Stems 0.3-1 dm L; herbage green, densely glandular-pubescent; lvs 2-7 cm L; lfts ± crowded, 1-3 mm L; infl open, few-fld, with filiform pedicels; fl-tube about 3 mm W; sepals 2-4 mm L; petals oblanceolate to oval, shorter than sepals; pistils usu 3. Gravelly and rocky places, 9000-13,000', Inyo to Placer Co., Jul-Aug.

8) *I. unguiculata*, Yosemite Ivesia: Stems 20-35 dm L; herbage silky-villous but green; basal lvs 8-14 cm L; lfts 14-25 pairs, crowded, 3-6 mm L; infl branched, the fls in dense clusters; fl-tube purplish, 2-3 mm W; sepals broadly lanceolate, 2-4 mm L; petals oblanceolate, about as long; pistils 3-9. Open slopes, 5000-8000', Fresno to Mariposa Co., Jun-Aug.

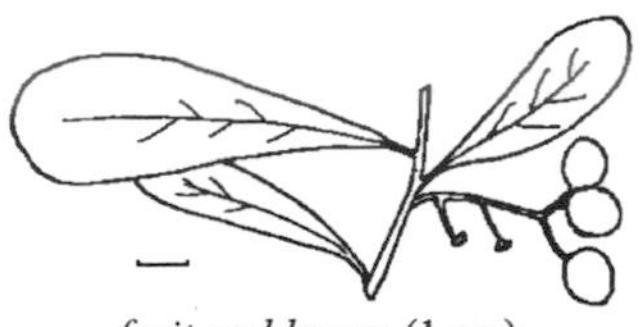
fruit and leaves (1 cm)

Oemleria

Oso Berry

branchlet cut longitudinally to show inner cross walls (1 cm)

Oe. cerasiformis: Shrub or tree 1-5 m H; lvs 5-10 cm L, deciduous, short-petioled, with small stipules; raceme 3-10 cm L; drupe about 1 cm L, black, glaucous. Canyons below 5600', SN (all), Apr-May. The berry is edible raw or cooked, although the taste is not appealing.

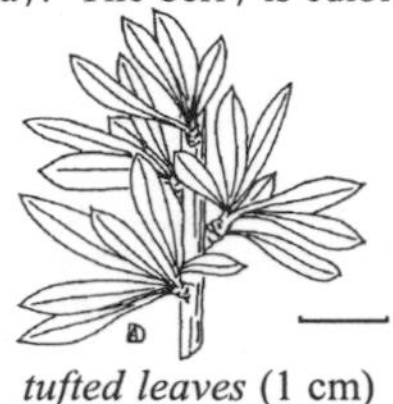
tufted leaves (1 cm)

Peraphyllum

Squaw-apple

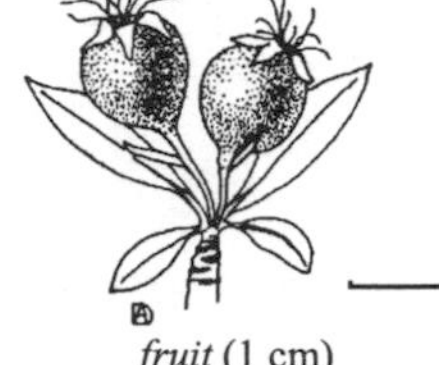
fruit (1 cm)

P. ramosissimum: Low shrub with grayish bark, 1-2 m H; lvs crowded at ends of spurlike branchlets, oblanceolate, entire or serrulate, 2-4 cm L; fls 1-3 on short spur-like branchlets; sepals about 3 mm L, reflexed, persistent; petals round, 7-8 mm L; fr a pome, yellowish, 8-10 mm in diameter, bitter. Dry slopes, 4000-8000', mly e. SN, Mono Co. n.

Petrophytum

Rock-spiraea

P. caespitosum ssp *acuminatum*: Depressed undershrub with prostrate branches forming dense mats 3-8 dm W, densely silky-pubescent; lvs in tufts, persistent, elliptic to oblong, 10-18 mm L; fls in dense elongate clusters; peduncles 3-10 cm H, bearing a spike 2-4 cm L. On limestone cliffs, 4000-7600', Tulare and Fresno cos., Aug-Sep. Edibility unknown.

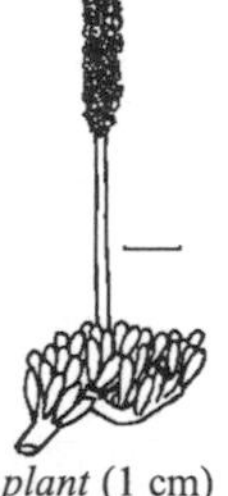
plant (1 cm)

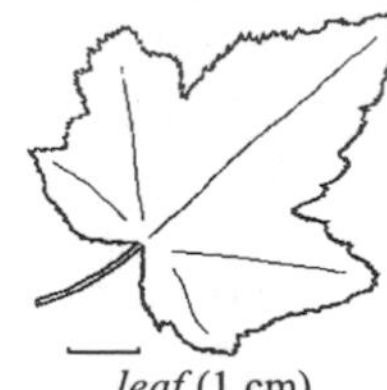
leaf (1 cm)

Physocarpus

Ninebark

inflorescence

P. capitatus: Deciduous shrubs, erect or spreading, 1-2.5 m H, with shredding bark; lvs 3-5-lobed, the middle lobe conspicuously longer, the lobes serrate, ± glabrous above, usu stellate-pubescent beneath, 3-7 cm L; petioles 1-2 cm L; fls in terminal clusters; pedicels and fl-tube densely stellate-pubescent; sepals and petals about 3 mm L. Moist banks and north slopes below 4500', SN (all), May-Jul. Edibility unknown.

flowers and fruit (1 cm)

Potentilla

Cinquefoil
Five-finger

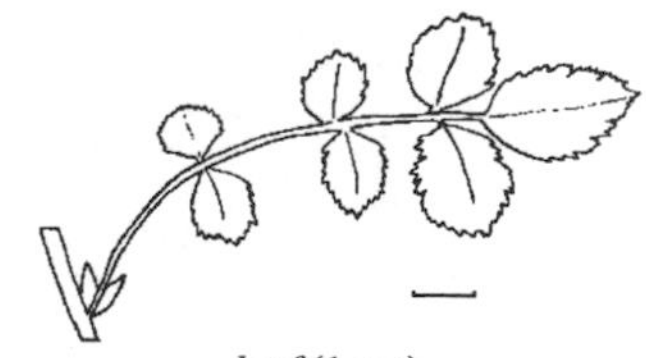
leaf (1 cm)

Lvs pinnately or digitately cmpd. Fls perfect, 5-merous. Fl-tube with 5 bractlets. Sepals ovate to lanceolate. Petals usu yellow, sometimes white or purple; usu in bloom from Jun to Sep. Intermediates and hybrids occur between several spp. The roots of *P. anserina* are edible after cooking. Tea can be made from lvs of *P. fruticosa*.

A Plants shrubs with woody branches 5) *P. fruticosa*
AA Plants herbaceous
B Basal lvs with 3 lfts
C Stems stout, 2-7 dm H; style torpedo-shaped 9) *P. norvegica*
CC Stems slender, 1-3 dm H; style thread-like 4) *P. flabellifolia*
BB Basal lvs with 5 to many lfts
C Basal lvs palmately cmpd; herbage mly silky-hairy
D Lfts dentate only near tips
E Lfts silky-hairy on both sides; stems hairy; Tulare Co. 13) *P. wheeleri*
EE Lfts hairy to glabrate; stems ± glabrous; widespread 2) *P. diversifolia*
DD Lfts serrate or lobed to base
E Lfts divided ± to midrib; stems to 1.5 dm H; above 10,500 12) *P. pseudosericea*
EE Lfts coarsely toothed; stem mly 4-7 dm H; below 11,000 7) *P. gracilis*
CC Basal lvs pinnately cmpd; pubescence various
D Lfts usu more than 12; usu e. SN
E Lfts serrate, usu whitish beneath 1) *P. anserina*
EE Lfts divided into ± linear divisions, green on both sides 8) *P. millefolia*
DD Lfts 5-12
E Style terminal and thread-like 3) *P. drummondii*
EE Style lateral or nearly basal; or if terminal then torpedo-shaped
F Fls large, purple; styles filiform 10) *P. palustris*
FF Fls yellow to white; styles fusiform
G Lfts deeply pinnately lobed, margins revolute; Inyo Co. 11) *P. pensylvanica*
GG Lfts serrate, not revolute 6) *P. glandulosa*

1) *P. anserina*, Silverweed: Stolons present; lvs 1-2 dm L, in basal rosettes; lfts 9-31, 1-4 cm L, ± oblong, green and subglabrous above, white silky-tomentose beneath; petioles 1-5 cm L; peduncles solitary, 1-fld; sepals 3-5 mm L; petals oval, 7-10 mm L. Moist, ± alkaline places, 4000-8000', e. SN (all).

2) *P. diversifolia*, Diverse-leaved Cinquefoil: Stems slender, 1-3-leaved; basal lvs digitate or crowded-pinnate; lfts 5-7; bractlets lanceolate, acute, 3-5 mm L; sepals acuminate, 5-6 mm L; petals obcordate, 6-10 mm L. Moist, rocky places, 8000-11,600', SN (all).

3) *P. drummondii*, Drummond's Cinquefoil: Stems decumbent or erect, 1-6 dm H, puberulent, few leaved, branched above; stipules about 2 cm L; lvs ovate-oblong in outline; lfts 2-6 pairs, crowded to moderately spaced, 2-6 cm L, deeply and sharply serrate or lobed; fls long-pedicelled; fl-tube hairy, 7-9 mm W; sepals 4-7 mm L; petals obcordate, usu exceeding sepals. Moist habitats, 4500-13,000', SN (all).

4) *P. flabellifolia*, Mount Rainier Cinquefoil: Stems 1-3 dm L, erect or ascending; lvs few, mly basal, on petioles 2-12 cm L; the lfts subsessile, sometimes the terminal one long-petiolulate, fan-shaped, 1-3.5 cm L; infl few-fld; sepals 5-6 mm L; petals obovate, deeply notched, 8-10 mm L. Moist places 5800-12,000', SN (all).

5) *P. fruticosa*, Bush Cinquefoil: Much-branched shrub, 2-12 dm H; lvs ± pinnate, to about 2 cm L; lfts 3-7, linear-oblong, entire, 0.5-2 cm L, revolute; fls 1 or few in small clusters; sepals 5-6 mm L; petals round, 5-15 mm L. Moist places, 6400-12,000', SN (all).

6) *P. glandulosa,* Sticky Cinquefoil: Stems erect, 1-8 dm H, leafy, glandular-villous, branching above, often reddish; lfts 5-9, mly 1-4 cm L; fl-tube usu 4-8 mm W; bractlets usu linear; sepals and petals various. Dryish to moist places, below 12,400', SN (all).

7) *P. gracilis* ssp *nuttallii*, Slender Cinquefoil: Stems slender, 4-7 dm H, erect or ascending; lfts 5-7, oblanceolate, 3-6 cm L; infl many-fld; sepals 6-7 mm L; petals obcordate, 8-11 mm L. Moist places, 2500-11,000', SN (all) .

8) *P. millefolia*, Many-leaved Cinquefoil: Stems several, 1-1.5 dm L, slender, spreading to prostrate; basal lvs oblong in outline, somewhat shorter or equal to stems; stem lvs few; infl open, few-fld; pedicels slender, often reflexed in fr; sepals about 5 mm L; petals obcordate, longer than sepals. Damp grassy places 2500-6000', Nevada Co. n.

9) *P. norvegica* ssp *monspeliensis*, Rough Cinquefoil: Stems erect or ascending, branched above, with stiff mly spreading hairs; lfts oblanceolate to obovate, coarsely serrate, 3-5 cm L; infl leafy; sepals 4-5 mm L, petals obovate, 3-5 mm L. Moist places, 4500-7500', scattered in SN.

10) *P. palustris,* Purple Cinquefoil: Stems stout, 1-6 dm H, glabrous below, soft-hairy and glandular above; lfts sharply serrate, 2-6 cm L; infl few-fld; sepals 10-15 mm L, purplish; petals wine-purple, about half as long as sepals. Swamps and bogs below 8000', Eldorado Co. n.

11) *P. pensylvanica* var. *strigosa*, Pennsylvania Cinquefoil: Stems several, ascending, 1-4 dm H, densely puberulent and with longer hairs; petioles 3-7 cm L; lfts 7-11, 1-5 dm L, paler beneath; infl dense; sepals 5-7 mm L; petals roundish, equal to sepals. Moist habitats, 9000-12,000'.

12) *P. pseudosericea*, Strigose Cinquefoil: Stems to 1.5 dm L, few-leaved, ascending; herbage grayish silky; lvs mly basal, congested; lfts 5-9, 1-2 cm L infl few-fld; sepals 4-5 mm L; petals obovate. Dry, rocky places, 10,500-13,000', Inyo and Tuolumne cos. n.

13) *P. wheeleri*, Wheeler's Cinquefoil: Stems spreading to prostrate, 0.5-2 dm L; basal lvs many, 1-8 cm L; lfts 5, 1-2.5 cm L; stem lvs few; infl open; sepals 2-3 mm L; petals obcordate. Edge of mdws, 6500-11,500'.

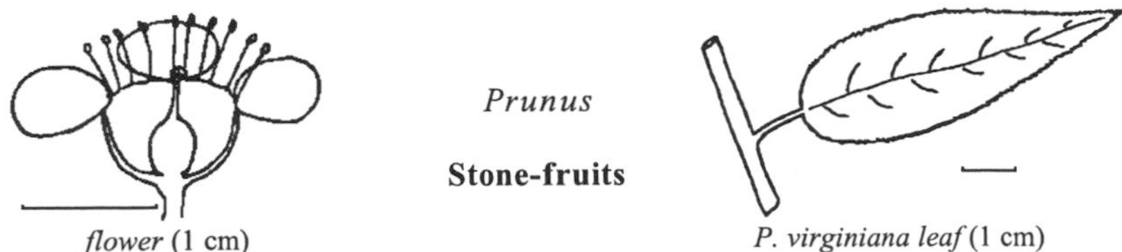

Prunus

Stone-fruits

flower (1 cm) *P. virginiana leaf* (1 cm)

Trees and shrubs with deciduous finely serrate lvs. Stipules small, deciduous. Fls 5-merous. Pistil 1. Fr edible (see individual sp). Lvs of most spp poisonous.

Branches with spinose branchlets — *P. subcordata*
Branches without spinose branchlets
  Young twigs shiny red; lvs 2-5 cm L; petioles short — *P. emarginata*
  Young twigs gray-brown; lvs 3-8 cm L; petioles 1 cm L — *P. virginiana*

P. emarginata, Bitter Cherry: Stems 1-6 m H, older bark smooth; lvs oblong-obovate, usu rounded at tip; infl 3-10-fld; sepals about 2 mm L, oblong; petals 5-7 mm L, obovate; drupe red, 6-8 mm in diameter. Rocky ridges to dampish slopes and canyons below 9000', SN (all), Apr-May. Fr bitter when red; juicy but insipid when dark.

P. subcordata, Sierra Plum: Stems 1-3(-6) m H, with stiff crooked branches; lvs elliptic to roundish, obtuse, 2-5 cm L; petioles 6-17 mm L; fls 2-4, white, pink in age; sepals 4-5 mm L; petals obovate, 4-6 mm L; fr 1.5-2 cm L, red-purple. Dry, rocky or moist slopes below 6000', SN (all), Apr-May. Fr is good eating although the amount of pulp varies with habitat.

P. virginiana var. *demissa,* Western Choke Cherry: Stems 1-5 m H; lvs oblong-ovate, usu long-pointed at tip, ± pubescent beneath, ± subcordate at base; infl many-fld, 5-10 cm L, at ends of short leafy branches; sepals obtuse, about 1 mm L; petals white, rounded, 5-6 mm W; fr round, 5-6 mm thick, dark red. Dampish places in woods and on brushy slopes and flats below 8200', SN (all), May-Jun. The immature fr and pits contain cyanogenic substances and should be avoided. The mature fruit is bitter but edible. Cooking removes the bitterness and the volatile cyanide located in the pits.

Purshia

Antelope Bush

fruit (1 mm) *P. virginiana leaf* (1 cm)

P. tridentata: Grayish shrub 1-3 m H; young twigs ± glandular and pubescent; lvs 3-toothed at apex, crowded, appearing fascicled, usu with revolute margins, white-woolly beneath; fls solitary at ends of short branches; sepals about 3 mm L; petals 7-8 mm L, cream to yellow. Dry slopes, 3000-10,400', mly e. SN (all), May-Jul. Fleshy covering of akene extremely bitter.

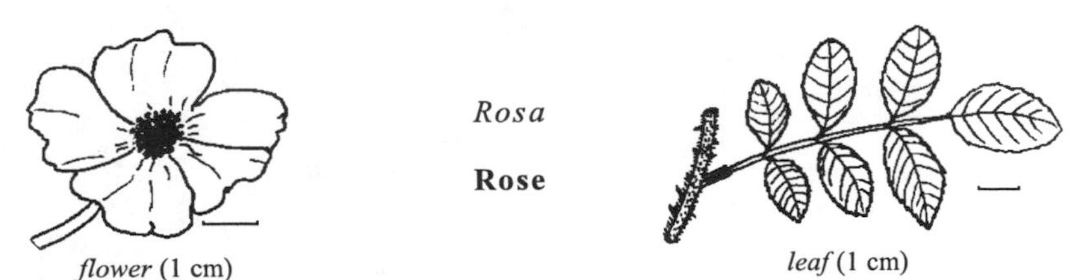

Rosa

Rose

flower (1 cm) *leaf* (1 cm)

Prickly shrubs with odd-pinnate lvs and stipules adnate to petiole. Fls many, rose-pink, 5-merous. Stamens many. Fr a fleshy hip. The hip may be eaten raw or cooked with some sugar added to improve flavor. The hips are an excellent source of vitamin C.

A Fls usu in corymbs; plants 1-3 m H; lvs usu once-serrate, the teeth not gland-tipped
B Stems armed with stout flattened recurved prickles; pedicels villous 1) *R. californica*
BB Stems armed with ± straight weak slender prickles; pedicels glabrous 5) *R. woodsii*
AA Fls usu solitary; plant usu less than 1 m H; lvs usu doubly serrate with gland-tipped teeth
B Plant 1-3 dm H; prickles rather stout, about 1 cm L 4) *R. spithamea*
BB Plant 5-10-(-30) dm H; prickles slender, usu shorter
C Sepals and styles deciduous in fr 2) *R. gymnocarpa*
CC Sepals and styles persistent in fr 3) *R. pinetorum*

1) *R. californica*, California Wild Rose: Stems erect, branched, usu purple; lfts 3-7, oval, 1-3.5 cm L, green above, white-hairy beneath; stipules narrow; sepals lanceolate, long-tipped; petals 1-2.5 cm L; fr globose, 8-16 mm across. Fairly moist places below 6000', SN (all), May-Aug.

2) *R. gymnocarpa*, Wood Rose: Stems slender; lfts 5-7, oval to roundish, 1-4 cm L, ± glabrous on both surfaces; stipules narrow, usu dentate; sepals ovate, glabrous on back, about 1 cm L; petals 8-12 mm L; fr red, glabrous, 5-10 mm L. Shaded woods below 6000', Fresno Co. n., May-Jul.

3) *R. pinetorum*, Pine Rose: Stems erect; lfts usu 5, elliptic to broadly oval, obtuse, 1-3 cm L, usu puberulent and glandular beneath; stipules pubescent, often glandular; sepals mly glandular on backs; petals 10-14 mm L; fr 6-12 mm in diameter. Open woods, 2000-6500', SN (all), May-Jul.

4) *R. spithamea*, Ground Rose: Lfts 3-7, ± oval, 1-3 cm L, subglabrous and green above, paler and glandular beneath; pedicels, sepals and fl-tube usu glandular-hairy; petals obcordate, 12-20 mm L; fr 7-8 mm across. Open woods below 5000', Tulare to Yuba Co., Jun-Aug.

5) *R. woodsii* var. *ultramontana*, Mountain Rose: Stems erect; lfts 5-7, oval, 1-4 cm L, glabrous above, puberulent beneath; sepals glabrous on back, long-tipped; petals 1.5-2 cm L; fr 7-10 mm across. Damp places, 3500-11,500', e. SN (all), Jun-Aug. Odor particularly pleasant

var. *gratissima*: Stems more densely armed, even floral branches usu heavily spiny; lfts 1-2 cm L; fls 1 to few; petals 1.5 cm L. Dry, stony slopes, 5000-9000', Fresno and Mono cos. s., Apr-Aug.

R. parviflorus fruit (1 cm)

Rubus

Berry

R. leucodermis leaf (1 cm)

Shrubs or vines. Stems in their first year shooting up and sterile, in second year usu flowering and with different foliage. Stipules adnate to petiole. Fl white, carpels many, crowded on an elevated receptacle, becoming drupelets which coalesce and form a thimble-shaped fr. All berries edible raw or in jams or pies.

Plants without prickles; lvs simple, palmately lobed *R. parviflorus*
Plants with prickles; lvs pinnately cmpd
Prickles many; sepals deflexed at anthesis *R. leucodermis*
Prickles few; sepals not deflexed *R. glaucifolius*

R. glaucifolius, Glaucous-leaved Raspberry: Plants with erect flowering shoots from decumbent or prostrate runner-like main stems; lvs green and ± glabrous above, paler and puberulent beneath, lfts 3, the terminal 4-8 cm L, oval to oblong, coarsely and unequally dentate; fls mly few, in umbel-like clusters overtopped by lvs; sepals 6-8 mm L; petals white, about as long; fr with few drupelets, less than 1 cm H. Dry slopes, 3000-7400', SN (all), Jun-Jul.

R. leucodermis, Western Raspberry: Stems to 2 m L, arched and branched, rooting at tips, whitish at least when young; prickles 4-6 mm L; lvs much like *R. glaucifolius*; fls mly 3-10 in rather compact clusters on lateral shoots, 7-10 mm W; sepals 5-8 mm L, long-tipped, exceeding petals; petals narrow-elliptic; fr dark purple to blackish, or yellow-red, to about 1.5 cm in diameter. Slopes and canyons below 7000', SN (all), Apr-Jul.

R. parviflorus, Thimbleberry: Stems erect, mly 1-2 m H, the bark shreddy in age; lvs deciduous, puberulent, ± cordate at base; fls few in terminal clusters, mly white, 2-5 cm W; sepals 10-15 mm L; petals elliptic, 1.5-2 cm L; fr red, 1-1.5 cm W. Open woods and in canyons below 8000', SN (all), May-Aug.

Sanguisorba

Western Burnet

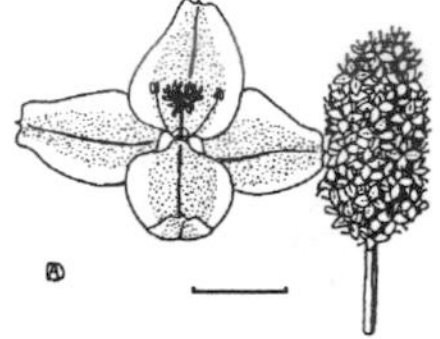

leaf (1 cm) *flower and inflorescence* (1 mm)

S. occidentalis: Glabrous annual with branching stems 1-4 dm H; stipules adherent to petiole, leafy, divided; lvs basal and cauline, lower cauline with 11-15 lfts, these 1-2 cm L, pinnatifid into linear segments; fls crowded in a dense spike on a long, scapose peduncle; spike 1-3 cm L, 10-50-fld; sepals usu 4, green, 2-3 mm L, ovate; petals none. Dry, open, often disturbed sites, 2500-6500', Nevada Co. n., May-Jul.

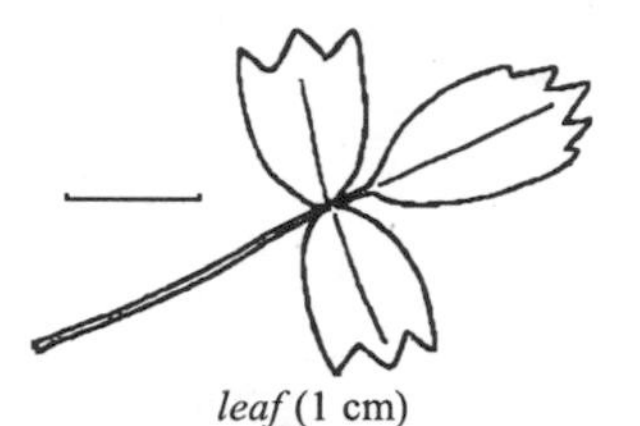

Sibbaldia

Sibbaldia

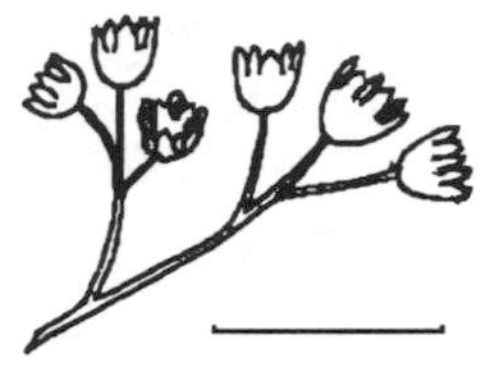

leaf (1 cm) *inflorescence* (1 cm)

S. procumbens: Low tufted perennial; petioles mly 1-4 cm L; stipules prominent; lfts of basal lvs obovate to oblanceolate, 1-2 cm L, 2-5-toothed at apex, green with silvery hairs; stem lvs smaller; fl-stems to about 8 cm H; infl 5-10-fld; sepals ± ovate, 2-3 mm L; petals narrow-elliptic, about 1.5 mm L. Dry, stony habitats, 6000-12,000', SN (all), Jun-Aug. Edibility unknown.

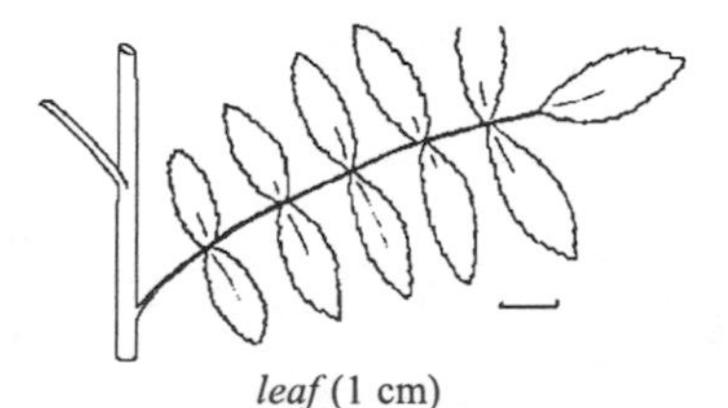

Sorbus

Mountain-ash

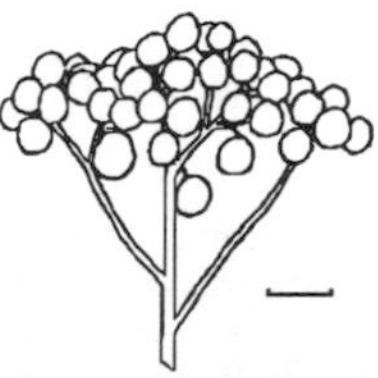

leaf (1 cm) *cluster of fruits* (1 cm)

Deciduous shrubs with alt, stipulate lvs. Fls white in a large terminal infl. Stamens 15-20. Pistils 1-5. The fr may be eaten raw, cooked or dried. It is best to wait until after the first few frosts to pick the fr for they are often quite bitter before then.

Lfts 7-11; petals 3-4 mm L; infl somewhat convex, 30-60-fld *S. californica*
Lfts 11-13; petals 5-6 mm L; infl flat-topped, 80-100-fld *S. scopulina*

S. californica, California Mountain-ash: Many-stemmed shrub, 1-2 m H; lfts oblong-oval, serrate, glabrous; sepals 2 mm L; petals round, 2-4 mm L; fr scarlet, ellipsoid to pear-shaped, 7-10 mm L. Moist, shady places, 5000-11,000', SN (all), May-Jun.

S. scopulina, Western Mountain-ash: Shrubby, 1-4 m H, with thick reddish bark; lateral lfts lanceolate, 3-6 cm L, finely and sharply singly or doubly serrate, glabrous; fls about 1 cm W; sepals hairy, 1.5 mm L; petals oval, 5-6 mm L; fr orange to scarlet, globose, 8-10 mm W. Occasional, in canyons and on wooded slopes, 4000-9000', SN (all), May-Jun.

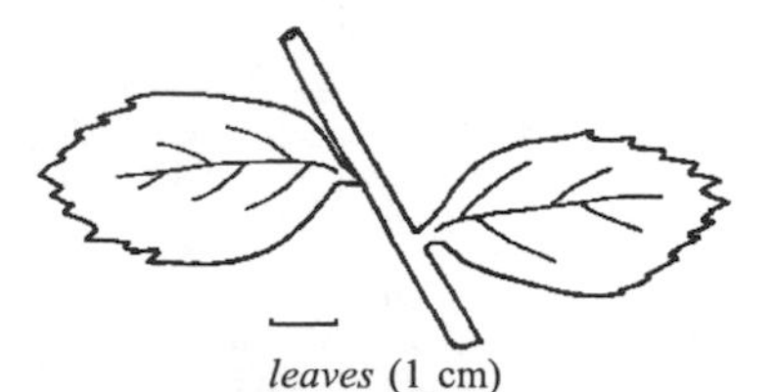

Spiraea

Spiraea

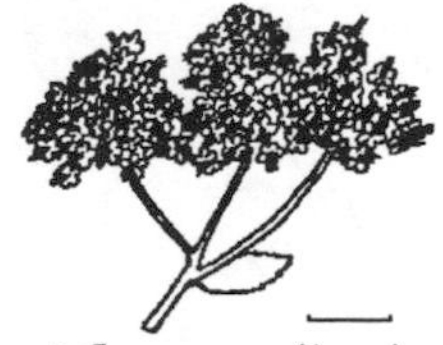

leaves (1 cm) *inflorescence* (1 cm)

Lvs deciduous, without stipules. Petals rose. Stamens 15 to many. A tea can be made from the lvs of a related sp.

Infl flat-topped to hemispheric; lvs ± oval, green beneath — *S. densiflora*
Infl elongate; lvs ± oblong, whitish beneath — *S. douglasii*

S. densiflora, Mountain Spiraea: Stems glabrous, ± ciliate, 2-9 dm H, young stems reddish aging gray; lvs 1.5-3 cm L, sharply serrate in upper 2/3; petioles to 3 mm L; infl 2-4 cm W; sepals 1 mm L; petals 1.5 mm L. Moist, rocky places, 5000-11,000', SN (all), Jul-Aug.

S. douglasii, Douglas Spiraea: Erect shrub, 1-2 m H; young growth puberulent; lvs serrate above the middle, 3-9 cm L; petioles 3-10 mm L; infl 1-1.5 dm L, congested; sepals to 1 mm L; petals 1.5 mm L. Moist habitats below 6400', Butte and Plumas Co. n., Jun-Sep.

RUBIACEAE - Madder Family

Lvs entire. Stamens as many as the corolla lobes and inserted on the tube. Calyx usu lacking. Fr of 2 indehiscent carpels separating when mature.

Lvs in whorls mly of 4 or more; stems 4-angled — *Galium*
Lvs opp; stems round — *Kelloggia*

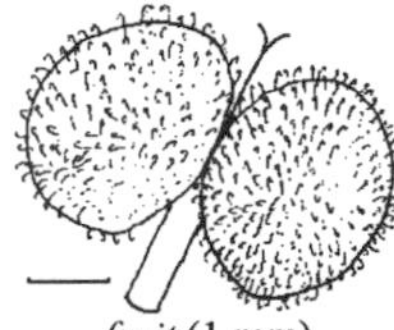
fruit (1 mm)

Galium

Bedstraw
Cleavers

whorl of leaves (1 cm)

Annual or perennial with slender stems and branches. Stems often with short stiff hairs on angles. Lvs in apparent whorls because of large leaf-like stipules. Plants often monoecious or dioecious. Fls small; sepals none. Corolla flat, 4-, rarely 3-parted. Stamens 4 or 3, short. Ovary 2-lobed; styles 2. The seeds of *G. aparine* and probably the other spp may be roasted and ground and used as a substitute for coffee. The young sprouts may be eaten in salads or as a potherb.

A Lvs mly 6-8 in a whorl (see also *G. trifidum*)
  B Plant annual; stems easily clinging to other vegetation — 1) *G. aparine*
  BB Plant perennial; stems not clinging to other vegetation
    C Fls several on each side-branch; lvs linear — 2) *G. asperrimum*
    CC Fls 2-3; lvs ovate-oblong to broadly obovate — 9) *G. triflorum*
AA Lvs mly 4 in a whorl, occasionally 2, 5 or 6
  B Ovary and fr glabrous; stems 2-5 dm L
    C Stems matted or interwoven, not woody at base; wet habitats to 10,500' — 8) *G. trifidum*
    CC Stems ± erect, woody at base; dryish habitats
      D. Corolla usu purplish red; lvs ± oblong; below 6000', Fresno Co. n. — 4) *G. bolanderi*
      DD Corolla greenish; lvs broader; 5500-8000', rare n. of Fresno Co. — 7) *G. sparsiflorum*
  BB Ovary and fr bristly-hairy; stems 0.5-1.5 dm L
    C Annual with slender base; lvs usu very unequal in whorl — 3) *G. bifolium*
    CC Perennial with woody base; lvs usu all similar in whorl
      D Corolla greenish white; lvs 5-10 mm L; Eldorado Co. n. — 5) *G. grayanum*
      DD Corolla yellowish to red; lvs 2-7 mm L; Plumas Co. s. — 6) *G. hypotrichium*

1) *G. aparine*, Goose-grass: Stem hairy at the nodes, 1-10 dm L; lvs mly linear-oblanceolate, 1.5-7 cm L, bristle-tipped, ± pubescent; fls 2 mm W, whitish, 2-5 in cymes in upper axils, the peduncles with a whorl of leaf-like bracts; frs bristly, 3-5 mm W; bristles often hooked at tips. Common on shaded banks below 5000', SN (all), May-Jul.

2) *G. asperrimum*, Tall Rough Bedstraw: Stems 3-8 dm L, freely branched above; lvs oblanceolate to broadly linear, 1.5-4 cm L, hairy on margins and midrib; fls in diffuse terminal ± leafy infl, on capillary pedicels; corolla white, 2.5-3.5 mm W; fr with very short hooked bristles. Shaded places, 1500-7300', Mariposa Co. n., Jun-Aug.

3) *G. bifolium*, Low Mountain Bedstraw: Stems erect, usu unbranched; lvs 2-4, 1-2.5 cm L, lanceolate to sublinear; pedicels 1-fld, recurved and equaling lvs in fr; fls whitish, about 1 mm W; fr about 2 mm W, with hooked hairs. Moist, partly shaded places, 5000-10,500', SN (all), Jun-Sep.

4) *G. bolanderi,* Bolander's Bedstraw: Stems erect or diffusely spreading, tufted, 1-4 dm H, glabrous except sometimes n the angles; lvs oblong to lance-oblong, 0.6-2.5 cm L; fls unisexual, the staminate fls in small terminal cymes, pistillate solitary in upper axils; corolla about 2 mm W; fr fleshy, dark purplish, about 3 mm W. Dry, rocky habitats, May-Aug.

5) *G. grayanum* and ssp, Gray's Bedstraw: Stems tufted; lvs broadly ovate, plant dioecious; infl narrow, leafy, 3-10 cm L; corolla 3-4 mm W; fr, including hairs, about 6-7 mm W, the hairs soft, tawny, about 3 mm L. Dry, rocky slopes, 6000-10,000', Jul-Aug.

6) *G. hypotrichium*, Alpine Bedstraw: Stems similar to *G. grayanum*, unbranched, leafy; lvs round-ovate to lance-ovate, mly 3-nerved, 2-7 mm L; plants mly dioecious; infl 2-5 cm L. Dry places about rocks, 7300-12,000', May-Aug.

7) *G. sparsiflorum*, Sequoia Bedstraw: Stems erect or spreading, 3-5 dm L, rough-textured; lvs oval to elliptic, 8-15 mm L; fls unisexual, the staminate few, in leafy clusters, pistillate solitary; corolla about 2 mm W; fr fleshy, about 4 mm W, on recurved pedicels. Plumas Co. s., but rare n. of Fresno Co., Jun-Aug.

8) *G. trifidum* vars, Trifid Bedstraw: Stems 0.5-4 dm L; lvs elliptic-oblong to linear-spatulate, sometimes fleshy, 0.4-2 cm L; fls 1 to few on axillary peduncles, pedicels ± glabrous; fr dry, 1-2 mm W. SN (all), Jun-Sep.

9) *G. triflorum*, Fragrant Bedstraw: Stems 2-8 dm L, often rough surfaced; lvs mly in 6's, thin, 1.5-6 cm L, the surfaces mly glabrous; peduncles axillary, 3-fld or 3-forked; corolla greenish white; fr densely bristly, scarcely 2 mm W. Moist, shaded canyons and wooded places below 8000', SN (all), May-Jul.

flower (1 mm) *upper stem* (1 cm)

K. galioides: Stems few, slender, 1-2.5 dm H; lvs 1-3 cm L, 3-8 cm W, ± glabrous, lanceolate; fls in loose-forking terminal infl, lavender to white; pedicels filiform, 1-12 cm L; calyx densely bristly-pubescent; corolla 3-4 mm L, mly 4-lobed, funnelform. Dry benches and slopes, 3000-9600, SN (all), May-Aug. Edibility unknown.

SALICACEAE - Willow Family

Lvs mly with stipules, usu finely serrate. Wood soft, light, mly pale. Winter buds scaly. Fls blooming early in spring, often before lvs.

Trees; young bark white — *Populus*
Shrubs or small trees; young bark green — *Salix*

P. tremuloides leaves (1 cm) *P. balsamifera leaf* (1 cm)

Bark pale. Winter-buds resinous. Lvs petioled. The inner bark of either sp may be eaten.

Petioles round; bark furrowed — *P. balsamifera*
Petioles flattened; bark smooth — *P. tremuloides*

P. balsamifera, Black Cottonwood: Tree 10-60 m H with broad open crown; lvs ovate, truncate or cordate at base, acute at apex, dark green above, pale and somewhat glaucous beneath, 3-7 cm L; petioles 2-6 cm L; catkins 4-8 cm L. Along streams below 9000', SN (all), Mar-Apr.

P. tremuloides, Quaking Aspen: Slender tree mly 3-20 m H; twigs slender; lvs round-ovate or wider, the blades 2-4 cm L; petioles about as long; catkins 3-6 cm L. Moist habitats, 6000-10,000', SN (all), Apr-Jun.

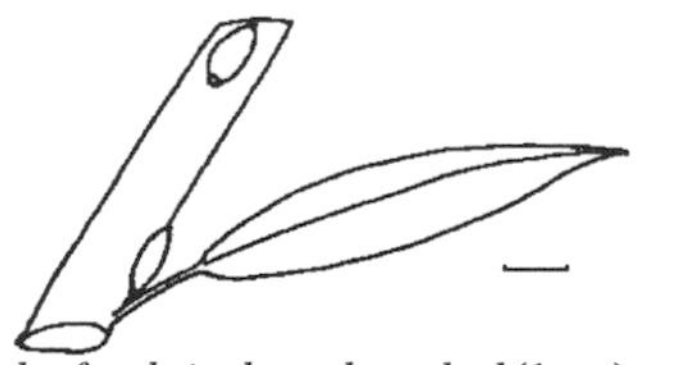

Salix

Willow

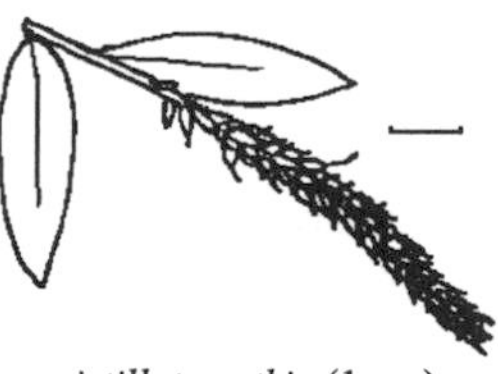

leaf and single scale on bud (1 cm)   *pistillate catkin* (1 cm)

Shrub or tree, dioecious. Twigs rather slender, often shiny and green. Lvs mly narrow, pinnately veined. Stamens mly 2-5; pistil 1, forming a capsule. Plant usu growing in moist to wet habitats. Hybridization and natural variation make identification of some specimens difficult. Inner bark may be eaten, and an extract of the bark is good for headaches..

A Plants mly less than 1 dm H; lvs entire, less than 4 cm L
  B Plants forming dense mats less than 0.5 dm H
    C Lvs 1-4 cm L; 8500-12,000', Lassen Co. s. — 1) *S. arctica*
    CC Lvs 0.7-1.2 cm L; 10,000-12,000', Mono Co. — 17) *S. reticulata*
  BB Plants erect, usu over 1 dm H
    C Lvs pubescent beneath; Convict Creek, Mono Co. — 3) *S. brachycarpa*
    CC Lvs glabrous when mature; 8000-12,500', Tuolumne Co. s. — 15) *S. planifolia*
AA Plants mly over 1 m H; lvs often serrate and longer
  B Lvs ± sessile, narrow, usu less than 1 cm W; below 8000' — 13) *S. melanopsis*
  BB Lvs on petioles 3-15 mm L
    C Petioles with wart-like glands at summit; lvs shiny dark green above, usu over 2 cm W — 11) *S. lucida*
    CC Petioles not glandular; lvs various
      D Lvs pubescent on both surfaces, often densely so, rarely glabrate in age
        E Lvs serrate, usu gray-pubescent; stipules ovate, small; 7000-10,500' — 5) *S. eastwoodiae*
        EE Lvs entire; stipules none or lanceolate
          F Stems 3-5 m H; petioles 4-10 mm L; lvs silvery on both surfaces — 6) *S. geyeriana*
          FF Stems 1-4 m H; petioles 2-6 mm L; lvs often only slightly hairy above
            G Capsule 7-8 mm L; stamens hairy at base; 7400-12,000' — 14) *S. orestra*
            GG Capsule 4-6 mm L; stamens without hairs at base
              H Catkin-scales black; 8400-9500', Fresno Co. s. — 4) *S. drummondiana*
              HH Catkin-scales brown; 5500-10,000', SN (all) — 7) *S. jepsonii*
      DD Lvs glabrous above, usu glabrous beneath when mature; below 10,000'
        E Catkins mly over 4 cm L; lvs entire, ± revolute; below 7000' — 8) *S. lasiolepis*
        EE Catkins mly less than 4 cm L; lvs various
          F Stipules none or lanceolate; above 5000', SN (all) — 9) *S. lemmonii*
          FF Stipules present, broader
            G Catkins on short peduncles; lvs glandular-serrulate
              H Pistillate catkins 3-6 cm L; occasional below 7000' — 16) *S. proliva*
              HH Pistillate catkins 1-3 cm L; above 6000' — 2) *S. boothii*
            GG Catkins ± sessile; lvs seldom glandular-serrulate
              H Lvs 1-2 cm W, about 5 times as long — 10) *S. ligulifolia*
              HH Lvs 1.5-4 cm W, about 3 times as long
                I Lvs usu hairy beneath, entire; catkin-scales black — 18) *S. scouleriana*
                II Lvs never hairy beneath, often serrulate; catkin-scales yellow — 12) *S. lutea*

1) *S. arctica*, Arctic Willow: Branchlets erect; lvs entire, broadly lanceolate; petioles 3-10 mm L; catkins 1-6 cm L. Moist habitats, Jul-Aug.

2) *S. boothii*, Booth's Willow: Stems to 4 m H; twigs ± shiny; lvs ± elliptic, 3-8 cm L, thickish, veiny; stipules narrow-ovate; catkins 1.5-4 cm L, the scales dark brown. To 11,300', SN (all), Jun-Jul.

3) *S. brachycarpa*, Short-fruited Willow: Lvs entire, ± elliptic, 2-3 cm L. At 9900-10,600'.

4) *S. drummondiana* var. *subcoerulea*, Drummond's Willow: Twigs slender, subglabrous; lvs on fl-stems narrowly oblong, rather short, lvs on summer shoots ± oval-lanceolate, 3-7 cm L; stipules usu none; catkins 1-4 cm L. May-Jun.

5) *S. eastwoodiae*, Eastwood's Willow: Stems 0.5-2 m H; twigs dark, pubescent; lvs ± elliptical, 4-6 cm L, sometimes glabrate in age; catkins 1-4 cm L; catkin-scales brownish, pubescent. SN (all), Jun-Jul.

6) *S. geyeriana* var. *argentea*, Geyer's Willow: Twigs ± pubescent; lvs 2-7 cm L, 0.6-1.5 cm W; stipules none. Wet habitats, 5000-10,500', SN (all), May-Jun.

7) *S. jepsonii*, Jepson's Willow: Twigs ± pubescent when young, glabrous and shining in age; lvs oblanceolate, 3-7 cm L, firm; staminate catkins 1-2 cm L, pistillate 2-4 cm L. May-Jun.

8) *S. lasiolepis*, Arroyo Willow: Erect shrub or small tree 2-10 m H; bark smooth; twigs yellowish to brown, usu pubescent; lvs ± oblanceolate, 6-10 cm L; catkins mly appearing before lvs, 3-7 cm L, the scales dark, densely hairy. Common, SN (all), Apr.

9) *S. lemmonii*, Lemmon's Willow: Stems 1-5 m H; twigs subglabrous, yellowish; lvs oblanceolate to lanceolate, 3-10 cm L, deep green and shiny above, ± glaucous beneath; catkins 1-6 cm L, the scales dark. SN (all), May-Jun.

10) *S. ligulifolia*, Ligulate Willow: Stems 1-5 m H with dark gray bark; young twigs yellowish, puberulent; lvs strap-shaped, 5-10 cm L, subentire to glandular-serrulate; catkins 2-4 cm L, the scales white-woolly. At 3000-9500', Plumas Co. s., Apr-May.

11) *S. lucida*, Shining Willow: Tree with rough brown bark; twigs reddish, shining, glabrous; lvs ± lanceolate, glandular-serrate, 6-10 cm L; catkin-scales yellow. SN (all), Apr-May.

12) *S. lutea*, Yellow Willow: Stems 2-5 m H; twigs yellowish or brownish, glabrous; lvs yellowish green above, paler beneath, 4-10 cm L; catkins 2-4 cm L. At 5000-9500', c. and s. SN, especially e. slope.

13) *S. melanopsis*, Dusky Willow: Shrub or small tree 3-5 m H; twigs dark; lvs oblanceolate to elliptic, 4-7 cm L, denticulate, shiny above, paler beneath; catkins 3-4 cm L. SN (all), Apr-May.

14) *S. orestra*, Sierra Willow: Twigs brown, hairy; lvs oblanceolate or broader, 4-6.5 cm L; stipules none or lanceolate; catkins 1-4 cm L, the scales dark brown, silky. SN (all), Jun-Jul.

15) *S. planifolia* var. *monica*, Mono Willow: Stems ascending to erect, 0.3-1.5 m H; twigs glabrous; lvs ± elliptic to obovate, bright green above, paler beneath; petioles 2-8 mm L; catkins 1-3 cm L. Tuolumne, Fresno, Mono and Inyo cos., Jun-Aug.

16) *S. proliva*, MacKenzie's Willow: Shrub or small tree to 6 m H; lvs 6-10 cm L, lanceolate or broader; catkins 2.5-6 cm L, the scales dark brown. SN (all), May.

17) *S. reticulata* var. *nivalis*, Snow Willow: Lvs oblong-obovate, shining above, ± revolute; catkins to 3 cm L. Uncommon, Jul-Aug.

18) *S. scouleriana*, Scouler's Willow: Shrub or small tree to 10 m H, ± pubescent; twigs stoutish, yellowish to brownish; lvs mly oblanceolate to obovate, 3-10 cm L, mly entire; stipules broad; catkins appearing before the lvs. Below 10,000', SN (all), Apr-Jun.

SANTALACEAE - Sandalwood Family

Comandra

Bastard Toad-flax

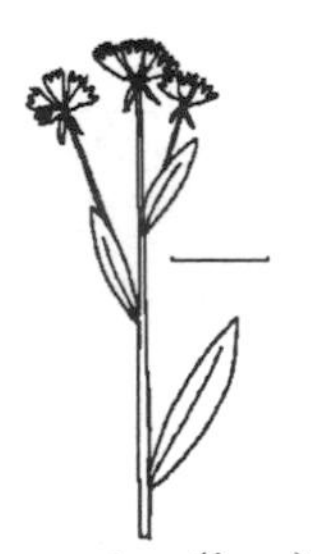

upper stem (1 cm)

C. umbellata ssp *californica*: Perennial; stems many, 1-3 dm H; lvs alt, entire, oblanceolate, 1-2.5 cm L, 6-11 mm W, subsessile; fls perfect, about 5 mm L, 5-merous; sepals green at base, lighter at tips; fr drupe-like, crowned by a persistent calyx. Occasional as a root parasite, below 9000', SN (all), Apr-Aug. The fleshy fr may be eaten raw. It tastes best when still slightly green, though it is still edible when brown and completely mature.

SARRACENIACEAE - Pitcher-plant Family

Darlingtonia

California Pitcher-plant

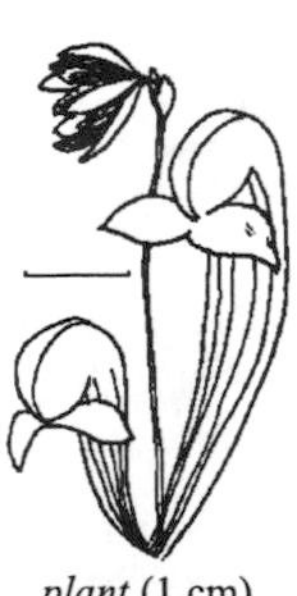

plant (1 cm)

D. californica: Lvs greenish yellow, 2-6 dm H; conspicuously net-veined, enlarged upward into a globose, irregularly dotted hood with opening underneath the hood; fl-stems equaling or exceeding lvs with several yellow scales and 1 fl; fls 5-merous, nodding; sepals yellow-green with purplish lines, 4-6 cm L; petals dark purple, 2-4 cm L. Below 6000', Plumas and Nevada cos. Apr-Jun. Plant too rare to consider edibility.

SAXIFRAGACEAE - Saxifrage Family

Mly perennial herbs. Lvs often without stipules. Fls perfect, perigynous. Sepals and petals mly 5. Styles, stigmas, and ovaries commonly 2.

A Fertile stamens 10
  B Lvs peltate, 1-4 dm across — *Darmera* p. 189
  BB Lvs not peltate, mly smaller
    C Lvs 5-10 cm W; petals ± 3-cleft at apex, reflexed — *Tellima* p. 191
    CC Lvs usu 1-4 cm W; petals not reflexed
      D Petals entire; styles 2; basal lvs rarely lobed — *Saxifraga* p. 190
      DD Petals usu cleft; styles 3; basal lvs usu deeply lobed — *Lithophragma* p. 189
AA Fertile stamens 5 (sterile staminodia also present in *Parnassia*)
  B Fls solitary at the ends of long stalks; petals usu yellow, conspicuous — *Parnassia* p. 190
  BB Fls variously clustered; petals often small, white or greenish
    C Fl-stems leafless, or if leafy then basal lvs shallowly lobed
      D Petals cleft; pedicels short; infl narrow, 1-2 cm W — *Mitella* p. 190
      DD Petals entire; pedicels usu over 1 cm L; infl usu wider — *Heuchera* p. 189
    CC Fl -stems usu leafy
      D Lower lvs ternately divided; lfts 1-2 cm L; Plumas Co. n. — *Suksdorfia* p. 191
      DD Lower lvs simple, 5-7-lobed or -cleft
        E Stems 1.5-3 dm H; lower lvs 2-3 cm W; fl-tube 5-8 mm W — *Bolandra* p. 188
        EE Stems 2-9 dm H; lower lvs 5-15 cm W; fl-tube 2-3 mm W — *Boykinia* p. 188

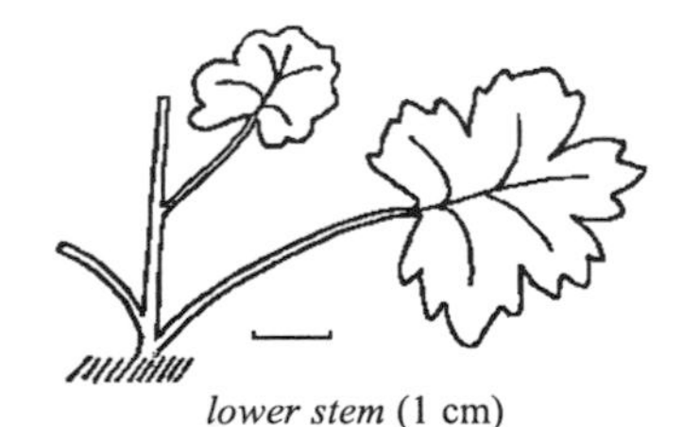

lower stem (1 cm)

Bolandra

Sierra Bolandra

inflorescence (1 cm)

B. californica: Stems slender; lvs thin, the blades round-cordate, glabrous; petioles 2-10 cm L; upper lvs reduced; fls few, loosely clustered, purplish, sepals 3-4 mm L; petals greenish with purplish tips and edges, about 5 mm L; carpels 2, united part way. Uncommon, wet rocks, 3300-8000', Mariposa to Eldorado Co., Jun-Jul. Edibility unknown.

B. major flower (1 cm)

Boykinia

Boykinia

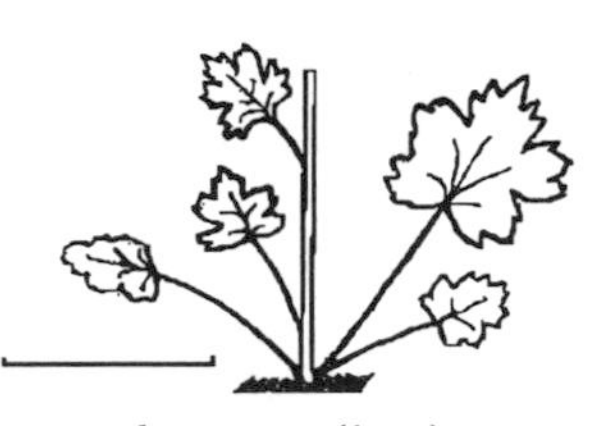

lower stem (1 cm)

Stems unbranched with small brown gland-tipped hairs. Lvs large, several, alt and basal, kidney-shaped, long-petioled, with stipules. Infl many-fld, densely glandular-puberulent. Petals white, usu early deciduous. Edibility unknown.

Upper lvs with large green leafy stipules; petals 5-7 mm L — *B. major*
Upper lvs with brownish stipules often reduced to bristles; petals 3-4 mm L — *B. occidentalis*

B. major, Mountain Boykinia: Stems stout, 3-9 dm H; lower lvs on petioles 1-2 dm L, the blades 5-7-cleft and with gland-tipped teeth; upper lvs reduced; infl dense, many-fld; sepals 3 mm L; petals broadly ovate to obovate. Moist rocky habitats below 7500', Madera Co. n., Jun-Sep.

B. occidentalis, Brook Foam: Stems erect, 2-6 dm H; lower lvs thin, 2-8 cm W, incised into 5-7 lobes with bristle-pointed teeth; petioles 5-15 cm L; upper lvs reduced; sepals about 2 mm L; petals ± oblanceolate. Shaded, moist habitats below 5000', Amador Co. n., Jun-Jul.

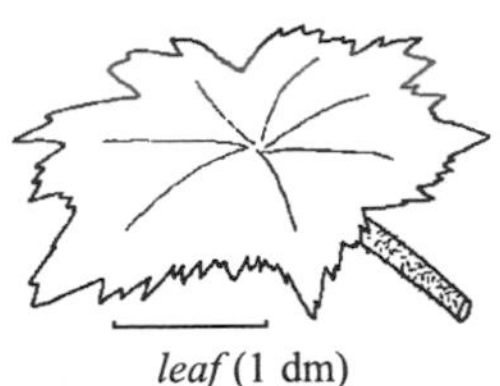
leaf (1 dm)

Darmera

Indian-rhubarb
Umbrella Plant

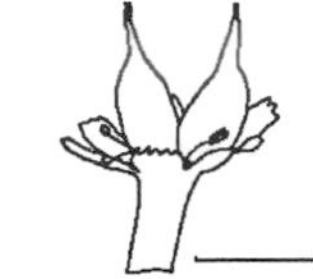
dissected flower showing the two ovaries (5 mm)

D. peltata: Stout perennial; lvs basal, 9-15-lobed, mly glabrous; petioles 3-10 dm L; fl-stems 3-10 dm H; sepals 4-5 mm L; petals broad, 5-7 mm L. Banks of streams, below 6000', SN (all), Apr-Jul. The fleshy leafstalks may be peeled and eaten as is or put in a salad.

Heuchera

Alum-root

Lvs long-petioled; stipules united with petiole at base. Fl-stem lateral with a terminal paniculate infl. The spring lvs are edible boiled or steamed. The root eaten raw will usu cure diarrhea.

plant (1 cm)

Basal lvs 3-8 cm L; petals white, 2 mm L; mly below 7000' — *H. micrantha*
Basal lvs 1-2.4 cm L; petals pink, 3-5 mm L; above 6000' — *H. rubescens*

H. micrantha, Small-flowered Heuchera: Basal lvs 5-7-lobed, cordate at base; petioles soft long-hairy, 5-18 cm L; fl-stems stoutish, 3-10 dm H, 0-4-leaved; infl much-branched, open conical, puberulent. Moist banks of humus and rocks, SN (all), May-Jul.

H. rubescens vars. Pink Heuchera: Lvs roundish, often subcordate at base, 3-5-lobed, dentate; petioles slender, 1-5 cm L, glandular-puberulent; fl-stems 1-3 dm H, leafless; infl 3-20 cm L, usu racemose without secondary branching. Dry, rocky places, 6000-12,000', Plumas Co. s., May-Aug.

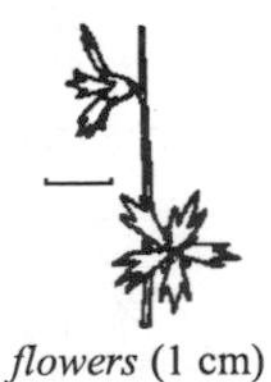
flowers (1 cm)

Lithophragma

Woodland Star

leaf (1 cm)

Perennials with slender, unbranched flowering shoots. Lvs mly basal, petioled, usu deeply 3-5-cleft; petioles with stipule-like dilated bases. Racemes simple, few-fld. Edibility unknown.

A Basal lvs shallowly lobed; petals ± entire — 1) *L. bolanderi*
AA Basal lvs lobed almost to their base; petals lobed
  B Stem lvs reduced; fls usu few; 4500-11,000' — 2) *L. glabrum*
  BB Stem lvs well developed; fls 3-14; below 7000'
    C Stem lvs similar to basal; petals 7-16 mm L — 3) *L. parviflorum*
    CC Stem lvs mly pinnatifid; petals 3-7 mm L — 4) *L. tenellum*

1) *L. bolanderi*, Sierra Star: Fl-stems several, 2-8 dm H; herbage pubescent; basal lvs 3-5-lobed; stem lvs 2-3; fls 3 to many; pedicels short; corolla wide-spreading, the petals ovate-elliptic, 4-7 mm L. Dry habitats mly below 7000', w. SN (all), May-Jul.

2) *L. glabrum*, Rock Star: Stems slender, 1-3.5 dm H, subglabrous glandular-pubescent; basal lvs about 2 cm W, usu 3-parted, the segments many-cleft or round-lobed; stem lvs similar but slightly smaller; fls 1-6, 4-5-merous, the petals usu exceeding the fl-tube, pink or rarely white, 3-7 mm L, palmately 5-parted. Dry to moist open places, SN (all), May-Jul.

3) *L. parviflorum*, Small-flowered Lithophragma: Stems slender, 2-5 dm H, nearly glabrous to densely pubescent; basal lvs 3-parted, about 3 cm W; stem lvs 2-3, about 1.5-3 cm W; pedicels about 7 mm L; fl-tube glandular; petals white or pink, obovate, always 3-cleft. Open slopes, 2000-6000', SN (all), Apr-May.

4) *L. tenellum*, Modoc Lithophragma: Stems 1.5-3 dm H; herbage light green and sparsely pubescent; basal lvs irregularly 3-5-lobed; stem lvs 2; petals mly pink, palmately 5-parted. Occasional, Butte and Plumas cos., May-Jul.

Mitella

Bishop's-cap Mitrewort

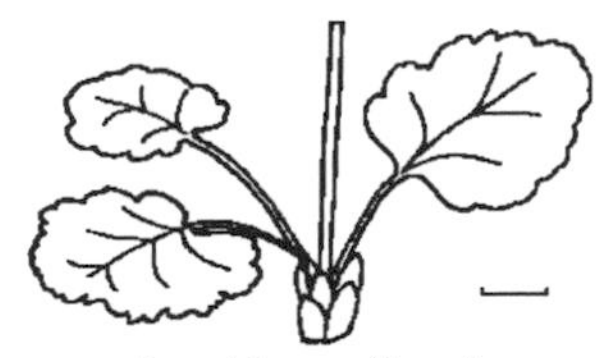

M. breweri flower (1 cm) *basal leaves* (1 cm)

From rootstocks. Lvs round-cordate, slender-petioled, basal. Fl-stems slender, leafless, with terminal spike-like raceme of small fls. Edibility unknown.

Stamens opp the sepals *M breweri*
Stamens opp the petals *M. pentandra*

M. breweri, Brewer's Mitrewort: Fl-stem slender, 1-2 dm H, minutely puberulent; lvs 3-6 cm W, with 6-10 shallow lobes, these again toothed or lobed; petioles 3-10 cm L, purplish below, green above, sparsely long white- or brown-hairy; fls racemose, the pedicels 6-10 mm L; fl-tube about 3-4 mm W; sepals 1 mm L, the tips reflexed; petals about 5 mm L, green with 5-7 filiform lobes off a linear rachis. Shady, moist habitats, 6000-11,500', Jun-Aug.

M. pentandra, Alpine Mitrewort: Much like *M. breweri* except as in key. Uncommon, shaded, rocky or wooded places, 4500-8000', May-Jul.

Parnassia

Grass-of-parnassus

P. fimbriata: Glabrous perennials; basal lvs ovate, entire, 2.5-4 cm L, with 5-7 principal veins; petioles 2-10 cm L; fl-stem 2.5-5 dm H with an ovate bract above the middle or sometimes lacking; sepals 4-6 mm L; petals oblong-ovate to rounded, about 8-14 mm L. Wet mdws below 11,000', SN (all), Jul-Oct. Edibility unknown.

plant (1 cm)

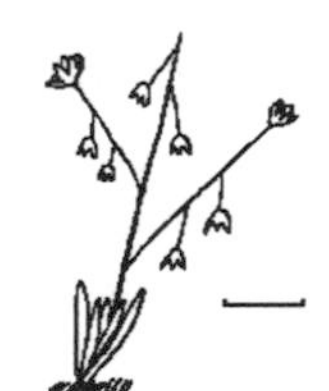

Saxifraga

Saxifrage

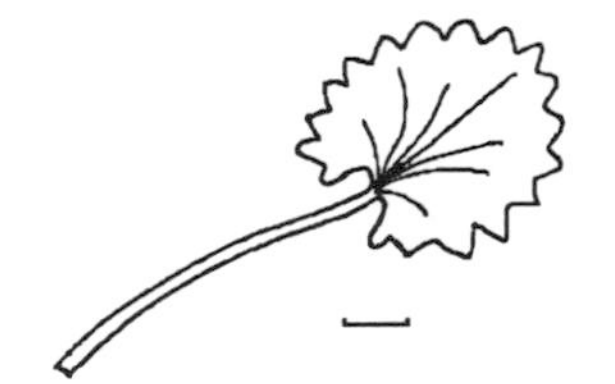

S. bryophora lower stem (1 cm) *S. punctata leaf* (1 cm)

Mly perennials. Lvs mly basal, usu glabrous, entire to dentate, rarely lobed. Fls 5-merous; petals deciduous. Related sp has edible young lvs.

A Lf-blades roundish to broader than long
  B Lvs 0.6-1.4 cm W, mly deeply lobed; plants mly less than 1 dm H 8) *S. rivularis*
  BB Lvs 2-8 cm W, shallowly toothed; plants 1-4 dm H
    C Lvs without stipules; above 6500', SN (all) 6) *S. odontoloma*
    CC Lvs with stipular expansions; 4000-5000', Mariposa and Nevada cos. 4) *S. mertensiana*
AA Lf-blades definitely longer than broad
  B Plants annual; petals often dissimilar; lvs about 5 mm W, in basal tuft 2) *S. bryophora*
  BB Plants perennial; petals equal; lvs wider or not in basal tuft
    C Stems usu less than 1 dm H, often reddish or purplish
      D Lower branches densely covered with ± linear revolute lvs; plants woody at base 9) *S. tolmiei*
      DD Lvs in a basal tuft; plants not woody at base 1) *S. aprica*
    CC Stems usu over 1 dm H, not reddish; lvs in basal tuft
      D Lvs coarsely dentate, abruptly narrowed to petiole 3) *S. fallax*
      DD Lvs entire to remotely dentate, usu gradually narrowed at base
        E Lvs 4-12 cm L; fl-stem stout, 3-8 dm H 7) *S. oregana*
        EE Lvs 1-4 cm L; fl-stem slender, 1-3 dm H 5) *S. nidifica*

1) *S. aprica*, Sierra Saxifrage: Lvs ovate to oblong or spatulate, glabrous, 1-4 cm L, on shorter petioles, entire to dentate; fl-stem solitary, unbranched, subglabrous, 0.3-1.2 dm H; infl subcapitate; petals white, elliptic, about 2 mm L. Moist, gravely and stony places, 5500-12,000', SN (all), May-Aug.

2) *S. bryophora*, Bud Saxifrage: Herbage glandular-pubescent; lvs ± oblong, sessile or nearly so, 0.5-1.5 cm L, mly entire; fl-stems slender, 0.5-2 dm H, openly branched above; pedicels filiform, all but the terminal soon deflexed; sepals mly reflexed; petals white with 2 yellow spots, 2-3 mm L. Moist, gravely places, 7000-11,200', Tulare to Plumas Co., Jul-Aug.

3) *S. fallax*, Greene's Saxifrage: Lvs oblong-oval, 1-5 cm L, on petioles as long or longer; fl-stems 1-3 dm H, glandular-pubescent, loosely and widely branched above; fls not clustered; sepals usu not reflexed; petals oblong, about 3-4 mm L. Rocky places, 3000-7500', Fresno to Plumas Co., Apr-May.

4) *S. mertensiana*, Wood Saxifrage: Lvs mly basal, orbicular, usu cordate at base, doubly serrate, on petioles 3-12 cm L; fl-stems slender, 1-3 dm H, finely glandular-pubescent, with 1 to few reduced lvs in lower part; infl open; sepals reflexed; petals white to pink, 3-5 mm L, ovate to oblong. Moist, rocky places, Jun-Jul.

5) *S. nidifica*, Peak Saxifrage: Lvs ovate, 1-4 cm L, on shorter petioles; fl-stems glandular-pubescent, mly few-branched above; petals white, roundish-obovate, about 2 mm L. Moist places, 2500-11,000', SN (all), Jun-Aug.

6) *S. odontoloma*, Brook Saxifrage: Lvs basal, roundish, 1.5-7 cm W, coarsely dentate, on petioles 2-20 cm L; fl-stems 2-4 dm H, slender, glandular-pubescent upward; infl open, with spreading to reflexed rather few-fld clusters; sepals reflexed, usu purplish; petals white with 2 yellow dots at base, rounded, 2-5 mm L. Moist stream banks, 6500-11,200', SN (all), Jun-Aug.

7) *S. oregana*, Bog Saxifrage: Lvs oblanceolate-spatulate, 3-12 cm L, narrowed to a broadly margined petiole or scarcely petiolate; fl-stems glandular-pubescent; fls in several small clusters; petals white, obovate, 2-4 mm L. Wet mdws and boggy places, 3500-11,000', SN (all), May-Aug.

8) *S. rivularis*, Lobed-leaved Saxifrage: Glabrous to glandular-puberulent, tufted, perennials; stems slender, leafy, 1- to few-branched, each branch with 1 terminal fls; lower lvs thin, reniform in outline, on slender petioles 1-3 cm L; upper lvs smaller; petals white, oblong-spatulate, 3-6 mm L. Damp places in shade of overhanging rocks, 11,000-12,000', Inyo, Tulare, Madera and Tuolumne cos., Jul-Aug.

9) *S. tolmiei*, Alpine Saxifrage: Stems densely tufted, diffusely branched, ± prostrate; lvs thickish, 0.8-1.5 cm L; fl-stems slender, 3-12 cm H, finely glandular-pubescent, few-branched at summit; petals white, elliptic-spatulate, 4-5 mm L. Moist, rocky places, 8500-11,800', SN (all), Jul-Aug.

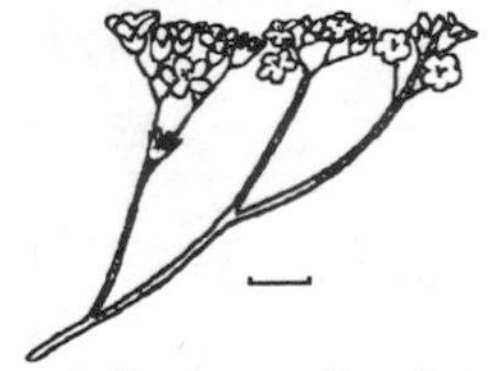

inflorescence (1 cm)

Suksdorfia

Suksdorfia

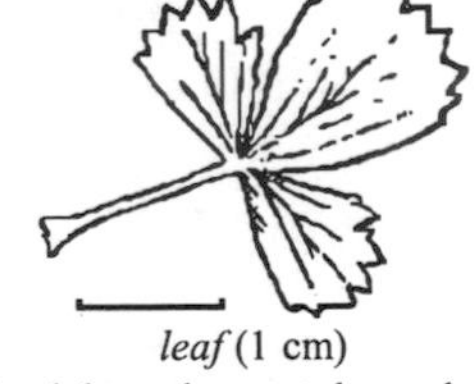

leaf (1 cm)

S. ranunculifolia: Stems 1-3 dm H, slender, unbranched, ± glandular-pubescent; lower lvs with 3 wedge-shaped lfts, the lfts 3-4-lobed; petioles 2-10 cm L; stipules green and leafy; upper lvs smaller; infl compact, often flat-topped; sepals tinged purplish, about 2 mm L; petals white, 4-6 mm L, entire. Wet rocks, 5000-6000', Jun-Aug. Edibility unknown.

Tellima

Fringe-cups

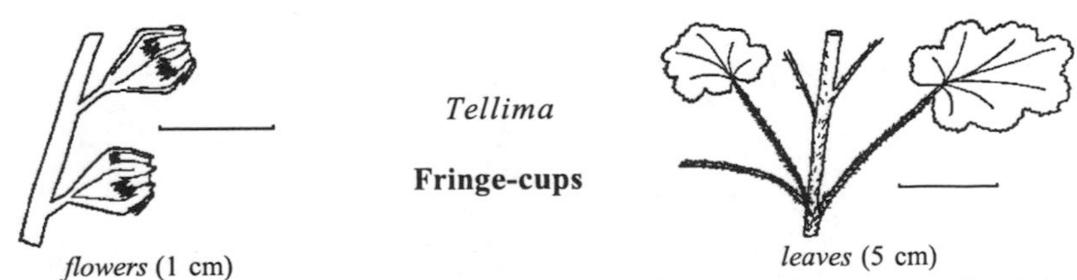

flowers (1 cm) *leaves* (5 cm)

T. grandiflora: Perennial with numerous basal lvs; fl-stems unbranched, stout, spreading-hairy; basal lvs hairy, shallowly 3-7-lobed, the lobes sharp-toothed, the lvs cordate at base, 5-10 cm W, on petioles, 5-20 cm L; stem lvs few, smaller; infl glandular-puberulent, few- to many-fld; fl-tube 3-5 mm L, ± urn-shaped; sepals about 2 mm L; petals at first whitish, later red, 4-6 mm L. Moist woods and rocky places below 5000', Eldorado Co. n., Apr-Jun. Edibility unknown.

SCROPHULARIACEAE - Figwort Family

Herbs, occasionally shrubs. Lvs simple, mly entire, without stipules. Fr a capsule.

A Lvs alt or basal, often pinnatifid or dissected
 B Fls highly irregular; stamens enclosed in narrow upper corolla-lip;
  C Lvs usu pinnatifid or fern-like with many narrow segments, rarely merely dentate — *Pedicularis* p. 198
  CC Lvs entire or few-lobed at base
   D Lower corolla lip as long as upper, 3-lobed; lvs linear or with linear lobes — *Cordylanthus* p. 194
   DD Lower corolla lip much shorter than upper lip; usu reduced to 3 teeth
    E Stigma dot-like; tip of corolla closed — *Orthocarpus* p. 198
    EE Stigma expanded, entire to 2-lobed; tip of corolla open — *Castilleja* p. 192
 BB Fls ± regular; stamens not hidden in corolla folds
  C Stems tufted less than 10 cm H; lvs all basal, 1-1.5 cm L; wet habitats — *Limosella* p. 195
  CC Stems solitary, 50-180 cm H; lvs alt, 15-40 cm L; dry, disturbed areas — *Verbascum* p. 201
AA Lvs (at least lower) opp, usu simple
 B Fls usu less than 5 mm W, or if larger, appearing 4-lobed
  C. Upper lvs tripartite; corolla 5-lobed, 2-4 mm L; stamens 4; Butte Co. n. — *Tonella* p. 201
  CC Upper lvs simple, sometimes clasping or bract-like; corolla appearing 4-lobed
   D Upper corolla-lip appearing irregularly 1-lobed; stamens 2 — *Veronica* p. 201
   DD Upper corolla-lip 2-lobed, lower appearing 2-lobed with the stamens enclosed in the folded middle "hidden" lobe — *Collinsia* p. 193
 BB Corolla 5-lobed, usu 2-lipped
  C Sepals separate to base; corolla white; fertile stamens 2 — *Gratiola* p. 195
  CC Sepals united into a ± cylindrical tube; corolla usu colored; fertile stamens 4
   D Stigmas flattened forming a 2-lipped surface that folds together when touched — *Mimulus* p. 195
   DD Stigmas united and capitate
    E Corolla brownish, mly 5-7 mm L; lvs coarsely toothed — *Scrophularia* p. 201
    EE Corolla rarely brownish, 10-40 mm L; lvs mly entire
     F Fls whitish; base of fertile filaments densely hairy — *Keckellia* p. 195
     FF Fls red to violet; base of fertile filaments glabrous — *Penstemon* p. 199

Castilleja

Indian Paint-brush
Owl's-clover

Herbs or slightly woody plants, partially root-parasites. Lvs sessile, entire to divided into narrow lobes, passing above into usu more incised and colored bracts of the terminal spike-like infl. Calyx tubular, 4-lobed or seemingly compressed, the upper lip (galea) elongated, entire, enclosing the style and stamens; lower lip often rudimentary, 3-toothed. Many, perhaps all spp have fls that may be eaten raw. The plant has the ability to concentrate selenium should it be in the soil; the fls should therefore be eaten only in moderate quantities.

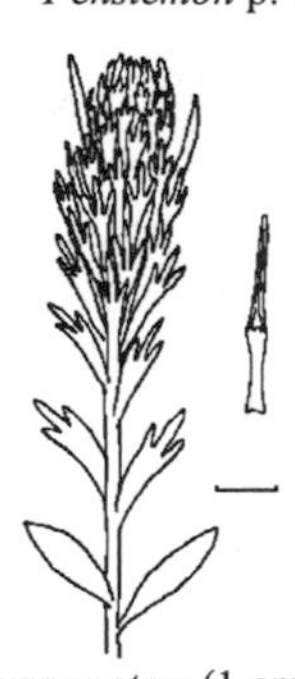

upper stem (1 cm)

A Corolla mly 2-3 cm L, the lower lip less than half as long as galea; infl mly red to orange
 B Plants often hairy but not glandular
  C Calyx mly 20-30 mm L; wet habitats — 5) *C. miniata*
  CC Calyx mly 12-15 mm L; dry habitats — 10) *C. pruinosa*
 BB Plants glandular-pubescent at least below infl
  C Corolla beak 1/2 length of tube; stigma clearly 2-lobed — 7) *C. parviflora*
  CC Corolla beak as long as tube; stigma faintly 2-lobed — 1) *C. applegatei*
AA Corolla mly 1-2 cm L, the lower lip about half to almost as long as galea; infl purplish to yellow or green
 B Annuals; bracts green; lower lip deeply 3-saccate (Owl's-clover group)
  C Bracts entire; lower herbage glabrous — 2) *C. campestris*
  CC Bracts palmately 3-7-cleft; herbage pubescent below
   D Sacs of lower lip 3-5 mm deep — 3) *C. lacerus*
   DD Sacs of lower lip about 2 mm deep — 11) *C. tenuis*

BB Perennials; bracts mly tipped purplish or yellow; lower lip with shallow sacs
C Herbage glandular; moist habitats — 4) *C. lemmonii*
CC Herbage not glandular; dry habitats
D Calyx divided 1/2 length on front and back, 1/8 length on sides — 9) *C. praeterita*
DD Calyx divided ± 1/2 length on each side
E Bracts rounded distally; corolla 17-22 mm L; Alpine Co. n. — 8) *C. pilosa*
EE Bracts sharp-pointed; corolla 13-16 mm L; Eldorado Co. s. — 6) *C. nana*

1) *C. applegatei*, Wavy-leaved Paint-brush: Stems clustered, often branched, 2-5 dm H from a woody base; lvs 2-3.5 cm L, wavy-margined, entire or 3-lobed; bracts and sepals usu red towards tip, occasionally orange or yellow; calyx 10-20 mm L, cleft about half way to base on top and bottom,, about 1/4 of the way on the sides; corolla 2-3 cm L, the galea 10-15 mm L with red margins, the lower lip about 2 mm L, greenish. Common in dry habitats below 11,000', SN (all), May-Aug.

2) *C. campestris*, Field Owl's-clover: Stems 1-2.5 dm H; lvs entire, lance-linear, 1.5-4 cm L; spike densely- to few-fld; calyx 2-cleft to middle, each half 2-lobed; corolla bright yellow, sometimes whitish, 1.5-2.5 cm L, long-exserted; galea slightly longer than lower lip. Moist habitats below 5000', Fresno Co. n., Apr-Jul.

3) *C. lacerus*, Cut-leaved Owl's-clover: Much like *C. tenuis*; spike broader; corolla bright yellow with 2 brown dots at base of lower lip. Open habitats below 7400', Fresno Co. n. May-Jul.

4) *C. lemmonii*, Lemmon's Paint-brush: Plants glandular-pubescent below infl; stems 1-2 dm H, often several to many; at least upper bracts with 1-2 pairs of lobes; calyx 16-18 mm L, cleft halfway on top and bottom, the lateral clefts only 1-2 mm L; corolla 17-20 mm L, the galea 8-10 mm L with thin, purple margins. Moist mdws, 7000-11,500', SN (all) Jul-Aug.

5) *C. miniata*, Great Red Paint-brush: Stems few, erect, sometimes branched above, 4-10 dm H; lvs lanceolate, mly 2-5 cm L, entire or the upper occasionally lobed; infl hairy; bracts and sepals mly scarlet towards tip; calyx mly 2-3 cm L, medianly cleft 1/2-2/3 its length, laterally cleft 3-7 mm; corolla mly 2.5-3.5 cm L, the galea as long as the tube; lower lip 1-2 mm L, green, ± exserted. Below 11,000', SN (all), May-Sep.

6) *C. nana*, Alpine Paint-brush: Stems many, clustered, 0.5-2.5 dm H, mly unbranched; lvs ± linear, mly 3-5-lobed; bracts and sepals yellow or purplish red at tips; calyx about 15 mm L, deeply cleft into linear lobes; corolla white or pinkish, the galea about 6 mm L. Rocky habitats, 6400-12,000', Jul-Aug.

7) *C. parviflora*, Peirson's Paint-brush: Much like *C. applegatei*; stems 1-2.5 dm H. Moist, rocky areas, 7800-11,000', Eldorado Co. s., Jul-Aug.

8) *C. pilosa*, Hairy Paint-brush: Much like *C. nana*; stems several, 1.5-3.5 dm H; herbage long-hairy, the hairs scarcely glandular; galea about 8 mm L. Dryish habitats, 5000-10,000', mly e. SN, Jun-Aug.

9) *C. praeterita*: Stems highly branched, 1-4.5 dm H, stiff-hairy; lvs mly linear, 3-5 cm L; infl 8-15 cm L; bracts 15-25 mm L, pale green, tipped yellow or reddish, the lobes usu 3; calyx about 15 mm L; corolla about size of calyx; galea about 5 mm L; lower lip 2 mm L with green teeth. Alpine sagebrush fields, 7500-12,000', Inyo and Tulare cos.

10) *C. pruinosa*, Gray Paint-brush: Stems often branched, 3-7 dm H; lvs linear-lanceolate, entire or the upper with a pair of lobes; bracts and sepals red at tips; calyx yellow in the middle, cleft about 1/3 its length on top and bottom, cleft laterally 3-5 mm; corolla 2.5-3 cm L, the galea 15-20 mm L; lower lip about 2 mm L, included. Rocky habitats below 8000', Tuolumne Co. n., May-Aug.

11) *C. tenuis*, Hairy Owl's-clover: Stems slender, 1-4 dm H; lvs narrowly linear-lanceolate, 1-4 cm L, the upper 3-5-cleft; spike slender; bracts 1-2.5 cm L; calyx 8-10 mm L, subequally 4-lobed, corolla yellow or white, 12-20 mm L. Mdws, 3000-8000', Tulare Co. n. May-Aug.

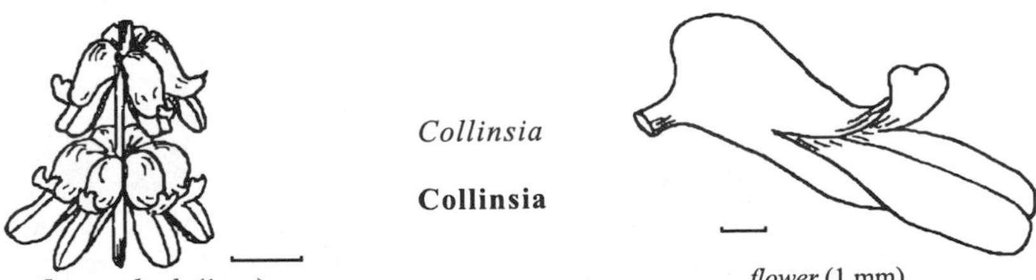

Collinsia

Collinsia

flower whorls (1 cm) — *flower* (1 mm)

Lvs simple, mly opp, entire to scalloped, the upper sessile or clasping, passing into ± leafy bracts. Fls in axils of uppermost lvs. Upper corolla-lip 2-lobed, the lower 3-lobed. The middle lobe keel-shaped and enclosing the 4 declined stamens and style. Corolla swollen at base on upper side. Edibility unknown.

A Stems unbranched; fls in 2-6 many-fld whorls; pedicels 0-2 mm L — 4) *C. tinctoria*
AA Stems usu branched; fls usu 2-4 at upper nodes; pedicels 3-30 mm L
B Infl glandular-puberulent
C Both corolla-lips white to pale violet, subequal — 1) *C. childii*
CC Lower corolla-lip longer, darker blue — 5) *C. torreyi*
BB Infl ± glabrous
C Fls mly 2 at lower nodes, 4 in upper; stems usu over 1 dm H — 2) *C. parviflora*
CC Fls solitary at lower nodes, 2 in upper; stem usu less than 1 dm H — 3) *C. sparsiflora*

1) *C. childii*, Child's Blue-eyed Mary: Stems erect, 1-4 dm H; lf-blades oblong-lanceolate, subentire to serrulate, 1-4 dm L, short-petioled; fls whorled in upper lf and bract axils; pedicels 5-25 mm L; calyx-lobes 3-4 mm L; corolla 6-8 mm L. Dry, shaded places, 3000-7000', Mariposa Co. s., May-Jun.

2) *C. parviflora*, Small-flowered Blue-eyed Mary: Stems ascending to erect, puberulent, 0.5-4 dm H; lf-blades lance-oblong, mly entire, glabrous, 2-4 cm L, sessile or the lower short-petioled; pedicels 3-15 mm L; calyx 5-7 mm L; corolla 4-7 mm L, the upper lip white to violet-blue at the tips, the lower longer, violet-blue. Moist, ± shaded places, 2500-11,150', SN (all), Apr-Jul.

3) *C. sparsiflora* var. *collina*, Few-flowered Blue-eyed Mary: Stems 0.5-2.5 dm H; lf-blades glabrous to sparsely puberulent, narrow-oblong, 1-3 cm L; calyx 5-6 mm L; corolla 5-8 mm L, purple, the upper lip whitish toward base. Grassy places and open woods, below 5000', Fresno to Eldorado Co., Apr-Jun.

4) *C. tinctoria*, Tincture Plant: Stems erect, 2-6 dm H, occasionally sparsely branched, glandular-pubescent in infl; lvs thin, ovate to lance-oblong, 2-8 cm L, the upper clasping, the lower short-petioled; calyx 5-8 mm L, the lobes parted almost to base; corolla yellow to greenish white, with purple dots or lines, 12-17 mm L, the upper lip less than half as long as lower. Dry or moist, mly stony habitats, 2000-7500', SN (all), May-Aug.

5) *C. torreyi*, Torrey's Collinsia: Stems erect, 0.5-2 dm H; lvs 1.5-4 cm L, 2-7 mm W, sessile or short-petioled; infl open, leafy; pedicels 5-10 mm L; calyx 3-4 mm L, lobed ± halfway; corolla 7-10 mm L, the upper lip pale with yellow base having purple dots. Damp or half-dry sandy banks and flats below 10,000', SN (all), May-Aug.

var. *wrightii*: Plants 0.5-3 dm H; fls 4-6 mm L. Granitic sand, mly 7000-11,000', SN (all).

C. tenuis inflorescence (1 cm)

Cordylanthus

Bird's-beak

flower in normal position (1 cm)

Stems branched; roots yellow. Lvs linear and entire or pinnatifid into linear segments. Fls dull yellow or purple, in heads or scattered. Calyx forming a single piece, split almost to base ventrally, extending dorsally into a tongue-like structure, entire or 2-cleft apically. Calyx proper narrow, enclosing corolla at base only (the bract below the fl often confused with calyx). Galea not or scarcely longer than lower lip. Corolla itself more than twice as long as wide. Related sp has been shown to contain several alkaloids.

Infl a dense head of 3-10 fls, subtended by several 3- or 5-lobed bracts — *C. rigidus*
Infl of racemosely arranged clusters of 1-3 fls, subtended by 1 to few entire bracts — *C. tenuis*

C. rigidus, Stiffly-branched Bird's-beak: Stems 3-7 dm H; lvs mly 1-2 cm L, linear and entire or with a pair of linear lobes; heads mly 5-6-fld; bracts several; fl-bract 15-17 mm L, lance-oblong, purplish; calyx 12-15 mm L, lanceolate, ± entire; corolla 12-15 mm L. Dry, granitic slopes, 3000-6000', Mariposa Co. s., Jul-Aug.

C. tenuis, Slender Bird's-beak: Stems 3-6 dm H; lvs narrow-linear, entire, 1-3 cm L; fl-bract 13-14 mm L, purplish, oblong; calyx about 15 mm L, linear-lanceolate, purplish, entire or bidentate; corolla 12-13 mm L, greenish yellow, the brown galea purple apically. Dry, open slopes, 4500-8500', Fresno to Butte and Plumas cos. Jul-Sep.

Gratiola

Hedge-hyssop

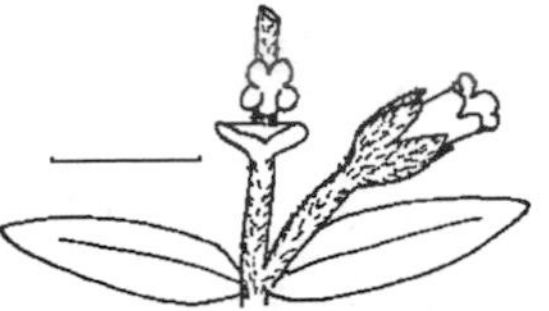

G. neglecta axillary flower (1 cm) *G. ebracteata axillary flower* (1 cm)

Stems erect or diffuse. Lvs opp, entire to denticulate, sessile. Fls white with yellow tube, loosely racemose, in axils of leafy bracts. Lvs of related sp edible.

Pedicels stout, without bracts; below 5000' *G. ebracteata*
Pedicels slender, with 2 bracts just beneath calyx; 5000-6500' *G. neglecta*

G. ebracteata, Bractless Hedge-hyssop: Stems mly glabrous; lvs lanceolate-attenuate, slightly clasping at base, 0.5-2.5 cm L; pedicels 10-20 mm L; corolla 5-7 mm L, the upper lobes united less than half their length. Tuolumne Co. n., Apr-Aug.

G. neglecta, Common Hedge-hyssop: Annual; lvs oblong-lanceolate to ovate, round-clasping at base, 0.5-5 cm L; pedicels 10-20 mm L, spreading; corolla about 10 mm L, the upper lobes joined almost to apex. Tuolumne Co. n., May-Aug.

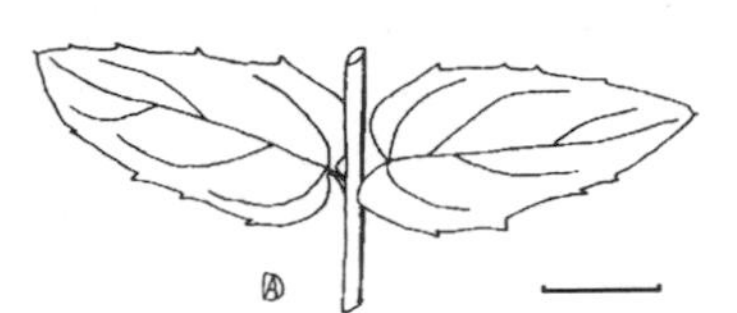

Keckellia

Gaping Penstemon

leaves (1 cm) *inflorescence* (1 cm)

K. breviflora: Shrub with many slender, glaucous stems 5-20 dm H; lvs opp, ± oblong, subsessile, 1-5 cm L, serrulate; infl 1-5 dm L, many-fld; pedicels glandular-pubescent; calyx 5-10 mm L; corolla white often tinged with pink, 15-18 mm L, the upper lip arched, galeate, over half the corolla length, the lower lip reflexed. Dry, rocky slopes below 8000', SN (all), May-Aug. Edibility unknown.

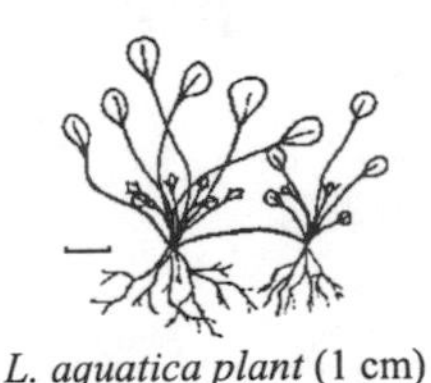

Limosella

Mudwort

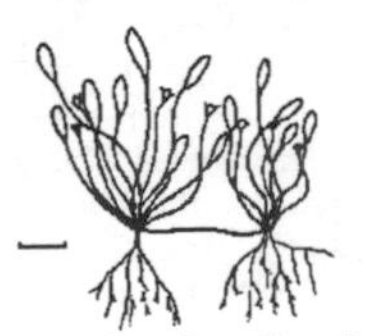

L. aquatica plant (1 cm) *L. acaulis plant* (1 cm)

Aquatic glabrous herbs with stolons. Lvs long-petioled, entire, palmately veined. Fls solitary on long pedicels. Corolla ± flat, 5-lobed, 1-2 mm W. Edibility unknown.

Lvs 0.5-2 mm W; corolla-lobes rounded *L. acaulis*
Lvs 2-8 mm W; corolla-lobes sharp-pointed *L. aquatica*

L. acaulis, Southern Mudwort: Lvs linear-oblanceolate, 0.6-1.2 cm L; petioles and pedicels several times as long. Muddy shores mly below 8000', Plumas Co. s., May-Oct.

L. aquatica, Northern Mudwort: Lvs elliptic to oblong, 1-1.5 cm L; petioles 2-4 times as long; pedicels about half as long as petioles. Occasional, below 10,500', SN (all), Jun-Sep.

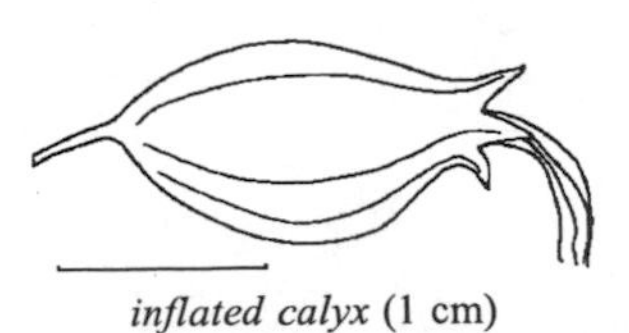

Mimulus

Monkey-flower

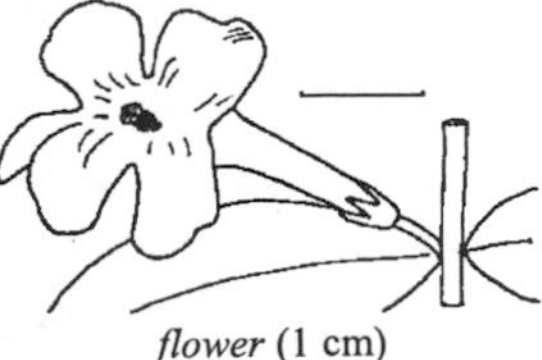

inflated calyx (1 cm) *flower* (1 cm)

Annual or perennial herbs or low shrubs. Lvs opp, entire to dentate or even divided. Fls in axils of foliate lvs or bracts, often in open racemes. Stamens 4. Style 1 with two flattened stigmas that fold together if touched. The young herbage may be eaten in salads; the lvs grow bitter with age but are still edible.

A Plant perennial
  B Fls red or magenta; 4-5 cm L; capsule 13-18 mm L
    C Stamens included; lvs lanceolate and entire — 13) *M. lewisii*
    CC Stamens exserted; lvs ± ovate, coarsely serrate — 5) *M. cardinalis*
  BB Fls yellow to buff; sometimes dotted red
    C Plant ± woody; corolla 5.5-6.5 cm L; capsule 15-17 mm L — 1) *M. aurantiacus*
    CC Plant herbaceous only; corolla mly 1-4 cm L; capsule 6-9 mm L
      D Calyx 1.5-2.5 cm L, often purple-dotted; corolla 2-3.5 cm L
        E Calyx scarious and strongly inflated when mature — 8) *M. guttatus*
        EE Calyx herbaceous, not strongly inflated; rootstocks yellow — 21) *M. tilingii*
      DD Calyx 0.6-1.2 cm L, not purple-dotted; corolla 1.5-2.6 cm L
        E Pedicels 1-3 cm L, equal to lvs; lvs pinnately veined — 16) *M. moschatus*
        EE Pedicels 2-12 cm L, erect, longer than lvs; lvs 3-5-veined from base — 18) *M. primuloides*
AA Plant annual
  B Pedicels conspicuous, mly longer than calyx; corolla mly deciduous, dropping before shriveling, leaving the styles exposed
    C Fls predominately yellow, sometimes partly white or with reddish spots
      D Lvs ± petioled except sometimes the uppermost, often palmately veined; stems usu at least 1 dm L
        E Stems glabrous to pubescent, never glandular; mature calyx strongly inflated — 10) *M. lanciniatus*
        EE Stems glandular-pubescent; mature calyx usu not inflated
          F Corolla 4-5 mm L; Lassen Co. n. — 3) *M. breviflorus*
          FF Corolla 7-20 mm L — 7) *M. floribundus*
      DD Lvs mly sessile except sometimes lowest, rarely palmately veined; stems often less than 1 dm L
        E Corolla less than 1 cm L
          F Stems 0.1-0.6 dm H, ± reddish, glandular-pubescent — 20) *M. suksdorfii*
          FF Stems 0.5-3.5 dm H, densely white-hairy, slightly sticky — 17) *M. pilosus*
        EE Corolla more than 1 cm L — 15) *M. montioides*
    CC Fls predominantly purple, red or magenta; mly Mariposa Co. s.
      D Plants glabrous; lvs 3-5-veined from base, serrulate — 9) *M. inconspicuous*
      DD Plants glandular-pubescent
        E Calyx-lobes mly 1-2 mm L; corolla 15-20 mm L — 6) *M. filicaulis*
        EE Calyx-lobes mly about 0.5 mm L; corolla 6-15 mm L — 4) *M. breweri*
  BB Pedicels mly shorter than calyces; corolla mly shriveling and persistent on the developing capsule
    C Corolla more than 20 mm L
      D Corolla-tube ± equal to calyx; stems mly over 10 cm H — 2) *M. bolanderi*
      DD Corolla-tube about 5 times as long as and exserted from calyx; stems 1-3 cm H — 19) *M. pulchellus*
    CC Corolla usu less than 20 mm L; stems mly glandular-pubescent
      D Stigmas and usu the longer stamens exserted from the corolla-throat; herbage strong-scented — 14) *M. mephiticus*
      DD Stigmas and anthers included in corolla-throat
        E Calyx 3-4 mm L; corolla 6-9 mm L — 12) *M. leptaleus*
        EE Calyx 5-8 mm L; corolla 10-22 mm L (see also *M. breweri*)
          F Corolla-lobes unequal; anthers glabrous — 22) *M. torreyi*
          FF Corolla-lobes subequal; anthers ciliate or pubescent
            G Herbage with strong odor; stems 1-2 dm H — 11) *M. layneae*
            GG Herbage without strong odor; stems 0.1-0.6 dm H — 23) *M. whitneyi*

1) *M. aurantiacus*, Notch-petaled Bush Monkey-flower: Stems 4-7 dm H, glutinous; lvs 2.5-6 cm L, oblong-elliptic, dark green above, entire to slightly toothed, ± revolute; pedicels 6-8 mm L; calyx 2.5-3 cm L, the lobes 5-10 mm L; corolla buff or cream. Rocky places below 5000', Placer to Plumas Co., May-Jul.

2) *M. bolanderi*, Bolander's Monkey-flower: Stems 2-6 dm H, erect; lvs glandular-pubescent, oblong to obovate, often serrulate toward tips, 2-6 cm L, with tobacco-like odor, sessile; pedicels 2-3 mm L; corolla mly 2-3 cm L, pink to red-purple, white with purple spots within, the lobes rounded. Dry, open places such as burns, below 6500', Fresno to Calaveras Co., May-Jul.

3) *M. breviflorus*, Short-flowered Monkey flower: Stems slender, often much-branched below, 0.3-1.5 dm H; lvs elliptic-lanceolate, denticulate, palmately 3-veined, 0.5-2 cm L,

subsessile to short-petioled; pedicels 5-13 mm L; calyx 5-8 mm L. Moist places, 5000-7100', May-Jun.

4) *M. breweri*, Brewer's Monkey-flower: Stems often branched, 0.3-1.8 dm H; lvs narrowly oblong-lanceolate, entire or nearly so, 0.5-3 cm L, sessile; pedicels becoming 3-10 mm L; calyx 5-7 mm L, the lobes about 1 mm L; corolla funnelform, pale pink to purplish, 6-10 mm L, the narrow throat with finely pubescent yellow ventral ridges, the subequal lobes notched. Damp, sandy places, 4000-11,000', SN (all), Jun-Aug.

5) *M. cardinalis*, Scarlet Monkey-flower: Stems freely branched, sticky-hairy, erect or decumbent, 2.5-8 dm L; lvs 2-8 cm L, sessile, longitudinally 3-5-veined, at least the upper with broad clasping bases; pedicels 5-8 cm L; calyx 2-3 cm L, the teeth 4-5 mm L; corolla sometimes yellowish, the throat yellowish with hairy ridges; capsule 16-18 mm L. Common in wet habitats below 8000', SN (all), May-Sep.

6) *M. filicaulis*, Slender-stemmed Monkey-flower: Stems erect, 0.5-0.9 dm H; lvs few, elliptic to oblong, less than 1 cm L; pedicels 10-25 mm L; calyx 5-6 mm L, often dotted red, inflated in fr; corolla with 2 broad yellow patches at the base of the lower lip. Moist places, Mariposa Co., May-Jun.

7) *M. floribundus*, Floriferous Monkey-flower: Stems much-branched, erect to decumbent, 1-5 dm L; herbage somewhat slimy; thin, ovate to lance-ovate, sharply dentate, scattered;, 1.5-4 cm L; pedicels 5-25 mm L; corolla-lobes unequal, rounded. Moist places, mly below 8000', SN (all), May-Aug.

8) *M. guttatus*, Common Large Monkey-flower: Stems glabrous below the infl, rooting at nodes, 0.5-10 dm H, mly unbranched; lvs oval, 1-8 mm L, the upper sessile, the lower long-petioled; infl ± racemose; pedicels 2-6 cm L; calyx-lobes acute and the lower sharply up-curved so as to partly close the orifice, the upper tooth mly less than 3 times the length of the others. Common in wet places below 10,000', SN (all), May-Aug.

9) *M. inconspicuous*, Small-flowered Monkey-flower: Stems erect, 0.5-2 dm H, often branched from base, with long internodes; lvs few, broadly ovate, 0.7-1.8 cm L, subsessile; pedicels 5-10 mm L; calyx 8-10 mm L, ridge-angled, the lobes 0.5 mm L; corolla 13-16 mm L, rose to rose-purple, with white and yellow throat. Moist places below 9500', May-Jul.

10) *M. lanciniatus*, Cut-leaved Monkey-flower: Stems 0.5-3 dm H, usu unbranched; lvs remote, oblong to oval in outline, 0.5-2.5 cm L; pedicels 1.5-4.5 cm L; calyx 5-15 mm L, ± spotted, the lower lobes up-curved against the much longer upper tooth; corolla 7-15 mm L, usu with large red-brown spot on lower lip. Damp sandy places, 3300-8700', Tuolumne Co. s., May-Jul.

11) *M. layneae,* Layne's Monkey-flower: Stems erect, often branched, 0.5-2.5 dm H; lvs oblong to linear, ± entire, 1-2.5 cm L, subsessile; calyx 5-9 mm L, somewhat inflated and strongly ribbed in age, the teeth 1-2 mm L; corolla 13-20 mm L, distally white with purple spots, the lobes with a purple median line. Dry, sandy and disturbed places below 7500', Fresno Co. n., May-Aug.

12) *M. leptaleus*, Least-flowered Monkey-flower: Stems less than 1 dm H, sometimes branched; lvs obovate to sublinear, entire, 0.5-1.5 cm L, the lower short-petioled; calyx 3-4 mm L, scarcely ridged, the teeth about 1 mm L; corolla slender, red-purple, ventrally yellow, with dark purple lines and spots, sometimes the lobes yellow. Open, gravely places, 6400-11,000', Tulare to Plumas Co., Jun-Aug.

13) *M lewisii*, Lewis' Monkey-flower: Stems erect, mly unbranched, 3-8 dm H; herbage ± sticky-pubescent; lvs oblong-elliptic, 2-7 cm L, 3-5-veined from base, sessile or the lower short-petioled; pedicels 3-6 cm L; calyx 15-25 mm L, sharply angled, the teeth subequal, 4-6 mm L; corolla rose to lavender with darker lines down the throat and with 2 yellow hairy ridges. Stream banks, 4000-10,000', SN (all), Jun-Sep.

14) *M. mephiticus*, Skunky Monkey-flower: Plants with strong skunky odor; stems erect, often branched, 0.2-1.2 dm H; lvs oblanceolate to sublinear, entire, subsessile; pedicels 1-5 mm L; calyx ridge-angled, the teeth lanceolate, the lobes about 2 mm L; corolla purplish-red to yellow, 12-17 mm L, the lobes rounded, subequal.

15) *M. montioides*, Montia-like Monkey-flower: Stems 2-6 cm H, usu profusely branched; lvs entire, ± linear, 0.6-1.3 cm L; pedicels 0-.5-2.5 cm L; calyx 5-7 mm L, the calyx-lobes roundish, about 0.5 mm L; anthers glabrous, stigmas fringed. Moist habitats, 3000-7000', May-Jul.

16) *M. moschatus*, Musk Flower: Stems 0.5-3 dm L, creeping or decumbent, often diffusely branched, glabrous to glandular-pubescent; lvs ovate to somewhat oblong, 1-5 cm L, subentire to denticulate, short petioled or the upper sometimes sessile; calyx 9-12 mm L, the teeth 2-3 mm L, often slightly unequal; corolla-tube purple veined. Moist to wet habitats, below 8800', SN (all), Jun-Aug.

17) *M. pilosus*, Downy Mimetanthe: Stems erect or ascending, 0.5-3.5 dm H; lvs lanceolate to oblong, entire, 1-3 cm L, sessile; pedicels 5-15 mm L; calyx 6-7 mm L, deeply lobed, the lobes unequal; corolla 7-8 mm L, the lower lip usu with maroon spots. Common in moist, sandy and gravelly places below 8500', SN (all), May-Sep.

18) *M. primuloides*, Primrose Monkey-flower: Stems 1-5 cm L, often with stolons; lvs obovate to oblong, often long-hairy above, dentate to entire, 1-4 cm L, usu in several pairs and closely set; pedicels slender, extending far above lvs; calyx 6-8 mm L, the lobes 1-2 mm L. Common in mdws and on wet grassy banks, 4000-11,200', SN (all), Jun-Aug.

19) *M. pulchellus*, Pansy Monkey-flower: Lvs sometimes petioled; calyx-lobes very unequal; corolla 2-5 cm L, the lower lip yellow, the yellowish tube 2.5-4 times as long as calyx, the throat purple except yellow below, the upper lobe shorter and purple, the lateral purple or yellow.

20) *M. suksdorfii*, Suksdorf's Monkey-flower: Stems usu densely branched; lvs oblong to oblanceolate, entire, 0.5-1.2 cm L, subsessile, often reddish below; pedicels 2-7 mm L; calyx 4-6 mm L; corolla 5-6 mm L, the ventral ridges faintly brown-spotted, the lobes mly notched. Moist, sandy places, mly 5000-13,000', SN (all), May-Aug.

21) *M. tilingii*, Larger Mountain Monkey-flower: Stems glabrous or nearly so, ± branched, 2-4 dm H, lvs few, often slimy, broadly ovate to subelliptic, usu dentate, 1-3 cm L, the lower short-petioled, the upper sessile, not usu connate; fls few; pedicels 2.5-5 cm L; calyx 1.5-2 cm L in fr, strongly angled, the lowermost pair of teeth curved up against the median upper which is 3-5 mm L. Wet banks, 6400-12,000', SN (all), Jul-Sep.

22) *M. torreyi*, Torrey's Monkey-flower: Stems erect, often branched, 0.5-3.5 dm H; lvs obovate to narrow-elliptic or oblong, 1.5-3.5 cm L, entire or nearly so, usu petioled; calyx 7-8 mm L, the lobes unequal, 1 mm L; corolla 15-22 mm L, the throat yellow-ridged, the lobes rounded. Dry, disturbed places below 8000', w. SN, Plumas Co. s., May-Aug.

23) *M. whitneyi*, Varicolored Monkey-flower: Lvs linear to elliptical, 1-2 cm L, entire, subsessile to short-petioled; calyx 5-6 mm in fr, the lobes unequal, 1-2 mm L; corolla 1-1.5 cm L, pale yellow with maroon blotches and lines, or purple with similar paler areas. Rare, gravelly places, 6000-11,000', Fresno and Tulare cos., Jun-Aug.

Orthocarpus

Owl's-clover

O. copelandii inflorescence (1 cm) *flower* (1 cm)

O. cuspidatus: Stems 1-3.5 dm H; lvs 1-5 cm L, entire or the upper with a pair of lateral lobes; spike dense, 2-10 cm L; bracts abruptly different from lvs, about 1 cm L, purplish, with a broad middle and small side lobes; calyx about 1 cm L, deeply 2-cleft and each division 2-lobed; corolla 1-2 cm L, nearly hidden by bracts. Dry habitats often in sagebrush, 6800-9000', Jun-Aug.

Pedicularis

Lousewort

P. semibarbata plant (5 cm) *P. groenlandica flower* (1 mm)

Stems usu unbranched. Lvs alt or basal, usu pinnatifid or pinnate. Fls in a usu spike-like raceme. Capsule flattened, glabrous. Lvs of related sp have been used to make a tea.

A Lvs serrate-dentate; calyx seemingly 2-cleft; Nevada Co. n. — 4) *P. racemosa*
AA Lvs deeply pinnatifid or pinnate; calyx 4-5-cleft
  B Upper lip of corolla slender ± tubular and curved like an elephant's trunk; lvs pinnate
    C Infl hairy; trunk-like beak 2-6 mm L — 1) *P. attollens*
    CC Infl glabrous; trunk-like beak 6-12 mm L — 3) *P. groenlandica*
  BB Upper lip of corolla blunt; lvs bipinnate
    C Stem less than 1 dm H, exceeded by basal lvs; 5000-11,000' — 5) *P. semibarbata*
    CC Stem 1-5 dm H, exceeding lvs; below 6000' — 2) *P. densiflora*

1) *P. attollens*, Little Elephant Heads: Stems erect, glabrous below the infl, 1.5-4 dm H; lvs basal and on lower stem, the blades 3-9 cm L; petioles shorter; corolla body about 7 mm L, glabrous, lavender or pink. Common in mdws and moist places, 5000-12,000', SN (all), Jun-Sep.

2) *P. densiflora* ssp *aurantiaca*, Indian Warrior: Stems several, 1-5 dm H, pubescent; lvs mly in a basal rosette or well distributed, the blades 5-15 cm L on shorter petioles; spike dense,

oblong, the bracts about as long as fls; calyx 8-12 mm L, deep red; corolla 2-2.5 cm L, the galea 12-18 mm L, orange to yellow. Dry slopes, Sierra to Plumas Co., May-Jun.

3) *P. groenlandica*, Elephant Heads: Stems 3-6 dm H; lvs basal and on lower stems, the blades 3-10 cm L, purplish when young; petioles about 5-20 mm L; infl 1-2 dm L; corolla-body 8-10 mm L, glabrous, red-purple. Occasional, mdws and wet places, 6000-11,200', SN (all), Jun-Aug.

4) *P. racemosa*, Leafy Lousewort: Stems 3-5 dm H, glabrous below the finely pubescent infl; lvs all cauline, 3-7 cm L, short-petioled, glabrous, lanceolate; pedicels 2-4 mm L; calyx 5-8 mm L; corolla 10-12 mm L, glabrous, pink or whitish; upper lip erect, strongly incurved, prolonged into a hooked beak.; lower lip 10-12 mm W. Dry slopes, 4000-7000', Jun-Aug.

5) *P. semibarbata*, Pine-woods Lousewort: Stems few, mly underground, not over 1 dm L; lvs in a rosette, 5-15 cm L, the primary segments ovate in outline usu irregularly pinnately lobed; petioles to about as long as blades; pedicels 4-5 mm L; calyx about 10 mm L; corolla 15-20 mm L, yellowish with purplish tips, the lower lip shorter than the upper. Common on dry forest floor, SN (all), May-Jul.

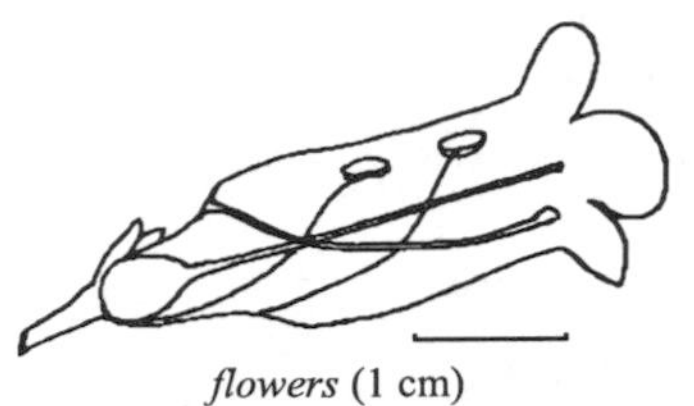

flowers (1 cm)

Penstemon

Beard-tongue

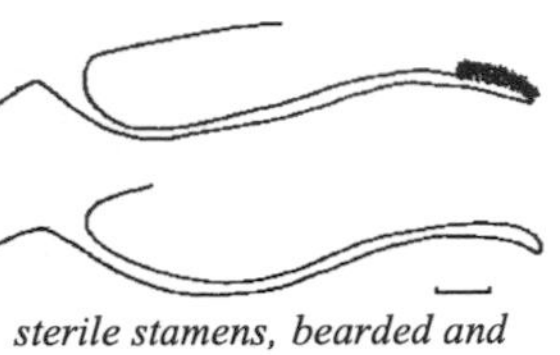

sterile stamens, bearded and glabrous (1 mm)

Herbs or shrubs. Lvs opp, rarely whorled or the upper alt, the lower usu petioled, the upper sessile. Fls usu showy in panicles, 5-merous, the 5th stamen sterile, often bearded. Edibility unknown.

A Corolla white to yellow with purple guidelines; plants woody — 4) *P. deustus*
AA Corolla red to violet
  B Staminode usu glabrous
    C Corolla scarlet, 20-35 mm L; Alpine Co. s. — 14) *P. rostriflorus*
    CC Corolla bluish or purplish
      D Lvs spatulate to round with a long petiole, often clustered near base — 2) *P. caesius*
      DD Lvs linear to lanceolate or oblanceolate, if broader then sessile
        E Corolla less than 1 cm L; stems 4-15 cm H; above 6500' — 13) *P. procerus*
        EE Corolla mly 2 cm L; stems 10-50 cm H
          F Infl glandular-pubescent
            G Herbage green, usu hairy — 8) *P. laetus*
            GG Herbage glaucous, glabrous — 9) *P. neotericus*
          FF Infl not glandular-pubescent
            G Lvs just below infl clasping stem at base
              H Corolla 14-20 mm L; anthers about 1.5 mm L — 12) *P. parvulus*
              HH Corolla 20-30 mm L; anthers mly 2-3 mm L — 1) *P. azureus*
            GG Lvs just below infl narrowed at base
              H Lvs linear; below 5000' — 7) *P. heterophyllus*
              HH Lvs usu wider, thickish; mly above 5000' — 16) *P. speciosus*
  BB Staminode bearded, usu densely so
    C Corolla 20-35 mm L; plants usu woody
      D Stems erect, 2-8 dm L; stems herbaceous; 7000-9500', e. SN — 11) *P. papillatus*
      DD Stems decumbent or ascending, usu less than 3 dm H
        E Stems 1.5-3 dm H; corolla rose-red to rose-purple — 10) *P. newberryi*
        EE Stems less than 1 dm H; corolla purple-violet — 3) *P. davidsonii*
    CC Corolla 7-20 mm L; plants mly herbaceous
      D Infl not glandular-puberulent — 15) *P. rydbergii*
      DD Infl, calyx and pedicels glandular-puberulent
        E Stems mly 1-2 dm H; staminode included; mly 8000-12,000', Plumas Co. s. — 6) *P. heterodoxus*
        EE Stems 2-7 dm H; staminode reaching orifice; 4200-8300', Lake Tahoe n. — 5) *P. gracilentus*

1) *P. azureus*, Azure Penstemon: Stems woody at base; the herbage blue-glaucous and glabrous throughout; basal lvs oblanceolate to obovate, 1-5 cm L, short-petiolate, 5-18 mm W; calyx 3.5-6 mm L; corolla deep blue-purple, 7-12 mm W. Dry slopes, 3500-8000', Fresno Co. n., May-Aug.

2) *P. caesius*, Cushion Penstemon: Stems decumbent to erect, 1.5-4.5 dm L, matted at base; herbage ± glaucous, glabrous below; lower lvs 1-2 cm L, on petioles as long; infl glandular-pubescent, few-fld; calyx 4-7 mm L; corolla purplish blue, ± bell-shaped, about 20 mm L. Dry, rocky slopes, 6700-11,300', Tulare Co., Jun-Aug.

3) *P. davidsonii*, Timberline Penstemon: Stems forming creeping mats; lvs elliptic to round, thick, ± glandular-punctate, entire, 0.5-1.5 cm L; infl few-fld, glandular-pubescent; calyx 7-11 mm L; corolla 20-35 mm L, bell-shaped; anthers included, arched along sides of corolla; staminode barely 1/2 as long as fertile white-hairy filaments. Rocky places, 9000-12,300', SN (all), Jul-Aug.

4) *P. deustus*, Hot-rock Penstemon: Stems woody and much-branched below, forming clumps 2-6 dm H; lvs 10-45 mm L, coarsely serrate; infl elongate, of about 10 distinct whorls of fls; corolla 10-15 mm L, the upper lip shorter than lower. Dry, rocky places below 8500', Alpine Co. n., May-Jul.

5) *P. gracilentus*, Slender Penstemon: Stems many from compact crown, glabrous below the ± glandular-pubescent infl; lvs entire, thin, mly basal, the lowest oblanceolate, 1-5 cm L, with petioles equally long; infl open, with 3-5 nodes; calyx 3-5 mm L; corolla purplish blue to red-purple, 13-16 mm L. Rather dry places, 4200-8300', Jun-Aug.

6) *P. heterodoxus*, Sierra Penstemon: Lvs thin, glabrous, the basal linear-oblanceolate to spatulate, 1.5-6 cm L; petioles as long as blades; stem lvs oblanceolate to broadly lanceolate, sessile; infl of 2-4 rather distinct many-fld clusters; calyx 3-6 mm L; corolla blue-purple, rarely paler, 10-16 mm L, the lips equal; staminode occasionally glabrous. Common, rocky slopes and alpine mdws, Jul-Aug.

7) *P. heterophyllus* ssp *purdyi*, Foothill Penstemon: Stems 2.5-7 dm H, puberulent throughout, woody at base; lvs 2.5-9 cm L; infl glabrous; calyx 4-6 mm L; corolla rose-violet, with blue or lilac lobes, 25-35 mm L. Dry hillsides, Madera to Butte Co., May-Jun.

8) *P. laetus* vars., Gay Penstemon: Stems woody at base, 2-8 dm H, the herbage mly puberulent; lvs linear to oblanceolate, the upper lanceolate, 1-6 cm L; infl narrow but open; calyx 5-15 mm L; corolla 15-30 mm L. Dry, rocky and disturbed slopes, below 8500', SN (all), May-Jul.

9) *P. neotericus*, Plumas County Beardtongue: Stems many, erect, 2-6 dm H, green; herbage glabrous, blue-glaucous; lvs crowded near base, more remote above, leathery, the lower narrowly oblanceolate to spatulate, 1-4 cm L; infl usu elongate, the divergent peduncles 1-2-fld; calyx 4-7 mm L corolla usu tricolored, 25-35 mm L. Dry places, 3500-6,000', Sierra Co. n., May-Aug.

10) *P. newberryi*, Mountain Pride: Herbage green or glaucous; lvs leathery, elliptic to ovate, obtuse, mly short-petioled, serrate, 1-3.5 cm L; infl short, ± 1-sided; calyx 7-12 mm L; corolla 20-30 mm L; anthers exserted; staminode 2/3 as long as fertile filaments. Rocky and gravely places, 5000-11,000', SN (all), Jun-Aug.

11) *P. papillatus*, Inyo Penstemon: Stems few, erect, 2-4 dm H; herbage gray-green, puberulent; basal lvs elliptic to spatulate-orbicular, 1-3 cm L, on petioles about as long; infl glandular-pubescent, compact, of 3-6 nodes; calyx 7-10 mm L; corolla blue, 24-30 mm L. Rocky, open slopes, Jun-Jul.

12) *P. parvulus*, Small Azure Penstemon: Near to *P. azureus*; stems 2-3.5 dm L; basal lvs 1-3 cm L; infl narrow, short. Dry slopes, 2400-9500', Tulare and Fresno cos., and perhaps at Donner Lake and locations further n., Jun-Aug.

13) *P. procerus*, Small-flowered Penstemon: Stems slender, tufted, glabrous throughout; lvs deep green, rather firm, the basal elliptic to linear, the blades about 1 cm L, sometimes folded, the petioles about as long; infl of 1-2 dense clusters; calyx 2-3 mm L; corolla deep blue-purple, 7-11 mm L; staminode glabrous to lightly bearded. Rocky, sometimes moist slopes 6500-11,000', Alpine, Mono and Tuolumne cos., Jul-Aug.

14) *P. rostriflorus*, Beaked Penstemon: Stems woody at base, 3-10 dm H; herbage yellow-green, glabrous or puberulent; lower lvs 2-6 cm L on somewhat shorter petioles, linear-oblanceolate to spatulate; infl ± 1-sided, glandular-pubescent; calyx 4-8 mm L; corolla 4-6 mm W, the lower lip sharply reflexed, the upper sometimes galeate. Dry slopes, 5000-10,700', Jun-Aug.

15) *P. rydbergii*, Rydberg's Meadow Penstemon: Stems 2-5(-7) dm H, the herbage bright green, glabrous, stem lvs 6-15 mm W; lvs thin, the basal linear-oblanceolate to elliptic, 2.5-8 cm L, short-petioled, the uppermost clasping stem; infl of 1-6 rather distinct many-fld clusters; calyx 3-5(-7) mm L; corolla blue-purple, staminode reaching orifice, rarely glabrous, 10-14 mm L. Dry to wet mdws, 4600-8500', Inyo and Fresno cos. n., May-Aug.

16) *P. speciosus*, Showy Penstemon: Stems erect, tufted, 2-8 dm H; lvs entire, thickish, the basal 3-8 cm L, the stem lvs linear-lanceolate; infl elongate, of many obscurely interrupted showy clusters; calyx 4-6 mm L; corolla bright blue-purple, 25-35 mm L, abruptly flaring into the wide throat. Dry habitats below 8000', SN (all), Jun-Aug.

ssp *kennedyi*: Calyx 8-12 mm L; 8000-10,400'.

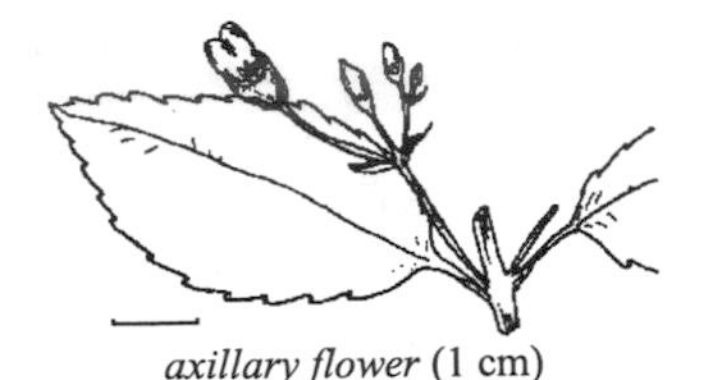

Scrophularia

Bee Plant

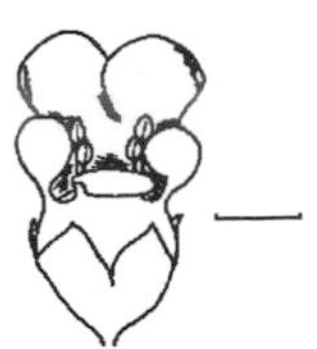

axillary flower (1 cm) *flower*(1 cm)

S. californica: Stems coarse, 10-20 dm H, finely pubescent; lvs opp, ± triangular, 3-12 cm L; petioles 2-6 cm L; infl mly 1.5-3 dm L; calyx 2-3 mm L; corolla greenish purple to maroon. Dryish slopes below 5000', w. SN (var *floribunda*) and 5000-10,000, mly s.e. SN (var. *desertorum*), Jun-Aug. Lvs and roots of similar sp very bitter and inedible.

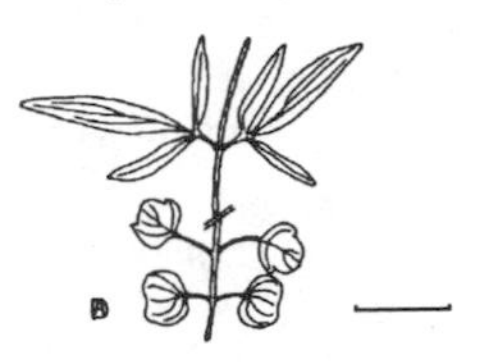

Tonella

Tonella

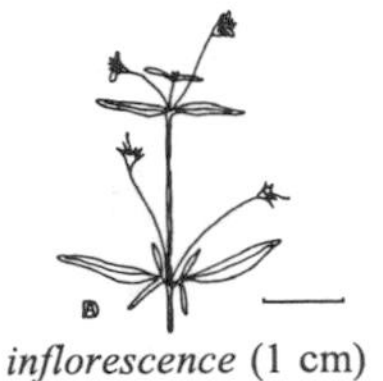

lower and upper leaves (1 cm) *inflorescence* (1 cm)

T. tenella: Slender annual, stem erect, 1-3 dm H, ± glabrous; lvs 1-1.5 cm L, ovate, entire or 3-lobed or -parted, petioled; fls in 2's or 3's in axils of upper lvs, pedicels filiform, 10-15 mm L; fls 2-4 mm L, violet. Mly in shade below 4500', Apr-May. Edibility unknown.

Verbascum

Common Mullein

V. thapsus: Stem stout, unbranched, 1-18 dm H; herbage woolly; basal lvs in rosettes; stem lvs gradually reduced upward, sessile, entire; infl dense, cylindric, 2-4 cm thick; calyx 7-9 mm L; corolla 20-25 mm W, yellow. Common, mly above 4000', SN (all), Jun-Sep. Mullein has astringent properties and has been used in a variety of ways for medical treatments. The herbage is not pleasant to taste and related spp have been shown to contain alkaloids.

plant (1 cm)

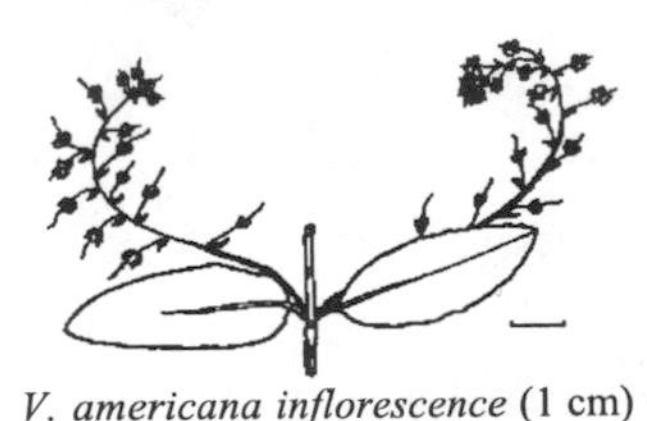

Veronica

Speedwell

V. americana inflorescence (1 cm) *V. serpyllifolia fruit* (1 mm)

Lvs opp or the upper bract-like, alt. Calyx 4- to 5-parted. Capsule flattened. The lvs and stems of *V. americana* and probably the others may be eaten in salads or as potherbs and are a source of vitamin C.

A Stems glabrous; main stem with lateral racemes below tip
  B Lvs on main stem short-petioled; corolla 7-10 mm W — 1) *V. americana*
  BB Lvs on main stem sessile; corolla 5-7 mm W — 4) *V. scutellata*
AA Stems pubescent; main stem ending in a single raceme-like infl
  B Plant annual; corolla 2-3 mm W, white — 3) *V. peregrina*
  BB Plant perennial; corolla 5-13 mm W
    C Corolla 10-13 mm W; stem 0.5-1.2 dm H; Alpine and Placer cos. — 2) *V. cusickii*
    CC Corolla 5-8 mm W; stems 1-4 dm H; widespread
      D Capsule 5-7 mm L, shallowly or not notched; corolla glabrous within — 6) *V. alpina*
      DD Capsule 4-5 mm L, deeply notched; corolla pubescent in the tube — 5) *V. serpyllifolia*

1) *V. americana*, Brooklime: Stems 1-10 dm L, ± succulent; principal lvs lance-ovate to lanceolate, serrate, 1-9 cm L, short-petioled; racemes open, few- to many-fld, the lower pedicels

to about 12 mm L; corolla violet-blue to lilac; capsule turgid, suborbicular. Wet places along streams below 10,500', SN (all), May-Aug.

2) *V. cusickii*, Ornamental Speedwell: Stems branched; lvs elliptic-oval, mly glabrous, entire, sessile, 0.5-1.8 cm L; pedicels 5-8 mm L; capsule deeply notched. Rare, mdws and moist places, 8200-9200', Jul-Aug.

3) *V. peregrina* ssp *xalapensis*, Purslane Speedwell: Stems erect, 1 to few, glandular-pubescent; lvs linear-oblong to spatulate, entire to ± dentate, sessile or the lower petioled, 1-2.5 cm L; raceme open, leafy-bracted; pedicels 1-2 mm L; capsule about 3 mm L, shallowly notched. Moist places below 10,200', SN (all), May-Aug.

4) *V. scutellata*, Marsh Speedwell: Stems weak, slender, 1-6 dm H; lvs sessile, linear to narrow-lanceolate, ± entire, 1.5-8 cm L; racemes divergent, open, small-bracted, rather few-fld; pedicels 5-15 mm L; corolla lilac to blue-lavender; capsule deeply obcordate. Wet habitats, 3500-7000', Fresno Co. n., May-Aug.

5) *V. serpyllifolia* and var. *humifusa*, Thyme-leaved Speedwell: Stems much-branched, pubescent, sometimes glandular; lvs ovate or oblong, ± entire, 0.5-1.5 cm L, the lower petioled; pedicels 4-5 mm L; corolla bright blue to whitish. Common in moist places, mly 6000-11,000', SN (all), May-Aug.

6) *V. wormskjoldii*, Alpine Speedwell: Stems erect, loosely pilose on stems and lvs, glandular-pubescent in infl; lvs elliptic or oblong-oval, 5-15 mm L, of 4-7 pairs, the upper often alt; racemes to 1 dm L in fr; pedicels 3-8 mm L, often alt; calyx 2-4 mm L; corolla violet with darker veins, the tube yellow. Wet places, mly 7000-11,500', SN (all), Jun-Aug.

SIMAROUBACEAE - Quassia Family

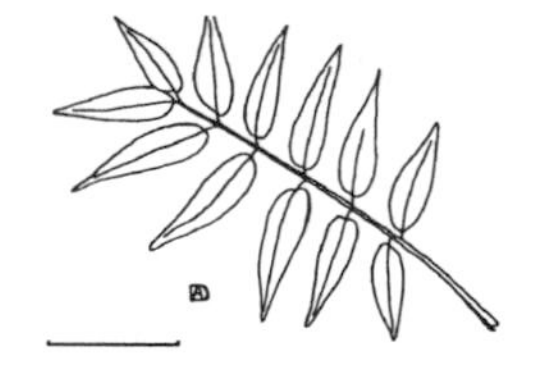

Ailanthus

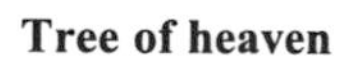

Tree of heaven

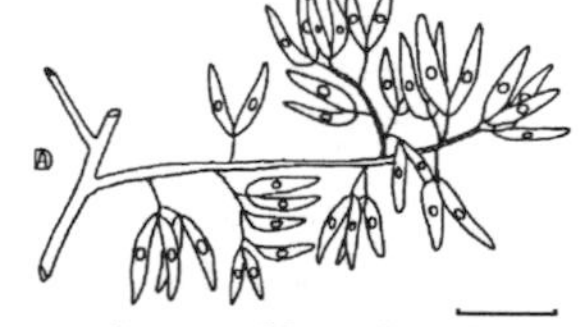

leaf (1 cm) *clusters of fruits* (1 cm)

A. altissima: Small tree 3-20 m H; lvs deciduous, odd-pinnate, 3-9 dm L, lfts 13-25, 8-13 cm L, lanceolate, ill-smelling when crushed; infl a large terminal panicle, the individual fls inconspicuous, 5-6 merous; fr a winged akene. Locally common in disturbed habitats, especially near old Chinese settlements, below 3500'.

SOLANACEAE - Nightshade Family

Lvs alt. Fls perfect, regular, 5-merous. Corolla radially symmetric, usu lobed.

Stems 3-20 dm H; fr a capsule; corolla ± tubular; annual — *Nicotiana*
Stems 0.5-3 dm H; fr a berry; corolla flat or bowl-shaped; perennial
  Corolla whitish, shallowly lobed and bowl-shaped; stems erect, rough-pubescent — *Chamaesaracha*
  Corolla deep violet with distinct spreading lobes; stem often in prostrate mats, often woody at base — *Solanum*

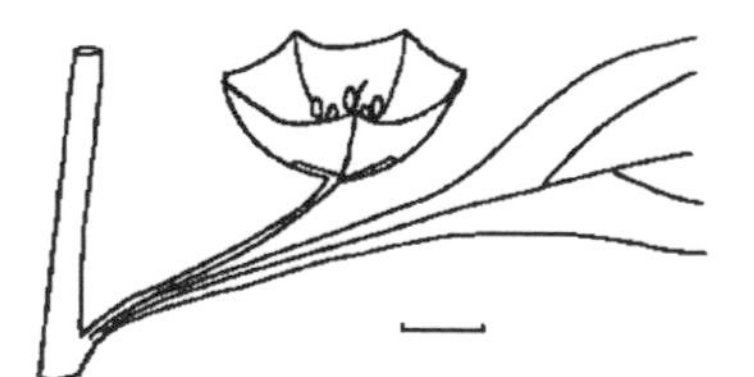

Chamaesaracha

Dwarf Chamaesaracha

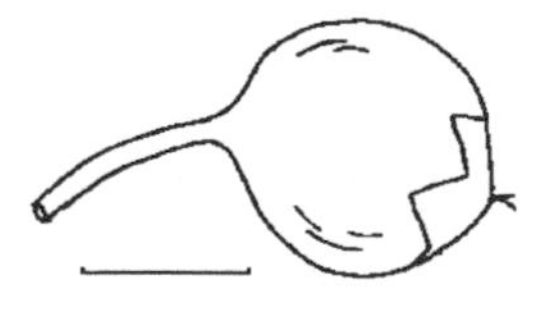

axillary flower (1 cm) *fruit* (1 cm)

C. nana: Stems 1 to few, leafy, 0.5-2.5 dm H; lvs ovate, 1.5-5 cm L, ± entire, white-pubescent, abruptly narrowed into equally long, narrowly winged petioles; calyx 6-10 mm L in fr; corolla with 5 basal green spots, 15-25 mm L; berry dull white to yellowish, about 1 cm in diameter. Sandy flats, 5000-9000', Mono Co. n., May-Jul. The berries of a similar sp are edible raw, cooked, or dried.

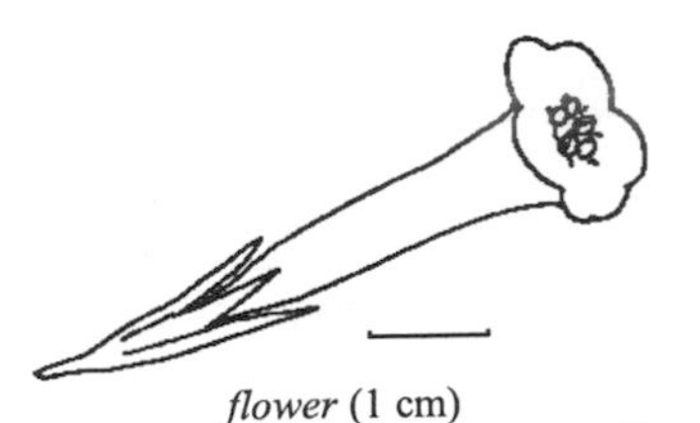

flower (1 cm)

Nicotiana

Coyote Tobacco

fruit (1 cm)

N. attenuata: Stems erect, heavily-scented; lvs 5-15 cm L, ovate to lanceolate, mly petioled; infl terminal, open, several- to many-fld; calyx 6-8 mm L; corolla white, 2.5-3 cm L, 1 cm W; capsule about 8-12 mm L. Disturbed places below 10,000', SN (all), May-Oct. Lvs may be used as an inferior smoking tobacco. The not unpleasant tasting lvs and stems contain nicotine which may cause irregular breathing, weakness and evidence of pain if eaten.

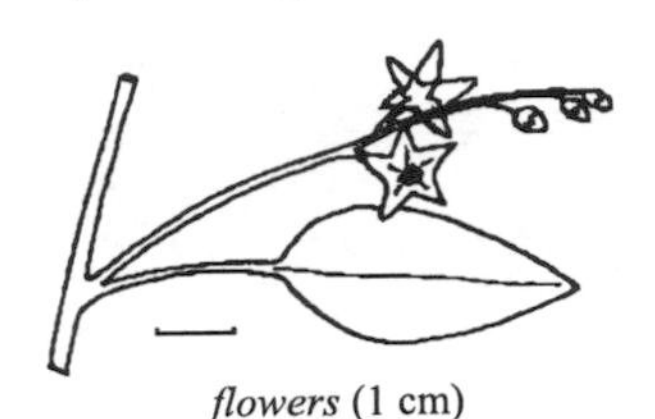

flowers (1 cm)

Solanum

Nightshade

leaves (1 cm)

S. xantii: Stems 1-4 dm L; lvs ovate, subentire, sometimes lobed at base, 2-4(-7) cm L, on petioles 5-20 mm L; fls mly 6-10 in lateral subumbellate clusters; calyx 5-6 mm L; corolla 1.5-2.5 cm in diameter; anthers bright yellow, about 5 mm L; berry greenish, round, about 10 mm in diameter. Dry places, 5000-9000', Nevada Co. s., May-Sep. All parts of the plant contain the poison solanine. Symptoms of solanine poisoning include trembling, drowsiness and weakness leading to unconsciousness and sometimes death. Digestive upset may or may not occur.

STAPHYLEACEAE - Bladdernut Family

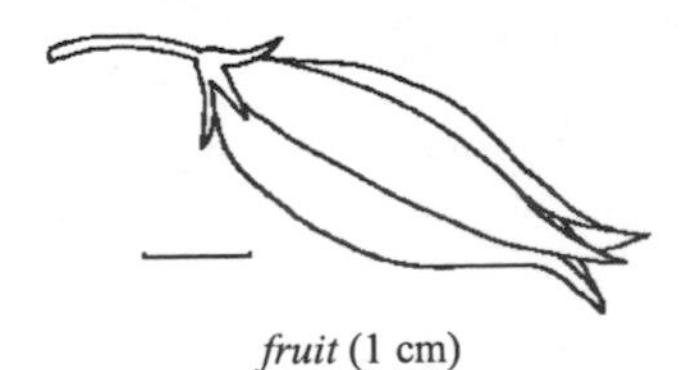

fruit (1 cm)

Staphylea

Bladdernut

leaves and axillary flowers (1 cm)

S. bolanderi: Stems erect, shrubby or arborescent, 2-6 m H, glabrous; lvs opp, deciduous, 3-foliolate; lfts broadly ovate to orbicular, 2.5-6 cm L, serrulate; petiole slender, 2-6 cm L; fls 5-merous in drooping axillary infl; calyx white, 8-10 mm L; petals white, 10-12 mm L; capsule 2.5-5 cm L, 3-horned. Occasional on canyon walls, below 4500', SN (all), Apr-May. Edibility unknown.

STERCULIACEAE - Cacao Family

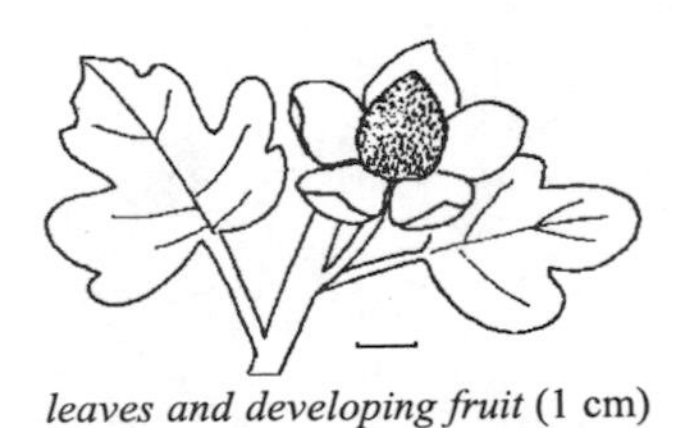

leaves and developing fruit (1 cm)

Fremontodendron

Fremontia
Flannel Bush

flower (1 cm)

F. californicum: Spreading evergreen shrub, 1.5-4 m H; lvs and fls mly on short lateral spur-like branchlets; lvs ± 3-lobed, dull green and sparsely stellate-pubescent above, densely tawny-stellate beneath, 1-5 cm L; petioles 1-3 cm L; fls showy, solitary, opp the lvs; calyx clear yellow, flat, 3.5-6 cm in diameter; capsule ovoid, 2.5-3.5 cm L. Dry, mly granitic slopes, 3000-6000', SN (all), Apr-Jun. Edibility unknown.

URTICACEAE - Nettle Family

Urtica

Stinging Nettle

upper stem (5 cm)

U. dioica: Stem 5-10 dm H, with stinging hairs (sting caused by injection of formic acid); lvs opp, lanceolate to narrow-ovate, 5-12 cm L, coarsely serrate, pubescent; petioles 1-4 cm L; stipules narrow-oblong, 5-10 mm L;. fls small, in dense axillary clusters, the pistillate and staminate in separate clusters. The young stems and lvs are edible after boiling, being a good spinach substitute. The older stems are too fibrous to be good eating.

VALERIANACEAE - Valerian Family

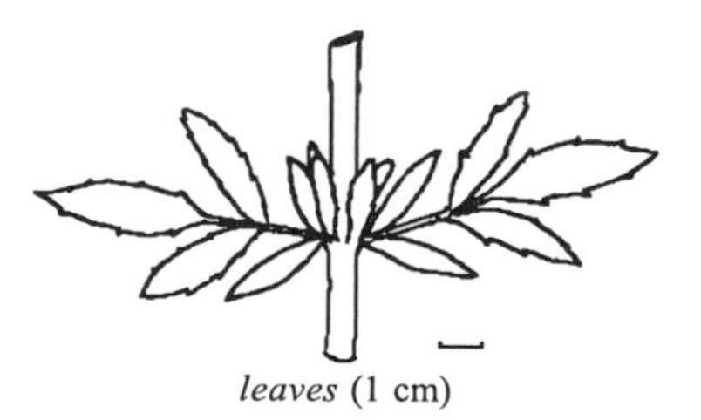
leaves (1 cm)

Valeriana

California Valerian

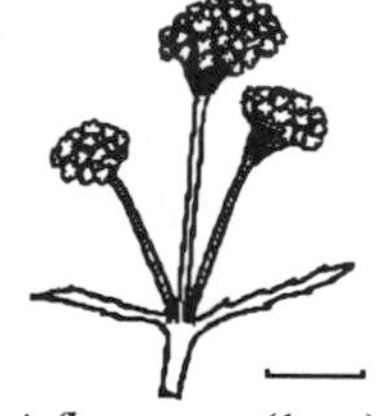
inflorescence (1 cm)

V. californica: Perennial from thickened strong-scented rhizome; stems 2-6 dm H, ± pubescent; lvs mly basal, usu forming a rosette, mly undivided, sometimes pinnatifid, 5-15 cm L including petiole; stem lvs opp, mly pinnate or pinnatifid, 3-8 cm L; infl 1.5-3 cm W at anthesis; corolla about 3 mm L, ± funnelform, the tube gibbous; stamens 3; stigma 3-cleft. Usu moist habitats, 5000-11,200', SN (all), Jul-Sep. The herbage of two related spp is edible raw; the root is edible cooked.

VERBENACEAE - Vervain Family

Verbena

Verbena Vervain

upper stem (1 cm)

V. lasiostachys: Stems much-branched, 3-8 dm L, pubescent; lvs ± ovate, opp, coarsely serrate to cut and lobed, 2-6 cm L, short-petioled; fls in slender terminal spikes, the spikes usu in 3's, 5-10 cm L; calyx 4-5 mm L; corolla mly purple, slightly two-lipped, the tube 4-5 mm L, the limb 3-4 mm W; fr dry, separating into 4 narrow nutlets. Dry to moist habitats below 8000', SN (all), Jun-Sep. Nutlets probably edible. Herbage of related spp slightly toxic..

VIOLACEAE - Violet Family

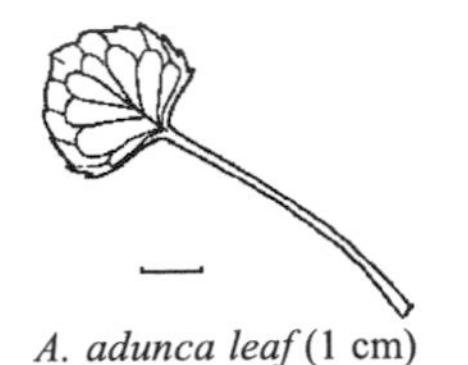

A. adunca leaf (1 cm)

Viola

Violet

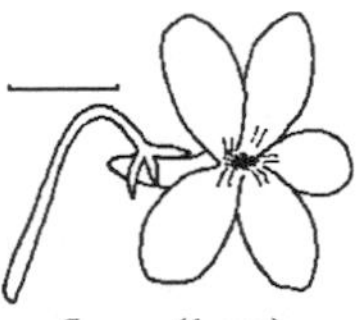

flower (1 cm)

Annual to perennial herbs. Lvs alt or basal, with stipules. Fls axillary, nodding, 5-merous, usu of 2 kinds - those of early season with showy petals and those of later season, cleistogamous - both usu fertile. Lower petal spurred, the other 4 in an upper, usu larger, pair and a lateral pair. Fr a capsule. Lvs and stems edible as greens.

A Petals mly white or violet with little or no yellow; plants usu of moist habitats
  B Lvs palmately 3-parted then bipinnately parted into linear segments — 3) *V. beckwithii*
  BB Lvs simple, entire to serrate
    C Plants lacking lf-bearing stems; spur 2-3 mm L — 8) *V. macloskeyi*
    CC Plants with evident lf-bearing stems
      D Petals white, deep red-violet on back; spur 3-5 mm L — 4) *V. cuneata*
      DD Petals violet, rarely whitish; spur 5-12 mm L — 1) *V. adunca*
AA Petals mly yellow; plants of dry habitats (except *V. glabella*)
  B Lf-blade dissected or deeply lobed
    C Lvs bipinnately 3-5-parted — 5) *V. douglasii*
    CC Lvs palmately lobed or divided
      D Lvs with 3-7 lobes, the lobes entire, ± oblong — 7) *V. lobata*
      DD Lvs 3-parted, these divisions again palmately 3-7-parted — 10) *V. sheltonii*
  BB Lf-blades ± entire to serrate, not dissected
    C Stems usu leafy only near apex, naked below
      D Back side of upper petals yellowish; stipules almost entire — 6) *V. glabella*
      DD Back side of upper petals brown; stipules deeply toothed — 7) *V. lobata* var. *integrifolia*
    CC Stems leafy throughout
      D Petals with some purple on backs; stipules coarsely toothed — 9) *V. purpurea*
      DD Petals not purplish on backs; stipules ± entire
        E Plants ± pubescent but not woolly; petals 8-10 mm L — 2) *V. bakeri*
        EE Plants gray-woolly; petals 6-7 mm L — 11) *V. tomentosa*

1) *V. adunca*, Western Dog Violet: Stems becoming 4-20 cm L; lvs round-ovate, ± cordate, obscurely scalloped, 1-4 cm L; petioles 1-6 cm L; sepals 5-8 mm L; petals 8-13 mm L, the 3 lower white at base and veined purple, the lateral white-bearded; spur rather broad, obtuse, often hooked at tip. Damp banks and edge of mdws, 5000-8000', SN (all), May-Jul.

2) *V. bakeri*, Baker's Violet: Stems 2-5 cm L; lvs lanceolate to narrow-ovate, entire, 2.5-4 cm L; petioles as long or longer; sepals 4-5 mm L; petals light yellow, 8-10 mm L, the lower 3 with brownish veins; spur about 1 mm L. Forest floor, 4500-8000', Fresno Co. n., May-Jul.

ssp *grandis*: Stems to 15 cm L; lf-blades 4-6 cm L; petioles 5-14 cm L. At 7000-8000', Placer and Plumas cos.

3) *V. beckwithii*, Great Basin Violet: Stems almost underground; lvs 2.5-3 cm L, the ultimate segments oblong to linear; petioles 1-5 cm L; peduncles 5-7 cm L; sepals 7-8 mm L; petals 8-14 mm L, the 2 upper dark red-violet, the 3 lower lilac with yellow area at base and veined dark violet. Dry, gravely places, often among shrubs, 3000-6000', mly e. SN, Inyo Co. n., Mar-May.

4) *V. cuneata*, Wedge-leaved Violet: Stems 7-20 cm H; basal lvs round-ovate to deltoid, scalloped, 1-2.5 cm L; petioles 2-7 cm L; stem lvs smaller; sepals 4-5 mm L; petals 8-10 mm L, the upper 2 may have a purplish base, the lateral with purple eye-spot near the base, the lower veined purple; spur yellowish. Moist, open forests below 5000', Nevada Co., Apr-Jun.

5) *V. douglasii*, Golden Violet: Stems 5-10 cm L; lvs ovate in outline; petioles 2-6 cm L; peduncles 5-12 cm L; sepals 6-12 mm L; petals mly 8-16 mm L. Grassy slopes and flats moist early in the season, mly 3500-7000', SN (all), Apr-May.

6) *V. glabella*, Smooth Yellow Violet: Herbage bright green; stems suberect, 1-3 dm H; basal lvs 2-3, reniform-cordate, toothed, 3-9 cm W, on petioles 5-25 cm L; stem lvs similar but smaller; stipules 4-10 mm L; petals 5-15 mm L, the lateral and lower with purple veins, the lateral bearded; spur 12 mm L. Wet, shaded places in woods below 8000', SN (all), Apr-Jul.

7) *V. lobata*, Pine Violet: Stems erect, 1-3 dm H; lvs mly at summit of stem, 2.5-10 cm W; petioles mly 1-5 cm L; stipules green, 10-15 mm L; sepals 6-10 mm L; petals 8-15 mm L, the 2

upper deep purple on back, others slightly purple, all or lower 3 with purple-brown veins toward base; spur 1-2 mm L. Rather dry slopes, in open woods, below 6500', SN (all), Apr-Jul.

var. *integrifolia*: Like the sp but lvs not lobed, irregularly scalloped.

8) *V. macloskeyi*, Macloskey's Violet: Plant with stolons, often forming dense patches; lvs thin, 1-3 cm L; petioles slender, 1-10 cm L; peduncles 2-15 cm L, reflexed; sepals 3-4 mm L, the lower 3 petals with purple veins. Wet banks and mdws, 3500-11,000', SN (all), May-Aug.

9) *V. purpurea*, Mountain Violet complex: Stems erect to prostrate or buried; lvs various; stipules membranous on basal lvs, greenish on stem lvs; spur 1-2 mm L. Dry habitats below 11,000', SN(all), Apr-Jun. Plants with an oval to round leaf blade 10-50 mm L are typical *V. purpurea*; those with narrower leaf blades have been referred to as either *V. pinetorum* (blade 3-25 mm W) or *V. praemorsa* (blade 20-45 mm W).

10) *V. sheltonii*, Fan Violet: Stems rising only slightly above ground; lvs 2-6 cm W, dark blue-green; petioles 3-10 cm L; stipules membranous; peduncles mly 8-15 mm L; sepals 6-8 mm L; petals 10-15 mm L, the 3 lower veined brown-purple, the 2 upper brown-purple on back. Rich loam, shade of open woods or brushy places, 2500-8000', SN (all), May-Jul.

11) *V. tomentosa*, Woolly Violet: Stems 3-15 cm L, prostrate in sun, suberect in shade; basal lvs 2-5, erect, entire or nearly so, 1.5-5 cm L on petioles 2-5 cm L; stem lvs smaller; stipules 5-15 mm L; peduncles 1-4 cm L; sepals 5-6 mm L; upper petals sometimes with darkening on back; spur scarcely 1 mm L. Local in dry, gravely places, 5000-6500', Eldorado to Plumas Co., Jun-Aug.

VITACEAE - Grape Family

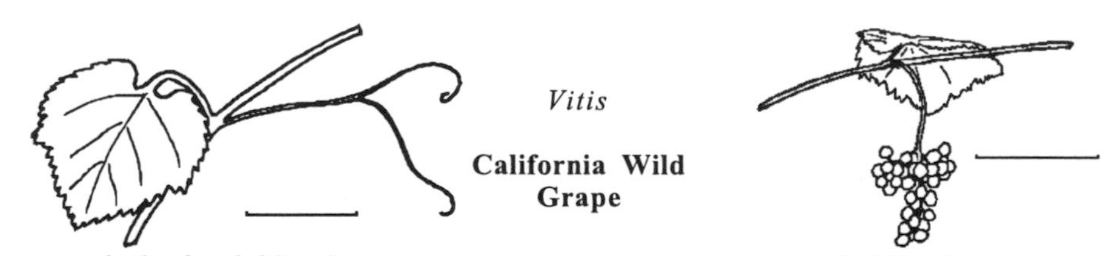

leaf and tendril (1 cm) *fruit* (5 cm)

V. californica: Stem to 15 m L, climbing by means of tendrils opp the lvs; lvs roundish, ± 3-lobed, usu deeply cordate at base, 7-14 cm W; petioles 3-12 cm L; infl 5-15 cm L; fls greenish yellow, 5-merous; fr spherical, 6-10 mm in diameter, purplish with a white bloom. Shady canyons, mly below 5500', w. SN (all), May-Jun. The fr is edible though inferior to the cultivated grape.

VISCACEAE - Mistletoe Family

Parasitic herbs. Branches swollen at the nodes. Lvs opp, without stipules, simple, entire, sometimes reduced to scales. Fr a berry.

Berry on a recurved pedicel, compressed; plants yellow — *Arceuthobium*
Berry sessile, globose; plants mly green — *Phoradendron*

Arceuthobium

Dwarf-mistletoe

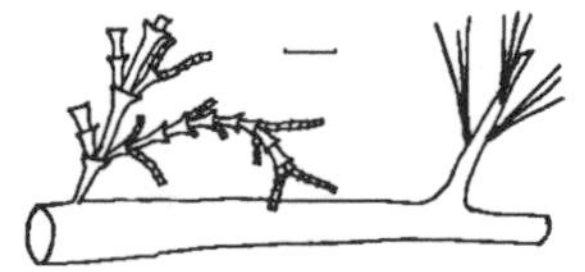

plan growing on host (1 cm)

Yellowish or brownish plants with fragile jointed stems, the segments glabrous, often ± 4-angled. Lvs reduced to connate scales. Parasitic on conifers. Possibly poisonous.

Fls blooming Aug-Sep; on *Pinus*, *Abies*, or *Tsuga* — *A. campylopodon*
Fls blooming Apr-Jun
  Plant on *Pinus* — *A. americanum*
  Plant on *Pseudotsuga* — *A. douglasii*

A. americanum, American Dwarf-mistletoe: Stem-segments usu greenish yellow, 12-15 times as long as thick; shoots 3-6(-10) cm L. Rare in SN.

A. campylopodon, Western Dwarf-mistletoe: Stem-segments yellowish to olive-green or brown, mly 5-10 times longer than thick; shoots 2-15 cm L. Most common sp in SN.

A. douglasii, Douglas' Dwarf-mistletoe: Stem-segments usu olive-green, 5-6 times longer than thick; shoots mly 0.8-2 cm L.

Phoradendron

Mistletoe

Stems ± woody, much-branched, brittle. Lvs entire and faintly nerved, or reduced to connate scales. Fls sunk in the jointed rachis. Berries of *P. flavescens* poisonous, causing intestinal inflammation, cardiovascular collapse and death.

plant (1 cm)

Lvs reduced to connate scales; on *Juniperus occidentalis* — *P. juniperinum*
Lvs broad and leaf-like
  On conifers — *P. bolleanum*
  On broad-leaved trees and shrubs — *P. flavescens*

P. bolleanum, Fir Mistletoe: Stems much-branched, dense to open; lvs 1-3 cm L, 3-10 mm W, usu oblanceolate. SN (all).

P. flavescens, Common Mistletoe: Plants stout, 3-6 dm H, internodes 3-5 cm L; lvs thickish.

P. juniperinum var. *ligatum*, Constricted Mistletoe: Stems glabrous, rather stout; internodes about 1 cm L

KEY TO THE MONOCOTYLEDON FAMILIES

A Petals and sepals lacking or reduced to bristles or scales
  B Fls in the axils of thin, dry scales and ± concealed by them
    C Lf-sheaths split lengthwise on the side opp the blade; stems mostly hollow and round; successive lvs forming an angle of 180 degrees along axis of stem POACEAE p. 228
    CC Lf-sheaths continuous around stem or ruptured by age; stems triangular, usu solid; successive lvs mly forming an angle of 120 degrees along axis of stem CYPERACEAE p. 209
  BB Fls not in the axils of thin, dry bracts or if subtended by bracts then exceeding or equaling them and not concealed
    C Plants floating or submersed
      D Fls in spikes or heads; lvs mly alt POTAMOGETONACEAE p. 242
      DD Fls axillary and few; lvs opp ZANNICHELLIACEAE p. 243
    CC Plants of land or shallow water; usu well emergent
      D Plant 1-3 m H; fls in a dense cylindrical spike TYPHACEAE p. 243
      DD Plant under 1 m H; fls not as above
        E Fls individually subtended by bracts or in dense clusters JUNCACEAE p. 213
        EE Fls in an open bractless raceme JUNCAGINACEAE p. 215
AA Petals present, conspicuous; sepals green or petaloid
  B Plants aquatic with ± floating or submersed lvs and submersed stems
    C Lvs ± linear, about 1 mm W; rare HYDROCHARITACEAE p. 212
    CC Lvs ± arrowhead-shaped, wider; widespread ALISMATACEAE p. 208
  BB Plants terrestrial, often growing in moist habitats
    C Ovary superior; fls regular LILIACEAE p. 215
    CC Ovary inferior; fl-stem leafless or fls irregular
      D Fls regular; lvs usu basal; infl 1-2-fld IRIDACEAE p. 212
      DD Fls irregular; stem lvs present or if not infl with many fls ORCHIDACEAE p. 225

ALISMATACEAE - Water-plantain Family

Perennials with leafless stems. Lvs basal, long-petioled, sheathing. Fls whorled, subtended by a whorl of bracts; petals white, 3; sepals 3.

Plants with clear sap; fls perfect *Alisma*
Plants with milky juice; fls unisexual *Sagittaria*

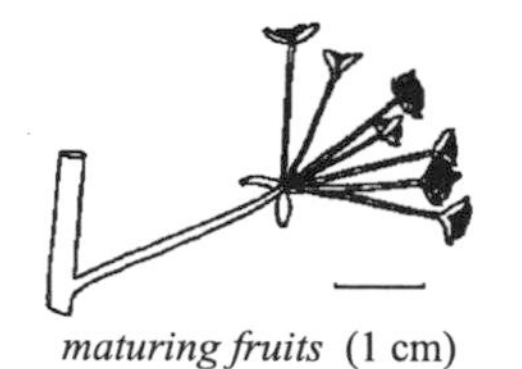

Alisma

Water-plantain

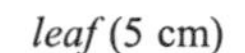

maturing fruits (1 cm)

leaf (5 cm)

A. plantago-aquatica: Fl-stems 6-12 dm H, exceeding lvs; lf-blades oblong to ovate, 3-15 cm L; pedicels slender, 2-4 cm L; fls many, small, in a pyramidal infl; sepals 2-4 mm L; petals 3-6 mm L. Margins of ponds and other wet habitats, below 5000', SN (all), Jun-Jul. The starchy bulbous base of the plant is edible after drying (removal of the strong flavor).

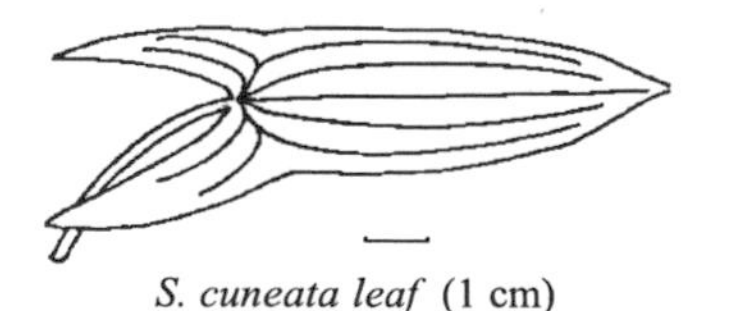

Sagittaria

Arrowhead

S. cuneata leaf (1 cm)

flowers (1 cm)

At least outer most lvs without distinct blades, the lf-blades arrow-shaped or lanceolate. Fls conspicuous, pedicelled, in whorls of 3. Stamens and pistils many. The tubers are an excellent starch source. They are found at the end of rhizomes perhaps 1 m or more from the parent plant. Best collected by wading barefoot and finding them with ones toes. Roast or boil for 15-30 minutes in salted water. The stem and lvs contain an alkaloid and may be poisonous.

Lf-blades 3-17 cm L; petals 6-10 mm L; fr heads 1-1.5 cm W — *S. cuneata*
Lf-blades 10-40 cm L; petals 10-20 mm L; fr heads 1.5-3 cm W — *S. latifolia*

S. cuneata, Arum-leaved Arrowhead: Lvs extending above water surface and 2-4 dm H or submersed and longer, arrow-shaped. Shallow ponds and swampy places below 7500', SN (all), Jun-Aug.

S. latifolia, Wappato, Tule-potato: Plants 2-12 dm H, glabrous; lvs variable. Edge of ponds or slow streams, below 7000', SN (all), Jul-Aug.

CYPERACEAE - Sedge Family

Stalks solid or rarely hollow, terete to variously angled. Lvs mainly basal, alt, the blades narrow and the sheaths closed. Spikelets composed of a series of scales (bracts) subtending individual fls. Perianth represented by several bristles or by an inner membrane or absent. Stamens 1-3. Pistil 1, the ovary superior.

A Fls all unisexual; akene surrounded by a sac-like bractlet
  B Pistillate fls all enclosed in perigynia — *Carex* p. 209
  BB Pistillate fls closely subtended by an inner glume; Convict Lake Basin — *Kobresia* p. 211
AA Fls perfect or perfect and staminate; akene naked
  B Scales on the spikelet in 2 columns
    C Stems 3-10 dm H, solitary, jointed; lvs about 10 cm L — *Dulichium* p. 210
    CC Stems less than 1 dm H, usu tufted; lvs 2-5 cm L, celery-scented — *Cyperus* p. 210
  BB Scales of spikelet spirally imbricate
    C Annuals from thread-like roots; bristles absent; invol. lvs present
      D Stalks 0.3-4 cm H; spikelets about 2 mm L — *Lipocarpha* p. 211
      DD Stalks 5-15 cm H; spikelets oblong, 5-8 mm L — *Bulbostylis* p. 209
    CC Perennials from rhizomes; bristles present, or if not then invol. lvs also absent
      D Bristles conspicuous, soft, smooth, white — *Eriophorum* p. 210
      DD Bristles hidden within scales, or if exserted then not as above
        E Spikelet solitary, terminal, erect; invol. lvs none — *Eleocharis* p. 210
        EE Spikelet various, but never as above — *Scirpus* p. 211

Bulbostylis

Slender Bulbostylis

B. capillaris: Stalks filiform; lvs basal, the blades filiform, shorter than stalks, the sheaths pubescent with long hairs; invol. lvs 1-3, short and bristly; infl a terminal umbel of 2 to several spikelets, or sometimes a solitary spikelet; scales puberulent, with deep brown sides and a green midrib. Mountain mdws, 4000-6000', Mariposa and Tuolumne cos., Jun-Aug. Edibility unknown. Related spp a source of fiber, the stems being too tough to eat..

upper stem (5 mm)

Carex

Sedge

spikelet (1 cm)

section through pistillate flower (1 mm)

Stalks solid, sharply triangular to nearly terete, leafy at base or the lowest lvs scale-like. Infl variable as to shape. Fls without perianth, each staminate fl consisting of 3 (or 2) stamens; each pistillate fl consisting of a 2 or 3 (or rarely 4) stigmatic pistil enclosed in an urn- or flask-shaped bractlet (perigynium) through the small orifice of which the style protrudes. Akene contained by and falling with the perigynium at maturity. Edibility unknown. This genus is almost impossible to distinguish to sp in the field. There exists a very large number of spp in the Sierra and hybridization between spp is common. No attempt will be made to describe the individual spp.

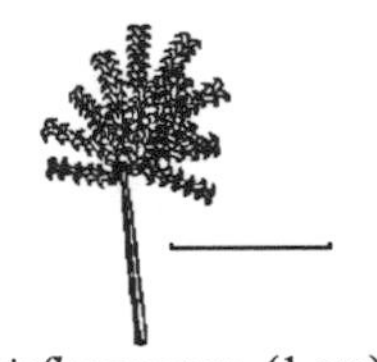

inflorescence (1 cm)

Cyperus

Umbrella-sedge

plant (1 cm)

C. aristatus: Stalks slender, smooth, 1-20 cm H; lvs 2 or 3 on stalk, flat, 0.5-3 mm W; invol. lvs extending much beyond the infl; infl umbellate, the umbels capitate or the rays 2 cm L or longer, bearing capitate clusters of spikelets; spikelets linear-oblong, 4-10 mm L, compressed. Wet ground up to 8500', SN (all). Rhizome of related sp edible, but rhizome of Sierran sp would be so small as to have negligible food value.

Dulichium

Dulichium

D. arundinaceum: Stalks stout, erect; lvs numerous, the basal lvs reduced to brown sheaths, the stalk lvs green, flat, linear, 2-8 cm L, 4-8 mm W; peduncles of the axillary racemes 5-25 mm L; spikelets linear, flat, about 6 to a raceme, 1-2.5 cm L, 6- to 12-fld; scales greenish brown, the lowermost scale of each spikelet sterile. Occasional in swamps, 5000-7500', Fresno to Plumas Co., Jul-Oct. Rhizome probably edible, but the plant is relatively rare and is usu growing among other plants of higher food value.

leaves and spikelets (5 mm)

Eleocharis

Spike-rush

Annual or perennial herbs with rhizomes, stolons or fibrous roots. Stalks unbranched, roundish, usu striate. Lvs reduced to basal lf-sheaths. Spikelets several- to many-fld. All grow in mdws or wet places. Rhizome of related spp edible. The general appearance of the various spp is very similar making them difficult to distinguish in the field.

plant (5 cm)

A Plants annual with fibrous roots
  B Plants 2-6 cm H; 3000-8000', Tulare Co. n. — *E. acicularis*
  BB Plants 10-40 cm H; Mariposa to Plumas cos. — *E. englemannii*
AA Plants perennial with rhizomes or stolons, these stolons filiform
  B Style 2-parted; seeds lens-shaped; widespread and common — *E. macrostachya*
  BB Style 3-parted; seeds triangular to obovoid
    C Spikelets many-fld; below 7000' — *E. kolanderi*
    CC Spikelets 4-10(-12)-fld; to 12,000'
      D Seeds with several longitudinal ridges; below 8000' — *E. acicularis*
      DD Seeds not longitudinally ridged; usu 5000-10,000' — *E. pauciflora*

Eriophorum

Slender Cotton-grass

E. gracile: Stalks slender, smooth, ± round, 30-60 cm H; lf-blades triangular, channeled, 2-30 cm L; invol. lf solitary, 1-2 cm L; infl a cluster of 2-5 spikelets; spikelets 6-8 mm L; scales gray to almost black; bristles numerous, 1-2 cm L. Mountain mdws and bogs. To 7000', Calaveras Co. n., May-Jul. Lower stems of similar sp edible.

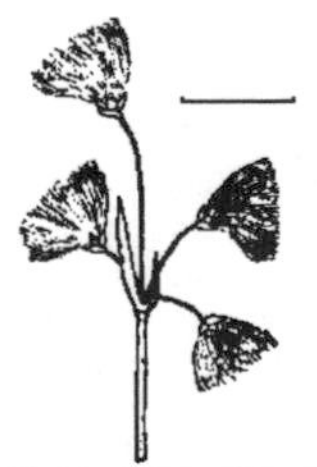

inflorescence (1 cm)

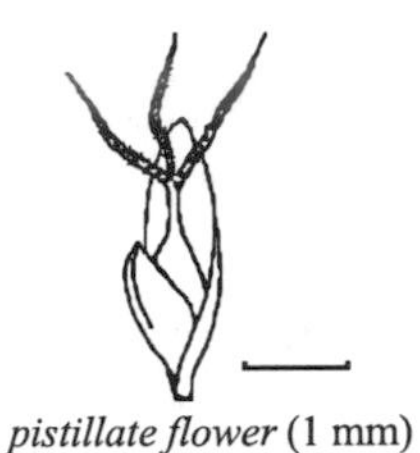

Kobresia

Mouse-tail Kobresia

pistillate flower (1 mm) *plant* (1 cm)

K. bellardii: Stalks very slender, 1-4.5 dm H, erect, leafy below; lvs filiform, the margins ± revolute; spike subtended by a short bract or bractlets, 1.5-3 cm L, 3-4 mm W, usu densely fld with 1 fl per scale; the lower fls usu pistillate, the upper staminate. Moist habitats, 9700-10,600'. Plant too rare to consider edibility.

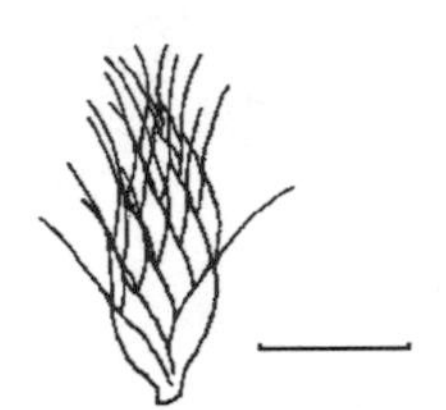

Lipocarpha

Lipocarpha

L. occidentalis spikelet (1 mm) *L. micrantha plant* (5 mm)

Slender annuals usu less than 4 cm H with grooved filiform lvs. Infl of 1-3 reddish brown, many-fld spikelets. Pollen and seeds edible

Scales of spikelets not awned; widespread below 5000' *L. micrantha*
Scales of spikelets awned; local in central SN *L occidentalis*

L. micrantha var *minor*, Common Lipocarpha: Stalks erect, surpassing the basal lvs; invol. lvs 1-3, 5-20 mm L; spikelets ovate. Moist habitats, Fresno to Plumas Co., Aug-Oct

L. occidentalis, Western Lipocarpha: Much like *L. micrantha* except as in key. At 4000-6000', Jun-Aug.

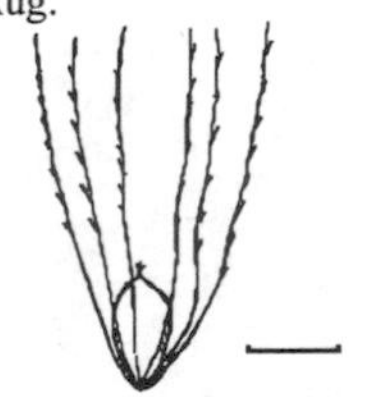

Scirpus

Bulrush Club-rush

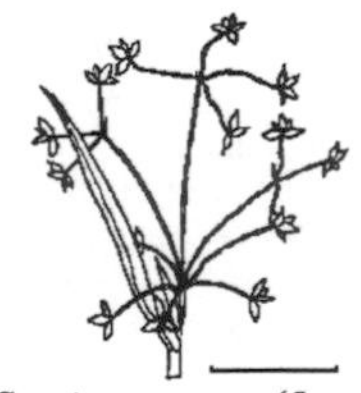

S. criniger floret (1 mm) *S. microcarpus* (5 cm)

Perennials with rhizomes or stolons. Stems erect, triangular to round The rhizomes of all spp are edible raw, cooked or dried and ground into a flour. The pollen and seeds and young stem-tips are also edible and can be prepared in a variety of ways.

A Infl open, conspicuously and diffusely branched
  B Akenes lens-shaped; stalks ± round, often over 5 dm H — 5) *S. microcarpus*
  BB Akenes triangular; stalks triangular, mly 3-5 dm H
    C Perianth bristles 2-4 mm L, usu exserted beyond scales — 3) *S. congdonii*
    CC Perianth bristles 1-2 mm L, usu included — 4) *S. diffusus*
AA Infl ± capitate, without conspicuous branches
  B Plant submersed except spikelets — 8) *S. subterminalis*
  BB Plant terrestrial
    C Low alpine dwarfs, mly less than 1 dm H; infl few-fld
      D Plant without stolons; bristles 6; Tuolumne Co. s. — 2) *S. clementis*
      DD Plants with stolons; bristles none; Convict Lake, Mono Co. — 6) *S. pumilis*
    CC Plants stout, mly over 3 dm H; infl usu many-fld
      D Stem round — 1) *S. acutus*
      DD Stem triangular — 7) *S. pungens*

1) *S. acutus* ssp *occidentalis*, Tule: Lvs usu basal, the blades much smaller than the sheaths; spikelets 3-many, 7-10 mm L, usu clustered. Wet places below 7500', SN (all).

2) *S. clementis*, Clement's Bulrush: Stalks slender, tufted; basal sheath bearing a linear blade 0.5-3 cm L; true invol. lf lacking, but lowermost scale of spikelet has green midrib extended as an awn 2-4 cm L is often confused as a lf; spikelets 2-4-fld. Alpine mdws, 8000-12,000', Jul-Aug.

3) *S. congdonii*, Congdon's Bulrush: Stalks slender; basal lvs to 8 mm W; invol. lvs 3-7 cm L, often shorter than infl; bristles white, longer than akenes. Mountain mdws, 4000-9000', Jun-Aug.

4) *S. diffusus*, Diffuse Bulrush: Much like *S. congdonii*; the infl more branched with primary rays mly again divided. Mly below 6500', SN (all).

5) *S. microcarpus*, Small-fruited Bulrush: Stems stoutish, leafy; lvs flat, 1-2 cm W, often reaching above stalk; longer invol. lvs usu extending beyond the heads; infl a loose cmpd umbel, the primary rays to 10 cm L. Wet habitats to 9000', SN (all), May-Aug.

6) *S. pumilis*, Rolland's Bulrush: Stalks in tufts, 5-15 cm H; spikelet solitary, 3-4 mm L. On limestone at 10,200-10,600'.

7) *S. pungens*, Common Threesquare: Stalks to 2 m H, strongly arched or erect, sharply 3-angled; lvs mly basal, usu as long as stem; blades 2-6, usu longer than sheaths; spikelets 1-5, 7-14 mm L, 4-5 mm W. Shores of lakes and marches below 7000', SN (all).

8) *S. subterminalis*, Water Bulrush: Stems 20-100 cm H, usu erect, about 1 mm W; lvs with blade much longer than sheath; spikelet 1, 5-10 mm L; bract 1, 1-6 cm L, erect, ± stem-like. Rare, in lakes or marshes below 7000', Nevada Co. n.

HYDROCHARITACEAE - Frogbit Family

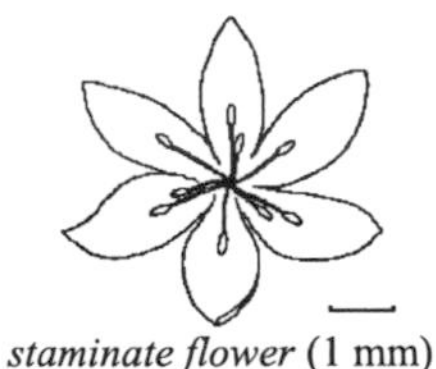

staminate flower (1 mm)

Elodea

Elodea

stem and leaves (1 cm)

E. nuttallii: Plants submersed, forming large masses of branching elongate stems with occasional slender roots; stems 3-10 dm H; lvs linear-oblong, 7-15 mm L; fls solitary. Rare, in slow streams and ponds below 9200', SN (all), Jul-Aug. Plant too rare to consider edibility.

IRIDACEAE - Iris Family

Perennials. Lvs mly basal, linear or sword-shaped. Fls terminal, showy, subtended by 1 or more green or membranous bracts, called a spathe. Perianth of 6 parts in 2 series, all united into a tube. Ovary inferior; stamens 3; pistil 1. Fr a few- to many-seeded capsule.

Infl not umbellate; perianth with 3 erect and 3 drooping segments, 4-7 cm L — *Iris*
Infl umbellate; perianth segments all alike, 2 cm L — *Sisyrinchium*

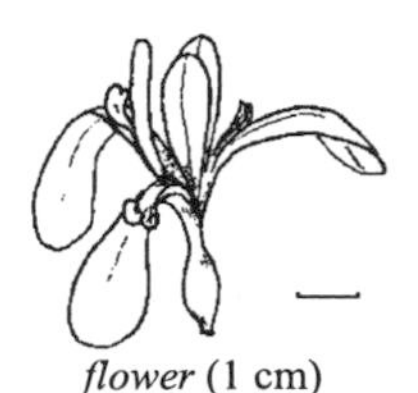

flower (1 cm)

Iris

Iris

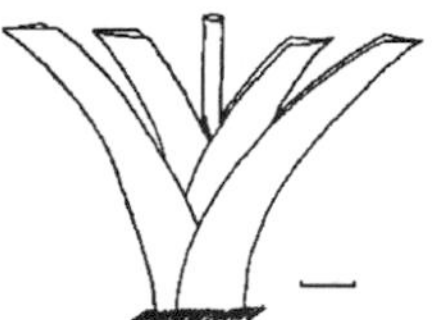

equitant basal leaves (1 cm)

Plants of moist or dry areas with large linear acute lvs and fls similar to the cultivated iris. The plant, especially the rootstock, produces a burning sensation when chewed and if eaten in quantity will cause vomiting and diarrhea.

Rhizome about 20-30 mm thick; moist places, usu e. SN — *I. missouriensis*
Rhizome less than 10 mm thick; dry places, usu w. SN
    Spathe-bracts separate, bases about 1-4 cm apart on stem, usu divergent — *I. hartwegii*
    Spathe-bracts opp, like closed lobster-claws — *I. tenuissima*

I. hartwegii, Hartweg's Iris: Lvs few, 2-6 mm W, up to 4.5 dm L; fl-stem slender, 0.5-3 dm H; fl color variable, usu pale yellow to cream, sometimes lavender or deep yellow. Wooded slopes, 2000-6000', Plumas Co. s., May-Jun.

I. missouriensis, Western Blue Flag: Lvs light green, glaucous, to about 4.5 dm L and 3-6 mm W, sometimes purplish at base; fl-stem slender, 2-5 dm H; spathe-bracts opp, membranous, with green parts at base, 4-7 cm L; sepals about 6 cm L, mly pale lilac to whitish with darker veins. Moist areas, 3000-11,000', May-Jun.

I. tenuissima, Slender Iris: Lvs gray-green, sometimes slightly glaucous, to 4 dm L and 6 mm W, the bases often pinkish or red; fl-stem with 2-3 stem lvs; perianth usu pale cream with purple veins. Dry, sunny woods, Sierra Co. n., May-Jun.

flowers (1 cm)

Sisyrinchium

Blue-eyed Grass

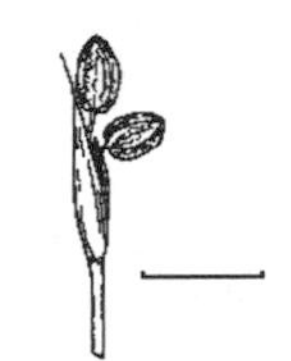

mature capsules (5 mm)

Fl-stems slender, compressed, ± winged. Fls ephemeral, opening in sun, subtended by 2 bracts. Herbage and rootstocks probably contain toxic quantities of alkaloids and should be avoided

Fls yellow — *S. elmeri*
Fls blue to purplish — *S. idahoense*

S. elmeri, Elmer's Blue-eyed Grass: Fl-stems less than 2 dm H; lvs about 1-3 mm W, half as long as stems; outer bract of spathe 2.3-3.5 cm L, scarcely longer than inner; perianth segments 8-12 mm L, orange-yellow, with 5 dark veins. Boggy and wet places, mly 4000-8500', Plumas Co. s., Jul-Aug.

S. idahoense, Idaho Blue-eyed Grass: Plants pale green, glaucous, 1-4.5 dm H; lvs about half as long as stems, 1-3 mm W, entire; spathe 1, green or faintly purplish; outer bract 2-6 cm L, the inner shorter; pedicels glabrous, 1.5-3 cm L; perianth segments 10-15 mm L, with darker veins. Wet mdws, 4500-10,800', SN (all), Jul-Aug.

JUNCACEAE - Rush Family

Mly perennial herbs usu of moist places. Stems rarely branching, terete or compressed. Lvs alt, sheathing, grass-like, flat to terete. Infl usu branched, rarely reduced to 1 fl. Fls borne singly and each subtended by a bractlet, or in a head-like or spike-like cluster and not individually subtended; infl usu subtended by 1 or more bracts. Perianth small, regular, usu 6-parted. Stamens usu 3 or 6. Ovary superior; style 1; stigmas 3.

Lf-sheaths open; lvs stiff, terete to flat — *Juncus*
Lf-sheaths closed; lvs soft, flat — *Luzula*

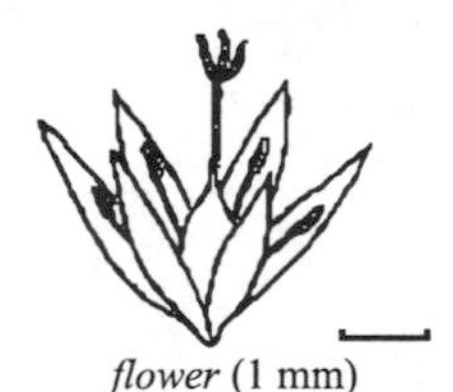

flower (1 mm)

Juncus

Rush
Wire-grass

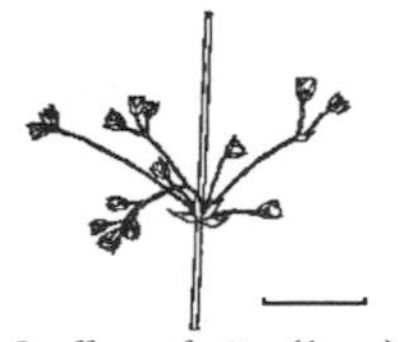

J. effusus fruits (1 cm)

Perennial or annual, glabrous. stems leafy or lvs all basal. Fls greenish to brownish or purplish. Stems tough, ± fibrous and inedible. Large genus which is often difficult to determine to sp in the field. Individual descriptions of the spp are not presented.

A Small annuals mly under 15 cm H
  B Infl ± open, forming the upper half of plant; stems to 20 cm H — *J. bufonius*
  BB Infl capitate
    C Plant 5-12 cm H; mly below 7500' — *J. triformis*
    CC Plants 0.5-5 cm H; widespread below 11,000'
      D Fls 1-7 per stem; bracts pointed
        E Fr shorter than perianth and lighter colored — *J. capillaris*
        EE Fr usu longer than perianth and similar in color — *J. tiehmii*
      DD Fls 1 per stem; bracts blunt at tip or lacking
        E Plants 0.5-1.5 cm H; fr mly 1 mm L. — *J. bryoides*
        E Plants mly over 1 cm H; fr 2-3 mm L — *J. hemiendytus*
AA Perennials mly over 15 cm H; stems tufted or from rhizomes

B Infl appearing lateral, with the bracts extending the stem above the infl; fls all individually subtended by bractlets
C Stems densely tufted, mly less than 30 cm H; fls usu 1-3; 6000-12,000', SN (all)
D Lf-blade 3-8 cm L — *J. parryi*
DD Lf-blade bristle-like, poorly developed — *J. drummondii*
CC Stems mly over 30 cm H; fls many
D Stems densely tufted, 6-13 dm H; mly below 7000', SN (all) — *J. effusus*
DD Stems scarcely tufted, 1-6 dm H; below 11,000'
E Lf-blades usu well developed; stem flattened, often twisted; Eldorado Co. s. — *J. mexicanus*
EE Lf-blades mly lacking; stems round, wiry; SN (all) — *J. balticus*
BB Infl appearing terminal, the bracts not forming an extension identical to the stem but usu flattened or grooved
C Lf-blades flat, the surface facing stem, without internal cross-walls
D Lf-blades 2-6 mm W; stems from stout rhizomes
E Infl of 2-10 heads; 3000-7500', Butte Co. n. — *J. howellii*
EE Infl of 1 to several heads; 4000-11,000', SN (all) — *J. orthophyllus*
DD Lf-blades mly less than 2 mm W; stems mly tufted; plants not common
E Each fl subtended by 2 bracts
F Infl ± open; perianth usu over 4 mm L; mly below 7000', Madera Co. n. — *J. occidentalis*
FF Infl dense; perianth 3-4 mm L; 4000-8000', Tuolumne Co. n — *J. confusus*
EE Fls in heads, the heads subtended by 2 bracts
F Stems 10-25 cm H; perianth segments 2-4 mm L; below 8000' — *J. covillei*
FF Stems 20-50 cm H; perianth segments 5-6 mm L; 5000-9500', Inyo and Mariposa cos. n — *J. longistylis*
CC Lf-blades round, or if flat then with edge toward stem, with ± evident cross-walls
D Lf-blades ± round, the cross-walls complete; below 11,000'
E Stems mly less than 25 cm H; perianth segments brown to black; SN (all)
F Heads usu solitary, purplish black; lvs soft — *J. mertensianus*
FF Heads usu more than 1, brown; lvs stiff — *J. nevadensis*
EE Stems mly over 25 cm H; c. SN
F Perianth segments light green, 4 mm L; stem lvs 4-10 cm L — *J. chlorocephalus*
FF Perianth segments brownish, 2-3 mm L; stem lvs 10-40 cm L — *J. dubius*
DD Lf-blades flat, edgewise to stem, cross-walls usu only partial; mly below 7000'
E Anthers longer than filaments; heads many, few-fld
F Perianth usu over 4 mm L; mly Nevada Co. n. — *J. phaeocephalus*
FF Perianth usu less than 4 mm L
G Perianth purplish brown; lvs 2-3 mm W — *J. macrandus*
GG Perianth greenish brown; lvs 3-6 mm W — *J. oxymeris*
EE Anthers much shorter than filaments; heads often few with many fls
F Heads few; auricles conspicuous; occasional at 6000-7500' — *J. saximontanus*
FF Heads usu many; auricles lacking
G Stem 5-9 dm H; lf-blades 10-40 cm L, 3-12 mm W; below 7000' — *J. xiphioides*
GG Stem 2-5 dm H; lf-blades 7-15 cm L, 2-5 mm W; below 9000' — *J. ensifolius*

L parviflora inflorescence (1 cm)

Luzula

Wood Rush

leaf with closed sheath (1 cm)

Tufted perennials with slender unbranched stems. Infl umbellate, paniculate or congested, bracteolate. Edibility unknown.

A Fls solitary or few at the end of the infl-branches
B Infl 5-10 cm L, the branches drooping; moist habitats — 4) *L. parviflora*
BB Infl 8-16 cm L, the branches inclined; mly dry habitats — 2) *L. divaricata*
AA Fls in dense clusters
B Infl with distinct branches
C Lvs with long white hairs on margins; infl-branches mly less than 2 cm L — 1) *L. comosa*
CC Lvs ± glabrous; infl-branches 2-5 cm L — 6) *L. subcongesta*
BB Infl ± dense, spike-like or subcapitate

C Infl nodding; lf-blades channeled — 5) *L. spicata*
CC Infl erect, subcapitate; lf-blades flat — 3) *L. orestera*

1) *L. comosa*, Hairy Wood Rush: Stems 1-6 dm H; lvs light green, 5-15 cm L, 3-6 mm W; spikes ± oblong, about 5-10 mm L; lowest bract leafy, 2-5 cm L. Mdws and open woods, 3000-10,500', SN (all), May-Aug.
2) *L. divaricata*, Forked Wood Rush: Stems 2-5 dm H; lvs 5-15 cm L, 3-10 mm W, shining; infl much branched; perianth about 2 mm L. At 7000-11,000, SN (all), Jul-Aug.
3) *L. orestera*, Mountain Wood Rush: Stems mly 5-20 cm H, green with purplish stalks and infls; lvs mly basal, 3-8 cm L, 2-3 mm W, occasionally with hairs on margin; infl 5-12 mm H, of 2-4 heads, subtended by 1-2 longer leafy bracts. Moist habitats, 9000-11,500', Tuolumne and Mono cos. s., Jul-Aug.
4) *L. parviflora*, Small-flowered Wood Rush: Stems 2-7 dm H; lvs glabrous, the blades 4-15 cm L, 4-12 mm W; infl much branched. Moist habitats, 3500-11,000', SN (all), Jun-Aug.
5) *L. spicata*, Spiked Wood Rush: Stems 1-3.5 dm H; lvs mly basal, 3-15 cm L, 1-4 mm W, stiffly erect; infl often interrupted, 1-2.5 cm L. Moist habitats, 8000-12,500', SN (all), Jul-Aug.
6) *L. subcongesta*, Donner Wood Rush: Stems 2-5 dm H; basal lvs 8-20 cm L, 4-8 mm W; stem lvs 2-4; infl ± congested; fls in subcapitate clusters at the ends of branches. Moist habitats, 7000-11,200', SN (all), Jul-Aug.

JUNCAGINACEAE - Arrow-weed Family

inflorescence (1 cm)

Triglochin

Arrow-grass

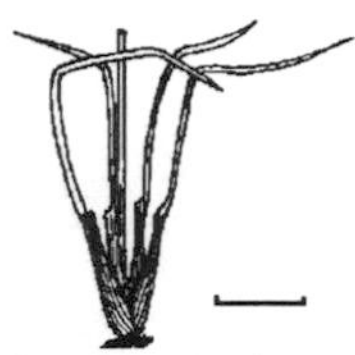
basal leaves (1 cm)

Marsh herbs with rhizomes. Lvs mly basal, linear, sheathing at base, fleshy. Fl-stems long, smooth. Fls small in terminal bractless racemes. Young lvs edible raw. Seeds and older herbage cyanogenic if eaten raw, but edible after cooking.

Carpels 6; fr oblong-ovoid, 3-4 mm L; below 7500' — *T. maritima*
Carpels 3; fr elongate, 5-7 mm L; above 7500' — *T. palustre*

T. maritima, Seaside Arrow-grass: Plant densely tufted, 3-7 dm H; lvs half-cylindric, about 2 mm W; infl 1-4 dm L; pedicels ascending, 2-3 mm L, 4-5 mm L in fr. SN (all), Jun-Aug.
T. palustris, Marsh Arrow-grass: Fl-stems slender, 1-6 dm H; lvs 5-30 cm L, linear; infl 0.5-3 dm L; pedicels appressed, becoming 4-6 mm L in fr. Mud flats and springy habitats, 7500-11,500', Tulare and Inyo cos., Jul-Aug.

LILIACEAE - Lily Family

Perennial herbs. Fls mly showy, sometimes small and greenish, but then usu many in clusters. Perianth usu of 6 distinct segments, these often all petaloid. Fr a capsule or berry.

A Stems leafless or stem lvs much reduced
B Lvs lanceolate, thick, spine-tipped — *Yucca* p. 225
BB Lvs not thick or spine-tipped
C Lvs linear, usu many
D Infl umbellate
E Perianth-segments distinct or near so; plant with onion-like odor — *Allium* p. 216
EE Perianth segments united into a ± conspicuous basal tube; plants without odor
F Perianth yellow, white or occasionally blue; fertile stamens 6 — *Triteleia* p. 224
FF Perianth blue to violet; fertile stamens 3
G Fls 3-10 in open umbel; perianth over 20 mm L — *Brodiaea* p. 217
GG Fls 10-35 in dense umbel; perianth 8-10 mm L — *Dichelostemma* p. 219
DD Infl usu a panicle or raceme; fls sometimes solitary
E Lvs folded and sheathing each other at base
F Fls 6-10 mm L; stamens densely yellow-hairy — *Narthecium* p. 222
FF Fls 3-5 mm L; stamens glabrous — *Tofieldia* p. 224

EE Lvs not equitant, usu over 2 dm L; fls blue or white
F Petals 3-6 mm L; infl a dense spicate raceme *Hastingsia* p. 221
FF Petals over 15 mm L; infl variable
G Infl a simple raceme, 2-5 cm W; petals dark blue *Camassia* p. 219
GG Infl a branching panicle over 5 cm W; petals white with green midveins *Chlorogalum* p. 219

CC Lvs broader, usu only 2-3
D Lvs broad; fls not nodding; fr a berry *Clintonia* p. 219
DD Lvs narrow; fls nodding; fr a capsule *Erythronium* p. 220
AA Stems usu leafy
B Stem lvs in distinct whorls, occasionally some also scattered
C Stem lvs in a whorl of 3 *Trillium* p. 223
CC Stems with several whorls of more than 3 lvs
D Fls white to yellow or orange; anthers usu attached in middle and versatile; capsule not winged *Lilium* p. 221
DD Fls usu purplish (occasionally yellow); anthers ± attached at base; capsule usu winged *Fritillaria* p. 220
BB Stem lvs alt
C Lvs broad, deltoid to elliptic
D Stems climbing or straggling, vine-like, bearing tendrils *Smilax* p. 222
DD Stems not vine-like or tendril-bearing
E Lvs ± elliptic, 20-40 cm L; stem 2-4 cm thick at base *Veratrum* p. 224
EE Lvs smaller; stems usu not over 1 cm thick at base
F Fls in terminal panicles or racemes *Smilacina* p. 222
FF Fls solitary or umbellate
G Fls terminal, solitary or in a small umbel *Disporum* p. 220
GG Fls axillary, 1 or 2 together; Plumas Co. n. *Streptopus* p. 223
CC Lvs lanceolate to linear, mly less than 2 cm W
D Perianth segments unlike, distinct sepals and petals *Calochortus* p. 218
DD Perianth segments alike, petaloid
E Lvs mly cauline, linear to lanceolate; fls usu purplish *Fritillary* p. 220
EE Lvs mly basal, ± grass-like; fls white to yellowish
F Infl open; pedicels less than 2 cm L; lvs 5-30 mm W *Zigadenus* p. 225
FF Infl dense, racemose; pedicels 2-5 cm L; lvs 2-4 mm W above base *Xerophyllum* p. 225

Allium

Wild Onion

Petals 1-nerved. Stamens 6, usu attached to base of perianth. All spp produce edible bulbs which may be used like cultivated onions and keep well stored. Overeating of any onion may cause poisoning. Difficult genus to work with in the field. Only abbreviated descriptions of the individual spp will be presented.

plant (1 cm)

A Fl-stem usu 1 dm H or less; lvs mly sickle-shaped, usu much longer than fl-stem
B Lf 1
C Lf round above the tubular sheath; petals purple to rose, 7-15 mm L 1) *A. abramsii*
CC Lf flat; petals greenish white, mly 5-8 mm L
D Pedicels 6-10 mm L; stamens equal to or exerted beyond perianth 3) *A. burlewii*
DD Pedicels 3-5 mm L; stamens included within perianth 8) *A. obtusum*
BB Lvs 2
C Lvs 2-6 mm W; fls mly 10-20
D Bracts usu 3; perianth segments thin, flat, the midvein not thickened; rare 11) *A. tribracteatum*
DD Bracts 2; perianth segments thicker, keeled with thickened midvein; common 9) *A. parvum*
CC Lvs 8-15 mm W; fls 30 or more
D Bracts with spine-like tip, 5 mm L; Yosemite National Park 13) *A. yosemitense*
DD Bracts with broader tip; Placer Co. n. 10) *A. platycaule*
AA Fl-stem more than 2 dm H; lvs rarely sickle-shaped, usu shorter than fl-stem
B Fl-stem stout, 5-10 dm H; lvs 3-6, 5-12 mm W 12) *A. validum*

BB Fl-stem less than 5 dm H; lvs usu 2-3, usu less than 5 mm W
C Ovary prominently 6-crested
D Petals mly 9-10 mm L; below 4500' — 7) *A. membranaceum*
DD Petals mly 5-8 mm L
E Infl ± open; fls on pedicels 1-3 cm L — 4) *A. campanulatum*
EE Infl dense; fls sessile to pedicelled — 2) *A. amplectens*
CC Ovary without crests or crests inconspicuous
D Below 5000', Eldorado Co. s. — 5) *A. hyalinum*
DD Above 5000', Monitor Pass, Fresno Co. n. — 6) *A. lemmonii*

1) *A. abramsii*, Fringed Onion: Fl-stem 3-10 cm H; bracts 2-3, sharp-pointed at apex, 6-12 mm L; fls 8-40; pedicels 4-15 mm L, stoutish; petals narrow, recurved-spreading; anthers yellow. Granitic gravel, 4500-10,000', Madera Co. s., May-Jul.

2) *A. amplectens*, Narrow-leaved Onion: Fl-stem 2-5 dm H; lvs 2-4, narrow, flattened, but convolute-filiform in age; bracts 2-3, broadly ovate; infl subglobose, many-fld; petals white to pinkish; anthers yellow or purplish. Dry slopes, mly below 6000', w. SN (all), Apr-Jun.

3) *A. burlewii*: Fl-stem 2-8 cm H; lf less than twice as long as stem; fls 8-20; perianth parts ovate, erect, entire, dull purple. Dry slopes, 6000-9000', s. SN.

4) *A. campanulatum*, Sierra Onion: Fl-stems 1-2, 1-3 dm H; bracts 2, ovate, abruptly long-pointed at tip; fls 15-40; petals pale rose; anthers reddish. Dry slopes in woods, 2000-8900', SN (all), May-Jul.

5) *A. hyalinum*, Paper-flowered Onion: Fl-stem 1.5-3 dm H; lvs 2, sometimes 1 or 3, often convolute; bracts 2, 1-2 cm L; fls 6-15; pedicels slender, 2-2.5 cm L; petals white or pinkish; anthers pale. Rather moist places, grassy and rocky slopes below 5000', Apr-Jun.

6) *A. lemmonii*, Lemmon's Onion: Fl-stem 1-2 dm H, narrowly 2-edged; lvs 2; bracts 2-4, ovate, 10-17 mm L; fls numerous; petals white to pale rose, erect; anthers yellow. Heavy soil, May-Jun.

7) *A. membranaceum*, Membranous Onion: Fl-stems 1-2, 1.5-3.5 dm H; lvs usu 2, as long as fl-stem; bracts 2, 1-2 cm L; petals whitish to pink, slightly saccate at base; anthers yellow. Wooded slopes, w. SN, Mariposa to Plumas Co., May-Jul.

8) *A. obtusum*, Red Sierra Onion: Fl-stem 2-8 cm H; bracts 2-3, 5-8 mm L; fls several to many; pedicels 3-5 mm L; petals elliptic; anthers yellowish or purplish. Sandy or gravely slopes and benches, 7000-12,000', Plumas Co. s., May-Jul.

9) *A. parvum*: Dry rocky ridges and slopes, 4000-8000', SN (all), Apr-Jul.

10) *A. platycaule*, Broad-stemmed Onion: Fl-stem 4-12 cm L, flattened; bracts 3-5; pedicels 12-20 mm L; petals deep rose with pale tips; anthers dark. Gravely slopes and knolls, 4000-9000', May-Aug.

11) *A. tribracteatum*, Three-bracted Onion: Fl-stem 3-12 cm L; bracts 2-3; pedicels 4-16 mm L; petals pale rose with dark purplish midveins; anthers yellow or purple. Volcanic soil, Yosemite National Park.

12) *A. validum*, Swamp Onion: Fl-stem angled, somewhat compressed; lvs channeled; fls many; petals purple-violet to almost white, saccate at base, anthers dark. Wet mdws, 4000-11,000', SN (all), Jul-Sep.

13) *A. yosemitense*, Yosemite Onion: Fl-stem 5-10 cm L, somewhat flat; bracts 3, purplish; pedicels purplish, 1-2 cm L; petals pale rose, linear oblong, about 10 mm L. Open forests, Yosemite National Park, Jun-Jul.

Brodiaea

Brodiaea

Perennial from corm. Lvs basal, grass-like, dying by anthesis. Infl umbel-like, scapose, open, the bracts membranous. Fls erect; outer surface of perianth shiny; perianth 6-lobed in 2 petal-like whorls. Stamens 3, alternating with staminodes. Fr a capsule

plant (1 cm)

Pedicels 1-5 cm L; staminodes held close to stamens — *B. coronaria*
Pedicels 5-10 cm L; staminodes distant from stamens — *B. elegans*

B. coronaria: Scape 5-25 cm H; fl blue to violet or rose, the tube 6-12 cm; the lobes ascending, 10-25 mm L; the tips recurved. Open, dry habitats below 6000', Tuolumne Co. n.

B. elegans, Harvest Brodiaea: Fl-stems mly 1-4 dm H; lvs about same length; pedicels 1-8 cm L; perianth ± funnelform, violet to deep blue-purple, rarely pink; staminodia erect, about 10 mm L; anthers 7-10 mm L, yellow. Common, usu in heavy soils, dry or occasionally moist flats and slopes up to 7000', w. Sierra, Tehama to n. Tulare Co., Apr-Jul.

Calochortus

Mariposa-lily

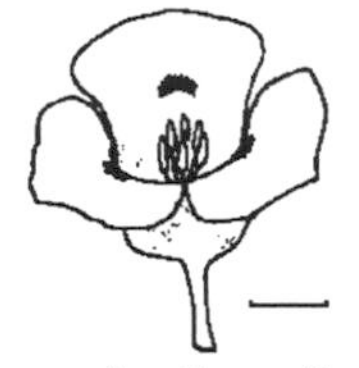

C. albus flowers (1 cm) *C. superbus flower* (1 cm)

Glabrous herbs. Lvs usu linear, the basal solitary, often long. Petals with gland near base and usu bearded on inner face. The bulbs are edible raw or cooked or dried and ground into flour. The seeds are also edible. These plants possess one of the most beautiful flowers in the Sierra and should be eaten only in emergencies.

A Corolla narrow or barely widened at apex; fls often nodding
  B Petals deep rose; fls narrow bell-shaped — 2) *C. amoenus*
  BB Petals white; fls subglobose — 1) *C. albus*
AA Corolla wide-spreading at petal tips; fls ± erect
  B Stems leafless; petals without central dark blotch
    C Petals bearded above gland, bluish; Amador Co. n. — 3) *C. coeruleus*
    CC Petals glabrous or nearly so, white (rarely pinkish) — 6) *C. minimus*
  BB Stems leafy (lvs often reduced); petals usu with dark blotch
    C Sepals 15-25 mm L
      D Base of anthers with divergent lobes 1-3 mm L — 5) *C. leichtlinii*
      DD Base of anthers notched to minutely parallel-lobed — 4) *C. invenustus*
    CC Sepals usu over 25 mm L
      D Petals with dark red blotch, occasionally a second paler blotch above first — 8) *C. venustus*
      DD Petal-blotch surrounded by bright yellow — 7) *C. superbus*

1) *C. albus*, Fairy Lantern, Globe-lily: Stem rather slender, erect, 2-5 dm H, branched; basal lf 3-5 dm L, 1-4 cm W; stem lvs 2-6; sepals 1-1.5 cm L; petals 2-2.5 cm L; capsule 3-winged, nodding, 2.5-4 cm L. Shaded, often rocky places in open woods or brush below 5000', Madera to Butte Co., May-Jun.

2) *C. amoenus*, Rosy Fairy Lantern: Stem slender, erect, 2-5 dm H; basal lf 2-5 dm L, 5-25 mm W; stem lvs 2-5; sepals 1-1.5 cm L; petals elliptic-obovate, 1.6-2.5 cm L; capsule narrowly 3-winged, nodding, 2-3 cm L. Leafy loam of grassy slopes below 5000', Madera Co. s., May-Jun.

3) *C. coeruleus*, Beavertail-grass: Stems usu slender, ± erect, 3-15 cm H; basal lf 1-2 dm L, 2-10 mm W; fls 1-8, subumbellate; sepals lance-oblong, about 10 mm L; petals obovate, 8-12 mm L; capsule nodding, 1-1.5 cm L. Open, gravely places in woods, 3500-7500', May-Jul.

4) *C. invenustus*, Plain Mariposa-lily: Stem 1.5-5 dm H; lvs linear, 1-2 dm L, becoming involute; fls 1-5, subumbellate, white to purplish, sometimes with purplish spot below the gland; petals wedge-shaped, 2-3.5 cm L; capsule lance-linear, angled, erect, 5-7 cm L. Dry soil, mly granitic, usu in woods, 4500-9000', Tuolumne Co. s., May-Aug.

5) *C. leichtlinii*, Leichtlin's Mariposa-lily: Much like *C. invenustus*; stem erect; fls often tinged pink, each petal with a red to dark spot above gland. Open gravely places, 4000-11,000', SN (all), Jun-Aug.

6) *C. minimus,* Lesser Star-tulip: Stem low, occasionally branched; basal lf 1, 5-15 cm L, 1-2 cm W; infl of 1-2 fls subtended by a pair of lanceolate herbaceous bracts; bracts 6-10 mm L, 3-5-veined; sepals 8-10 mm L; petals roundish, about 10 mm in diameter; capsule winged, nodding, 15-20 mm L. Moist, grassy places, 4000-9500', e. Eldorado Co. s., May-Aug. An apparent hybrid with *C. nudus* which displays pink to lavender-tinged petals occurs in n. SN.

7) *C. superbus,* Superb Mariposa-lily: Stems erect, 4-6 dm H; lvs linear; 1.5-2.5 dm L, 4-6 mm W; fls 1-3, bell-shaped, subumbellate, white to yellowish to lavender, usu purple at base; sepals 2-3.5 cm L; petals obovate, 2.5-4 cm L; capsule linear, angled, erect, 5-6 cm L. Open or wooded slopes, mly below 5000', SN (all), May-Jul.

8) *C. venustus*, Butterfly Mariposa-lily: Stem erect, usu branched; basal lvs linear, 1-2 dm L; fls 1-3, white to yellow, purple or dark red; sepals reflexed at tip, 2.5-3 cm L; petals obovate, 3-4.5 cm L; capsule linear, angled, erect, 5-6 cm L. Light, sandy soil, often decomposed granite, below 8000', Eldorado Co. s., May-Jul.

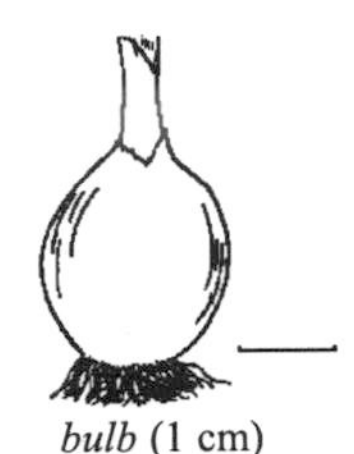

bulb (1 cm)

Camassia

Common Camas

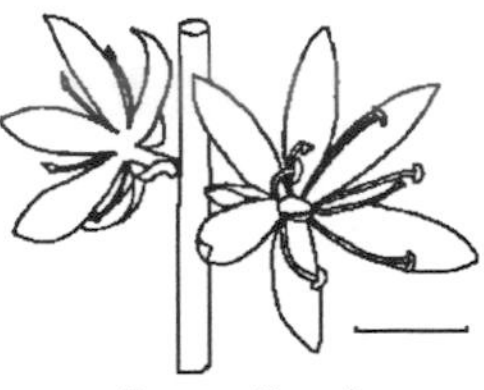

flowers (1 cm)

C. quamash ssp *breviflora*:: Lvs in a basal whorl, keeled, 1.5-6 dm L, 5-25 mm W; glaucous above; fl stem 2-13 dm H; pedicels 1.5-15 mm L; perianth-segments 6, somewhat spreading, 3- to 9-veined, persistent; stamens 6; capsule ovoid, sometimes oblong, 8-30 mm L. Wet mdw, 2000-8000', SN (all), May-Aug. The bulbs are edible and are best prepared by steaming in a fire pit for 24 hours. Care should be taken in collecting the bulbs, for death-camas (*Zigadenus*) will often be found in the same locations.

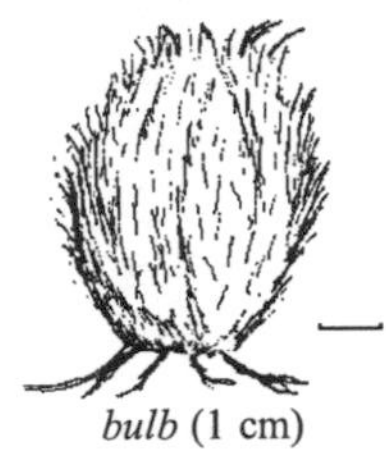

bulb (1 cm)

Chlorogalum

Soap Plant

flowers (1 cm)

C. pomeridianum: Bulb 7-15 cm L, heavily coated with persistent dark brown fibers of old coats; basal lvs several, tufted, linear, 2-7 dm L, 6-15 mm W, very wavy; stem lvs reduced; stem glaucous, stout, 6-25 dm H, branched above; pedicels slender, 5-25 mm L; petals linear, white with green or purple midvein, 15-25 mm L. Dry, open habitats, below 5000', May-Aug. The bulbs may be roasted and eaten; uncooked they have lather-producing qualities.

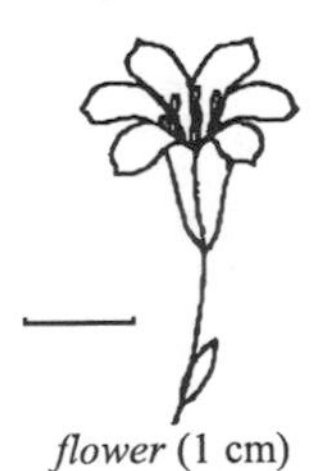

flower (1 cm)

Clintonia

Bride's Bonnet
Queen Cup

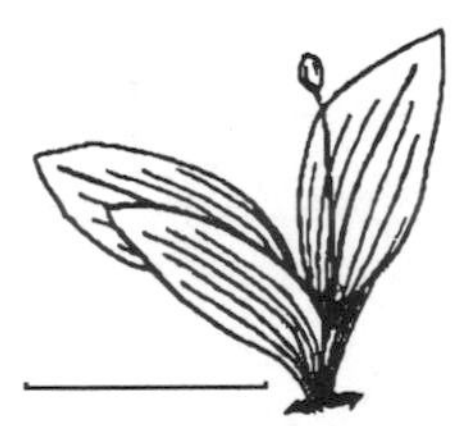

leaves and fruit (1 cm)

C. uniflora: Stem slender, erect, unbranched; lvs usu 2-3, obovate to oblanceolate, 7-15 cm L, 2.5-6 cm W, petioled; fls white, terminal, 1-2, the petals oblanceolate, spreading, about 20 mm L; berry about 10 mm in diameter. Shaded woods, 3500-6000', SN (all), May-Jul. Related sp has edible young lvs.

Dichelostemma

Wild Hyacinth

D. multiflorum: Fl-stems 3-8 dm H; lvs 3 or more, glaucous, equaling or exceeding fl-stems; infl ± spherical; bracts ovate, purplish, to 12 mm L; pedicels stiff, 3-15 mm L; perianth-tube pale, 8-10 mm L, the lobes 8-10 mm L; staminodia white to violet, 5-6 mm L; anthers 4-5 mm L. Open and wooded slopes below 5000', Mariposa Co. n., May-Jun.

plant (1 cm)

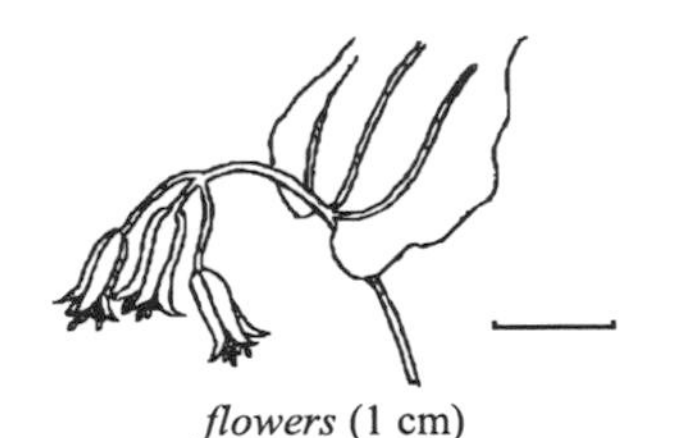

Disporum

Fairy Bells

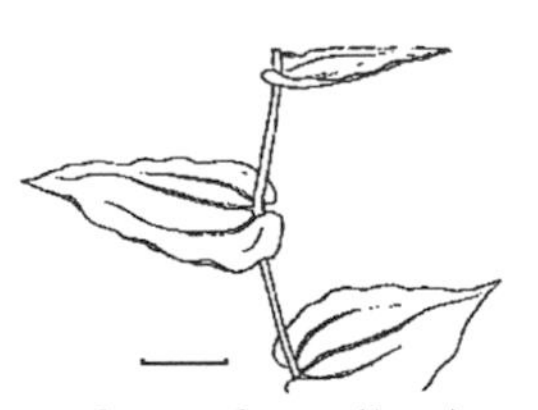

flowers (1 cm) *clasping leaves* (1 cm)

D. hookeri var. *trachyandrum*: Stems branched, scaly below, leafy above, usu 3-8 dm H; lvs ovate to oblong-ovate, somewhat asymmetric, at least the lower cordate-clasping; fls creamy-white to greenish, 9-12 mm L; berry scarlet, about 8 mm L. Dry, shaded slopes and benches below 5000', SN (all), Apr-Jun. Berries edible, sweet.

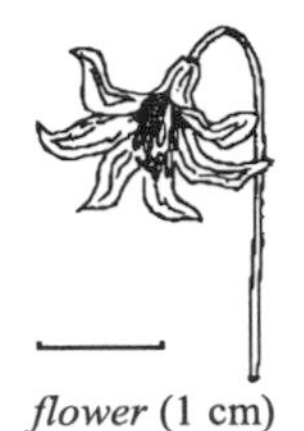

Erythronium

Adder's Tongue
Fawn Lily

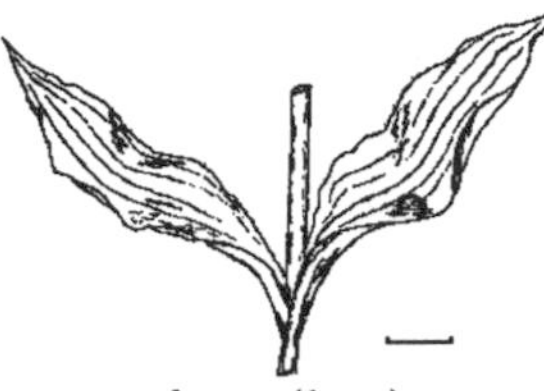

flower (1 cm) *leaves* (1 cm)

Perennial from bulb. Lvs 2, basal, lanceolate, green to ovate, ± wavy-margined, petioled. Infl a bractless raceme. Fls showy, nodding. Perianth segments 6, similar, ± lanceolate, recurved. Stamens 6, style 1. Corm and herbage of related spp toxic when raw, edible after cooking.

Perianth yellow throughout, aging bronze; Madera Co. — *E. pluriflorum*
Perianth white with yellow base, aging pink or purple
  Perianth segments 10-20 mm L; widespread — *E. purpurascens*
  Perianth segments 25-45 mm L, Tulare Co. — *E. pusaterii*

E. pluriflorum: Lf 5-30 cm L; scape 8-35 cm H; fls 1-10; perianth segments 15-30 mm L; stamens 8-12 mm L, anthers and filaments yellow. Open, rocky areas at about 7500'.

E. purpurascens: Stem 8-20 cm H, unbranched; corm 3-5 cm L, 5-6 mm W; lvs narrow to oblong-lanceolate, 10-15 cm L, 1-2.5 cm W, crisped along margin, yellow-green; fls 1-8, pedicels unequal, 5-40 mm L; petals white with yellow base, tinged purple with age, lance-linear, 10-15 mm L, spreading, slightly recurved. Usu moist habitats, 4000-8000', SN (all), May-Aug.

E. pusaterii, Kaweah Lakes Fawn Lily: Lf 10-35 cm; scape 10-40 cm; fls 1-8; stamens 8-15 mm; filaments white, anthers yellow. Rare, about 7000'.

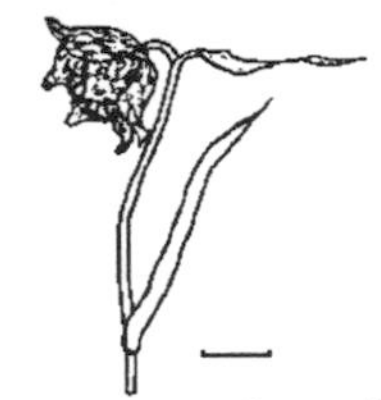

Fritillaria

Fritillary

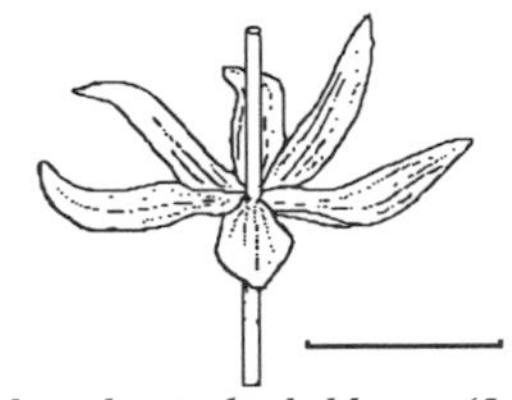

F. atropurpurea flower (1 cm) *F. brandegei whorled leaves* (5 cm)

Bulb of one or more fleshy scales, with or without rice-grain bulblets. Stems erect, unbranched. Lvs sessile. Bulbs of all Sierran spp are edible, however they are relatively rare and should be eaten only in emergency.

A Style barely cleft at apex
  B Petals yellow to orange; Sierra Co. n. — 5) *F. pudica*
  BB Petals pink to purplish; Tulare and Kern cos. — 2) *F. brandegei*
AA Style obviously 3-cleft at apex
  B Fls indistinctly mottled — 3) *F. micrantha*
  BB Fls conspicuously yellowish mottled
    C Fls scarlet; petals usu over 2 cm L; Nevada Co. n. — 6) *F. recurva*
    CC Fls purplish; petals usu less than 2 cm L; widespread
      D Fls nodding; stems slender, solid; bulb without rice-grain bulblets — 1) *F. atropurpurea*
      DD Fls ± erect; stems hollow above; bulb with rice-grain bulblets — 4) *F. pinetorum*

1) *F. atropurpurea*, Purple Fritillary: Stem 1.5-6 dm H; lvs 7-14, linear to lanceolate, alt or ± whorled on upper half of stem, 5-9 cm L, 2-6 mm W; fls nodding, purplish brown spotted with yellow and white; petals 1-2 cm L, 4-8 mm W, thick; capsule obovoid, 10-13 mm L, acutely angled. Leaf mold under trees. 6000-10,500', SN (all), Apr-Jun.

2) *F. brandegei*, Brandegee's Fritillary: Stem 4-10 dm H; lvs on upper stem, in whorls of about 5, lanceolate, 5-10 cm L; fls 4-12, nodding, pink to purplish; petals 12-17 mm L, 2-3 mm W. Granitic soils, open forests, 5000-7000', Apr-Jun.

3) *F. micrantha*, Brown Bells: Stem 4-9 dm H, light green; lvs on upper part of stem, in whorls of 4-6, linear, 5-15 cm L; fls 4-10, purplish or greenish white; petals 12-20 mm L, 4-5 mm W, apically white-tufted; capsule broadly winged, slightly wider than long. Dry benches and slopes, below 6000', Plumas Co. s., Apr-Jun.

4) *F. pinetorum*, Davidson's Fritillary: Stem glaucous, 1-3 dm H; lvs glaucous, 12-20, somewhat whorled, linear, 5-15 cm L; fls 3-9, erect or nearly so, purplish, mottled with greenish yellow; petals 14-20 mm L, 2-6 mm W; capsule 12-15 mm L, angled, with short horn-like processes at base and summit of each wing. Usu shaded granitic slopes, 6000-10,500', Alpine Co. s., May-Jul.

5) *F. pudica*, Yellow Bell: Stem 7-30 cm H; lvs alt, 3-8, linear to lanceolate, 6-20 cm L, 2-11 mm W; fls 1-3, nodding, bell-shaped; petals 15-20 mm L, 4-7 mm W. Grassy and brushy or wooded slopes below 5000', May-Jun.

6) *F. recurva*, Scarlet Fritillary: Stem 3-9 dm H; lvs usu 8-10 in 2-3 whorls near the middle of stem, 3-10 cm L, 3-14 mm W; fls nodding, mottled with yellow within, tinged purple without; petals recurved at tips, 2-3.5 cm L, 5-7 mm W; capsule winged, about 10 mm L. Dry hillsides in brush or woods, 2000-6000', May-Jul.

Hastingsia

White-flowered Schoenolirion

plant (1 cm)

H. alba: Stem 2.5-15 dm H, unbranched, from a coated bulb 2-3 cm L; lvs 1.5-6 dm L, 4-12 mm W, flat; raceme occasionally with 1-2 short racemose branches at base; pedicels mly 1-4 mm L; petals white, tinged with green or pink. Mdws and swampy places, 1500-8000', Nevada Co. n., Jun-Jul. Edibility unknown.

Lilium

Lily

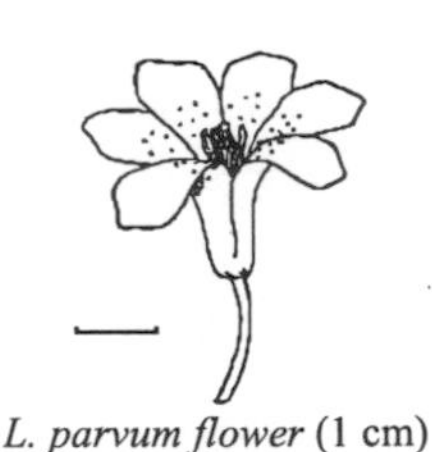
L. parvum flower (1 cm)

L. pardalinum upper stem (1 cm)

Stems unbranched, tall, from scaly bulbs or scaly rootstocks. Lvs narrow, sessile. Fls showy, large. Perianth deciduous, the segments 6, alike, each with a nectar-bearing gland near base. The bulbs of all spp are edible raw or cooked.

A Plants of rather dry habitats; petals 7-10 cm L
  B Fls white with a few reddish dots — 5) *L. washingtonianum*
  BB Fls orange-yellow spotted maroon or purple — 1) *L. humboldtii*
AA Plants of wet habitats
  B Fls 5-8 cm L; anthers 10-15 mm L — 3) *L. pardalinum*
  BB Fls 3-5 cm L; anthers less than 6 mm L
    C Fls nodding; petals recurved about half their length — 2) *L. kelleyanum*
    CC Fls usu horizontal or ascending; petals recurved only at tips — 4) *L. parvum*

1) *L. humboldtii*, Humboldt Lily: Stems 9-20 dm H, stout; lvs in 4-8 whorls of 10-20 each, 9-12 cm L, oblanceolate, green to purplish; fls nodding, few to many, petals 7-9 cm L, 12-25 mm W, recurved to near base. Dry, open habitats and forests below 4500', Fresno to Butte Co., Jun-Jul.

2) *L. kelleyanum*, Kelley's Lily: Stems 6-18 dm H; lvs oblong-lanceolate, 5-15 cm L, the lower alt, middle and upper in whorls of 3-8; fls fragrant, few to 25, orange toward tips or yellow throughout with minute maroon dots. At 4000-10,500', SN (all), Jul-Aug.

3) *L. pardalinum*, Leopard Lily, Panther Lily: Exceedingly variable; stems stout, 10-25 dm H; lvs linear to lanceolate, 1-2 dm L, in 3-4 whorls of 9-15 and some scattered; fls nodding, often not fragrant. Forming large colonies on stream banks and near springs, up to about 6000', SN (all), May-Jul.

4) *L. parvum*, Alpine Lily: Stems 4-15 dm H; slender; lvs light green, broadly lanceolate to linear, 5-12 cm L, mly scattered but usu the lower in whorls; fls erect, few to many, orange to dark red, spotted maroon; petals 3.5-4 cm L. Boggy habitats, often among alders or willows, 4000-8000', SN (all), Jul-Sep.

5) *L. washingtonianum*, Washington Lily: Stem 6-18 dm H; lvs light green 4-20 cm L, 1-4 cm W, mly horizontal in several whorls of 6-12; fls up to 20 or more, trumpet-shaped, fragrant, 8-10 cm L. Among bushes, dry, granitic and loamy soils, 4000-7400', Fresno Co. n., Jul-Aug.

inflorescence (5 cm) *flower* (1 cm)

N. californicum: Stem slender, solitary, erect, 4-5 dm H; basal lvs 1-3 dm L, 3-6 mm W, grass-like; stem lvs reduced; raceme 8-15 cm L; pedicels 6-15 mm L; petals yellow, 6-10 mm L, spreading; anthers densely yellow-hairy. Wet mdws and banks to 8500', Fresno Co. n., Jul-Aug. Edibility unknown.

S. stellata inflorescence (1 cm) *S. stellata leaves* (5 cm)

Stem unbranched, scaly below, leafy above. Lvs short-petioled or sessile. Fls small, usu white; perianth of 6 equal segments. The starchy aromatic roots may be eaten after soaking overnight in lye to remove the bitter acids and subsequent removal of the lye. The berries are also edible, although they may cause diarrhea if eaten in quantity. Cooking the bitter-sweet berries removes most of the purgative element.

Fls panicled, numerous; petals 1-2 mm L ..... *S. racemosa*
Fls racemose, few to several; petals 5-7 mm L ..... *S. stellata*

S. racemosa, Racemose False Solomon's-seal: Stems trailing to erect, 3-9 dm H; lvs 7-20 cm L; infl 3-18 cm H, dense; berry about 5 mm L, mly red. Shaded woods below 8200', SN (all), Apr-Jun.

S. stellata, Panicled False Solomon's-seal: Stems 3-6 dm H, mly erect; lvs lanceolate, 5-15 cm L, ascending, evenly many-veined, often ± folded: infl 3-15 fld; berry red-purple, becoming black, 7-10 mm in diameter. Wet places, often in brush, mly 4000-8000', SN (all), Apr-Jun.

umbel in fruit (1 cm) *leaves* (5 cm)

S. californica: Stems woody, rarely prickly, 1-3 m L; lvs ovate, 5-10 cm L, on petioles 10-15 mm L; peduncles 2-5 cm L; fls 8-20, greenish or yellowish, in umbels; perianth about 5 mm L;

berries black, 6 mm L. Thickets and stream banks below 5000', Butte Co. n., May-Jun. The roots may be used in soups or stews or dried and ground into flour. The flour mixed with water and sugar makes a good drink. The young shoots in early season may be eaten raw or cooked. Berry edible raw or cooked.

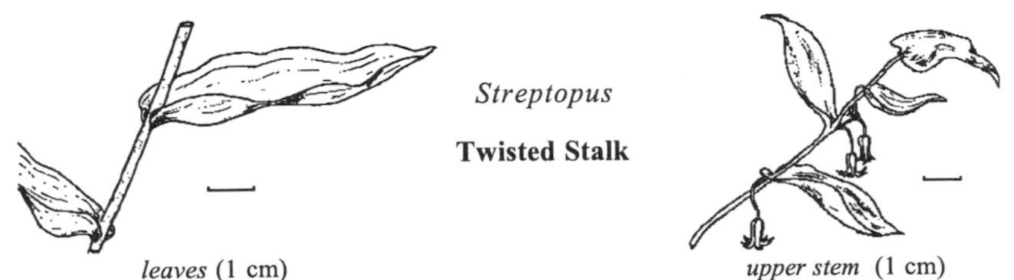

Streptopus

Twisted Stalk

leaves (1 cm) *upper stem* (1 cm)

S. amplexifolius var. *denticulatus*: Stem 3-8 dm H, glabrous, often branched, from a stout horizontal rootstock; lvs ± ovate, thin, cordate-clasping, 5-15 cm L, minutely serrulate; peduncles 1.5-3 cm L, sharply bent and twisted at the joint with the 1 or 2 pedicels; fls greenish or yellowish, 1-1.5 cm L; berry 1-1.5 cm across. Moist woods below 5500', May-Jun. The berry is edible raw or cooked, however in large quantities it may have a purgative effect.

Tofieldia

Tofieldia

inflorescence (1 cm) *basal equitant leaves* (1 cm)

T. occidentalis: Stems 2-7 dm H, solitary, glandular; lvs grass-like, 5-20 cm L; pedicels 2-10 mm L; fls usu in 3's in a subcapitate panicle; pedicels bracted at base and with 3 membranous ± united bractlets below each fl; petals ± oblong, 3-6 mm L, light yellow. Boggy places and mdws to 10,200', SN (all), Jul-Aug. Herbage of a related sp contains alkaloids.

Trillium

Wake-robin
Trillium

inflorescence (1 cm) *flower showing 3 styles* (1 cm)

Erect perennial from rhizome. Lvs 3 in a single whorl, ± ovate, subtending fl. Fl solitary, terminal; sepals 3, greenish; petals 3 distinct, withering. Stamens 6, styles 3. Stem and lvs may be boiled as greens. The root and possibly the berries are powerful emetics.

Petals oblanceolate, white to pink; Tuolumne Co. n. — *T. album*
Petals linear, purplish; widespread — *T. angustipetalum*

T. album: Stem 2-7 dm H; lvs sessile, 7-20 cm L, rounded; fl usu with a sweet, spicy scent; sepals 3-8 cm L, petals 4-10 cm L. Common below 6500'.

T. angustipetalum: Stem 2-7 dm H; lvs subsessile, 10-25 cm L; rounded, sometimes spotted; fl usu with a musty odor, sepals 3-6 cm; petals 5-10 cm, dark purple. Dry to moist areas below 6500', SN (all).

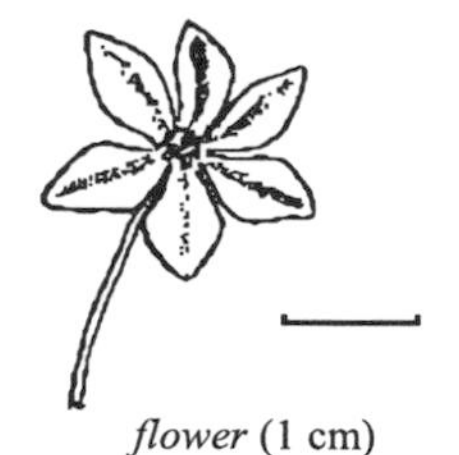

flower (1 cm)

Triteleia

Triteleia

T. laxa inflorescence (1 cm)

Perennials from corm. Lvs few, grass-like, basal, withering by anthesis. Infl an umbel with ± erect pedicels and several to many fls. Stamens 6; style 1. Corms best eaten after cooking.

A Perianth white to purple; forked appendages none
  B Perianth usu blue to purple, 20-40 mm L — 4) *T. laxa*
  BB Perianth white to light purple, 10-15 mm L — 2) *T. hyacinthina*
AA Perianth ± yellow; forked appendages present outside anthers
  B Perianth tube ± equal to lobes; mly above 9000' — 1) *T. dudleyi*
  BB Perianth tube much shorter than lobes; mly below 9000' — 3) *T. ixioides*

1) *T. dudleyi*, Dudley's Triteleia: Fls-stem 10-30 cm H; lvs 1-3 dm L, 4-8 mm W; pedicels 15-35 mm L; fls pale yellow, drying purplish, perianth about 2 cm L. Rare in subalpine forests, 9500-11,500', Tulare Co, Jun-Aug.

2) *T. hyacinthina*, White Brodiaea: Fl-stems 3-6 dm H; lvs 1-4 dm L, 5-20 mm W; pedicels 5-50 mm L; fl-tube cup-shaped, the lobes spreading, 7-12 mm L; anthers white to blue. Common, moist to dryish habitats below 7200', w. SN (all), May-Aug.

3) *T. ixioides*, Golden Brodiaea, Pretty Face: Fl-stems 1-4 dm H; lvs 2-4 dm L, 3-14 mm W; perianth-tube 4-7 mm L, the lobes 7-12 mm L. Common in dryish habitats, SN (all), May-Jul.

4) *T. laxa*, Grass Nut, Ithuriel's Spear: Fl-stems 1-7 dm H; lvs 2-4 dm L; pedicels 2-9 cm L, usu slightly bent at apex; perianth ± horizontal with pistil on lower side and filaments curved upward, lobes 8-20 mm L. Common in heavy soils below 4600', Tehama Co. s., Apr-Jun.

Veratrum

Corn-lily

V. californicum: Coarse erect herbs 1-2 m H; stems unbranched, leafy throughout; lvs broadly oval to lanceolate, entire, sheathing, 2.5-4 dm L, pleated; panicles 2-5 dm L; pedicels 2-6 mm L; petals white, 9-15 mm L. Common, wet mdws below 11,000', SN (all), Jul-Aug. Toxic alkaloids are found throughout the plant; symptoms include depressed heart rate, salivation, burning sensation in mouth, and headache.

plant (1 dm)

inflorescence (1 dm)

Xerophyllum

Bear-grass
Indian Basket-grass

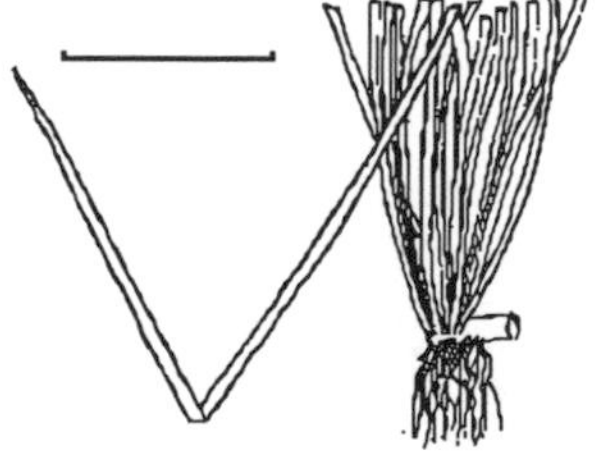

base of plant (5 cm)

X. tenax: Stem 3-15 dm H, stout, unbranched; basal lvs densely clustered, rigid, 2-4 mm W above the base; stem lvs passing into long, linear bracts; raceme 1-6 dm L; fls whitish; petals linear-oblong, 6-10 mm L. Open, dry slopes and ridges below 6000', Placer Co. n., May-Aug. The fibrous root is best eaten roasted or boiled.

Yucca

Yucca
Spanish Bayonet

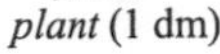
plant (1 dm)

Y. whipplei ssp *caespitosa*: Lvs all basal, in compact tuft, 3-8 dm L, 4-10 cm W, stiff, fleshy, sharp-tipped; fl-stalk unbranched, 1-3 m H, stout; infl 5-8 dm L; fls large, pendent. Dry, rocky slopes, Middle Fork of Kings River s., below 4000', May-Jun. The fls, buds and young fl-stalks may be eaten raw, roasted or boiled. If the stalk is cut into slices and cooked the outer rind comes off easily.

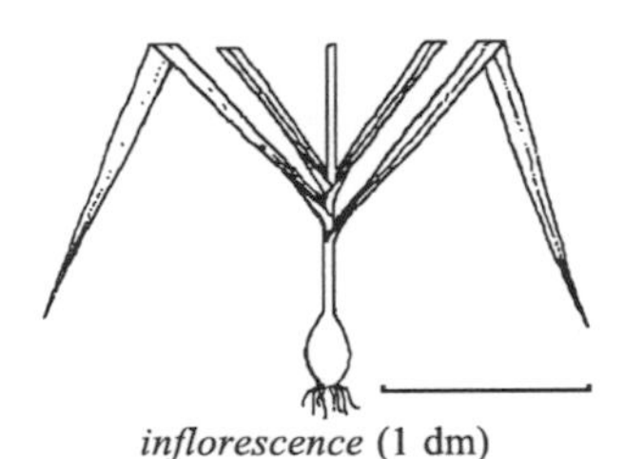
inflorescence (1 dm)

Zigadenus

Zygadene

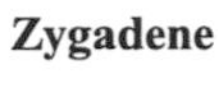

Z. venenosus inflorescence (1 cm)

Stems unbranched, leafy below; lvs glabrous. Fls greenish or yellowish white, in terminal racemes or panicles. Perianth withering-persistent. The bulbs and fls of *Z. venenosus* are very toxic, causing severe indigestion and circulatory collapse. The other species are less toxic but definitely should be avoided.

Fls normally racemose — *Z. venenosus*
Fls mly paniculate
  Stem 2-6 dm H; Nevada Co. n., dry places — *Z. paniculatus*
  Stem 6-10 dm H; Butte Co. s., wooded slopes — *Z. exaltatus*

Z. exaltatus, Giant Zygadene: Bulb 5-8 cm L; stem stout, smooth; basal lvs 4-7 dm L; infl 2-4 dm L; petals 7-8 mm L; capsule 2-3 cm L. Below 4000', May-Jul.

Z. paniculatus, Sand-corn: Bulb ovoid, 3-4 cm L; stems smooth; basal lvs 2-5 dm L, 5-20 mm W; infl 5-25 cm L; pedicels 1-2 cm L; petals yellowish white, broadly ovate, 4 mm L; stamens 4-5 mm L; capsule 10-12 mm L. Mly e. SN, 4000-7000', May-Jun.

Z. venenosus, Death-camas: Bulbs oblong-ovoid, 1.5-2.5 cm L with dark outer coats; stems 2.5-6 dm H, glabrous, slender; basal lvs 1.5-6 dm L, usu pleated; raceme 5-20 cm L; the lower branches and each upper fl subtended by a lance-linear bract; petals whitish, 2-5 mm L, spreading, yellow at base; capsule cylindric, 1-1.5 cm L. Moist habitats below 9200', SN (all), May-Jul.

ORCHIDACEAE-Orchid Family

Perennials from short or elongate rhizome with fibrous to fleshy roots. Lvs sheathing, often reduced to scales. Fls perfect, irregular, bracted.

A. Plants white to brown, not green; lvs scale-like
  B Plant white; perianth 12-15 mm L, not striped — *Cephalanthera* p. 226
  BB Plant brown, yellow or purple; perianth 6-8 mm L or longitudinally striped if longer — *Corallorhiza* p. 226

AA Plants green; lvs with conspicuous, flattened blade
  B Fls few, with leafy bracts
    C Fls 1-3; lip an inflated pouch, 2-4 cm L — *Cypripedium* p. 226
    CC Fls 3-15; lip saccate at base, about 1 cm L — *Epipactis* p. 227
  BB Fls many; bracts not leafy
    C Lvs $\pm$ ovate
      D Lvs 2, subopposite, sessile, green throughout — *Listera* p. 227

DD Lvs several, basal, petioled, with white stripe along midvein — *Goodyera* p. 227
CC Lvs oblong to linear
D Fls spiraled up axis of spike; lip not spurred — *Spiranthes* p. 228
DD Fls in ± vertical rows; lip conspicuously spurred
E Stem ± scapose; lvs often withered in fl — *Piperia* p. 227
EE Stem leafy; lvs persistent in fl — *Platanthera* p. 228

Cephalanthera

Phantom Orchid

inflorescence (1 cm)

C. austinae: Stem stout, erect, 2-5 dm H; lvs 2-4 cm L, mly sheathing stem; infl ± spicate, 5-15 cm L; perianth white with yellowish tinge. Dry woods below 6000', Fresno Co. n., May. Probably toxic.

Corallorhiza

Coralroot

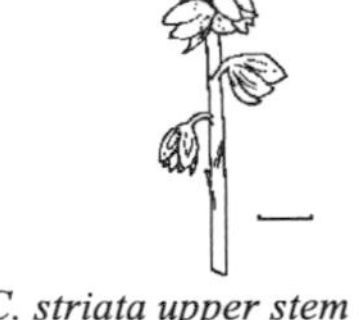

C. striata upper stem (1 cm)

Rhizomes much branched. Infl a loose terminal *raceme*. Rhizome toxic, causing fever and profuse perspiration.

Perianth segments 8-15 mm L, with purplish longitudinal stripes — *C. striata*
Perianth segments usu shorter, lacking stripes
Perianth segment 5-6 mm L, 1 veined; rare in Plumas Co. — *C. trifida*
Perianth segments 6-9 mm L, 3-veined; below 9000', SN (all) — *C. maculata*

C. maculata, Spotted Coralroot: Stems 2-7 dm H, glabrous, the sheaths whitish; fls few to many; pedicels 2-3 mm L; sepals and petals crimson-purple to greenish, 3-nerved; petals ± oblong; lip white, spotted and veined with crimson, unequally 3-lobed, 6-8 mm L; column yellow, curved, compressed; capsule nodding, 1.5-2.5 cm L. Woods below 9000', SN (all), Jun-Aug.

C. striata, Striped Coralroot: Stems glabrous, erect, 1.5-5 dm H; lvs tubular sheaths, whitish to purplish; fls few to many, pinkish yellow or whitish, tinged and striped with purple; lip white with purple veins; capsule 1.2-2 cm L. Woods below 7500', Sierra Co. n., May-Jul.

C. trifida, Northern Coralroot: Stems 1-3 dm H, yellow; sepals white to yellow; lip 3-4 mm W, white or red-spotted. Whet habitats in coniferous forests, 5000-6000'.

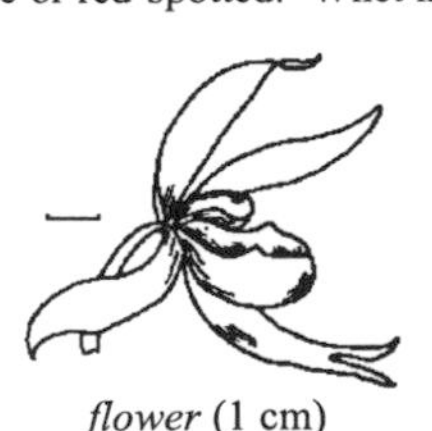

flower (1 cm)

Cypripedium

Lady-slipper

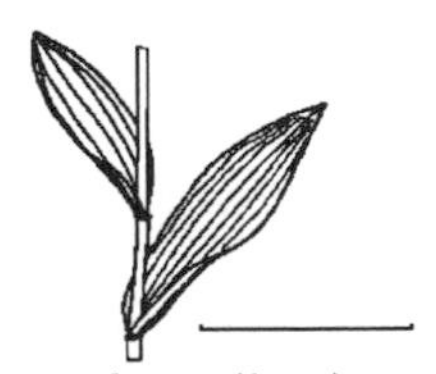

leaves (1 cm)

Stems with 4 or more lvs; basal lvs 1-2. Lvs alt. Fls 1-few with leafy bracts. Lower sepals fused and descending behind lip; lateral petals descending, ± like upper sepal. Fertile stamens 2. Rhizome somewhat toxic but has been used medicinally as a stimulant and an antispasmodic.

Petals 10-20 mm L, oblong to lanceolate, flat — *C. californicum*
Petals 30-60 mm L narrower, twisted — *C. montanum*

C. californicum, California Lady's-slipper: Stems 20-100 cm H; lvs 5-12, 5-15 cm L; lower ± elliptic, the upper lanceolate; lip white. Moist habitats in coniferous forests, n. SN, May-Aug.

C. montanum, Mountain Lady-slipper: Stems stout, 3-6 dm H, glandular-puberulent; lvs 4-7, elliptic-ovate or narrower, 8-15 cm L, 4-7 cm W, clasping; fls showy; sepals brown-purple; petals similar to sepals but crisped and narrower, 25-60 mm L; lip white with purple veins, 2-3 cm L; capsule ascending, 2 cm L. Moist woods below 5000', Mariposa Co. n., May-Aug.

Epipactis

Giant Helleborine

upper stem (1 cm)

E. gigantea: Stems stout, 3-9 dm H, unbranched, leafy; lower lvs ovate, 5-15 cm L; upper lanceolate and gradually reduced; fls racemose; pedicels 4-8 mm L; sepals 12-20 mm L, greenish; petals shorter, purplish to reddish; lip strongly veined and marked with red or purple, constricted above saccate base; capsule pendant. Moist stream banks below 7500', SN (all), May-Aug. Herbage slightly toxic if eaten.

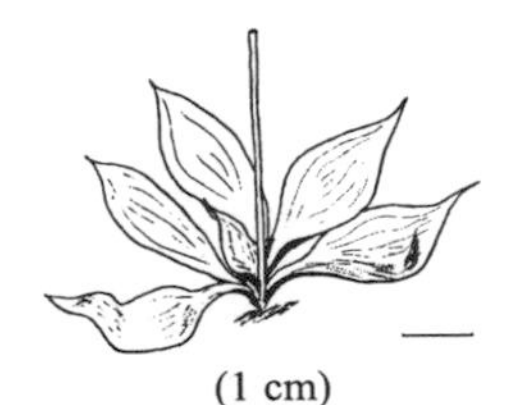

(1 cm)

Goodyera

Rattlesnake-plantain

(1 cm)

G. oblongifolia: Fl-stem stout, 2-4 dm H; lvs basal, oblong-ovate, 3-6 cm L; winged petioles about half as long; perianth about 8 mm L; lateral sepals free, the dorsal adnate on petals forming a galea; capsule about 1 cm L. Dry forest floor below 5500', Mariposa Co. n., Jul-Aug. Herbage slightly toxic if eaten.

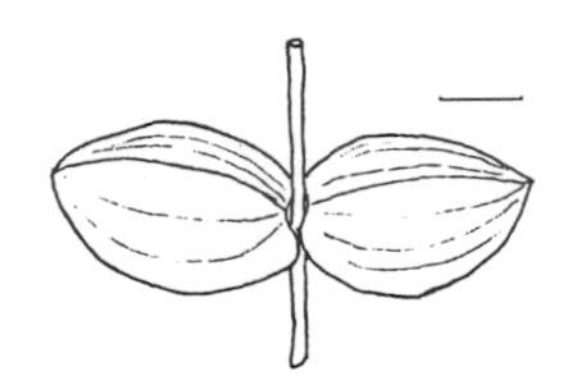

leaves (1 cm)

Listera

Broad-lipped Twayblade

inflorescence (5 cm), *flower* (1 cm)

L. convallarioides. Stem 1-2 dm H, slender, usu ± glandular-pubescent above the 2 lvs; lvs sessile, broadly ovate, 3-5 cm L, rounded at base; raceme 3-15 fld; pedicels 5-8 mm L; sepals and petals yellow-green, reflexed in flowering, 4-5 mm L; lip 7-10 mm L, shallowly 2-lobed at tip. Moist to wet places in shade below 8000', Fresno Co. n., Jun-Aug. Herbage probably slightly toxic.

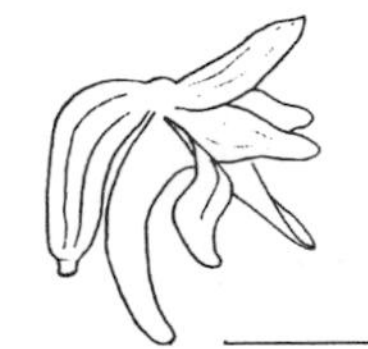

P. unalascensis flower (1 cm)

Piperia

Rein Orchid

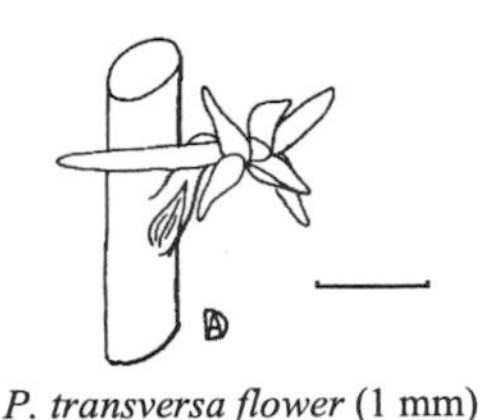

P. transversa flower (1 mm)

Perennials with erect, unbranched stems. Lvs basal, 2-5; cauline lvs much reduced. Infl a spike or raceme. Usu found in wet habitats.

Spur 6-10 mm, usu straight; sepals white to yellowish *P. transversa*
Spur 2-7 mm L, at least upper sepal green

Spur curved; lateral petals recurved — *P. unalascensis*
Spur straight,; lateral petals ± pointing forward — *P. leptopetala*

P. leptopetala: Stem 15-70 cm H; basal lvs 5-15 cm L; infl 5-40 cm L, open; perianth green. Uncommon. Dry places below 7000', SN (all).

P. transversa: Stem 15-50 cm H; basal lvs 5-20 cm L; petals whitish with green midvein. Mly dry sites below 8500', SN (all).

P. unalascensis: Stem 3-6 dm H, ribbed; lower lvs oblanceolate, 1-2 dm L, the upper reduced, long-pointed; spike slender, 1-3 dm L; fls 10-12 mm L; lip linear to lance-linear, 6-8 mm L. Mly along streams or boggy places 4000-11,000', SN (all), Jun-Aug.

inflorescence (1 cm)

Platanthera

Bog Orchid

Perennials with erect, unbranched stems. Lvs basal and cauline, mly sessile. Infl a terminal spike. Petals shorter than sepals; lip with a basal spur. Wet sites. The roots of most spp may be eaten raw or cooked; however, orchids are rare and should be used for food only in need.

Fls white; stems leafy; infl dense — *P. leucostachys*
Fls greenish; stems often naked; infl usu open
  Spur 2-4 mm L, cylindric; Convict Creek Lake Basin — *P. hyperborea*
  Spur 6-8 mm L, filiform; widespread — *P. sparsiflora*

P. hyperborea, Green -flowered Bog Orchid: Stem 2-5 dm H; lvs 6-15 cm L; spike 8-20 cm L, with divergent bracts; fls 10-12 mm L; lip lanceolate, 4-5 mm L. Bogs to 10,000', Jun-Aug.

P. leucostachys, Sierra Rein Orchid: Stems 4-8 dm H; lvs lanceolate, 8-20 cm L, gradually reduced above; spikes 1-2 dm L; fls 10-14 mm L; lip 7-8 mm L; spur 7-20 mm L. Wet places below 11,000', SN (all), May-Aug.

P. sparsiflora: Sparsely-flowered Bog Orchid: Stem 3-6 dm H, ribbed; lower lvs oblanceolate, 1-2 dm L, the upper reduced, long-pointed; spike slender, 1-3 dm L; fls 10-12 mm L; lip linear to lance-linear, 6-8 mm L. Mly along streams or boggy places, 4000-11,000', SN (all), Jun-Aug.

Spiranthes

Hooded Ladies-Tresses

S. romanzoffiana: Stem stout, erect, glabrous, leafy only below, 1-5 dm H; lvs 3-5, linear to lance-linear, 0.5-3 dm L; spike 5-12 cm L; bracts lanceolate, membranous, exceeding fls; perianth segments greenish white, 6-8 mm L; lip oblong, constricted below the dilated crisped apex. Wet banks and mdws below 10,000', SN (all), Jun-Aug. Another sp, *S. porrifolia*, does not have the lip constricted below apex. A related sp has strong diuretic properties, making it undesirable as a food source.

plant (1 cm)

POACEAE - GRASS FAMILY

Annual or perennial herbs. Stalks solid at nodes, otherwise usu hollow. Lvs consisting of a ± linear blade and a sheath that wraps around the stalk for some distance below the base of the blade. Infl an open to dense cluster of spikelets; the spikelets 1- to many-fld, consisting of a short axis (rachilla), usu 2 lower bracts (glumes) that do not contain fls, and 1 to many upper bracts. Each upper bract (lemma) usu contains a second 1- to 2-nerved bract (palea) and the fl (positioned between the two bracts). The fl is usu bisexual with 3 stamens and 1 style. The seeds of all Sierran grasses are edible but are often small and tedious to collect.

A Spikelets sessile or infl appearing solid, without easily discernible branches
  B Spikelets many, in a dense, usu uninterrupted infl; pedicels present but inconspicuous
    C Awns conspicuous, 5-10 mm L; infl often plume-like

D Awns twisted or bent, not producing a fur-like texture *Trisetum aristata* p. 241
DD Awns straight, soft and often fur-like *Polypogon* p. 240
CC Awns 0-2 mm L
D Spikelets 2-fld; infl often interrupted below *Koeleria* p. 237
DD Spikelets 1-fld; infl continuous or disintegrating above
E Awns inconspicuous; infl disintegrating leaving a naked rachis *Alopecurus* p. 232
EE Awns conspicuous, purple-tipped; rachis disintegrating with infl *Phleum* p. 239
BB Spikelets sessile or nearly so, borne in ± 2 rows
C Spikes linear or consisting of only 1 to 2 spikelets
D Spikes of only 1 to 2 spikelets; high montane *Danthonia unispicata* p. 234
DD Spikes linear, 2-10 cm L, of several 1-fld spikelets *Scribneria* p. 240
CC Spikes broader, usu with several to many spikelets
D Axis of infl breaking apart at the nodes in fr; spikelets 3 per node, dimorphic *Hordeum* p. 237
DD Axis of infl not usu breaking at nodes; usu spikelets 1-4 per node, all alike
E Spikelet 1 per node; internodes on infl 1-3 mm L *Agropyron* p. 231
EE Spikelets 2 to many per node, or if 1, the internodes longer
F Glumes ± elliptic; lemma awns 1-25 (-90) mm L *Elymus* p. 235
FF Glumes usu narrow; lemma awns 1-3 mm L *Leymus* p. 237
AA Infl branched or spikelets on easily observed pedicels
B Spikelets with 1 sterile floret below fertile floret; infl openly and diffusely branched *Panicum* p. 239
BB Lowest floret of spikelet fertile; infl usu not both openly and diffusely branched
C Spikelets 1-fld, mly small and hard to examine without hand lens
D Glumes over 4 mm L
E Panicle dense, spike-like, glumes 4-8 mm L *Calamagrostis purpurescens* p. 233
EE Panicle ± open, or if dense, glumes 8-30 mm L
F Awns mly 5-15 cm L, smoothly curving, not bent *Hesperostipa* p. 237
FF Awns shorter, usu bent and twisted *Achnatherum* p. 230
DD Glumes 1-4 mm L
E Glumes falling with spikelets; stalks 5-15 dm H *Cinna* p. 233
EE Glumes remaining on pedicel; stalks usu shorter
F Lemma as long as or longer than glumes and exserted beyond them
G Lemma with twisted awn 10-20 mm L; moist habitats above 9000' *Ptilagrostis* p. 240
GG Lemma awn lacking or shorter; if similar then plants of dry habitats
H Infl mly less than 4 cm W; secondary branches not appressed to primary *Muhlenbergia* p. 238
HH Infl usu over 5 cm W; secondary branches appressed to primary *Sporobolus* p. 241
FF Glumes longer than and covering lemma
G. Fl with tuft of hairs at base; palea usu well developed *Calamagrostis* p. 233
GG Fl without tuft of hairs; palea usu small or none *Agrostis* p. 231
CC Spikelets mly with 2 or more fls
D Lemma not awned (see also *Bromus inerminus*, *Festuca kingii* and *F. viridula*)
E Veins on lemma converging at tip; blades narrow to hair-like *Poa* p. 239
EE Veins on lemma parallel throughout; blades usu flat
F Lf-sheath connate at apex; below 8500' *Glyceria* p. 236
FF Lf-sheath open at apex
G Lower glume 1-veined *Torreyochloa* p. 241
GG Lower glume 3 to 5-veined *Melica* p. 238
DD Lemma awned
E At last upper glume longer than lowest lemma and usu longer than remainder of spikelet awn on lemma dorsal and usu bent
F Spikelets mly 5-fld *Danthonia* p. 234
FF Spikelets 2-fld (occasionally 3-fld in *Trisetum*)
G Awn (usu present) from above middle of lemma, conspicuously exerted *Trisetum* p. 241
GG Awn from below middle of lemma, usu inconspicuous in intact spikelet *Deschampsia* p. 234
EE Neither glume longer than lowest lemma; awn on lemma terminal and straight (rarely dorsal in *Bromus*)
F Glumes membranous on margins; lemma 7 to 9-nerved, entire at apex *Melica* p. 238

FF Glumes green or membranous throughout
G Lemma 2-cleft at tip; nerves prominent, 5 to 9 — *Bromus* p. 232
GG Lemma entire at tip; nerves usu faint, 5 or less
H Annuals; stamen usu 1 — *Vulpia* p. 242
HH Perennials; stamens usu 3 — *Festuca* p. 235

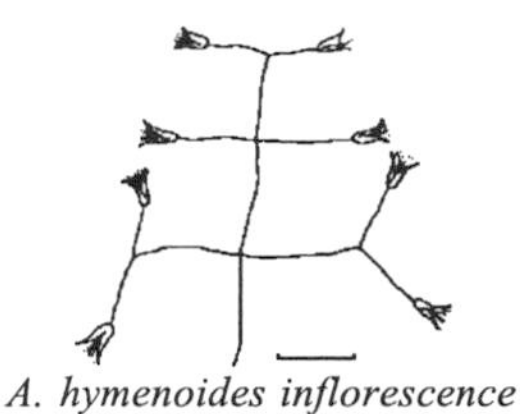

A. hymenoides inflorescence (1 cm)

Achnatherum

Needlegrass

A. californica inflorescence (1 cm)

Slender tufted perennials, with involute or filiform blades. Glumes subequal, often papery. Lemma about as long as glumes, broad, pubescent. The large seeds of some spp are worth collecting for food.

A Lemma awnless at maturity
B Infl open, the branches and pedicels widely spreading — 1) *A. hymenoides*
BB Infl compact, branches and pedicels ascending — 11) *A. webberi*
AA Lemma awned, the awn usu 10-40 mm L
B Lower part of awn hairy, the lowest hairs longer than lemma hairs
C Lowest hairs on awn shorter than hairs on lemma — 6) *A. occidentalis*
CC Lowest hairs on awn longer than hairs on lemma
D Awn with 1 bend; uppermost ligule about 1 mm L, ciliate — 8) *A. speciosum*
DD Awn with 2 bends; uppermost ligule 2-4 mm L, glabrous
E Infl 15-30 cm L, Tuolumne Co. s. — 2) *A. latiglumis*
EE Infl 5-15 cm L; mly e. slope — 10) *A. thurberianum*
BB Lower part of awn without hairs
C Glumes about 15 mm L; lemma bilobed at tip, the lobes extending into 2 lateral awns 2-3 mm L; rare — 9) *A. stillmanii*
CC Glumes mly shorter; lemma entire or obscurely lobed
D Lemma densely appressed-hairy, with hairs 3-4 mm L and extending beyond tip of lemma — 7) *A. pinetorum*
DD Lemma pubescent but never long-hairy
E Palea 1/3-2/3 lemma length — 5) *A. nelsonii*
EE Palea more than 3/4 lemma length
F Stalks 3-8 dm H; below 7500' — 3) *A. lemmonii*
FF Stalks 2-3 dm H; mly e. SN — 4) *A. lettermanii*

1) *A. hymenoides*, Indian Ricegrass: Blades almost as long as stalks; infl 8-15 cm L, open, with capillary pedicels: glumes 6-7 mm L, long-pointed. Common in dry, sandy habitats, below 10,400', mly e. SN.

2) *A. latiglumis*, Wide-glumed Stipa: Stalks slender, 5-10 dm H; infl narrow, 1.5-3 dm L, loosely-fld; glumes subequal, 13-15 mm L; lemma densely pubescent, 8-9 mm L. Tuolumne Co. s.

3) *A. lemmonii*, Lemmon's Stipa: Stalks 3-8 dm H; blades 1-2 dm L, 1-2 mm W; infl 5-12 cm L, narrow, pale or purplish; glumes 8-10 mm L; lemmas 6-7 mm L, pubescent; awn 2-3.5 cm L, twice bent. SN (all).

4) *A. lettermanii*, Letterman's Stipa: Stalks in large tufts; blades filiform, involute, 2-3 dm L; infl narrow, 1-1.5 dm L; glumes 7-9 mm L; lemmas 4-6 mm L. Mly e. SN (all).

5) *A. nelsonii*: Stalks 4-15 dm H; blades 1-5 mm W; infl 10-35 cm L; glumes 6-12 mm L subequal; awn 15-45 mm L. Open, habitats below 11,500', SN (all).

6) *A. occidentalis*, Western Needlegrass: Stalks to 12 dm H; blade to 1 mm W, usu involute; infl 8-30 cm L; glumes 10-15 mm L, subequal; awn 15-40 mm L, bent twice. Open, dry habitats below 12,000', SN (all).

7) *A. pinetorum*, Pine Stipa: Stalks tufted, 1-5 dm H; lvs mly basal, blades filiform, involute, 5-12 cm L; infl narrow, 8-10 cm L; glumes about 9 mm L; lemmas about 5 mm L. Rocky habitats, 7000-12,500', mly e. SN, Eldorado Co. s.

8) *A. speciosum*, Desert Needlegrass: Stalks 3-6 dm H; blade involute; infl partly enclosed by upper leaf sheath, 10-15 cm L; glumes 15-20 mm L. Rocky habitats below 7500', s. SN.

9) *A. stillmanii*, Stillman's Stipa: Stalks stout, 6-10 dm H; blades folded or involute; infl narrow, dense, 1-2 dm L; glumes papery; lemmas 9 mm L; awn about 2.5 cm L, once or twice bent. Rocky habitats, 4000-6000', Placer Co. and probably n.

10) *A. thurberianum*: Stalks pubescent, 3-7 dm H; blade to 2 mm W; infl 5-15 cm L; glumes 10-15 mm L, the lower longer than the upper; awn 30-60 mm L. Dry, open habitats 5000-10,000'.

11) *A. webberi*, Webber's Ricegrass: Stalks erect, densely tufted; infl narrow, 2.5-5 cm L; glumes about 8 mm L, narrow, often purplish. Rocky slopes, 5000-10,500', e. SN(all), w. SN Tuolumne Co. s.

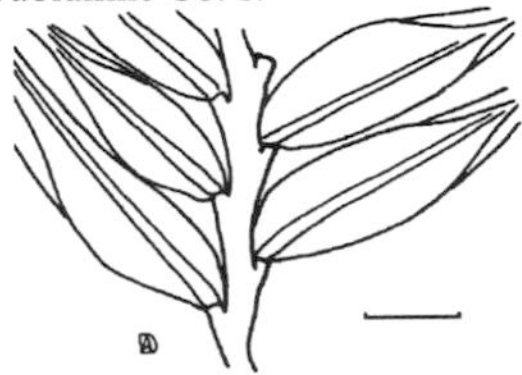

rachis and spikelets (1 mm)

Agropyron

Crested Wheatgrass

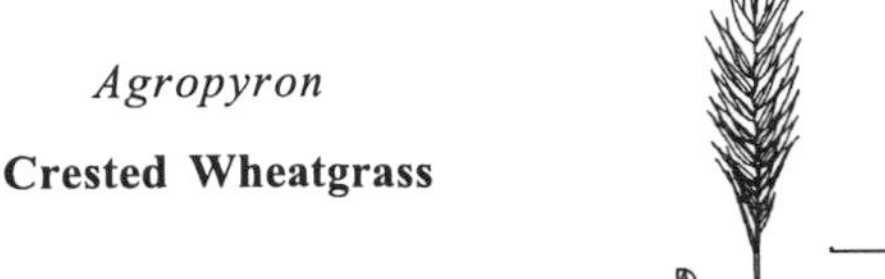

inflorescence (1 cm)

A. desertorum: Perennial, stalk usu erect. Spikes usu erect, green or purplish. Glumes 2, equal. Lemma convex on back, rather firm, mostly acute or awned from apex; palea shorter than lemma. Rootstock can be dried and ground into flour.

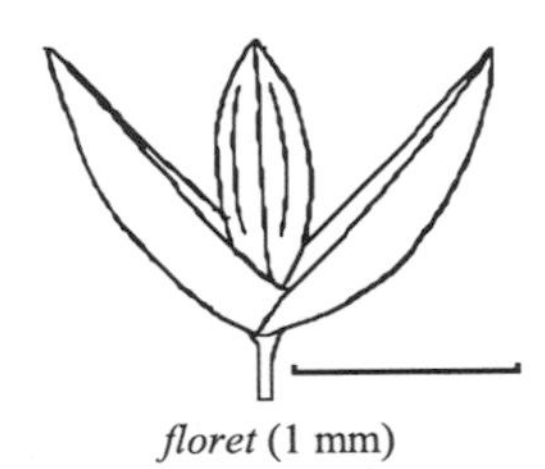

floret (1 mm)

Agrostis

Bentgrass

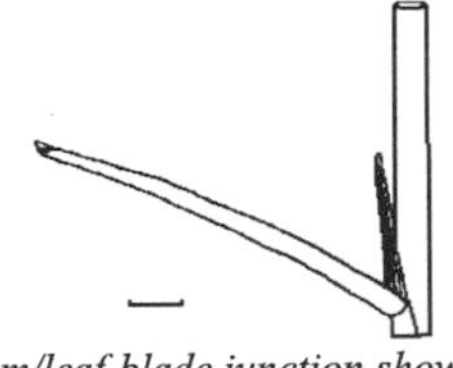

stem/leaf-blade junction showing ligule (shaded) (1 cm)

Perennials with flattish to filiform lvs and open to narrow delicate panicles of many small spikelets. Glumes subequal, usu purplish, occasionally long-tipped but never conspicuously awned.

A Lf-blades 2-10 mm W; stalks mly 3-12 dm H
  B Ligule 4-6 mm L; infl branches short, erect; to 10,000' ... 1) *A. exarata*
  BB Ligule 1-3 mm L; infl branches often lax; below 7500'
    C Plant with rhizomes; infl narrow with ascending branches; ... 4) *A. pallens*
    CC Plant from fibrous roots; infl wide-spreading ... 3) *A. oregonensis*
AA Lf-blades 1-3 mm W; stalks slender, 1-4 dm H
  B Palea evident, about 2/3 length of lemma; stalk lvs usu well developed ... 6) *A. thurberiana*
  BB Palea minute or obsolete; stalk lvs often filiform
    C Infl narrow, ± dense; rocky slopes ... 7) *A. variabilis*
    CC Infl open, sometimes widespreading; moist habitats
      D Upper branches of infl mly erect, not capillary; Tulare and Inyo cos. ... 4) *A. pallens*
      DD Upper branches of infl spreading, capillary; SN (all)
        E Stems 20-60 cm H; panicles 15-25 cm L ... 5) *A. scabra*
        EE Stems 10-30 cm H; panicles 5-10 cm L ... 3) *A. idahoensis*

1) *A. exarata*, Western Bentgrass: Stalks tufted; blades 2-10 mm W, infl 5-30 cm L, often interrupted at base. Moist, open habitats, SN (all), Jun-Aug.

2) *A. idahoensis*, Idaho Bentgrass: Stalks tufted, spikelets about 2 mm L. Mountain mdws, 5000-11,500', SN (all), Jul-Aug.

3) *A. oregonensis*, Oregon Bentgrass: Stalks 6-9 dm H; blades 2-4 mm W; infl branches whorled, to 10 cm L. Wet habitats below 7000', SN (all), Jun-Aug.

4) *A. pallens*: Stalks tufted, 1-7 dm H; ligule to 4 mm L; blades flat or folded, 2-6 mm W; infl often purple, erect, 1-2 dm L, the lower branches 2-5 cm L, spreading. Mdws, below 10,500', SN (all), May-Aug.

5) *A. scabra*, Ticklegrass: Blades flat, 1-3 mm W, 8-20 cm L; infl 15-25 cm L. Moist habitats, 3500-10,000', SN (all), Jul-Sep.

6) *A. thurberiana*, Thurber's Bentgrass: Stalks very slender; lvs ± crowded near base; infl rather narrow, drooping, 5-7 cm L; spikelets green to purplish. Moist habitats, 3500-10,000', SN (all), Jul-Sep.

7) *A. variabilis*, Variable Bentgrass: Stalks densely tufted, 1-2.5 dm H; blades flat, to 1 mm W; infl 2-6 cm L; spikelets purplish. At 5000-12,000', SN (all), Jul-Aug.

Alopecurus

Foxtail

Perennials with conspicuous brownish nodes and a unique cylindrical, dense infl. Ligule 3-5 mm L. Spikelets strongly compressed laterally.

Awn straight, usu less than 2 mm L — *A. aequalis*
Awn bent, usu 2-4 mm L — *A. geniculatus*

inflorescence (1 cm)

A. aequalis, Short-awned Foxtail: Stem 1-5 dm H; blades 1-4 mm W; infl soft 2-7 cm L, 4-5 mm W. Common in moist to wet habitats to 11,500', SN (all).

A. geniculatus, Water Foxtail: Stem 1-6 dm H; blades 2-8 cm L; infl 4-8 mm W. Moist habitats below 6000', SN (all).

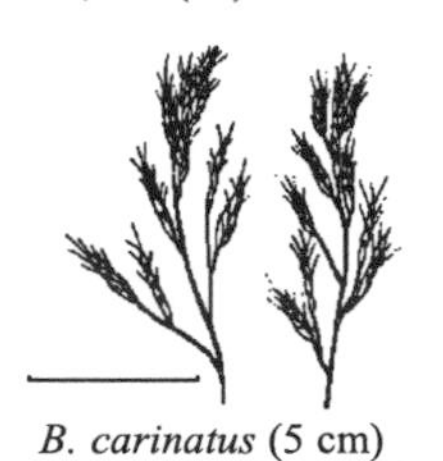

B. carinatus (5 cm)

Bromus

Bromegrass

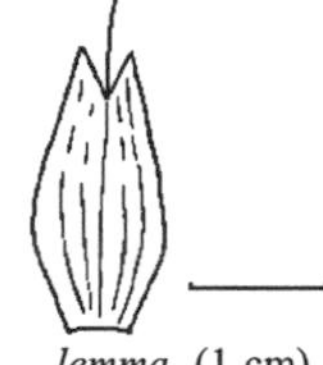

lemma (1 cm)

Usu perennials with closed sheaths and usu flat blades. Spikelets usu 1 to several cm L. Glumes unequal, acute. Lemmas rounded or keeled on back, 5-9-nerved, mly with a 5-10 mm L awn. Palea usu shorter than lemma.

A First glume mly 1-nerved
  B Stalks 3-6 dm H; blades 1-4 mm W; awn 12-14 mm L — 10) *B. tectorum*
  BB Stalks 5-15 dm H; blades 4-15 mm W; awn 0-6 mm L
    C Blades glabrous — 6) *B. inermis*
    CC Blades pubescent above — 4) *B. ciliatus*
AA First glume 3-5-nerved
  B Spikelets strongly flattened — 3) *B. carinatus*
  BB Spikelets cylindrical or slightly flattened
    C Infl narrow, mly less than 2 cm W, the branches erect — 9) *B. suksdorfii*
    CC Infl wide-spreading, over 2 cm W, the branches often drooping
      D Awns 0-1 mm L — 2) *B. brizaformis*
      DD Awns 2-10 mm L
        E Awns 2-3 mm L; stalks 3-8 dm H; rare, 7000-10,500' — 1) *B. anomalus*
        EE Awn 4-8 mm L; stalks 6-15 dm H; widespread below 8000'
          F Second glume 5-nerved; stalks decumbent at base and rooting at lower nodes — 7) *B. laevipes*
          FF Second glume 3-nerved; stalks erect from base
            G Infl mly 15-20 cm L, the branches spreading to nodding — 5) *B. grandis*
            GG Infl mly 7-15 cm L, the branches spreading to ascending — 8) *B. orcuttianus*

1) *B. anomalus*: Much like *B. laevipes*; blade 1-2.5 dm L, 2-5 mm W, usu erect; infl 7-15 cm L; spikelets 13-15 mm L; lemmas 8-13 mm L. Dry habitats, Eldorado, Tulare and Inyo cos., Jul-Aug.

2) *B. brizaformis*, Rattlesnake Brome: Annual to 7 dm H; sheath densely soft-hairy, blade to 10 mm W; infl open, nodding to 15 cm L; awn 0-1 mm L. Open slopes, 5000-6000', n. SN.

3) *B. carinatus*, California Brome: Blades mly 2-3 dm L, 3-10 mm W; infl 1.5-3 dm L, with spreading or drooping branches; spikelets 2-3 cm L, mly 6-10-fld; first glume 6-9 mm L, the second 10-15 mm L. Common in dry, open habitats below 10,500', SN (all), May-Aug.

4) *B. ciliatus*, Fringed Brome: Blades 5-10 mm W; infl 1-2 dm L, open; spikelets 1.5-2.5 cm L, 4-9-fld; lemmas 10-12 mm L; awn 3-5 mm L. Moist habitats, 7500-9600', known from Tulare and Inyo cos., but probably extending further n., Jul-Aug.

5) *B. grandis*: Perennial, 6-15 dm H; blade 5-10 mm W; infl 15-20 cm L, open, the branches spreading to nodding; awn 3-6 mm L. Dry habitats below 8000', Mariposa Co. s.

6) *B. inermis*, Smooth Brome: Much like *B. ciliatus*; stalks smooth; first glume 6-8 mm L, second 7-10 mm L; lemmas often awnless. Occasional in disturbed places below 8000', SN (all), May-Aug.

7) *B. laevipes*, Woodland Bromegrass: Sheaths and blades glabrous; blades 4-10 mm W; infl drooping, 1-2 dm L; spikelet 2.5-3.5 cm L, 5-11-fld; first glume 6-9 mm L, second 10-12 mm L; lemmas 12-15 mm L. Cool, moist habitats below 8600', SN (all), Jun-Aug.

8) *B. orcuttianus*, Orcutt's Bromegrass: Much like *B. laevipes*: sheaths and blades occasionally pubescent; infl erect. Various habitats, up to 10,000', in Fresno and Tulare cos., below 8000' further n., Jun-Sep.

9) *B. suksdorfii*, Suksdorf's Bromegrass: Stalks 3-10 dm H; sheaths and blades glabrous; infl 7-13 cm L; spikelets 1.5-3 cm L, 5-7-fld; first glume 7-11 mm L, second 9-12 mm L; awn 2-5 mm L. Many habitats, 4000-11,000', SN (all), Jul-Aug.

10) *B. tectorum*, Downy Cheat: Annual; sheaths and blades pubescent; infl broad, drooping, 5-12 cm L with slender reddish branchlets; spikelets nodding, 1-2 cm L; first glume 4-6 mm L, second 8-10 mm L; lemmas about 10 mm L. Common at mid-altitudes throughout the Sierra, May-Jun.

C. purpurascens spike (1 cm)

Calamagrostis

Reedgrass

C. canadensis spikelet (1 mm)

Perennial, usu tall grasses, mostly with creeping rhizomes. Glumes subequal. Lemma 5-veined, the midvein exserted as an awn; infl often purplish.

A Awns exserted beyond glumes; lf-blades usu less than 4 mm W
  B Stalks 1.5-3 dm H; infl open; glume 3-4 mm L — 1) *C. breweri*
  BB Stalks 4-6(-10) dm H; infl spike-like; glume 6-8 mm L — 3) *C. purpurascens*
AA Awns mly included within glumes; lf-blades mly 3-8 mm W
  B Infl nodding, dense to open, 1-2.5 dm L — 2) *C. canadensis*
  BB Infl ± erect, dense, 0.5-1.5 dm L — 4) *C. stricta*

1) *C. breweri*, Brewer's Reedgrass: Stalks densely tufted, slender; lvs mly basal, usu involute; infl ovoid, 3-8 cm L, the lower branches 1-2 cm L. Mdws, 6200-12,200', SN (all), Jul-Sep.

2) *C. canadensis*, Blue-joint: Stalks 6-15 dm H, tufted; blades flat; glumes 3-4 mm L. Moist habitats, 5000-12,000', SN (all), Jul-Sep.

3) *C. purpurascens,* Purple Reedgrass: Stalks tufted; blades ± involute, 5-12 cm L; glumes 6-8 mm L. Rocky habitats, 9500-13,000', SN (all), Jul-Sep.

4) *C. stricta*, Narrow-spiked Reedgrass: Stalks 4-12 dm H; blades flat or ± involute; infl narrow; glumes 3-4 mm L. Moist habitats, 4500-11,000', SN (all), Jun-Aug.

Cinna

Woodreed

inflorescence (1 cm)

Perennial with erect stems. Infl open, the branches spreading to ascending. Spikelet breaking below glumes; glumes 1-veined.

Spikelet 4-6 mm L; lemma faintly 5-veined — *C. bolanderi*
Spikelet 2-4 mm L; lemma usu 3-veined — *C. latifolia*

C. bolanderi, Scribner Woodreed: Stem 10-20 dm H; blade to 40 cm L and 20 mm W; infl 7-40 cm L, spreading; spikelets 4-6 mm L; awn inconspicuous or lacking. Moist to wet habitats 5500-8000', Mariposa Co. s.

C. latifolia, Drooping Woodreed: Stalks 5-15 dm H; blades flat, 5-15 cm L, 6-15 mm W; infl with spreading capillary branches naked at the base; spikelets about 4 mm L; glumes subequal;

lemma resembling glumes, almost as long with a straight, 1 mm L awn just below apex. Moist places, 4500-9500', SN (all), Jul-Aug.

D. californica inflorescence (1 cm)

Danthonia

Oatgrass

D. unispicata spikelet (1 cm)

Tufted perennials with an open or spike-like infl of rather large spikelets. Glumes subequal, broad, papery, mly exceeding the upper floret. Lemmas bifid at tip, the teeth mostly slender-awned and a stout, twisted, bent awn arising from between the teeth.

Infl with appressed branches; spikelets several to many — *D. intermedia*
Infl with wide spreading or reflexed branches; spikelets 1 to few
  Infl usu of one spikelet; mly above 5000' — *D. unispicata*
  Infl of several spikelets; mly below 5000' — *D. californica*

D. californica, California Oatgrass: Stalks 3-10 dm H, glabrous; blades 1-2 dm L, often flat, glabrous; pedicels 1-2 cm L; glumes 15-20 mm L; lemmas 8-10 mm L; awns 5-10 mm L. Common in dry habitats below 6800', SN (all), May-Jul.

D. intermedia, Mountain Oatgrass: Stalks 1.5-4 dm H; blades ± involute, subglabrous; infl purplish, 2-5 cm L; glumes about 12 mm L; lemmas 7-8 mm L; awns 6-8 mm L. Moist banks 8200-11,000', Eldorado Co. s., Jul-Sep.

D. unispicata, Few-flowered Oatgrass: Much like and may not be distinct from *D. californica*: stalks 1-2 cm L, in spreading tufts. Rocky habitats above 4500', SN (all), May-Jul.

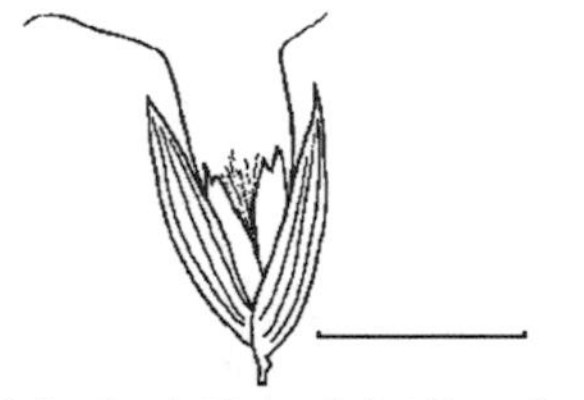

D danthonioides spikelet (5 mm)

Deschampsia

Hairgrass

D. caespitosa lemma (1 cm)

Annual or perennial grasses with flat or involute lvs. Spikelets 2-fld. Glumes subequal, keeled, membranous, shining. Lemmas thin, truncate, 2-4-toothed at summit, with a slender awn from or below middle.

Infl narrow, almost spike-like — *D. elongata*
Infl open, widespreading
  Stalks few, slender, 1-5 dm H — *D. danthonioides*
  Stalks densely tufted, 6-12 dm H — *D. caespitosa*

D. caespitosa, Tufted Hairgrass: Lvs mly basal, flat or folded; infl nodding, 1-2 dm L, the branches capillary; spikelets 4-5 mm L, green to purple. Wet mdws, 3300-12,500', SN (all), Jul-Aug.

D. danthonioides, Annual Hairgrass: Blades few, 2-8 cm L, about 1 mm W; infl 5-12 cm L; glumes 6-8 mm L; awns 4-6 mm L. Moist habitats below 9000', SN (all), May-Aug.

D. elongata, Slender Hairgrass: Stalks tufted, slender, 3-10 dm L; blades flat, about 1 mm W; infl 1-3 dm L; glumes 4-6 mm L; awns to 4 mm L. Wet places, 4500-10,000', SN (all), May-Aug.

Elymus

Wildrye

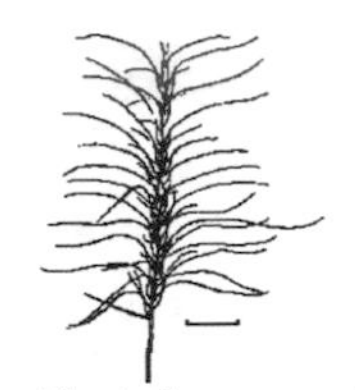

E. glaucous inflorescence (1 cm) *E. elymoides inflorescence* (1 cm)

Usu tufted perennials. Stems often bent at base. Infl spike-like, spikelets often 2 or more per node.

A Glumes awn-like, conspicuously extending into 1 or more awns, 25-100 mm L
B Glume usu entire and often shorter than awn on lemma 1) *E. elymoides*
BB Glume divided into 3 or more outwardly curving awns 4) *E. multisetus*
AA Glumes flat; awns usu less than 25 mm
B Spikelets usu 2 per node 2) *E. glaucous*
BB Spikelets 1 per node
C Lemmas with 15-25 mm L awn, curving out at maturity 5) *E. sierrae*
CC Lemmas with straight awn or lacking awn
D Lemma awn lacking; below 4000', Lassen Co. n. 3) *E. lanceolatus*
DD Lemma awn 1-30 mm L
E Internodes of infl 5-15 mm L 6) *E. stebbinsii*
EE Internodes of infl 12-25 mm L 7) *E. trachycaulus*

1) *E. elymoides*, Squirreltail: Stalks 1-5 dm H; blades flat or involute, 1-5 mm W; spike 2-8 cm L; glumes narrow; awns 2-7 cm L, widespreading. Common on dry open ground below 13,000', SN (all).

2) *E. glaucous*, Blue Wildrye: Stalks 5-15 dm H; blade 4-12 mm W, usu flat; infl 5-15 cm L (excluding awns), not breaking apart with age; internodes 4-8 mm L. Open habitats below 8000', SN (all).

3) *E. lanceolatus*, Thickspike Wheatgrass: Stalks 4-12 dm H; blade 1-5 mm W; infl 5-20 cm L; internodes 5-20 mm L; spikelet 10-15 mm L. Open forests below 4000'.

4) *E. multisetus*, Big Squirreltail: Stalks 2-6 dm H; blades flat or involute, 1-4 mm W; spike dense, 3-10 cm L. Dry habitats below 10,000', SN (all).

5) *E. sierrae:* Stalks 3-5 dm L, decumbent; lf sheath usu glabrous; blade 1-3 mm W; infl 4-7 cm L, flexible; middle internodes about 10 mm L; spikelet mly 3-7, 10-15 mm L; glumes with straight 5 mm L awn; lemma with longer awn. Rocky slopes, 7500-11,000', Sierra Co. s.

6) *E. stebbinsii*: Stalks 7-12 dm; lf sheath usu glabrous; blade 2-6 mm W, flat or involute; infl 10-25 cm L; spikelet 10-20 mm L; glumes 10-15 mm L; awn 1-5 mm L, straight. Dry slopes below 5500', SN (all).

7) *E. trachycaulus*, Slender Wheatgrass: Stalks 5-10 dm H, tufted; blades 2-8 mm W; spike 5-25 cm L, usu compact. Moist to dry habitats below 11,000', SN (all).

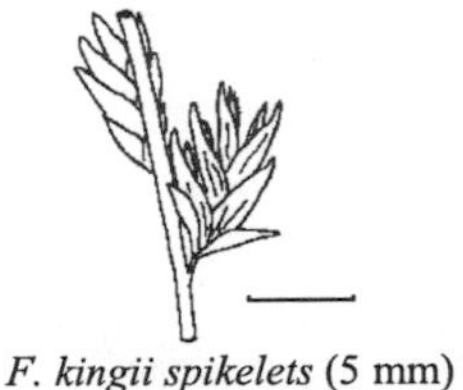

Festuca

Fescue

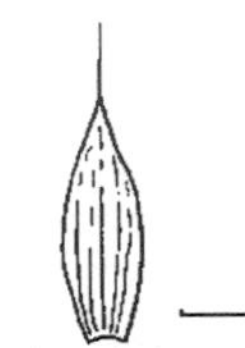

F. kingii spikelets (5 mm) *lemma with awn* (1 mm)

Usu glabrous perennials with erect stems. Lvs mly basal; blade often rolled. Glumes shorter than lowest floret, unequal. Axis breaking above glumes and between florets. Species difficult to distinguish.

A Lemmas awnless
B Plants dioecious; lf-blades without lobes 4) *F. kingii*
BB Plants bisexual; lf-blades with basal lobes clasping stem 6) *F. pratensis*
AA Lemmas awned, the awn short in *F. viridula*; plants bisexual; leaf blades not lobed at base
B Plants of dry rocky places above 8500'; about 1 dm H 1) *F. brachyphylla*
BB Plants of various habitats, below 9000'; taller
C Awns scarcely if at all evident; less than 1 mm L 9) *F. viridula*
CC Awns well-developed, 2 to many mm L
D Plant over 3 dm H; often with many dead leaf sheaths at base; mly below 6000'

E Lf-collar pubescent; blade flat or rolled, usu over 2 mm W — 2) *F. californica*
EE Lf-collar glabrous, blade rolled, less than 2 mm W
F Lf-sheath closed; base of stem decumbent and reddish — 7) *F. rubra*
FF Lf-sheath open; base of stem with dead leaf sheath — 3) *F. idahoensis*
DD Plant mly less than 3 dm H; lacking dead leaf sheaths at base; widespread
E Lf-blade ± flat, rather soft and lax, over 2 mm W
EE Lf-blade narrow or rolled, not lax, less than 2 mm W
F Awns on lemma 5-10 mm L — 5) *F. occidentalis*
FF Awns on lemma less than 3 mm L — 8) *F. trachyphylla*

1) *F. brachyphylla*, Alpine Fescue: Stalks tufted, slender; blades many, filiform, 2-6 dm L; infl short, narrow, 2-5 cm L; spikelets 2-5-fld; lemmas about 3 mm L. SN (all), Jul-Sep.

2) *F. californica*, California Fescue: Stalks ± stout, 6-12 dm H; blades flat; panicle open, few-branched; spikelets compressed, about 5 fld. Cooler, dry habitats below 6000', Jun-Jul.

3) *F. idahoensis*, Idaho Fescue: Stalks 3-10 dm H; blade 5-35 cm L, less than 2 mm W, involute, stiff; infl 6-20 cm L; branches ± appressed; spikelets 5-15 mm L; axis usu visible and zigzag.; awn 1-6 mm L. Dry habitats below 6000', Mariposa Co. n.

4) *F. kingii*, Spiked Hesperochloa: Stalks erect, in large dense clumps, 5-8 dm H; blades 5-30 cm L, 3-6 mm W, flat or loosely involute; panicle erect, 7-20 cm L, with short appressed branches; spikelets 5-15 mm L; glumes about 5 mm L. Dry slopes, 7000-12,000', s. SN, Jun-Aug.

5) *F. occidentalis*, Western Fescue: Stalks slender, 4-10 dm H; blades many, mly basal, narrow-involute; infl 7-20 cm L; spikelets loosely 3-5-fld, 6-10 mm L, lemmas 5-6 mm L. Dry, rocky, forested habitats below 6500', SN (all), May-Aug.

6) *F. pratensis*, Meadow Fescue: Stalks to 12 dm H, loosely clumped; blade 10-30 cm L, 2-7 mm W, flat or loosely involute; infl 10-25 cm L, narrow, branched only at lowest node; spikelet 10-15 mm L. Disturbed sites below 7000', SN (all).

7) *F. rubra*, Red Fescue: Much like *F. occidentalis*; infl narrow, with erect branches; spikelets often purplish. Moist habitats below 8500', SN (all), May-Jul.

8) *F. trachyphylla*, Hard Fescue: Stalks 2-7 dm H, clumped; blades 8-30 cm L, less than 1 mm W, folded; infl 5-10 cm L; branches ± appressed; spikelet 5-10 mm L; awn to 3 mm L. Open habitats below 10,000', n. SN.

9) *F. viridula*, Mountain Bunchgrass: Stalks loosely tufted, 6-10 dm H; blades erect, 1-2 mm W; infl open, 1-1.5 dm L, the branches spreading to ascending, ± remote; spikelets 10-12 mm L; first glume 5-6 mm L, the second 8-9 mm L. Mdws, 6000-10,000', Alpine to Sierra Co., Jul-Aug.

Glyceria

Mannagrass

G. elata branch of inflorescence (5 cm)

G. borealis spikelet (5 mm)

Aquatic or marsh perennials with creeping rhizomes and flat blades. Panicle open or contracted. Glumes unequal. Grains large and worth gathering for food.

Spikelets linear, 10-15 mm L; infl narrow, erect — *G. borealis*
Spikelets ovate or oblong, less than 6 mm L; infl open, usu nodding
Lf-blades mly 2-4 mm W, ± folded; stalks 3-12 dm H — *G. striata*
Lf-blades 6-12 mm W, flat; stalks 10-18 dm H — *G. elata*

G. borealis, Northern Mannagrass: Stalks 5-10 dm H; blades usu 2-4 mm W, erect; infl mly 2-4 dm L, the branches appressed; spikelets mly 6-12-fld, not purplish; lemmas 3-4 mm L. Below 7000', Mariposa Co. n.

G. elata, Tall Mannagrass: Stalks ± fleshy; infl much branched, 1.5-3 dm L; spikelets 6-8-fld, 4-6 mm L, often purplish. Below 8500', SN (all).

G. striata, Fowl Meadow Grass: Much like *G. elata*; stalks slender; infl 1-2 dm L; spikelets 3-7-fld, 3-4 mm L. At 5000-8000', Tuolumne Co. n.

Hesperostipa

Needle-and-thread

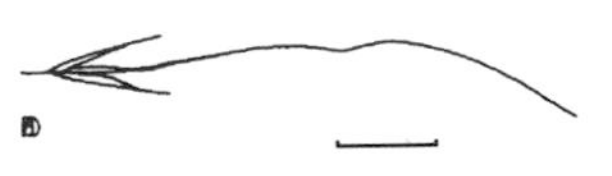

inflorescence (4 cm) *spikelet* (2 cm)

H. comata: Tufted perennial with erect, glabrous stalks 3-6 dm H; blade 1-3 dm L, involute, 1-2 mm W; infl 1-2 dm L, narrow; glumes 15-20 mm L; lemmas about 10 mm L with smoothly curving awns 5-15 cm L. Mly e. slope SN to 11,300'.

Hordeum

Barley

Perennial with flat blades and dense bristly spikes, these often appearing 4-angled. Glumes very narrow, subtending the 3 spikelets. Lemmas awned.

Awns less than 1.5 cm L *H. brachyantherum*
Awns 2-5 cm L *H. jubatum*

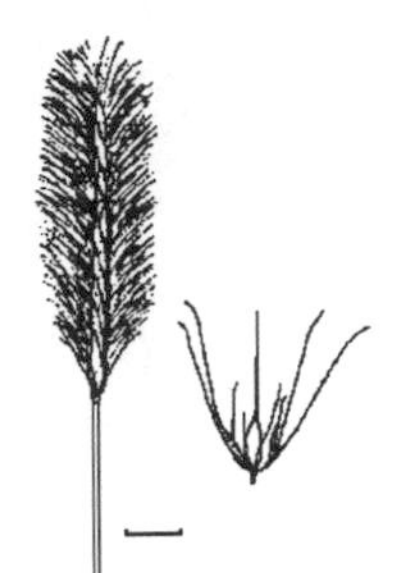

H. jubatum (1 cm)

H. brachyantherum, Meadow Barley: Stalks tufted, 2-7 dm H; spike 8-10 cm L, sometimes purplish; floret of central spikelet 7-10 mm L, the awn 10 mm L; lateral spikelets reduced. Moist habitats below 11,000', SN (all), May-Aug.

H. jubatum, Foxtail: Stalks 3-6 dm H; blades 2-5 mm W, spike nodding, 5-10 cm L, pale; lateral spikelets reduced to 1-3 awns. Common in moist habitats below 10,000', SN (all), May-Jul.

Koleria

Junegrass

inflorescence (2 cm) *spikelet* (2 mm)

K. macrantha: Tufted perennial with mly basal lvs, glabrous to puberulent; blades 3-10 mm L, usu involute, 4-6 mm L, laterally compressed; glumes unequal, the lower 3 mm L, the upper ± 5 mm L; lemmas 3-5 mm L, awn less. Dry sites throughout SN to 11,000'.

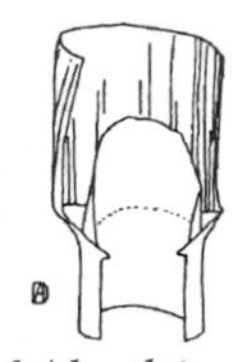

Leymus

Leymus

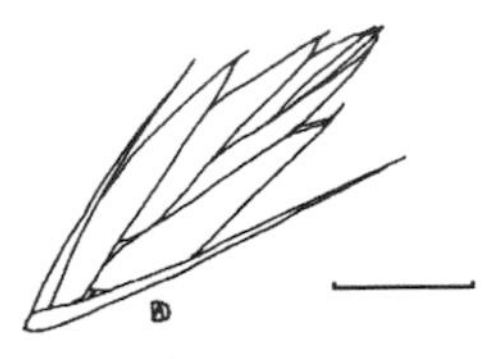

leaf blade/sheath junction (5 mm) *spikelet* (2 mm)

L. cinereus: Tufted perennial to 20 dm H, ± glabrous; blade 5-10 mm W, the upper surface rough; infl 10-20 cm L; glumes 8-15 mm L, awl-like; lemma 5-10 mm L, often tipped with an awn. Usu moist habitats below 10,000', SN (all). (*L. triticoides* is a spp of smaller stature with overlapping range and forming hybrids with *L. cinereus*).

Melica

Melic

M. bulbosa inflorescence (1 cm) *M. aristata spikelet* (1 cm)

Perennials, the stalks usu bulbous at base; sheaths closed; blades flat. Glumes less firm than lemmas, with papery or membranous margins. Lemmas firm, not keeled, with membranous tips and upper margins. Juice sweet.

A Lemmas awned
  B Awn short, less than 4 mm L; lf-blades mly 1-3 dm L — 5) *M. harfordii*
  BB Awn 5-12 mm L; lf-blades 0.6-1.4 dm L — 1) *M. aristata*
AA Lemmas not awned
  B Stalks 6-12 dm H; spikelets usu about 20 mm L; below 7000'
    C Lemma acuminate; infl branches usu ascending erect — 7) *M. subulata*
    CC Lemma ± blunt; infl branches often reflexed — 4) *M. geyeri*
  BB Stalks 2-6 dm H; spikelets usu shorter; to 11,600'
    C Spikelets reflexed; glumes falling with spikelet at maturity — 6) *M. stricta*
    CC Spikelets erect; glumes remaining on pedicel at maturity
      D First glume 3-5 mm L; 4000-7000' — 3) *M. fugax*
      DD First glume 6-10 mm L; mly 7000-11,000' — 2) *M. bulbosa*

1) *M. aristata*, Awned Melic: Stalks 6-12 dm L, not bulbous at base; blades 3-6 mm W; infl 1-2.3 dm L, mly narrow; spikelets 1-2 cm L, 2-3-fld; glumes 7-11 mm L. Dry habitats, 4500-10,000', SN (all), Jun-Aug.

2) *M. bulbosa*, Western Melic: Stalks usu bulbous at base; infl narrow; spikelets 0.6-2.4 cm L, 2-5-fld. Moist habitats, Fresno Co. n., Jul-Aug.

3) *M. fugax*, Small Oniongrass: Stalks bulbous at base; blades 2-4 mm W; infl 8-20 cm L, the branches appressed to spreading; spikelets 4-15 mm L, 2-4-fld. Dry habitats, SN (all), May-Jul.

4) *M. geyeri*: Stems 8-20 dm; blade 2-8 mm W; infl to 15 cm W; spikelet 1-25 mm L; fertile florets 2-6. Dry habitats below 7000', Tuolumne Co. n.

5) *M. harfordii*, Harford's Melic: Stalks 6-12 dm H, not bulbous at base; blades 2-6 mm W; infl 6-20 cm L, narrow; spikelets 7-20 mm L, 2-6-fld; glumes 5-10 mm L. Dry habitats mly below 7000', Tuolumne Co. n.

6) *M. stricta*, Nodding Melic: Stalks densely tufted, purplish and thickened near base but not bulbous; blades 2-5 mm W; infl 3-20 cm L, narrow; spikelets 6-20 mm L, 2-5-fld; glumes 5-15 mm L. Rocky habitats, 4000-11,600', SN (all).

7) *M. subulata*, Alaska Oniongrass: Stalks bulbous at base; blades 2-10 mm W; infl 1-2.5 dm L, mly narrow; spikelets 2-3 cm L, 2-5-fld. Shaded banks and woods, Eldorado Co..

Muhlenbergia

Muhly

M .jonesii spikelet (1 mm) *M. rigens inflorescence* (1 cm)

Lvs flat or involute. Infl narrow, sometimes spike-like, sometimes open. Glumes usu shorter than lemma. Lemma firm-membranous, 3-nerved

A Stalks capillary, mly 5-20 cm H; lemma mly 1 mm L
  B Infl narrow, less than 1 cm W — 2) *M. filiformis*
  BB Infl open, with branches 1.5-5 cm L — 4) *M. minutissima*
AA Stalks mly thicker, taller; lemma mly over 3 mm L
  B Awns conspicuous
    C Awns about 5 times longer than lemma, usu twisted — 5) *M. montana*
    CC Awns 1-2 times length of lemma, ± straight — 1) *M. andina*
  BB Awns not obvious in infl
    C Infl + open, 5-15 cm L, usu with several capillary branches — 3) *M. jonesii*
    CC Infl dense, narrow, rarely branched
      D Infl 25-50 cm L; stalks erect, 7-15 dm H — 7) *M. rigens*
      DD Infl 1-3 cm L; stalks ascending, 1-5 dm L — 6) *M. richardsonis*

1) *M. andina*, Hairy Muhly: Stalks 5-10 dm H, from rhizomes, pubescent at nodes; blades flat, 2-6 mm W; infl narrow, ± interrupted, 7-15 cm L; spikelets sessile. Open flats, 6500-10,000', SN (all), Jul-Sep.

2) *M. filiformis*, Slender Muhly: Blades flat, 1-3 cm L; infl very slender, interrupted, 1-3 cm L, few-fld; glumes 1 mm L. Open, moist habitats, 5000-11,000', SN (all), Jun-Aug.

3) *M. jonesii*, Jones' Muhly: Stalks tufted, 2-4 dm H; lvs mly basal; spikelets 3-4 mm L; glumes about 1 mm L. Dry habitats, 4000-6500', Placer Co. n., Jul-Aug.

4) *M. minutissima*: Annual to 3 dm H; blade to 4 cm L, 1-2 mm W, flat; infl 1-20 cm L, open with ascending branches. Open, disturbed places below 8000', Mariposa Co. n.

5) *M. montana*, Mountain Muhly: Stalks erect, densely tufted, 1.6-6 dm H; blades 1-2 mm W; sheaths mly basal; infl narrow, 5-15 cm L; lemma 3-4 mm L. Dry habitats, 4500-10,500', Nevada Co. s., Jun-Aug.

6) *M. richardsonis*, Richardson's Muhly: Stalks wiry, smooth; blades 2-5 cm L, flat or involute; spikelets 2-3 mm L. Dry to moist, open habitats, 5000-11,000', SN (all), Jun-Aug.

7) *M. rigens*, Deergrass: Stalks tufted, slender; blades 15-25 cm L, involute, with a long, slender point; glumes 2-3 mm L; lemmas awnless. Dry to moist habitats below 7000', SN (all), Jun-Sep.

upper stem (1 cm)

Panicum

Panicgrass

P. acuminatum: Perennials; stalks leafy, ascending to spreading, soft-hairy; blades 4-8 cm L, 5-7 mm W, appressed-hairy beneath; infl 4-7 cm L; spikelets ± compressed, about 2 mm L; glumes 2, green, nerved, mly unequal, the first often minute, the second typically equaling the sterile lemma, the latter like a third glume; fertile lemma and palea papery. Moist habitats, SN (all), Jun-Aug.

Phleum

Timothy

P. alpinum: Stalks 2-5 dm H, from tufted base; blades flat, mostly 4-6 mm W; infl mly 1.5-2.5 cm L; glumes equal, awned, the awns about 2 mm L; lemma shorter than glumes. Moist to wet habitats, 5000-11,500', SN (all), Jul-Aug.

upper stem (1 cm) *and spikelet*

Poa

Bluegrass

P. bolanderi spikelet (5 mm)

P. pratensis inflorescence (1 cm)

Blades narrow, flat, folded or involute, ending in a boat-shaped tip. Spikelets 2-8-fld. The uppermost floret often rudimentary. Glumes acute, keeled, the first 1-nerved, the second larger and usu 3-nerved. Lemmas somewhat keeled. Palea nearly equaling lemma. Difficult genus to determine to sp in field. Individual sp descriptions will not be presented.

A Florets mly formed into dark purple bulblets; stalks with a bulb-like base; rare, below 4500', Fresno Co. n. ... *P. bulbosa*

AA Florets normal, green to purple-tipped; stalks not bulbous at base

 B Infl ± open; pedicels or ultimate branchlets capillary

  C Infl few-fld; glumes 2-3 mm L; stalks leafy

D Spikelets 2-3-fld; common, in open forests, 5000-10,000', SN (all) — *P. bolanderi*
DD Spikelets 2-5-fld; rare, in moist habitats, 6000-10,500', Nevada Co. s. — *P. leptocoma*
CC Infl many-fld; glumes 4-6 mm L
D Creeping rhizomes present; stalks usu leafy
E Spikelets 5-9 mm L, ± open; common, 4500-12,500', Nevada Co. s. — *P. wheeleri*
EE Spikelets 4-6 mm L, compact; mly below 8000', SN (all) — *P. pratensis*
DD Rhizomes lacking; stalks usu naked above basal lvs
E Infl very open; lower branches wide-spreading; common, 6000-12,000', Butte Co. s. — *P. secunda*
EE Infl often dense above; branches ascending to erect; 5000-12,000', Nevada Co. s. — *P. cusickii*
BB Infl dense; branches mly erect; pedicels mly less than 1 cm L, usu erect
C Stalks barely exceeding basal tuft of lvs; lemmas glabrous on back; 11,000-13,700', s. SN
D Infl 1-2 cm L; spikelet purple, 3-4 mm L; lemmas 2-3 mm L; rare — *P. lettermanii*
DD Infl 2-3 cm L; spikelet green to purplish, 4-6 mm L; lemmas 3-5 mm L; common — *P. keckii*
CC Stalks usu well exceeding basal tuft, if not lemmas pubescent on back near base
D Stalk and nodes distinctly compressed, wiry — *P. compressa*
DD Stalk and nodes not compressed
E Spikelets not compressed, ± cylindric; common, SN (all) — *P. secunda*
EE Spikelets ± compressed
F Lemmas keeled; Mono Co. s.
G Spikelets 5-7-fld, 6-10 mm L, pale or greenish; 3200-10,200' — *P. fenleriana*
GG Spikelets 2-4 fld, 2.5-5 mm L, purplish; 11,000-13,000' — *P. glauca*
FF Lemmas rounded on back; stalks mly 1-2 dm H
G Lemmas less than 4 mm L; 7000-12,000', Eldorado Co. s. — *P. stebbinsii*
GG Lemmas mly 6 mm L; 6000-10,000', Placer Co. n. — *P. pringlei*

Polypogon

Annual Beard Grass

P. monspeliensis: Stalks 2-10 dm H with a terminal dense, soft infl; sheath open, glabrous; blade-flat, rough; infl 3-10 cm L, about 1 cm W. Common below 6000', SN (all).

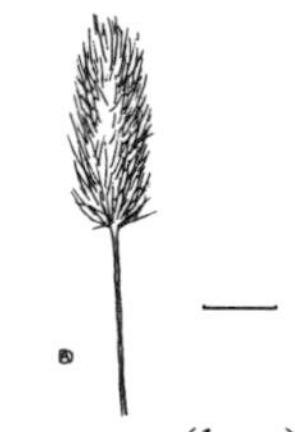

inflorescence (1 cm)

Ptilagrostis

King's Ricegrass

P. kingii: Tufted perennial; stalks slender, 2-4 dm H; lvs mly basal, 3-15 cm L, filiform; infl narrow; glumes broad, subequal; lemma about 3 mm L with a long twisted awn. Moist habitats, 9300-11,500', Tuolumne and Mono cos.

inflorescence (1.5 cm)

Scribneria

Scribneria

S. bolanderi: Annual; stalks 0.5-3 dm H, tufted, ascending to erect; lvs few with short, narrow blades; spike linear, cylindrical, 2-11 cm L; spikelets about 7 mm L, partly sunken into hollows in rachis; glumes equal, pointed; lemma shorter than glumes, membranous; palea about as long as lemma. Many habitats below 9000', SN (all), Mar-Jun.

inflorescence (1 cm)

inflorescence (3 cm)

Sporobolus

Sand Dropseed

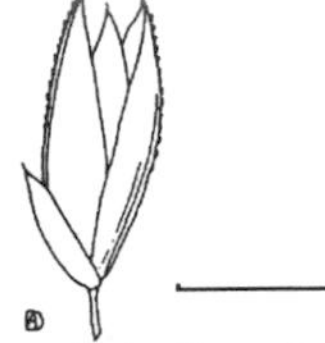

spikelet (1 mm)

S. cryptandrus: Tufted perennial; stalks 3-10 dm H; lvs mly basal; blade 3-15 cm L, 1-5 mm W; infl pyramidal, the primary branches usu over 3 cm L, spreading, spike-like, with secondary branches tightly appressed. Open slopes and washes below 9000', mly e. SN.

P. pallida leaf (1 cm)

Torreyochloa

Alkali Grass

P. erecta inflorescence (1 cm)

Tufted perennials. Lf-sheaths open. Spikelets roundish. Glumes shorter than first lemma, the first mly 1-nerved, the second 3-nerved. Lemmas usu firm, conspicuously 5-nerved, rounded on back. Plants of wet habitats.

Infl narrow in outline, length 5-15 times width, usu less than 1 cm W — *T. erecta*
Infl ovate to elliptic in outline, length 1-5 times width, 1-10 cm W — *T. pallida*

T. erecta, Upright Alkali Grass: Stalks mly 3-6 dm H; blades flat, mly 5-12 cm L. infl ± narrow; spikelets 4-6-fld, often purplish. At 7000-11,000', SN (all), Jul-Sep.

T. pallida, California Alkali Grass: Stalks 5-12 dm H; blades flat, 10-15 cm L; infl often open; spikelets mly 5-6-fld, 4-5 mm L, often purplish. Below 11,000', SN (all), Jul-Sep.

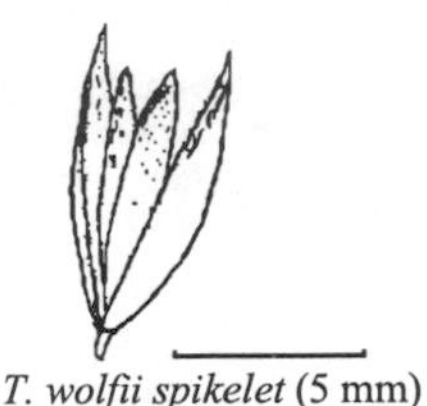

T. wolfii spikelet (5 mm)

Trisetum

Trisetum

T. spicatum spikelet (1 mm)

Tufted perennials with flat blades. The rachilla prolonged beyond the upper floret, usu hairy. Glumes ± unequal, acute, the second usu exceeding the first floret. Lemma 2-toothed at apex, the teeth often awned.

Awns included within the glumes or lacking — *T. wolfii*
Awns exerted
- Infl dense, spike-like; stalks 1-4 dm H — *T. spicatum*
- Infl open, drooping; stalks 6-12 dm H — *T. canescens*

T. canescens, Nodding Trisetum: Blades usu 5-15 cm L, 1-3 mm W; infl 1.5-3 cm L; spikelets 6-12 mm L, usu 3-fld; lemmas 5-6 mm L; awns 5-10 mm L, curved. Moist habitats, 4000-9000', SN (all), Jun-Aug.

T. spicatum, Spike Trisetum: Stalks densely tufted; infl pale or dark purple, 5-15 cm L; spikelets 4-6 m L; lemmas 5 mm L; awns 5-6 mm L, bent. Montane habitats, 7200-13,000', SN (all), Jul-Aug.

T. wolfii, Beardless Trisetum: Stalks erect, 3-10 dm H, loosely tufted; blades flat, 2-4 mm W; infl erect, ± dense, green or pale, 8-15 cm L; spikelets 5-7 mm L, 2-3-fld; glumes about 5 mm L; lemmas 4-5 mm L. Mdws, 7000-10,000', SN (all), Jul-Aug.

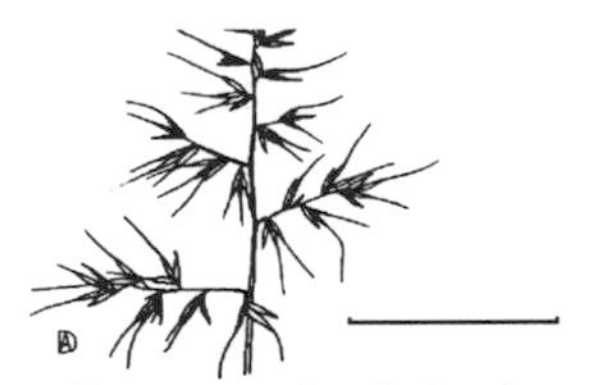

Vulpia

Annual Fescue

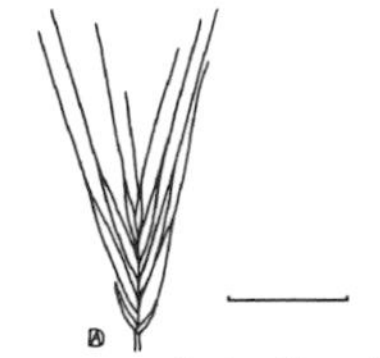

V. microstachys (2.5 cm) *V. myuros spikelet* (1 cm)

Annuals with solitary or loosely clumped, unbranched stalks.

Lower glume less than half length of upper glume — *V. myuros*
Lower glume over half length of upper glume
  Florets 1-5, loosely overlapping; awn 4-12 mm L — *V. microstachys*
  Florets 5-10, crowded; spikelet axis hidden; awn usu less than 4 mm L — *V. octoflora*

V. microstachys: Stalks erect or bent at base, 2-5 dm H, glabrous; blades ± involute; infl 5-12 cm L, the lower branches ± remote, spreading; glumes about 5 mm L; lemma 6-7 mm L. Below 5000', SN (all).

V. myuros: Stalks 2-6 dm H, glabrous; blades glabrous; infl 7-20 cm L; spikelets 4-5-fld, 8-10 mm L; lemma 4-6 mm L. Common below 5500', SN (all).

V. octoflora: Stalks less than 6 dm H, glabrous or hairy; infl to 15 cm L, dense; branches 1 per node; spikelet 5-10 mm L; lemma 3-5 mm L. Open habitats below 6500', SN (all).

POTAMOGETONACEAE-Pondweed Family

Potamogeton

Pondweed

P. amplifolius upper stem (1 cm) *P. pectinatus upper stem* (1 cm)

Perennial herbs growing from rhizomes. Stems jointed, leafy, fibrous-rooted at lower nodes. Lvs alt or opp, the submersed thin, the floating thicker, all sheathing at base. Fls perfect. The peduncles surrounded by a basal sheath formed by stipules. Bracts lacking. Perianth of 4 free, rounded, valvate segments. Stamens 4. Carpels 4, free, sessile. Fr drupe-like when fresh. Probably all spp have edible starchy rhizomes which make an excellent emergency food source. The herbage is an important food source for many wild fowl. Many species hybridize with each other, making identification difficult. Only a key to the species is presented.

A Lvs of 2 kinds, floating (thick) and submersed (thin)
  B Submersed lvs ± same shape as floating lvs; rare
    C Floating lvs ovate, with 30-50 veins; below 6000' — *P. amplifolius*
    CC Floating lvs lanceolate, with fewer veins, often reddish; above 5000' — *P. alpinus*
  BB Submersed lvs narrower than floating lvs
    C Stipules fused to lf base; infl of two types; n. SN below 8000' — *P. diversifolius*
    CC Stipules free from lf base; infls all emergent and similar; to 10,000'
      D Floating lvs truncate or lobed at base — *P. natans*
      DD Floating lvs tapering to petiole
        E Submersed lvs long petioled — *P. nodosus*
        EE Submersed lvs on short branches — *P. gramineus*
AA Lvs all submersed
  B Lvs linear, less than 3 mm W
    C Lvs with 2 prominent glands at base — *P. pusillus*
    CC Lvs lacking such glands; below 7000' — *P. pectinatus*
  BB Lvs wider, lanceolate to ovate; below 10,000; SN (all)
    C Lf short-petioled, the base not clasping — *P. illinoensis*
    CC Lf sessile; the base clasping, rare, Mariposa Co. n.
      D Lf blades 5-15 cm L, hood-like at tip — *P. praelongus*
      DD Lf blades 1-10 cm L, flat at tip — *P. richardsonii*

TYPHACEAE-Cattail Family

S. angustifolium upper stem (1 cm)

Sparganium

Bur-reed

Perennials from creeping rhizomes. Stems leafy. Lvs elongate, alt, sheathing at base. Infl unbranched. Fls in unisexual globose heads, the staminate heads further up the stem. Fr indehiscent, crowded, narrowed at base, nut-like. The thickened base of stem and the tubers on the rhizomes are edible.

Plants usu emergent; stems 2-10 dm H, stiff; lvs ± keeled — S. emersum
Plants usu submersed with floating lvs; lvs often rounded on back but not keeled
  Stems 3-10 cm L; lvs ± convex, 20-80 cm L — S. angustifolium
  Stems 8-40 cm L; lvs ± flat, 5-40 cm L — S. natans

S. angustifolium, Narrow-leaved Bur-reed: Staminate infls usu 2-3, crowded; pistillate infls 2-4, 5-20 mm in diameter. Common in water to 10' deep, 4000-12,000', SN (all).

S. emersum ssp. *emersum*, Emergent Bur-reed: Staminate infls usu 4-7, well separated; pistillate infl 3-4, 10-20 mm across in fr. Lakes and streams below 9000', SN (all).

S. natans, Small bur-reed: Staminate infl usu 1; pistillate infl 1-3, well separated, 5-10 mm in diameter. Rare, 7000-8000', Madera Co. n.

Typha

Cattail
Soft-flag

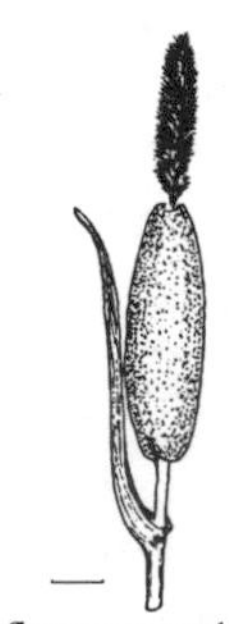
inflorescence (1 cm)

T. latifolia: Plant 1-2.5 m H; stems unbranched, usu submersed at base; lvs alt, 12-16, mly exceeding infl, 8-15 mm W; infl of 2 contiguous unisexual spikes with the staminate spike above the pistillate. Swampy habitats below 5000', SN (all), Jun-Jul. Rhizome can be dried and ground into flour; the young shoots are edible raw; the green infl is edible after cooking; the pollen may be eaten raw.

ZANNICHELLIACEAE - Horned Pondweed Family

upper stem (1 cm)

Zannichellia

Horned Pondweed

Z. palustris: Stems 3-5 dm L, capillary, branched; lvs 2-8 cm L, filiform but flat, 1-nerved; staminate and pistillate fls in same axil, staminate fls solitary, pistillate fls usu 4. Slow moving water, below 7000', SN (all), May-Sep. Edibility unknown.

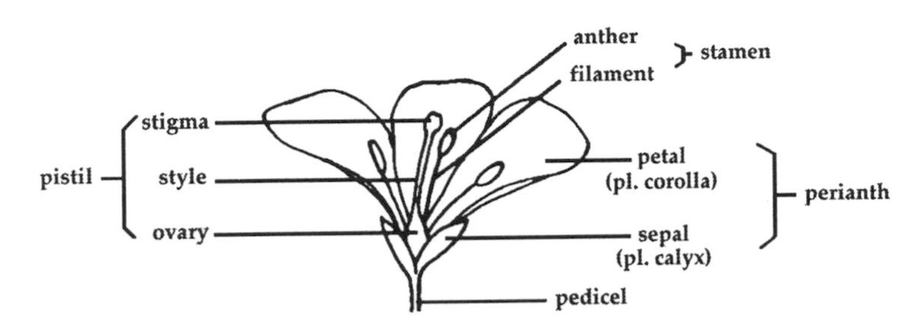

DIAGRAM OF FLOWER

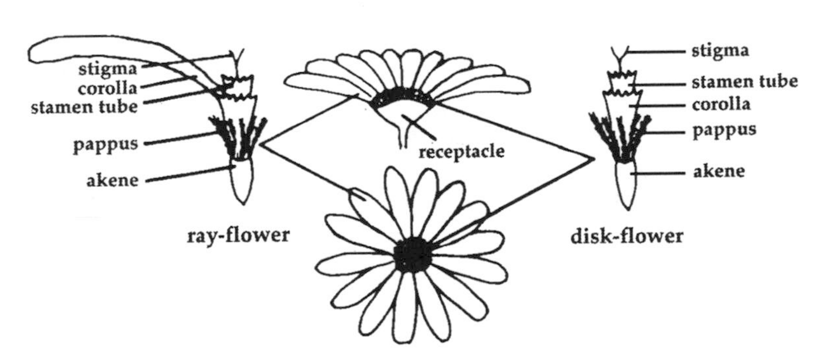

SUNFLOWER DIAGRAM

GLOSSARY

Acaulescent: plant without leafy stem; cf., caulescent (see *Primula*, p.165)
Acuminate: tapering to a long, slender point, as the tip of a leaf (see *Prunus*, p.181)
Acute: sharp-pointed but not long-pointed (see *Solanum*, p.203)
Adnate: attached or grown together
Aggregate: a dense collection of simple units
Aggregate fruit: a unified cluster of fruits developing from separate pistils (see *Rubus*, p.182)
Akene: a dry, hard, 1-seeded fruit not splitting at maturity (see *Hieracium*, p.60)
Alkaloid: nitrogen-containing slightly alkaline substances which are often poisonous
Alternate: arising between units, as in stamens alternate with petals; or in a staggered arrangement, as in alternate leaves (see *Holodiscus*, p.177)
Ament: see catkin
Amplexicaul: clasping the stem at base (see *Streptanthus*, p.82)
Annual: lasting only one season, growing from seed to maturity and dying in one year. A typical annual has filiform fibrous roots, a slender stem, and, in the case of dicotyledons, at least lowest leaves opposite.
Annular: ring-like
Anther: the pollen-containing, usually oblong structure, at the end of the filament (see Diagram of Flower, p.244)
Anthesis: the blooming period of a flower
Anthocyanous: containing anthocyanin, a purple or reddish pigment
Apetalous: without petals (see *Eremocarpus,* p.108)
Apical: located at the apex or tip
Appendage: an additional part
Appressed: flattened against another organ, as appressed hairs
Aquatic: living or growing in water
Arachnoid: pubescent with matted, soft, entangled hairs
Arborescent: having a large main trunk, tree-like
Arcuate: arched or smoothly curving like a bow (see *Arabis*, p.76)
Aristate: tipped with a bristle or awn (see *Festuca,* p.235)
Articulate: jointed
Ascending: curving upward, as the branches of an inflorescence (see *Hypericum*, p.132)
Attenuate: gradually narrowing to a long, slender tip (see *Asarum*, p.43)
Auricle: a small appendage at the base of the lf-blade
Awn: a $\pm$ stiff, slender bristle (see *Festuca,* p.235)
Axil: the angle formed by the upper surface of the base of a leaf or branch with the stem
Axillary: located in or arising from an axil (see *Navarettia*, p.155)
Axis: an imaginary line drawn along the center of an elongate structure
Banner: the upper petal of a flower in the Pea Family (see *Lathyrus*, p.111)
Basal: located at the base (see *Nothocalais*, p.64)
Beak: a narrowed, elongate, stiffened tip, as on an akene (see *Agoseris*, p.47)
Bearded: covered with conspicuous hairs (see *Penstemon*, p.199)
Berry: a fleshy fruit with several to many small seeds (see *Vaccinium*, p.108)
Bi-: a prefix meaning two
Bidentate: with two teeth (see *Bromus*, p.232)
Biennial: lasting for two years
Bifid: two-cleft (see *Draperia* style, p.127)
Bilabiate: two-lipped (see *Porterella*, p.86)
Bipinnate: pinnate and these segments again pinnately compound (see *Chamaebatia*, p.175)
Bipinnatifid: pinnately cleft and these divisions again pinnately cleft (see *Dicentra*, p.149)
Blade: the flattened, expanded portion of a leaf or petal (see *Cercis*, p.110)
Bract: a rudimentary leaf subtending a flower or flower-cluster (see *Zigadene*, p.225)
Bracteate: with bracts Bractlet: a secondary bract (see *Perideridia*, p.39) or sepal-like structures alternate with the sepals in some general of the Rose Family (see *Potentilla*, p.179)
Bur: a dry roundish fruit covered with spines (see *Castanopsis*, p.119)
Calyx: the outer, usually green whorl of the perianth; a collective term for the sepals (see Diagram of Flower, p.244)
Calyx-lobe: the sepal-like lobe in flowers with a calyx united at base (see *Mimulus*, p.195)
Campanulate: bell-shaped (see *Calochortus*, p.218)
Canescent: covered with fine gray-white hairs
Capillary: very thin and delicate; thread-like (see *Gayophytum*, p.146)
Capitate: in a + spherical or head-shaped form (see *Sphenosciaudium*, p.41)
Capsule: a dry, few to many-seeded fruit splitting open at maturity (see *Staphylea*, p.203)
Catkin: a dense, elongate cluster of apetalous, unisexual, minute flowers (see *Betula,* p.71)

Caudate: having a tail or tail-like appendage
Caudex: the thickened base of some perennial herbs
Caulescent: having a conspicuous and leafy stem (see *Veratrum*, p.224)
Cauline: related to or pertaining to the stem, such as cauline leaves
Cespitose: low and densely tufted or clumped
Ciliate: with a line of hairs along margins
Clasping: partly surrounding stem (see *Pholistoma*, p.132)
Claw: an elongate, narrowed base on a petal
Cleft: split or narrowly divided at least half-way to base (see *Lithophragma*, p. 189)
Cleistogamous: small flowers which do not open but fertilize themselves
Collar: outer side of the grass leaf at the junction of the sheath and blade
Compound: having two or more similar units in one organ (cf., palmately compound, pinnately compound)
Connate: forming a cone-like structure
Connate-perfoliate: an apparently single leaf encircling the stem. Actually two opposite leaves fused at their base (see *Claytonia*, p.161)
Contracted: narrowed or shortened at certain points
Convolute: rolled up longitudinally in bud
Cordate: heart-shaped (see *Smilax*, p.222)
Corm: a bulb-like underground thickening of the stem (see *Lewisia*, p.160)
Corolla: the inner usually brightly colored whorl of the perianth; a collective term for the petals (Diagram of Flower, p.244)
Corolla-lobe: the petaloid lobes in flowers with corolla united at base (see *Phlox*, p.155)
Cortex: the outer covering, bark
Corymb: an inflorescence with only primary branches, the lower (outer) branches being longer so that the flowers form a flat-topped or slightly convex formation
Cotyledon: the first leaf or leaves (opposite) of an embryo of a flowering plant
Crenate: scalloped (see *Thysanocarpus*, p.83)
Crested: having an elevated ridge or appendage
Crisped: slightly warped or curled (see *Disporum* leaves, p.220)
Crown: the upper leafy section of a tree; a whorl of appendages or scales at the apex of a structure (see *Microseris*, p.)
Cuneate: wedge-shaped; V-shaped at base (see *Suksdorfia*, p.191)
Cyme: a flat-topped or convex inflorescence with the primary branches again branching
Deciduous: falling off, especially referring to plants that lost their leaves each autumn
Declined: curved downward
Decumbent: lying on ground with the tip ascending (see *Rorippa*, p.82)
Decurrent: continuing down stem below a node (see *Helenium*, p.60)
Deflexed: turning down at tip (see *Pyrola* style, p.107)
Dehiscent: splitting open at maturity
Deltoid: triangular, with length of edge about equal to width of base (see *Balsamorhiza*, p.52)
Dentate: having margins cut into coarse + triangular teeth (see *Saxifraga*, p.)
Denticulate: finely dentate (see *Prunus*, p.181)
Dichotomous: repeatedly two-branched (see *Cerastium*, p.90)
Diffuse: widely and much-branched (see *Gayophytum*, p.146)
Dilated: expanded and usually flattened
Dimorphous: having two forms
Dioecious: having staminate and pistillate flowers on separate plants
Disarticulate: to break at joints
Disk-flower: flowers in the Sunflower family lacking a ray-like extension on the corolla (see Sunflower Diagram, p.244)
Discoid: a flowering head without ray-flowers (see *Antennaria*, p.48)
Dissected: much-divided into many small segments (see *Achillea*, p.46)
Distal: away from the midline or point of attachment
Distinct: separate to base and not connected to other segments of the same whorl of parts (see *Limnanthes*, p.138)
Divaricate: widely divergent:
Divergent: branching from stem at angle of about 45 degrees
Divided: cut to near base
Dominant: most common or characteristic plant of an area
Dorsal: referring to surface facing away from midline
Drupe: a fleshy 1-seeded fruit such as a cherry
Ebracteate: without bracts
Elliptic: slightly longer than wide, being widest near middle
Emarginate: slightly notched at apex (see *Veronica*, p.201)
Emersed: see emergent

Emergent: at least partly above water
Entire: with smooth edges (see *Arbutus*, p.103)
Equitant: a term for leaves sharply folded at base and enclosing the bases of other leaves so that the base of the plant is much wider from the front view than from the side (see *Iris*, p.212)
Erect: perpendicular to the ground or paralleling the axis from which the referred to plant part has branched (see *Linum* leaves, p.139)
Evergreen: retaining leaves through the winter
Exfoliate: to peel off in thin layers (see *Arbutus* bark, p.103)
Exserted: protruding beyond an opening, such as stamens exserted from corolla (see *Trichostema*, p.136)
Fascicle: a small cluster or bundle of leaves, flowers, etc. (see *Pinus*, p.16)
Fertile: capable of taking part in reproduction
Fertile stamens: stamens with anthers and pollen
Filament: the stalk on the anther (see Diagram of Flower, p.244)
Filamentous: very thin, hair-like
Filiform: see filamentous
Fleshy: thick and juicy; succulent
Floret: the individual flower in the Grass family; a small flower in a dense head (see *Calamagrostis*, p.233)
Flower: the reproductive structure of a large group of plants (see Diagram of Flower, p.244)
Flower-tube: a structure on some plants consisting of the united bases of the anthers, petals and sepals (see *Epilobium canum*, p.144)
Foliaceous: leaf-like; green, flattened
Foliolate: referring to the leaflets
Follicle: a dry, usually elongate fruit splitting along one seam (see *Asclepias*, p.43)
Frond: the leaf of a fern (see *Cystopteris*, p.10)
Fruit: in a strict sense the mature pistil, but more usually including such structures as a strawberry or raspberry
Funnelform: having the shape of a funnel (see *Calystegia*, p.98)
Fusiform: thickest near middle and tapering toward both ends, cigar-shaped
Galea: a very narrow upper lip of a corolla which is arched or folded along its long axis and often encloses the stamens (see *Castilleja*, p.192)
Galeate: having a galea
Gibbous: swollen or inflated (see *Collinsia*, p.193)
Glabrous: hairless
Glabrate: becoming glabrous
Glandular: bearing glands which usually secrete a sticky liquid
Glaucous: covered with a whitish powder
Globose: spherical, globe-shaped (see *Sparganium*, p.243)
Glomerate: growing in compact clusters or glomerules
Glomerule: a dense, roundish cluster
Glumes: the bracts (usually two) at the base of and occasionally enclosing the remainder of the spikelet of grasses (see *Agrostis*, p.231)
Glutinous: covered with a sticky fluid
Gynobase: a platform on which the ovary sits
Habitat: the general location and environs in which a plant grows
Hastate: widely arrowhead-shaped (see *Calystegia*, p.98)
Head: a dense rounded cluster of + sessile flowers (see *Monardella*, p.134)
Herb: a plant with non-woody stems that die to the ground each winter
Herbaceous: like a herb, not woody
Herbage: the non-woody parts of a plant
Hirsute: bearing long, shaggy hairs
Hyaline: glass-like, clear, translucent
Hypanthium: a cup-like structure formed by the fused bases of the sepals, petals, and stamens in flowers from the Rose and Saxifrage families (see *Prunus*, p.181)
Imbricate: overlapping, much like shingles on a roof (see *Juniperus*, p.16)
Incised: having margins with rather deep, narrow sinuses or cuts (see *Podistera*, p.40)
Included: not extended beyond the opening or enclosing structure
Indehiscent: not splitting open at maturity
Indusium: A structure often present in ferns that covers the sori (see *Dryopteris*, p.10)
Inferior: situated below
Inferior ovary: an ovary positioned below (proximal to) the base of the other flower parts (stamens, petals and sepals) along the floral axis
Inflated: appearing greatly expanded or puffed out (see *Mimulus*, p.195)
Inflorescence: the cluster of flowers on a plant. Often more than one per plant
Internode: the stem between successive nodes

Interrupted: with gaps or openings, not continuous (see *Potamogeton*, p.242)
Involucel: a whorl of bractlets subtending a portion of an infloresence such as an umbellet in the Carrot family (see *Tauschia*, p.41)
Involucrate: subtended by an involucre
Involucrate head: a type of infloresence mostly found in the Sunflower family (see *Helenium*, p.60)
Involucre: a whorl of several to many bracts subtending an inflorescence (see *Allium*, p.216)
Involute: having edges curled up and inward
Irregular: having different shaped petals or sepals on the same flower (see *Epipactis*, p.227)
Keel: a ridge running along the outside longitudinal axis of a structure; the lowest petaloid structure (actually the lower two petals which are fused together) on a flower in the Pea family
Labiate: lip-like in arrangement or form (see *Porterella*, p.86)
Laciniate: deeply incised into narrow lobes (see *Utricularia*, p.137)
Lanate: densely hairy with long, soft, entangled hairs (see *Gnaphalium,* p.59)
Lanceolate: elongate with the margins converging toward tip (see *Lysimachia,* p.164)
Leaflet: a leaf-like unit of a compound leaf (see *Rosa*, p.181)
Legume: a fruit of the Pea family similar to a pea pod (see *Cercis*, p.110)
Lemma: the outer bract enclosing the floret in grasses (see *Bromus*, p.232)
Lenticel: corky ventilating pores in the bark of certain woody plants
Ligule: in grasses the membranous appendage on a leaf originating at the blade/sheath junction and continuing up the stalk (see *Agrostis*, p.231)
Linear: very much longer than wide (usually over ten time)
Lip: two or three approximately parallel, more or less connected petals; usually united below with a similar but oppositely facing lip (see *Porterella*, p.86)
Lobe: a peripheral extension or partial segment of a leaf or other plant structure
Locule: a compartment such as in an ovary
Lunate: crescent-shaped
Membranous: resembling a membrane, thin, delicate, often translucent
Merous: parted, having sections, as a 5-merous flower has 5 petals and 5 sepals
Midrib: the predominant central vein of a leaf
Monoecious: having distinct male and female flowers on the same plant
Montane: of or pertaining to the mountains
Mucilaginous: slimy, sticky
Mucro: a short, fine tip on an otherwise blunt end of a structure
Mucronate: bearing a mucro
Nerve: a prominent unbranched vein (see *Arabis*, p.76)
Node: the joint of a stem where leaves, branches or flowers arise (see *Scutellaria*, p.135)
Nodding: drooping or hanging down (see *Erythronium*, p.220)
Nutlet: a small dry seed that does not split at maturity
Ob-: a prefix signifying the reverse of the term following
  Obcompressed: flattened perpendicularly to the axis
  Obcordate: heart-shaped with narrow end attached (see *Capsella*, p. 78)
  Oblanceolate: inversely lanceolate (see *Myrica*, p.141)
  Oblong: elongate with parallel sides (see *Umbellularia*, p.137)
  Obovate: inversely ovate (see *Vaccinium caespitosum*, p.108)
Obsolete: inconspicuous or lacking
Ocrea (pl. *ocreae*): a membranous structure surrounding stem above petiole base (see *Polygonum*, p.159)
Opposite: a leaf arrangement in which two leaves branch off the stem at each node (see *Pycnanthemum*, p.135); arising parallel to or directly in front of, as in stamens opposite petals
Orifice: an opening or mouth of a flower with petals united at base
Oval: almost round but being slightly longer than wide (see *Listera*, p.227)
Ovary: the seed-containing part of the pistil (see Diagram of Flower, p.244)
Ovate: egg-shaped, the widest part near the base (see *Listera*, p.227)
Palea: the inner bract immediately subtending the floret in grasses
Palmate: having lobes or divisions radiating from a central point
  Palmately compound: having leaflets all arising from a single point (see *Hippuris*, p.126)
  Palmately lobed: having lobes in a finger-like arrangement (see *Acer*, p.33)
Panicle: an infloresence with secondary branches, a branched raceme (see *Panicum*, p.239)
Paniculate: borne in a panicle
Pappus: the whorl of scales or bristles at the apex of the akene in the Sunflower family (see *Hieracium*, p.60)
Parted: cleft almost to base (see *Stellaria*, p.94)
Pedicel: the stalk on a single flower; or in the Grass family on a spikelet
Pedicellate: bearing a pedicel

Peduncle: the stalk on an inflorescence such as an umbel (see *Linnaea*, p.86)
Peltate: a leaf with the petiole intersecting the middle of the blade (see *Darmera*, p.189)
Pendulous: pendent, hanging (see *Fraxinus*, p.142)
Perennial: living for more than one year. Usually a perennial will have a rather thick base or enlarged root or storage organs
Perfect: a flower having both staminate and pistillate parts
Perfoliate: the stem transsecting the middle of leaf so that the leaf completely surrounds it (see *Claytonia*, p.161)
Perianth: a collective term for both the calyx and corolla (see Diagram of Flower, p.244)
Perigynous: a term used to describe the position of the ovary in many flowers in the Rose and Saxifrage families in which the bases of the sepals, petals and stamens are fused into a cup-like structure before intersecting central axis of flower just below ovary (see *Prunus*, p.181)
Petal: a usually brightly colored flower part located between sepals and stamens
Petaloid: brightly colored and petal-like
Petiole: the stalk on a leaf (see *Nuphar*, p.142)
Petiolule: the stalk on a leaflet
Phyllary: one of the bracts of the involucre in the Sunflower family (see *Machaeranthera*, p.62)
Pinna: the primary division of a 1- to many-pinnate leaf (see *Lastrea*, p.11)
Pinnate: having a main central axis with secondary branches or units arranged in two lines on either side of the central axis
Pinnately compound: having leaflets arranged in a pinnate fashion (see *Rosa*, p.181)
Pinnately lobed: having lobes arranged in a pinnate fashion (see *Crepis*, p.56)
Pinnatifid: pinnately cleft or lobed
Pinnule: a division of a pinna
Pistil: the female or seed bearing structure of a flower, consisting of the ovary, one or more styles and stigmas (see Diagram of Flower, p.244)
Pistillate: a flower with only one pistil, stamen absent
Plane: flat, not curled or undulate
Pleated: folded in a pleat-like manner (see *Calystegia*, p. 98)
Plumose: feathery, with soft, thin branches off a central axis (see *Geum*, p.176)
Polyploidy: a situation in which new species have been formed by the doubling, tripling, etc. of a basic number of chromosomes
Pome: an apple-like fruit (see *Peraphyllum*, p.179)
Procumbent: Prostrate on ground but not rooting at nodes
Proximal: toward or close to the central axis or point of attachment
Puberulent: the diminutive of pubescent
Pubescent: bearing soft hairs (see *Nama*, p.128)
Pungent: sharp-pointed at tip (see *Salsola*, p.97)
Raceme: an inflorescence with one main axis and subequal primary branches (pedicels) each bearing one flower (see *Lupinus*, p.113)
Racemose: in or of a raceme
Rachilla: the axis of a spikelet in grasses
Rachis: the longitudinal axis of a raceme, spike, or pinnately lobed or compound leaf
Radiate: in the Sunflower family, possessing ray-flowers (see *Helenium*, p.60)
Radical: pertaining to the root
Ray: one of the primary branches in an umbel or compound umbel (see *Cicuta*, p.36)
Ray-flower: a flower in the Sunflower family which bears a strap-like extension on one side of the corolla (see Sunflower Diagram, p.244)
Receptacle: in the Sunflower family the solid, fleshy base of the inflorescence on which all the flowers and maturing akenes sit (see Sunflower Diagram, p.244)
Recurved: bent back on itself (see *Lillium*, p.221)
Reflexed: sharply recurved (see *Dodecatheon*, p.164)
Regular: having identically shaped sepals so as to appear radially symmetric (see *Sagina*, p.92)
Remote: distant, well-spaced
Reniform: kidney-shaped (see *Oxyria*, p.158)
Retrorse: having processes (usually hairs) bent back toward base
Revolute: being curled under or having margins which are curled under (see *Kalmia*, p.105)
Rhizome: a horizontal underground stem rooting on its lower side
Rib: a prominent vein or ridge on a structure (see *Sphenosciadium*, p.41)
Ridge-angled: a cylindrical structure with its surface sharply angled to form 3 or more sharp longitudinal ridges, thus making the structure somewhat star-shaped in cross-section
Rootstock: see rhizome
Rosette: a cluster of whorl of leaves usually at base of stem (see *Sedum*, p.100)
Rosulate: in a rosette
Rotate: wide-spreading in one place, like the spokes on a wheel

Runner: a thin, naked stem travelling horizontally on the ground and rooting at nodes (see *Fragaria*, p.176)
Saccate: inflated or pouch-like (see *Cypripedium*, p.226)
Sagittate: arrowhead-shaped (see *Sagittaria*, p.208)
Samara: a winged fruit which does not split open at maturity (see *Acer*, p.33)
Scale: a small membranous bract (see *Wyethia* pappus, p.70)
Scape: a leafless flowering stalk usually arising from a cluster of basal leaves (see *Primula*, p.165)
Scarious: membranous; thin and dry
Scorpioid: with a coiled shape much like a scorpion's tail (see *Cryptantha*, p.72)
Sepal: a usually green segment of the outer whorl of flower parts or calyx (see Diagram of Flower, p.244)
Sepaloid: green and sepal-like
Seriate: in a row or whorl; in the Sunflower family the arrangement of the phyllaries is termed uniseriate, biseriate, etc., depending on how many rows of phyllaries make up the involucre
Serrate: having sharply toothed edges; a general term for a roughened but not lobed margin (see *Eryngium*, p.36)
Sessile: not on a stalk, situated directly on the stem or rachis (see *Streptopus* leaves, p.223)
Sheath: the lower part of the leaf in grasses, sedges, etc., that wraps around the stem (see *Leymus*, p.237)
Sheathing: partially or wholly enclosing the stem
Shrub: a woody plant with several branches from near the base
Simple: single, not further divided, not compound
Sinuate: conspicuously and often deeply undulating or wavy (see *Erythronium* leaves, p.220)
Sinus: a deep indentation between lobes of a leaf, leaflet, petal, etc.
Solitary: alone, borne singly
Sorus (pl. *sori*): the cluster of sporangia on the underside of fern leaves (see *Polypodium*, p.13)
Spatulate: spatula-shaped, with a long, narrow base and a widened, roundish tip (see *Ceanothus* petals, p.171)
Spicate: spike-like
Spike: an elongate, unbranched inflorescence in which the flowers are all sessile (see *Myosurus*, p.169); in grasses the infloresence consisting of one to many spikelets
Spikelet: a secondary spike found in grasses and sedges
Sporangium (pl. *sporangia*): a spore case or single cell giving rise to spores (see *Cystopteris*, p.10)
Spore: the reproductive unit of ferns with functions similar to a see
Sporocarp: a structure containing spores
Spur: a pouch-like to linear process near the base and extending away from the tip of a sepal or petal (see *Aquilegia*, p.167)
Stamen: the pollen-bearing structure of a flower, consisting of a filament, an anther and the pollen (see Diagram of Flower, p.244)
Staminate: a flower with only stamens, pistils absent
Staminode: a sterile stamen; a stamen usually lacking an anther and consisting only of a modified filament
Stellate: with several ray-like branches from a central point, star-shaped (see *Anelsonia*, p.76)
Stigma: the section of the pistil receptive to pollen (see Diagram of Flower, p.244)
Stipitate: situated on a stipe
Stipule: a bract- or leaf-like appendage located at the base of the petiole (see *Vicia*, p.119)
Stipulate: possessing stipules
Stolon: a horizontal stem running above ground and rooting at nodes (see *Fragaria*, p.119)
Stomate: a pore in the surface of a leaf or stem allowing for air exchange
Striate: marked with fine longitudinal lines or grooves (see *Lithospermum*, p.74)
Strict: stiffly erect
Strigose: bearing stiff appressed hairs
Strobilus: the cone-like structure on some fern relatives (see *Equisetum*, p.11)
Style: the thin stalk between the ovary and stigma on the pistil (see Diagram of Flower, p.244)
Subtend: to be situated immediately below
Subspecies: a taxon within a species
Subulate: awl-shaped (see *Arenaria* leaves, p.89)
Succulent: fleshy, thick and juicy (see *Sedum*, p.100)
Superior: above
Superior ovary: an ovary positioned above (distal to) the base of the other flower parts (stamens, petals and sepals) along the floral axis
Taxon: any category or classifying unit
Tendril: a slender, coiling appendage on a stem which is used for climbing and support (see *Vitis*, p.206)

Terete: cylindrical, circular in cross-section
Ternate: split into three small divisions (see *Suksdorfia*, p.191)
Tomentose: covered with a soft, dense, matted layer of hair (see *Gnaphalium*, p.59)
Toothed: having coarse, teeth-like serrations along the margin (see *Saxifraga*, p.190)
Truncate: abruptly cut off at tip or base
Tuber: an enlarged underground storage organ much like a potato (see *Sanicula*, p.40)
Tufted: in dense clusters
Turbinate: shaped like a top
Umbel: an inflorescence in which all the pedicels arise at the same point on the peduncle (see *Allium*, p.216)
Compound umbel: an inflorescence with several umbels themselves arranged in an umbel (see *Perideridia*, p.39)
Undulate: having a wavy margin (see *Streptopus*, p.223)
Uniseriate: arranged in one row or whorl (see *Eriophyllum*, p.58)
Variety: a taxon within a species
Vestigial: rudimentary, much reduced
Villous: densely long-hairy
Viscid: viscous: sticky
Whorl: three or more similar structures (leaves, bracts, petals, etc.) encircling a node (see *Hippuris*, p.126)
Wing: a thin, often membranous appendage bordering a structure (see *Acer*, p.33)
Woolly: bearing soft, curled or entangled hairs (see *Eriastrum*, p.152)

BIBLIOGRAPHY AND ADDITIONAL REFERENCES

Abrams, LeRoy, *Illustrated Flora of the Pacific States*, Stanford, Stanford University Press, Vol. I, 1940; Vol. II, 1944: Vol. III, 1951; Vol. IV with Roxanne S. Ferris, 1960.

Altschul, S., *Drugs and Food from Little-known Plants*, Cambridge, MA, Harvard University Press, 1973.

Butler, G.W., and R.W. Bailey, (eds.), *Chemistry and Biochemistry of Herbage*, Vol. I, London and New York, Academic Press, 1973.

Crowhurst, A., *The Weed Cookbook*, New York, Lancer Books, Inc., 1972.

Frye, T.C., *Ferns of the Northwest*, Portland, Metropolitan Press, 1934

Gillett, G.W. J.T. Howell, and H. Leschke, "A Flora of Lassen Volcanic National Park, California," *The Wasmann Journal of Biology*, 19(1): 1-185, 1961.

Griffin, J.R. and W.B. Critchfield, *The Distribution of Forest Trees in California*, USDA Forest Service Research Paper PSW-82/1972 (Reprinted with Supplement, 1976).

Grillos, S.J., *Ferns and Fern Allies of California*, Berkeley, University of California Press, 1966.

Hall, H.M. and C.C. Hall, *A Yosemite Flora*, San Francisco, Paul Elder and Company, 1912.

Harris, B.C., *Eat the Weeds*, Barre, MA, Barre Publishers, 1971.

Hedrick, U.P. (ed.), *Sturtevant's Edible Plants of the World*, New York, Dover Publications, 1972.

Hickman, J.C., *The Jepson Manual: Higher Plants of California,* Berkeley, University of California Press, 1993.

Hitchcock, A.S., *Manual of the Grasses of the United States*, 2nd edition revised by Agnes Chase, USDA, Misc. Publ. no. 200, Feb. 1951. Now published in two volumes by Dover Publications, New York, 1971.

Jepson, W.L., *A Manual of the Flowering Plants of California*, Berkeley, Associated Students Store, 1923-1925. Reprinted, 1963, by the University of California Press.

Kingsbury, J.M., *Poisonous Plants of the United States and Canada*, Englewood Cliffs, New Jersey, Prentice-Hall, Inc., 1964.

Kirk, D.R., *Wild Edible Plants of the Western States*, Healdsburg, CA, Naturegraph Publishers, 1970.

Mason, H.L., *A Flora of the Marshes of California*, Berkeley, University of California Press, 1969.

McMinn, H.E., *An Illustrated Manual of California Shrubs*, Berkeley, University of California Press, 1951.

Medsger, O.P., *Edible Wild Plants*, New York, The MacMillan Co., 1939. Reprinted 1969.

Muenschner, W.C., *Poisonous Plants of the United States*, New York, The MacMillan Co., 1939.

Munz, P.A., *Supplement to A California Flora*, Berkeley, University of California Press, 1968.

Munz, P.A., in collaboration with David D. Keck, *A California Flora*, Berkeley, University of California Press, 1959.

Niehaus, T.F., and D.L. Ripper, *A Field Guide to Pacific States Wildflowers*, Boston, Houghton Mifflin, 1976.

Parsons, M.E., *The Wild Flowers of California*, San Francisco, Cunningham, Curtiss and Welch, 1909.

Porter, C.L., *Taxonomy of Flowering Plants*, San Francisco, W.H. Freeman and Co., 1959.

Rockwell, J.A., and S.K. Stocking, *Checklist of the Flora*, Sequoia-Kings Canyon National Parks, Three Rivers, CA, Sequoia Natural History Assn., 1971.

Rodin, R.J., "Ferns of the Sierra," *Yosemite Nature Notes*, 39(4): 44-124, April, 1960.

Smith, G.L., "A Flora of the Tahoe Basin and Neighboring Areas," *The Wasmann Journal of Biology*, 31(1): 1-231, 1973.

——, "Supplement to A Flora of the Tahoe Basin and Neighboring Areas," *The Wasmann Journal of Biology*, 41(1-2): 1-46, 1983.

Storer, t.I., and R.L. Usinger, *Sierra Nevada Natural History*, Berkeley, University of California Press, 1963.

Sweet, M., *Common Edible and Useful Plants of the West*, Healdsburg, CA., Naturegraph Publishers, 1962.

Thomas. J.H., and D.R. Parnell, *Native Shrubs of the Sierra Nevada*, Berkeley, University of California Press, 1974.

Uphof, J.D., *Dictionary of Economic Plants*, New York, Hafner Service Agency, 1968.

INDEX